Emission and Control of Trace Elements from Coal-Derived Gas Streams

Woodhead Publishing Series in Energy

Emission and Control of Trace Elements from Coal-Derived Gas Streams

Edited by

Yongchun Zhao

Hailong Li

Jianping Yang

Junying Zhang

Chuguang Zheng

ELSEVIER

WP
WOODHEAD
PUBLISHING
An imprint of Elsevier

Woodhead Publishing is an imprint of Elsevier
The Officers' Mess Business Centre, Royston Road, Duxford, CB22 4QH, United Kingdom
50 Hampshire Street, 5th Floor, Cambridge, MA 02139, United States
The Boulevard, Langford Lane, Kidlington, OX5 1GB, United Kingdom

Library of Congress Cataloging-in-Publication Data
A catalog record for this book is available from the Library of Congress

British Library Cataloguing-in-Publication Data
A catalogue record for this book is available from the British Library

ISBN: 978-0-08-102591-8

For information on all Woodhead Publishing publications visit our website at https://www.elsevier.com/books-and-journals

Publisher: Joe Hayton
Acquisition Editor: Maria Convey
Editorial Project Manager: Mariana Kuhl
Production Project Manager: Poulouse Joseph
Cover Designer: Alan Studholme

Typeset by TNQ Technologies

Contents

List of contributors

Bengen Gong State Key Laboratory of Coal Combustion, Huazhong University of Science and Technology, Wuhan, China

Yingchao Hu School of Energy Science and Engineering, Central South University, Changsha, China

Hailong Li School of Energy Science and Engineering, Central South University, Changsha, China

Qin Li School of Energy Science and Engineering, Central South University, Changsha, China

Wenqi Qu School of Energy Science and Engineering, Central South University, Changsha, China

Jianping Yang School of Energy Science and Engineering, Central South University, Changsha, China

Zequn Yang Department of Civil Engineering, The University of Hong Kong, Hong Kong, China

Junying Zhang State Key Laboratory of Coal Combustion, Huazhong University of Science and Technology, Wuhan, China

Jiexia Zhao School of Energy Science and Engineering, Central South University, Changsha, China

Yongchun Zhao State Key Laboratory of Coal Combustion, Huazhong University of Science and Technology, Wuhan, China

Chuguang Zheng State Key Laboratory of Coal Combustion, Huazhong University of Science and Technology, Wuhan, China

Yuming Zhou State Key Laboratory of Coal Combustion, Huazhong University of Science and Technology, Wuhan, China

Introduction: background to trace elements

1

Jianping Yang [1], *Yongchun Zhao* [2], *Junying Zhang* [2], *Chuguang Zheng* [2]
[1]School of Energy Science and Engineering, Central South University, Changsha, China;
[2]State Key Laboratory of Coal Combustion, Huazhong University of Science and Technology, Wuhan, China

1.1 Introduction

Coal is the primary resource used for electric power generation, and supplied 41% of global electricity needs in 2017 [1]. Coal is also widely used in the field of metallurgical processes, gasification, cement industries, and many common industrial chemicals. The widespread use of coal has resulted in serious health problems in millions of people. Coal is an extraordinarily complex rock, which consists of a complex carbon matrix, almost all elements in the periodic table, scores of different minerals, water, gases, and oil. The formation conditions are beneficial to improving the toxic elements including arsenic (As), mercury (Hg), selenium (Se), lead (Pb), etc. Toxic elements are released during coal utilization, causing harm to the environment and human beings.

1.2 Typical trace elements in coal and its classification

1.2.1 Trace elements content in coal

Elements in coal can be classified into three broad groups on the basis of their concentration: (1) major elements (C, H, O, N, S), with concentrations above 1000 ppm; (2) minor elements, including coal–mineral matters (Si, Al, Ca, Mg, K, Na, Fe, Mn, Ti), and halogens (F, Cl, Br, I), with concentrations between 100 and 1000 ppm; and (3) trace elements (TEs), with concentrations below 100 ppm. TEs such as Hg, As, and Se present in coal are known to be of concern to public health. These TEs even when present in parts per million levels in coal can result in emissions of several tons of pollutants in the environment [2].

The concentration of TEs is significantly varied between coals from different sources and even between coals from the same seams [2–10]. The mean abundances of elements in coal can provide an important scientific basis for many geochemical comparisons and studies [11]. To clearly understand the release behavior of TEs during coal utilization processes, it is essential to investigate the variations of TE concentrations in different coals. The TE contents in coal samples from US, UK, Australian [3,12], and Chinese coals [13–15] were systematically investigated. Although the TE contents seem to vary strongly with coal type, the elements can be classified

Emission and Control of Trace Elements from Coal-Derived Gas Streams
https://doi.org/10.1016/B978-0-08-102591-8.00001-5

into four groups with concentrations >50 ppm, 10–50 ppm, 1–10 ppm, and <1 ppm. Most TEs are less concentrated than 50 ppm. The concentrations of Hg, As, and Se for most coals appear to vary in the ranges of 0.02–1 ppm, 0.5–80 ppm, and 0.2–4 ppm, respectively [16].

Ren et al. [17] estimated the concentration of 22 TEs in Chinese coals and Bai et al. [18] estimated the concentration of 31 TEs in coal based on 1123 samples collected from 504 coal mines in 26 provinces. Based on the work by Ren et al. [17] and Dai et al. [19,20], the background values of elements in Chinese coals are estimated [21]. Compared to US and world coals, concentrations of most TEs in Chinese coals are similar. However, Chinese coals have higher contents of Li (31.8 μg/g), Zr (89.5 μg/g), Nb (9.44 μg/g), Ta (0.62 μg/g), Hf (3.71 μg/g), Th (5.84 μg/g), and total rare earth elements ($\sum La - Lu + Y$, 136 μg/g) [21]. The higher concentration of these TEs contributes to high ash yields as well as intraseam alkali tonsteins in southwestern China [19]. In the past several years, toxic TEs like Hg, As, and F in Chinese coals have been widely studied. Some coals from Xingren, Guizhou [22–28], and Laochang, Yunnan [29], are significantly enriched in As and Hg. The world's highest As level (3.2%–3.5%) in coal was found in Xingren, Guizhou [23,25]. However, reserves of high-As coals in Xingren are extraordinarily small compared to abundant coal resources in western Guizhou [2]. A more detailed description of the content of elements in Chinese coals, as well as their comparisons with world coals, is presented in Chapter 2.

1.2.2 Modes of occurrence and association of trace elements in coal

The occurrence modes of TEs in coal include the specific mineral that an element forms, dispersion within a particular host mineral or maceral, the fraction that the element is associated with, the oxidation state of the element, and so on [2]. TEs could be associated with either the mineral or the organic parts of coal [2]. Fig. 1.1 illustrates the occurrence modes of TEs in coal [30], and Table 1.1 summarizes TE distribution in coal. Be, Sr, and Ge have an organic affinity in most cases, while Ba, Ce, Co, La, Mn, Ni, Rb, and Zr have an inorganic affinity, and other TEs occur variously in different coals [2]. The contents of TEs in coal vary significantly with coal type.

1.3 The classification of trace elements

TEs can be categorized into three main groups regarding their partitioning during coal combustion [31,32], as shown in Fig. 1.2:

1. The first group is the nonvolatile elements, which are enriched in coarse particles or between coarse and suspended particles evenly (such as Mn, Co, V). They rarely evaporate, so these elements are distributed in fly ash or bottom ash, and can usually be removed by conventional particle control systems.
2. The second group includes secondary volatile elements, which are more likely enriched in fine-grained particles (usually those particles less than 10 μm) such as As, Cd, Pb, Sb, and Se,

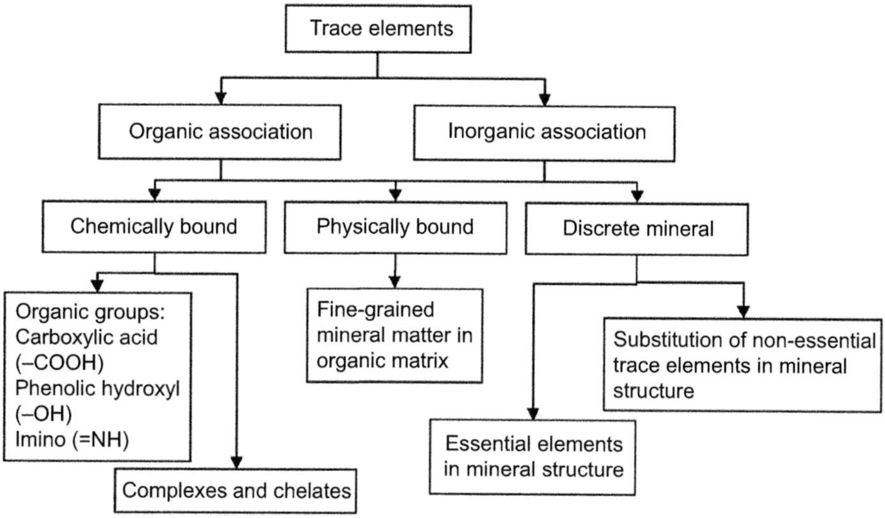

Figure 1.1 Modes of occurrence of trace elements in coal.
From Vejahati F, Xu Z, Gupta R. Trace elements in coal: associations with coal and minerals and their behavior during coal utilization – a review. Fuel 2010;89(4):904–11.

Table 1.1 Trace element distribution in coal.

Affinity	Mineral group	Mineral type	Elements
Inorganic	Clay minerals and feldspars	Kaolinite Montmorillonite	Al, Ba, Bi, Cr, Cs, Cu, Ga, K, Li, Mg, Na, Ni, P, Pb, Rb, Sn, Sr, Ta, Th, Ti, U, V, Y, and rare earth elements
	Iron sulfides	Pyrite Sphalerite	As, Cd, Co, Cu, Fe, Hg, Mo, Ni, Pb S, Sb, Se, Ti, W, and Zn
	Carbonates	Calcite, dolomite, ankerite	Ca, Co, and Mn
	Sulfates	–	Ba, Ca, Fe, and S
	Heavy minerals	Tourmaline	B
Organic	N, S, Be, B, Ge, V, W, and Zr (B exhibits partial association with tourmaline in the heavy fraction, and V with clay minerals)		

From Vejahati F, Xu Z, Gupta R. Trace elements in coal: associations with coal and minerals and their behavior during coal utilization – a review. Fuel 2010;89(4):904–11.

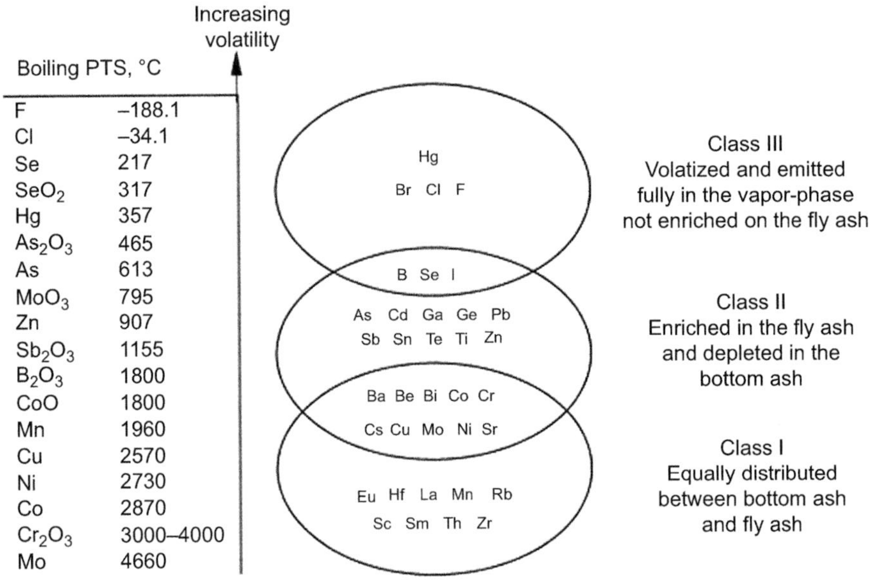

Figure 1.2 Classification of trace elements in coal combustion.
From Xu M, Yan R, Zheng C, Qiao Y. Status of trace element emission in a coal combustion process: a review. Fuel Process Technol 2004;85(2–3):215–37.

which may escape from particulate control systems. As the flue gas cools and the particle size decreases, these elements enriched in fine particles.

3. The third group is readily volatilized during coal combustion and can be enriched in the vapor or gas phase, such as the elements Hg, F, and Cl. In addition, there are also some elements between the groups, Cr and Ni are between groups I and II, which have intermediate partitioning behavior, e.g., Se shows a partitioning behavior between groups II and III.

The elements in group II are enriched with decreasing particle size [33,34]. These elements are enriched on in-stack suspended particles. Group I elements do not volatilize easily under coal combustion temperatures and form bottom ash or fly ash particles on which more volatile elements condense. The most volatile group III elements (such as Hg and the halogens) may remain in the gas phase during the coal utilization process.

1.4 Environmental effects and toxicity of trace elements

1.4.1 Mercury

Mercury (Hg) is a toxic element. In the 1960s, residents of Minamata, Japan, developed severe neuropathies from methylmercury (MeHg)-contaminated seafood. It is now called Minamata disease, and 2217 cases had been officially recognized in the city by 1989. Hg pollution is also a hazard in other countries like Iraq, Brazil, Indonesia, the United States, and China. Hg pollution has attracted attention globally due to its persistence, bioaccumulation, and toxicity. Mercury may persist and cycle in the environment for a long time. When mercury is released into the air, it is transported and eventually deposited back on the earth. In aquatic ecosystems, inorganic mercury is transformed into more toxic MeHg, which could be bioaccumulated in the food chain. After inhalation of Hg by human beings, elemental Hg (Hg^0) vapor rapidly diffuses through the alveolar membrane and is absorbed through the lungs. Exposure to more than 100,000–200,000 ng/m^3 of Hg^0 vapor for a few hours can cause acute chemical bronchiolitis and pneumonitis. The nervous system appears the most sensitive target organ, and neurological problems have been reported in adults. Exposure to high levels of MeHg can cause mental retardation, cerebral palsy, seizures, and ultimately death.

1.4.2 Arsenic

Arsenic is a noxious and carcinogenic element, which causes various adverse effects on human health. It accumulates in the body and generates reactive oxygen species, resulting in oxidative damage to vital organs [35]. The toxicity of arsenic depends on its chemical forms and oxidation states [36], and As^{3+} is 50 times more toxic than As^{5+}. Chronic arsenic accumulation eventually leads to psychological effects, decreased mental performance, peripheral neuropathy, hypertension, and increased risks of cardiovascular disorders, diabetes, and cancer [37]. Chronic arsenic toxicity impairs the same organs and tissues as acute toxicity although in cases of higher exposure levels, skin and nervous system disturbances are usually more pronounced [38]. Incidents of adverse health risks resulting from arsenic pollution occur in many countries. More than 3000 cases of arseniasis in villages in Guizhou, a typical area of arsenic contamination in China, were caused by the use of locally mined high-arsenic coal. In Bangladesh, about 57 million people are exposed to concentrations of arsenic in drinking water exceeding the World Health Organization (WHO)'s guideline of 10 μg/L.

1.4.3 Selenium

As a naturally occurring element, selenium acts as an essential micronutrient for humans and animals, and plays an important role in redox regulation of intracellular signaling, redox homeostasis, and thyroid hormone metabolism [39]. However, there is a narrow margin between beneficial and toxic effects [39]. High Se concentrations in drainage water caused gastrointestinal illness and hair loss among residents in Enshi, Hubei Province, China [40]. In southern California, USA, excessive Se also caused deformation, reproductive failure, or juvenile mortality in waterfowl [41] and fish [42]. In the endemic selenosis areas, human activities, such as mining of stone coal for use as a fuel or fertilizer, and discharging lime into cropland to improve soil, caused variable addition of Se to the soil and further accumulation of Se in the food chain. Selenium exists in four states, i.e., selenate (SeO_4^{2-}, +6), selenite (SeO_3^{2-}, +4), elemental selenium (Se^0, 0), and selenide (Se^{2-}, −2) [43]. Among these Se species, SeO_4^{2-} is highly water soluble and toxic to organisms at low concentrations [44].

1.4.4 Fluorine

Fluorine (F) is a TE in biological systems and has a reputation for causing harmful effects on health and the environment [45]. Fluorine is essential to animal life at low levels but is toxic at high levels [46]. Exposure to excess fluorine can result in mottling of tooth enamel and various forms of skeletal fluorosis, including limited movement of the joints and spinal curvature. Li et al. [47] conducted an investigation of fluorosis in Guizhou Province, China. They reported that a large amount of gas and dust containing fluoride was released by domestic coal combustion and quickly combined with vapor in the air to form an aerosol or a hydrofluoric acid mist, which adhered to the surface of corn and chili. They also pointed out that fluorine concentrations in the soils in the fluorosis area were higher than those in other areas. The authors demonstrated that fluorine concentrations in fresh corn from Zhijin County, Bijie County, and Guiyang City were 3.60, 2.75, and 5.35 mg/kg, respectively, higher than the world average value.

1.4.5 Lead

Lead is ubiquitous in nature and is harmful at high levels. Pb finds its way into the body through oral administration, inhalation, and other methods. It is not easily excreted and accumulates in the body. Pb in the body can cause blood disorders, anorexia, gastrointestinal symptoms, problems with the nervous system, and kidney malfunction [48]. Although the WHO set the maximum contamination concentration of Pb in drinking water at 10 μg/L, the presence of Pb in concentrations as low as 5 μg/L has been reported to cause cognitive impairment, aggression, and delinquency in children [49]. Children are easily exposed to various Pb sources and are more sensitive to Pb than adults [50]. Moreover, Pb(II) causes toxicity via disrupting functions of the brain, kidney, liver, and central nervous system, along with the inhibition of hemoglobin formation, anemia, weakness of muscles, dizziness, headache, irritability, and infertility.

1.5 The regulations and emission limits of trace elements during coal utilization

1.5.1 Mercury emission standards for coal-fired power plants

In recent years, with increasing environmental pollution, more and more attention has been paid to mercury emission and control in coal-fired power plants.

1.5.1.1 America

In December 2000, the US Environmental Protection Agency (US EPA) announced the control of regular mercury emissions from coal-fired power plants. On March 15, 2005, the US government issued the Clean Air Mercury Rule (CAMR). The power plants are allowed to remove mercury by existing air pollution control devices.

In February 2008, the US Federal Court of Appeal canceled CAMR, and asked the EPA to set stricter mercury emission standards for coal-fired power plants, requiring the use of "maximum achievable control technology" to develop mercury emission standards for coal-fired power plants. Specifically, the mercury emission standards are established according to the average mercury emission of the power plants with the least emission level, which accounts for 12% of all of the power plants. Since then, the United States has not issued a unified mercury emission standard, but several states have established mercury emission standards for coal-fired power station boilers, as shown in Table 1.2.

On March 16, 2011, the US EPA issued its first draft for public consultation on mercury and its toxic and harmful gases emission standards "Mercury and Air Toxics Standards" for power plants, and released standards for toxic gas emissions from coal-fired power plants, including mercury, on December 16, 2011. This standard stipulates that both existing and newly built coal-fired and oil-fired boilers should implement emission reduction. This is the first time that the United States has promulgated a national air pollution control law for coal-fired power plants. The new standard has been formally implemented since April 16, 2012. About 50% of mercury emissions

Table 1.2 Mercury emission reduction regulations in some states of the United States.

States	Implementation time	Regulations
Massachusetts	October 2012	Mercury removal efficiency ≥95% or ≤0.0025 lb/GWh
Connecticut	July 2008	Mercury removal efficiency ≥90%
Wisconsin	To 2015	Mercury removal efficiency ≥75%
	To 2018	Mercury removal efficiency ≥80%
	To 2020	Mercury removal efficiency ≥90%
Illinois	July 1, 2009	Mercury removal efficiency ≥90% or ≤0.0080 lb/GWh
Pennsylvania	To 2010	Mercury removal efficiency ≥80% or 0.024 lb Hg/GWh, prohibition of mercury emissions trading
	To 2015	Mercury removal efficiency ≥90% or 0.012 lb Hg/GWh, prohibition of mercury emissions trading
Indiana		Compliance with Clean Air Mercury Rule
South Carolina		Compliance with Clean Air Mercury Rule

Table 1.3 Mercury emission limits for existing and new coal-fired power plants in the United States (mg/MWh).

Category	Mercury emission standard (mg/MWh)
Existing unit—nonlow-rank raw coal combustion device	5.90
Existing unit—designed for low-rank raw coal unit	54.4
Existing unit—combustion device of integrated coal gasification combined cycle	18.16
New unit—nonlow-rank raw coal combustion device	1.36
New unit—designed for low-rank raw coal unit	18.16
New unit—combustion device of integrated coal gasification combined cycle	1.36

Note: Definition of low-rank coal is <19,305 kJ/kg, water content, free radicals; definition of nonlow-rank coal is >19,305 kJ/kg, water content, free radicals.

in the United States come from power plants. According to the Global Mercury Assessment 2018, the mercury emission to air from power plants in North America have decreased between 2010 and 2015 by 6.25 tonnes. The emission limits are shown in Table 1.3.

1.5.1.2 European Union

Mercury emissions are not limited in the directive (2001/80/EC) on the limitation of emissions of certain pollutants into the air from large combustion plants issued by the European Union in 2001. In the Reference Document on Best Available Technology for Large Combustion Plants published in 2006, no mandatory regulation on mercury emission was made. Only conventional flue gas pollutant removal devices, such as selective catalytic reduction (SCR), wet flue gas desulfurization (WFGD) and electrostatic precipitator (ESP)/fabric filter (FF), were recommended for collaborative mercury removal.

Article 13 of the German Federal Control Act (13 BlmSchv) requires a mercury emission limit of 0.01 mg/m^3 for coal-fired power plants, and all power plants are required to install continuous monitoring systems for mercury. VGB Power Technical Association formulated the association standard in 2010, requiring mercury emissions from coal-fired boilers to be controlled below 0.01 mg/m^3.

1.5.1.3 Canada

The federal government set environmental expectations in 2006, which control total mercury emissions from existing coal-fired power plants in provinces. The mercury emission from power plant in Canada was 2002.9 kg in 2006 and 718.7 kg in 2016

Table 1.4 Mercury emission reduction rates and emission limits for newly built coal-fired power plants in Canada.

Types of coal	Reduction rates (%)	Emission limits (kg/(tw h))
Bituminous coal	85	3 (About 0.0023 mg/m^3)
Sub-bituminous coal	75	8 (0.0062 mg/m^3)
Lignite	75	15 (0.0116 mg/m^3)
Blended coal	85	3 (0.0023 mg/m^3)

with a decrease of 64%. For new power plants, mercury emission reduction rates and emission limits are set according to coal types, as shown in Table 1.4.

1.5.1.4 China

The Ministry of Environmental Protection of the People's Republic of China promulgated the latest Air Pollutant Emission Standard for Thermal Power Plants (GB 13223-2011) in 2011. This standard put mercury on the emission limit of pollutants from coal-fired boilers for the first time and required that the emission limit of mercury from flue gas of coal-fired power plants be 0.03 mg/m^3 from January 1, 2015.

1.5.2 Emission standards and development of other trace elements in coal-fired power plants

TEs released into the atmosphere can exist for a long time and drift great distances due to the wind. Therefore TE pollution in the atmosphere is not a problem just for a region or country, it is an international problem. At present, many international agreements and national regulations have been formulated to limit the emission of TEs.

The United Nations Economic Commission for Europe has reached an agreement on the pollution of TEs in the atmosphere and promulgated a draft in 1998. The United States, Canada, Europe, and Russia also signed the draft. The draft requires member countries to reduce mercury, cadmium, and lead emissions from fixed sources. The draft considers that TEs are mainly emitted in the form of gaseous or particulate states, so it recommends the use of particulate matter emission control methods to control the emission of TEs. As a result, for fossil fuel combustion, the emission standard of a particulate matter is 50 mg/m^3 and 0.05 mg/m^3 for a toxic waste furnace, and 0.08 mg/m^3 for a municipal waste incinerator prior to more stringent emission restrictions being proposed. At present, these standard limits have been greatly reduced.

In an effort to reduce emissions of TEs, the member states of the European Community have three other agreements besides implementing the agreements of the United Nations Economic Commission for Europe. They are:

1. SPAR Oslo–Paris Agreement: reducing continental pollutant migration to the North Sea;
2. ELCOM Helsinki Protocol: protecting the Baltic Sea, similar to OSPAR;
3. Barcelona Agreement: reducing pollution in the Mediterranean.

In these agreements, European Community countries do not have emission standards for mercury or other TEs in coal-fired power plants. In 1989, the emission standard of waste incinerators over 3 t/h established by the European Community did not put forward the mercury limitation standard. Now the mercury emission standard is 0.05 mg/m^3. The emission standard of cadmium and thallium is 0.5 mg/m^3, and that of antimony, arsenic, chromium, lead, cobalt, copper, manganese, nickel, and vanadium is 0.5 mg/m^3. These standards are also suitable for mixed combustion of waste and other fuels.

Nordic plans to reduce emissions of pollutants such as mercury without changing the energy structure, improving the efficient supply, and utilizing energy make the best use of renewable energy, adjust the transport structure, and improve the Nordic energy management system. The Arctic Commission has established an Arctic environmental protection strategy, which considers mercury as the main pollutant in Arctic marine environmental protection, detection, and evaluation. Canada, Mexico, and the United States have reached a consensus on the North American Action Plan. The target and deadline for reducing mercury emissions were adopted in June 1999.

Nowadays, many countries have formulated regulations and plans to reduce the emission of heavy metals to conform with international agreements. The National Environmental Protection Agency has stipulated that: We should make full use of advanced technologies to control the emission of heavy metals in the coal utilization process, including coal-fired unit and integrated gasification combined cycle unit. The emission limits of heavy metals are listed in Tables 1.5 and 1.6. Some emission standards or guidelines for toxic TEs in coal combustion have also been established in Australia (Table 1.7).

Table 1.5 Emission limits for new or reconstructed coal-fired unit and integrated gasification combined cycle (IGCC) unit.

Subcategory	Pollutants	Emission limits
Coal-fired unit not low-rank virgin coal	Total non-Hg HAP metals	6.0E-2 lb/GWh
	OR	OR
	Individual HAP metals:	
	Antimony (Sb)	8.0E-3 lb/GWh
	Arsenic (As)	3.0E-3 lb/GWh
	Beryllium (Be)	6.0E-4 lb/GWh
	Cadmium (Cd)	4.0E-4 lb/GWh
	Chromium (Cr)	7.0E-3 lb/GWh
	Cobalt (Co)	2.0E-3 lb/GWh
	Lead (Pb)	2.0E-2 lb/GWh
	Manganese (Mn)	4.0E-3 lb/GWh
	Nickel (Ni)	4.0E-2 lb/GWh
	Selenium (Se)	5.0E-2 lb/GWh

Table 1.5 Emission limits for new or reconstructed coal-fired unit and integrated gasification combined cycle (IGCC) unit.—cont'd

Subcategory	Pollutants	Emission limits
Coal-fired units low-rank virgin coal	Total non-Hg HAP metals	6.0E-2 lb/GWh
	OR	OR
	Individual HAP metals:	
	Antimony (Sb)	8.0E-3 lb/GWh
	Arsenic (As)	3.0E-3 lb/GWh
	Beryllium (Be)	6.0E-4 lb/GWh
	Cadmium (Cd)	4.0E-4 lb/GWh
	Chromium (Cr)	7.0E-3 lb/GWh
	Cobalt (Co)	2.0E-3 lb/GWh
	Lead (Pb)	2.0E-2 lb/GWh
	Manganese (Mn)	4.0E-3 lb/GWh
	Nickel (Ni)	4.0E-2 lb/GWh
	Selenium (Se)	5.0E-2 lb/GWh
IGCC unit	Total non-Hg HAP metals	4.0E-1 lb/GWh
	OR	OR
	Individual HAP metals:	
	Antimony (Sb)	2.0E-2 lb/GWh
	Arsenic (As)	2.0E-2 lb/GWh
	Beryllium (Be)	1.0E-3 lb/GWh
	Cadmium (Cd)	2.0E-3 lb/GWh
	Chromium (Cr)	4.0E-2 lb/GWh
	Cobalt (Co)	4.0E-3 lb/GWh
	Lead (Pb)	9.0E-3 lb/GWh
	Manganese (Mn)	2.0E-2 lb/GWh
	Nickel (Ni)	7.0E-2 lb/GWh
	Selenium (Se)	3.0E-1 lb/GWh

HAP, Hazardous air pollution.

Table 1.6 Emission limits for existing coal-fired unit and integrated gasification combined cycle (IGCC) unit.

Subcategory	Pollutants	Emission limits
Coal-fired unit not low-rank virgin coal	Total non-Hg HAP metals	5.0E-5 lb/MMBtu or 5.0E-1 lb/GWh
	OR	OR
	Individual HAP metals:	
	Antimony (Sb)	8.0E-1 lb/TBtu or 8.0E-3 lb/GWh
	Arsenic (As)	1.1E0 lb/TBtu or 2.0E-2 lb/GWh
	Beryllium (Be)	2.0E-1 lb/TBtu or 2.0E-3 lb/GWh
	Cadmium (Cd)	3.0E-1 lb/TBtu or 3.0E-3 lb/GWh
	Chromium (Cr)	2.8E0 lb/TBtu or 3.0E-2 lb/GWh
	Cobalt (Co)	8.0E-1 lb/TBtu or 8.0E-3 lb/GWh
	Lead (Pb)	1.2E0 lb/TBtu or 2.0E-2 lb/GWh
	Manganese (Mn)	4.0E0 lb/TBtu or 5.0E-2 lb/GWh
	Nickel (Ni)	3.5E0 lb/TBtu or 4.0E-2 lb/GWh
	Selenium (Se)	5.0E0 lb/TBtu or 6.0E-2 lb/GWh
Coal-fired units low-rank virgin coal	Total non-Hg HAP metals	5.0E-5 lb/MMBtu or 5.0E-1 lb/GWh
	OR	OR
	Individual HAP metals:	
	Antimony (Sb)	8.0E-1 lb/TBtu or 8.0E-3 lb/GWh
	Arsenic (As)	1.1E0 lb/TBtu or 2.0E-2 lb/GWh
	Beryllium (Be)	2.0E-1 lb/TBtu or 2.0E-3 lb/GWh
	Cadmium (Cd)	3.0E-1 lb/TBtu or 3.0E-3 lb/GWh
	Chromium (Cr)	2.8E0 lb/TBtu or 3.0E-2 lb/GWh
	Cobalt (Co)	8.0E-1 lb/TBtu or 8.0E-3 lb/GWh
	Lead (Pb)	1.2E0 lb/TBtu or 2.0E-2 lb/GWh
	Manganese (Mn)	4.0E0 lb/TBtu or 5.0E-2 lb/GWh
	Nickel (Ni)	3.5E0 lb/TBtu or 4.0E-2 lb/GWh
	Selenium (Se)	5.0E0 lb/TBtu or 6.0E-2 lb/GWh

Table 1.6 Emission limits for existing coal-fired unit and integrated gasification combined cycle (IGCC) unit.—cont'd

Subcategory	Pollutants	Emission limits
IGCC unit	Total non-Hg HAP metals	6.0E-5 lb/MMBtu or 5.0E-1 lb/GWh
	OR	OR
	Individual HAP metals:	
	Antimony (Sb)	1.4E0 lb/TBtu or 2.0E-2 lb/GWh
	Arsenic (As)	1.5E0 lb/TBtu or 2.0E-2 lb/GWh
	Beryllium (Be)	1.0E-1 lb/TBtu or 1.0E-3 lb/GWh
	Cadmium (Cd)	1.5E-1 lb/TBtu or 2.0E-3 lb/GWh
	Chromium (Cr)	2.9E0 lb/TBtu or 3.0E-2 lb/GWh
	Cobalt (Co)	1.2E0 lb/TBtu or 2.0E-2 lb/GWh
	Lead (Pb)	1.9E + 2 lb/TBtu or 1.8E0 lb/GWh
	Manganese (Mn)	2.5E0 lb/TBtu or 3.0E-2 lb/GWh
	Nickel (Ni)	6.5E0 lb/TBtu or 7.0E-2 lb/GWh
	Selenium (Se)	2.2E + 1 lb/TBtu or 3.0E-1 lb/GWh

HAP, Hazardous air pollution.

Table 1.7 Guidelines and standards for emissions of trace elements from trade, industrial production, and industrial processes.

Countries or regions	Type	Limiting components	Limit value	Scope of application
Australia				
Nationwide	Guidance value	Hg and its compounds	3 mg/m^3	Trade, industrial production, and industrial processes
Nationwide	Guidance value	Cd and its compounds	3 mg/m^3	Trade, industrial production, and industrial processes
Nationwide	Guidance value	Ni and its compounds	20 mg/m^3	Power plant
Nationwide	Guidance value	Sb, As, Cd, Hg, Pb, Se, and its compounds	10 mg/m^3	Trade, industrial production, and industrial processes

Continued

Table 1.7 Guidelines and standards for emissions of trace elements from trade, industrial production, and industrial processes.—cont'd

Countries or regions	Type	Limiting components	Limit value	Scope of application
Capital region	Standard value	Hg and its compounds	3 mg/m^3	Any production process
Capital region	Standard value	Total Hg, As, Cd, Pb, Se and its compounds	10 mg/m^3	Any production process
New South Wales	Standard value	Hg and its compounds	1 mg/m^3	Any industry or production process (new power plant)
New South Wales	Standard value	Cd and its compounds	1 mg/m^3	Any industry or production process (new power plant)
New South Wales	Standard value	Total Sb, As, Cd, Pb, Hg, and its compounds	20 mg/m^3	Power plants built before 1986
New South Wales	Standard value	Total Sb, As, Cd, Pb, Hg, and its compounds	10 mg/m^3	Power plants built before 1986
New South Wales	Standard value	Total Sb, As, Be, Cd, Co, Pb, Mn, Hg, Ni, Se, V, and its compounds	5 mg/m^3	Trade, industrial production, and industrial processes
New South Wales	Standard value	Total Hg, As, Cd, Pb, Se, V, and its compounds	10 mg/m^3	Any production process
Queensland	Standard value	Total Hg, As, Cd, Pb, and its compounds	20 mg/m^3	Built power plant
Queensland	Standard value	Total Sb, As, Cd, Hg, Pb, and its compounds	10 mg/m^3	Trade, industry, and production processes (new)
Queensland	Standard value	Cd and its compounds	3 mg/m^3	Trade, industry, and production processes (new)
Queensland	Standard value	Hg and its compounds	3 mg/m^3	Trade, industry, and production processes

Table 1.7 Guidelines and standards for emissions of trace elements from trade, industrial production, and industrial processes.—cont'd

Countries or regions	Type	Limiting components	Limit value	Scope of application
Queensland	Standard value	Ni and its compounds (beside nickel carbonyl)	20 mg/m^3	Trade, industry, and production processes (new)
Queensland	Standard value	Nickel carbonyl	0.5 mg/m^3	Trade, industry, and production processes (new)
South Australia	Standard value	Sb and its compounds	10 mg/m^3	Trade, industry, and production processes
South Australia	Standard value	As and its compounds	10 mg/m^3	Trade, industry, and production processes
South Australia	Standard value	Pb and its compounds	10 mg/m^3	Combustion equipment
South Australia	Standard value	Cd and its compounds	3 mg/m^3	Trade, industry, and production processes
South Australia	Standard value	Hg and its compounds	3 mg/m^3	Combustion equipment
South Australia	Standard value	Total Hg, As, Cd, Pb, Se, and its compounds	10 mg/m^3	Combustion equipment
Tasmania	Standard value	Total Hg, As, Cd, Pb, Se, and its compounds	10 mg/m^3	Equipment after 1975
Victoria	Standard value	Total Hg, As, Cd, Pb, Se, and its compounds	10 mg/m^3	Power plant
Austria	Standard value	Hg	0.05 mg/m^3	Coal-fired power plant
Canada	Guidance value	As, Cu, Pb, Sb, and Zn (respective)	7 mg/m^3	New or recently renovated power plants
	Guidance value	Cd and its compounds	2 mg/m^3	New or recently renovated power plants
	Guidance value	Hg and its compounds	1 mg/m^3	New or recently renovated power plants

Continued

Table 1.7 Guidelines and standards for emissions of trace elements from trade, industrial production, and industrial processes.—cont'd

Countries or regions	Type	Limiting components	Limit value	Scope of application
Germany	Standard value	Class I: Cd, Hg, and its compounds	0.2 mg/m^3	Production process greater than 1 g/h
		Class II: Se and its compounds	2 mg/m^3	Production process greater than 5 g/h
		Class III: Sb, Cr, Mn, Pb, and its compounds	10 mg/m^3	Production process greater than 25 g/h
Italy	Standard value	Same as German classification and limitation criteria		
Switzerland	Standard value	Same as German classification and limitation criteria		

1.6 Summary

TEs in coal result in substantial pollutant emissions and cause serious damage to the ecological environment and human health. The emission and control of TEs during coal utilization are of significant concern. The objective of this book is to provide an overview of the emission characteristics and control strategies of TEs during coal utilization. The first section briefly summarizes the distribution of TEs in coals, including the average content of TEs in coals and the distribution of TEs in coals from different coal-forming periods, coal ranks, and coal-bearing regions. The second section summarizes TE emissions during the coal utilization process (combustion, gasification, industry furnace, power plant, etc.). Then, the sorbents used for TE removal are described. Finally, the recovery of TE resources from coal and coal utilization by-products is discussed. This book aims to help people understand the behavior of TEs and the control of their emission, and also provides basic information for the issue of emission standards and regulations of TEs.

Acknowledgments

This project was supported by the National Key Technologies R&D Program: [Grant Number 2018YFB0605104].

References

[1] World Coal Association, 2017. 2017. https://www.worldcoal.org/coal/uses-coal/coal-electricity.

[2] Vejahati F, Xu Z, Gupta R. Trace elements in coal: Associations with coal and minerals and their behavior during coal utilization – a review. Fuel 2010;89:904–11.

[3] Swaine D. Trace elements in coal. London: Butterworth; 1990.

[4] Smith I, IEA Coal Research, Canada Centre for Mineral and Energy Technology. Trace elements from coal combustion: emissions. London: IEA Coal Research; 1987. 87.

[5] Huggins F, Shah N, Huffman G, Kolker A, Crowley S, Palmer C, Finkelman R. Mode of occurrence of chromium in four US coals. Fuel Process Technol 2000;63(2–3):79–92.

[6] Kolker A, Huggins F, Palmer C, Shah N, Crowley S, Huffman G, Finkelman R. Mode of occurrence of arsenic in four US coals. Fuel Process Technol 2000;63(2–3):167–78.

[7] Spears D, Martinez-Tarazona M. Geochemical and mineralogical characteristics of a power station feed-coal, Eggborough, England. Int J Coal Geol 1993;22(1):1–20.

[8] Pires M, Teixeira E. Geochemical distribution of trace elements in Leao coal. Brazil Fuel 1992;71(10):1093–6.

[9] Querol X, Fernandez Turiel J, Lopez Soler A, Duran M. Trace elements in high-subbituminous coals from the teruel mining district, northeast Spain. Appl Geochem 1992;7(6):547.

[10] Solari J, Fiedler H, Schneider C. Modeling of the distribution of trace elements in coal. Fuel 1989;68(4):536–9.

[11] Ketris M, Yudovich Y. Estimations of Clarkes for carbonaceous biolithes: world average for trace element contents in black shales and coals. Int J Coal Geol 2009;78(2):135–48.

[12] Raask E. The mode of occurrence and concentration of trace elements in coal. Fuel 1985; 11(2):97–118.

[13] Yan R, Zeng H, Ouyang Z, Lu X, Xu T. The mode of occurrence and concentration of trace elements in several typical Chinese coals. Anal Lab 1995;14(4):43–7 [In Chinese].

[14] Lu X, Ali A, Liu H, Zeng H. Multivariable analysis and experimental researches on the distribution of trace elements in coal. Environ Chem 1995;14(6):494–9 [In Chinese].

[15] Lu X, Xu T, Liu H, Zeng H. Factor analysis method in study on trace element distribution in coal. J Fuel Chem Technol 1994;22(4):438 [In Chinese].

[16] Xu M, Yan R, Zheng C, Qiao Y, Han J, Sheng C. Status of trace element emission in a coal combustion process: a review. Fuel Process Technol 2004;85(2–3):215–37.

[17] Ren D, Zhao F, Dai S, Zhang J, Luo K. Geochemistry of trace elements in coal. Beijing: Science Press; 2006. p. 556 [in Chinese].

[18] Bai X, Li W, Chen Y, Jiang Y. The general distributions of trace elements in Chinese coals. Coal Quality Technol 2007;1:1–4 [in Chinese].

[19] Dai S, Zhou Y, Ren D, Wang X, Li D, Zhao L. Geochemistry and mineralogy of the late Permian coals from the Songzao coalfield, Chongqing, southwestern China. Sci China Ser D 2007;50(5):678–88.

[20] Dai S, Li D, Chou C-L, Zhao L, Zhang Y, Ren D, Ma Y, Sun Y. Mineralogy and geochemistry of boehmite-rich coals: new insights from the Haerwusu surface mine, Jungar coalfield, Inner Mongolia, China. Int J Coal Geol 2008;74:185–202.

[21] Dai S, Wang X, Zhou Y, Hower J, Li D, Chen W, Zhu X, Zou J. Chemical and miner alogical compositions of silicic, mafic, and alkali tonsteins in the late Permian coals from the Songzao coalfield, Chongqing. Southwest China. Chem Geol 2011;282(1–2):29–44.

[22] Belkin H, Finkelman R, Zheng B. Human health effects of domestic combustion of coal: a causal factor for arsenic poisoning and fluorosis in rural China. Trans Am Geophys Union 1999;80(17):377−8.
[23] Ding Z, Zheng B, Long J, Belkin H, Finkelman R, Chen C, Zhou D, Zhou Y. Geological and geochemical characteristics of high arsenic coals from endemic arsenosis areas in southwestern Guizhou Province, China. Appl Geochem 2001;16(11−12):1353−60.
[24] Zhao F, Ren D, Zheng B, Hu T, Liu T. Modes of occurrence of arsenic in high arsenic coal by extended X-ray absorption fine structure spectroscopy. Chin Sci Bull 1998;43:1660−3.
[25] Zheng B, Ding Z, Huang R, Zhu J, Yu X, Wang A, Zhou D, Mao D, Su H. Issues of health and disease relating to coal use in southwest China. Int J Coal Geol 1999;40:119−32.
[26] Zheng B, Wang B, Ding Z, Zhou D, Zhou Y, Zhou C, Chen C, Finkelman R. Endemic arsenosis caused by indoor combustion of high-As coal in Guizhou Province, P.R. China. Environ Geochem Health 2005;27:521−8.
[27] Dai S, Zeng R, Sun Y. Enrichment of arsenic, antimony, mercury, and thallium in a late Permian anthracite from Xingren, Guizhou, southwest China. Int J Coal Geol 2006;66(3): 217−26.
[28] Zhou R, Ren Y. Distribution of arsenic in coals of Yunnan Province, China, and its controlling factors. Int J Coal Geol 1992;20(1−2):85−98.
[29] Dai S, Ren D, Tang Y, Yue M, Hao L. Concentration and distribution of elements in late Permian coals from western Guizhou province, China. Int J Coal Geol 2005;61(1−2): 119−37.
[30] Querol X, Fernández-Turiel J, López-Soler A. Trace elements in coal and their behaviour during combustion in a large power station. Fuel 1995;74(3):331−43.
[31] Zhu W, Xu M, Sui J, Han J. Simulation of the vaporization of trace element during coal combustion. Coal Technol 2005;24:115−7.
[32] Chadwick M, Highton N, Lindman N. Environmental impacts of coal mining and utilization. Oxford, UK: Pergamon Press; 1987. p. 171−217.
[33] Germani M, Zoller W. Vapor-phase concentrations of arsenic, selenium, bromine, iodine and mercury in the stack of a coal-fired power plant. Environ Sci Technol 1988;22(9): 1079−85.
[34] Meij R, Van der Sioot H, Winkel B. Characterization of input and output streams of coal-fired power stations, Part IV. 1990. KEMA, report 98179/93663−MOC 90−3362, Arnhem, Netherlands.
[35] Abdul K, Jayasinghe S, Chandana E, Jayasumana C, De Silva P. Arsenic and human health effects: a review. Environ Toxicol Pharmacol 2015;40:828−46.
[36] Ma J, Sengupta M, Yuan D, Dasgupta P. Speciation and detection of arsenic in aqueous samples: a review of recent progress in non-atomic spectrometric methods. Anal Chim Acta 2014;831(11):1−23.
[37] Mazumder D, Dasgupta U. Chronic arsenic toxicity: studies in West Bengal, India. Kaohsiung J Med Sci 2011;27(9):360−70.
[38] Lagerkvist B, Zetterlund B. Assessment of exposure to arsenic among smelter workers: a five-year follow-up. Am J Ind Med 1994;25:477−88.
[39] El-Ramady H, Abdalla N, Alshaal T, Domokos-Szabolcsy É, Elhawat N, Prokisch J, Sztrik A, Fári M, El-Marsafawy S, Shams M. Selenium in soils under climate change, implication for human health. Environ Chem Lett 2014;13:1−19.
[40] Combs G. Selenium in global food systems. Br J Nutr 2001;85(5):517−47.
[41] Dennis Lemly A. A procedure for setting environmentally safe total maximum daily loads (TMDLs) for selenium. Ecotoxicol Environ Saf 2002;52:123−7.

[42] Saiki M, Lowe T. Selenium in aquatic organisms from subsurface agricultural drainage water, San Joaquin Valley, California. Arch Environ Contam Toxicol 1987;16:657–70.
[43] Holmes A, Gu F. Emerging nanomaterials for the application of selenium removal for wastewater treatment. Environ Sci 2016;3:982–96.
[44] Lai C, Wen L, Shi L, Zhao K, Wang Y, Yang X, Rittmann B, Zhou C, Tang Y, Zheng P, Zhao H. Selenate and nitrate bioreductions using methane as the electron donor in a membrane biofilm reactor. Environ Sci Technol 2016;50(18):10179–86.
[45] Zouari M, Ben Ahmed C, Fourati R, Delmail D, Ben Rouina B, Labrousse P, Ben Abdallah F. Soil fluoride spiking effects on olive trees (*Olea europaea* L. cv. Chemlali). Ecotoxicol Environ Saf 2014;108:78–83.
[46] Loganathan P, Hedley M, Grace N, Lee J, Cronin S, Bolan N, Zanders J. Fertiliser contaminants in New Zealand grazed pasture with special reference to cadmium and fluorine: a review. Aust J Soil Res 2003;41(3):501–32.
[47] Li R, Tan J, Wang W. Discussion on fluorosis induced by foodstuff in Guizhou Province. Chin J Med 1982;62(7):425–8.
[48] Reddy D, Harinath Y, Seshaiah K, Reddy A. Biosorption of Pb(II) from aqueous solutions using chemically modified *Moringa oleifera* tree leaves. Chem Eng J 2010;162(2): 626–34.
[49] Rahmani A, Mousavi H, Fazli M. Effect of nanostructure alumina on adsorption of heavy metals. Desalin 2010;253(1–3):94–100.
[50] Rossi E. Low level environmental lead exposure—a continuing challenge. Clin Biochem Rev 2008;29(2):63–70.

Trace elements in coals

2

Yuming Zhou, Yongchun Zhao, Junying Zhang, Chuguang Zheng
State Key Laboratory of Coal Combustion, Huazhong University of Science and Technology, Wuhan, China

2.1 Introduction

Trace elements (TEs) are one of the most important components in coal, and their healthy impact and potential value should be given special attention. As a result of different and distinct geological circumstances, the contents, occurrences, and distributions of TEs are various. In this chapter, the occurrence of TEs in coals, including the average content of TEs in coals, and the distribution of TEs in coals from different coal-forming periods, coal ranks, and coal-bearing regions and countries are summarized. According to these data on TEs, the change in the tendency of TEs around the world are summarized, and coal samples and regions that contain TEs in abnormal status are pointed out. These conclusions are conducive to evaluating the influence of TEs in coal during combustion procedures.

2.2 Average content of trace elements in coals

The average content of TEs in coals is one of the most significant statistics for the evaluation of potential hazards and can also be useful in assessing the geochemical characteristics of an element in regional coal. Large numbers of samples are the basic guarantee for accurate and precisely average contents. In addition, statistical methods are also important. There are four statistical methods that have been published in the literature to calculate the average concentration of TEs in coals: arithmetic mean, geometric mean, weighted mean, and bootstrap mean. A mathematical description of these methods is briefly introduced in the following [1]:

1. Arithmetic mean:

$$X_a = \frac{X_1 + X_2 + \cdots + X_n}{n} = \frac{1}{n}\sum_{i=1}^{n} X_i \tag{2.1}$$

where $\overline{X}_a$ is the arithmetic mean content of TEs in coal, X_i is the tested content value of TEs in each coal sample, and n is the number of these samples.

Emission and Control of Trace Elements from Coal-Derived Gas Streams
https://doi.org/10.1016/B978-0-08-102591-8.00002-7

2. Geometric mean:

$$\overline{X} = \sqrt[n]{X_1 \cdot X_2 \cdots X_n} = \sqrt[n]{\prod_{i=1}^{n} X_i} \tag{2.2}$$

where $\overline{X}$ is the geometric mean content of TEs in coal, X_i is the tested content value of TEs in each coal sample, and n is the number of samples.

3. Weighted mean:

$$X_w = \frac{W_1X_1 + W_2X_2 + \cdots + W_nX_n}{W_1 + W_2 + \cdots + W_n} = \frac{\sum_1^n W_iX_i}{\sum_1^n W_i} \tag{2.3}$$

where X_w is the weighted mean content of TEs in coal, X_i is the tested value of TEs in each coal sample, W_i is the replication times of X_i, and n is the number of samples.

4. Bootstrap simulation method:

A random sample $X = (x_1, x_2, \ldots, x_n)$ of size n is observed from a completely unspecified probability distribution F. The sampling distribution $R(X, F)$ is the function of X and F. Assuming that $\theta = \theta(F)$ is some parameter of F, F_n is the empirical distribution function of X, and $\widehat{\theta} = \widehat{\theta}(F_n)$ is the estimator of θ, the estimation error can be expressed as:

$$R(X, F) = \widehat{\theta}(F_n) - \theta(F) \tag{2.4}$$

The basic steps of computing the distribution $R(X, F)$ by bootstrap simulation are summarized as follows:

1. The value of observed sample $X = (x_1, x_2, \ldots, x_n)$ is the finite overall sample (called original sample) $x_i \sim F(x)$, $i = 1, 2, \ldots, n$. The empirical distribution function of the original sample is shown as:

$$F_n \begin{cases} 0 & x < x_{(1)} \\ \dfrac{k}{n} & x_{(1)} \leq x < x_{(k+1)} \\ 1 & x \geq x_{(1)} \end{cases} \tag{2.5}$$

where $x_{(1)} \leq x_{(2)} \leq \ldots \leq x_n$ is the statistics of $x_1, x_2, \ldots, x_n$ sorted in ascending order.

2. Monte Carlo simulation is used to randomly simulate N groups of samples $x^*_{(j)} = \left(x^*_{(1)}, x^*_{(2)}, \ldots, x^*_{(n)}, j = 1, 2, \ldots, N\right)$ (N is a very large number) from F_n, and this regeneration sample is called the bootstrap sample. The decision method of the empirical distribution function by Monte Carlo simulation is (a) generate random integer η, with independence and uniformity between 0 and M ($M \gg n$); (b) let $i = \eta\%\, n$, and i is the remainder of n divided by η; (c) find the sample x_i as the regeneration sample x^* in observed samples; and (d) ensure x^* is the needed random sample.

3. Calculate the statistics of bootstrap samples:

$$R^*(R^*, F_n) = \widehat{\theta}(F_n^*) - \widehat{\theta}(F_n) \rightarrow R_n \tag{2.6}$$

where F_n^* is the empirical distribution function of the bootstrap sample. A small sample cannot derive $\theta(F_n)$, but $\widehat{\theta}(F_n)$ is used to approximate it.

4. Use the distribution of R_n (under a given situation) to simulate the distribution of T_n. $\theta(F) \approx \widehat{\theta} - R_n$, which can receive N numbers of $\theta(F)$. Then, the distribution and eigenvalue of the unknown parameter θ can be obtained.

The weighted mean of TEs in Chinese coals was proposed by Ren et al. [2]. Ren classified coals based on the coal-forming periods and calculated the weighted mean of TEs according to the percentage of coal reserves from different coal-forming periods in national coal reserves. Tian et al. [1] evaluated the variability and uncertainty of the average concentration of TEs in Chinese coals by the bootstrap simulation method. Bootstrap simulation is a type of Monte Carlo simulation, which is a statistical processing method involving resampling and replacement; details of the bootstrap simulation method can be found in the literature [1]. Among the four statistical methods, the weighted mean and bootstrap mean are more accurate because the influence of very low or high data with low frequency is not significant. Dai et al. [3] reviewed the abundances of TEs in Chinese coals and concluded that the total TE concentration was 136.6 μg/g, which is higher than the average global content.

The average TE contents can be calculated based on the information of world coals. These data are meaningful for comparison study of TEs in coal. It is useful to evaluate the TE contents level in coals from different regions, ranks, or forming periods, and to conclude if the coal sample is normal, enriched, or impoverished in target TEs, so that the preliminary quality of studied coals can be obtained. The most important elements are toxic elements (such as Hg, Pb, As, and Se) and valuable elements (such as Ge, Ga, Re, and noble metals). Content evaluation and comparison of these elements are notably significant for avoiding potential environmental contamination and valuable elements extraction. In addition, through the analysis of TE contents in coals, the affinity between diverse elements and minerals can be researched. Many studies were focused on TE contents in coals from different countries.

According to TE contents in coals around the world, most elements' contents fluctuate to a certain and limited range. However, some elements' contents in coals change significantly in different countries. There are a number of abnormal enrichments of TEs widely distributed in different coals, the extremely high content is about 10 times the world average. A coal sample from Petrosani basin (Romania) shows W content at 18.43 μg/g, which is more than 16 times the world average content; another coal from Donetsk basin (Ukraine) has an extremely high Se content at 12.8 μg/g, which is almost 10 times of world average value [4]. P (1891.2 μg/g) in Pakistan coal has reached an exorbitant value; it is a result of a combination with pyrite, kaolinite, and calcite, which could influence the application of the coal [5]. These abnormal circumstances are useful to evaluate the coal quality of specific locations; the negative impact from enriched TEs should be paid particular attention during combustion or other processes. Other elements' contents change in a reasonable context.

As arithmetic average content of US coal can reach 24 μg/g, and UK coals have a higher content of 31.3 μg/g, which is almost three times the world average content; there is also an extremely high content of Cl in UK coals at 6120 μg/g [6,7]. As is also enriched in the North Bohemian Basin in the Czech Republic at 39.94 μg/g; the coal there also has an extremely high Ni content (94.07 μg/g) [8]. Li is enriched in southeastern Iran coal at 49.31 μg/g and in DPR Korea coal at 49.2 μg/g, which is more than four times the world average content value; Rb in southeastern Iran coal also has a high content at 71.12 μg/g [9,10]. Santa Catarina coal from Brazil is enriched with various TEs: Cs (8.49 μg/g), Ce (78.8 μg/g), Nb (29.2 μg/g), and Ta (5.21 μg/g) [11]. The values of most of TEs are more than world average contents; this condition is probably due to the relatively high mineral content. In the Dobrudza coal basin, Bulgaria, Ba is the most enriched element at 1384 μg/g. The other enriched elements are Cl, Br, Mn, and Pb; these elements are 7.1−3.5 times world average values [12]. A notably enriched Cs content was observed in South African at 85.2 μg/g [13]. Because coal is the most important fuel in South Africa, the its combustion could cause heavy metal pollution. Enrichments of valuable elements have been discovered, such as Co (36.14 μg/g) in India, Mn (657.7 μg/g) in Russia, Li (82.37 μg/g) in Iran, and Cr (118.7 μg/g) in Romania [4,14]. These results could be important information not only for the evaluation of coal quality and environmental impact but also for developing procedures for the concentration and extraction of TEs as valuable resources.

Table 2.1 shows the average concentration of TEs in coals globally, calculated by different researchers. The average concentration of Hg, Pb, F, and Se in Chinese coals is higher than that in global coals reported by Ketris and Yudovich [15]. Concentrations of elements Sb, Be, Ni, and Cr in Chinese coals are near those of the world average, while As in Chinese coals is lower than the world average. Considerating the toxicity and potential utilization of TEs, Hg, Se, and Pb should be paid special attention.

According to the analysis of TE contents in different countries around the world, TEs are distributed unevenly, impacted by various geological factors, and are always changing during the coal-forming periods. Thus analysis results of coal samples usually present a large discrepancy. However, the average contents of TEs are still limited. Hence investigation of average TE contents is indispensable. It should also be pointed out that there could be abnormal contents of TEs in any coalfields, so the given coal sample still needs to be studied thoroughly, especially from regions with complex geological circumstances.

2.3 Distribution of trace elements in coals from different coal-forming periods and coal ranks

The main coal-forming periods in China can be classified into six periods: Late Carboniferous and Early Permian (C_2−P_1), Late Permian (P_2), Late Triassic (T_3), Early to Middle Jurassic (J_1−J_2), Late Jurassic and Early Cretaceous (J_3−K_1), and

Paleogene and Neogene (E–N), which are essentially consistent with the global main coal-forming periods [2]. In addition, there is a minor but significant amount of Early Carboniferous (C_1) coal in southern China [3]. The reserves of coal formed in different coal-forming periods are quite different. The total recoverable coal reserves are made up of 38.1% C_2–P_1, 7.5% P_2, 0.4% T_3, 39.6% J_1–J_2, 12.1% J_3–K_1, and 2.3% E–N coals [20]. Different coal-forming periods, which vary in botanical composition, sedimentary environment, climate, and epigenetic geological factors that affect coalification, can significantly influence the distribution of TEs. Thus it is important to investigate the distribution of TEs in Chinese coals from different coal-forming periods.

Table 2.2 summarizes the average concentration of TEs in Chinese raw coals from different coal-forming periods. It can be clearly seen that most TEs are enriched in P_2 and T_3 coals, especially in T_3 coals, while the enrichment of TEs in J_1–J_2 and J_3–K_1 coals is relatively low. Elements of Ni, Co, As, F, and Cr are concentrated in E–N coals.

Based on the metamorphic stage, coals can be divided into three main coal ranks: lignite, bituminous coal, and anthracite. The distribution of TEs in coals of different ranks also varies. Shanxi Province is one of the most dominant coal-producing areas in China; the investigation of TEs in coal from Shanxi Province is a representative work. Zhang et al. [21] systematically analyzed the distribution of TEs in coals from Shanxi Province with different forming periods and ranks. Compared to bituminous coals and anthracite, Shanxi lignite is enriched with As, Ba, Cd, Cr, Cu, F, and Zn, Shanxi bituminous coals are highly enriched with Hg, B, and Cl, and Shanxi anthracite is enriched with Cl, Hg, U, and V. Compared with average contents in the earth's crust, on the basis of coal ranks, Shanxi lignite is enriched with As, Cd, Mo, and Se, Shanxi bituminous coals are enriched with Ce, Hg, and Se, and Shanxi anthracite is enriched with Hg, Mo, and Se. On the basis of coal-forming periods, Shanxi J_2 coal is slightly enriched with As and Se, P_1 coal is enriched only with Se, and C_3 coal is enriched with Cd, Hg, Mn, and Se. Comparing the average contents of US coals and world coals, Shanxi N coal is enriched with As and Se, and Shanxi J_2, P_2, and C_3 coal is enriched only with Hg. On the basis of coal ranks, Shanxi bituminous coals and anthracite are highly enriched with Hg; in addition, Shanxi bituminous coals have a high Cd, F, and Th concentration, and Shanxi anthracite is slightly enriched with Th. Detailed data are shown in Tables 2.3 and 2.4.

Tian et al. [1] investigated the distribution of eight TEs in coals of different ranks and forming periods on a nationwide level and concluded that the average concentration of Hg, Se, Pb, and Cr increases gradually with an increase in coal rank. Elements of Ni, As, and Sb are more enriched with lignite than with bituminous coal and minimally enriched with anthracite coal. In addition, all eight TEs were found to be enriched in Late Triassic (T_3) and Late Permian (P_2) coals, especially the former. However, the J_1–J_2 and J_3–K_1 coals had low concentrations of TEs. Detailed data are shown in Tables 2.2 and 2.5. Chen et al. [22] reported a Permian coal in Anhui Province enriched with Si, K, Se, Cd, Sn, Sb, and W, whereas Zn, As, and Lu are lower. The enrichment factor and concentration coefficient show that Se, Cd, Sn, Sb, and W are preferably enriched. A similar study of Permian coals in Anhui Province

Table 2.1 Trace element contents in different countries (μg/g).

Elements	World average [15]			China [3]	American [6,16]	UK [7]	Canada [17]	Iran [10]	Turkey [18]	South African [4,13]
	All	Brown	Hard							
Li	12	10	14	31.8	16	29.1	13.0	49.31		
Rb	14	10	18	9.25	21	9.8	8.7	71.12	14.8	(0.6)
Cs	1.0	0.98	1.1	1.13	1.1	0.79	0.7	7.13	0.97	(0.64)
Tl	0.63	0.68	0.58	0.47	1.2	0.73		0.41	0.37	
Sr	110	120	100	140	130	46.0	147	260	224	(770.1)
Ba	150	150	150	159	170	100	380	270	122	(597)
Be	1.6	1.2	2.0	2.11	2.2	1.65	0.9	1.81		
Sc	3.9	4.1	3.7	4.38	2.8	3.15	4.8	9.92		(5.3)
Y	8.4	8.6	8.2	18.2	8.5 (8.93)	5.84			6.74	
La	11	10	11	22.5	(11.70)		9.6	18.21	8.28	(50.1)
Ce	23	22	23	46.7	(23.69)		25.8	37.31	15.62	(85.2)
Pr	3.5	3.5	3.4	6.42	(10.21)		3.0		0.66	
Nd	12	11	12	22.3	(12.32)		6.2		1.63	(39.5)
Sm	2.0	1.9	2.2	4.07	(2.54)		1.5		7.73	(6.4)
Eu	0.47	0.50	0.43	0.84	(0.42)		0.3			(0.7)
Gd	2.7	2.6	2.7	4.65	(2.91)					(7.1)
Tb	0.32	0.32	0.31	0.62	(1.16)		0.3			(0.7)
Dy	2.1	2.0	2.1	3.74	(3.11)		2.0			(4.0)
Ho	0.54	0.50	0.57	0.96	(1.03)		0.2			
Er	0.93	0.85	1.00	1.79	(1.24)					
Tm	0.31	0.31	0.30	0.64	(0.63)		1.5			
Yb	1.0	1.0	1.0	2.08	(1.01)		0.9			
Lu	0.20	0.19	0.20	0.38	(0.37)		0.1			(0.28)
Ga	5.8	5.5	6.0	6.55	5.7	3.26		12.53	4.13	
Ge	2.2	2.0	2.4	2.78	5.7	9.31		0.43	0.59	
Ti	800	720	890		800		1200	1558	1200	(1428)
Zr	36	35	36	89.5	27				37.88	
Hf	1.2	1.2	1.2	3.71	0.73		1.2	1.89	2.08	(1.81)
Th	3.3	3.3	3.2	5.84	3.2		2.4	7.63	3.18	8.9 (8.25)
Sn	1.1	0.79	1.4	2.11	1.3	1.45		1.62	0.27	(1.9)
V	25	22	28	35.1	22	35.0	44.3	72.19		39.2 (14.7)

DPR Korea [9]	Czech [8]	Nigeria [19]	Brazil [11]	India [14]	Pakistan [5]	Bulgaria [12]	Australia [4]	Romania [4]	Russia [4]	Ukraine [4]	For most coals
49.2		25.5	39.9		45.67	27					1–80
27.1	14.71	21.7	44.3		93.72	23	8.7	0.7	1.1	0.4	2–50
	3.98	2.7	8.49			1.8	1.30	5.15	0.88	0.30	
0.3			1.69								0.2–1
132	87.50		69.3	127.6	137.26	112	84.4	93.5	271.1	85.5	15–500
	103.54	78.8	291.2		139.26	1384	67	317	332	44	20–1000
1.2	3.09	4.4	4.10	2.27	3.81	2.9					0.1–15
2.3	7.21	7.7	13.3		7.41	3.2	5.7	13.3	3.4	3.7	1–10
7.2	10.19	10.9	30.7			16					
14.5	15.17	18.2	28.9			14.5	3.4	22.8	13.2	8.9	
27.2	36.11	39	78.8	0.4		25	19.7	44.3	23.0	16.2	
2.9			7.69								
11.1	17.55		33.8				12.3	16.6	10.0	7.4	
2.3			7.40			2.4	2.6	4.6	2.4	1.8	
0.5	0.71		0.19			0.49	0.3	0.8	0.3	0.2	
1.4			6.92				2.6	5.6	2.2	2.1	
0.3	0.57		0.90			0.57	0.5	0.7	0.4	0.2	
2			4.94				2.0	3.1	1.6	1.1	
0.4			1.06								
1.1			2.50								
0.3			<0.1								
1	0.74		2.85			1.0					
			<0.8			0.22	0.21	0.44	0.20	0.19	
6.3	10.03		15.9			2.0					1–20
1.4	2.47		7.53			5.6					0.5–50
			2954		2200	1342	6.3	2456	747	382	10–2000
40.2	98.81	70.7	211.1			87					5–200
	1.61		5.22			1.3	1.84	1.95	1.57	0.74	
5.1	3.96		14.8			4.7	3.22	8.06	3.71	2.99	0.5–10
4.8	1.09		3.76				1.3	1.8	0.8	0.5	
32.2		54.7	104.0	86.00	116.61	42	23.6	64.0	8.8	8.5	2–100

Continued

Table 2.1 Trace element contents in different countries (μg/g).—cont'd

Elements	World average [15] All	Brown	Hard	China [3]	American [6,16]	UK [7]	Canada [17]	Iran [10]	Turkey [18]	South African [4,13]
Nb	3.7	3.3	4.0	9.44	2.9	1.17	6.0	5.4	4.95	
Ta	0.28	0.26	0.30		0.22		0.2	0.4		(0.64)
Mo	2.2	2.2	2.1	3.08	3.3	3.39	1.5	1.62	8.39	2.1 (0.38)
W	1.1	1.2	0.99	1.08	1.0		0.4		0.09	(7.68)
U	2.4	2.9	1.9	2.43	2.1	1.09	1.6		11.34	2.6 (2.01)
B	52	56	47	53	49		17.0	10.5		
P	230	200	250		430			378.5		
F	88	90	82	130	98					
Cl	180	120	340	255	614	6120	49.5			
Br	5.2	4.4	6.0		17		1.4			(1.6)
I	1.9	2.3	1.5		<1.0					
Cu	16	15	16	17.5	16	39.0	11.0	39.72	12.61	16.8 (80.5)
Ag	0.095	0.09	0.10		<0.1			0.08		
Au	0.037	3.0	4.4		<0.05			0.008		
Zn	23	18	28		53	8.44	22.0		30.06	17.9 (7.5)
Cd	0.22	0.24	0.20	0.25	0.47	0.05		0.19	0.02	0.3
Hg	0.10	0.10	0.10	0.163	0.17	0.174		0.24		0.3 (0.09)
In	0.031	0.021	0.04	0.047	<0.3			0.05		(0.03)
Pb	7.8	6.6	9.0	15.1	11	23.5	4.0	40.23	11.65	15.03
Bi	0.97	0.84	1.1	0.79	<1.0	0.12		0.31		
As	8.3	7.6	9.0	3.79	24	31.3	0.7	10.1	13.92	4.7 (1.3)
Sb	0.92	0.84	1.0	0.84	1.2	2.29	0.7	0.96	0.41	0.5 (0.11)
Se	1.3	1.0	1.6	2.47	2.8	2.33	2.3	1.55		1.2
Cr	16	15	17	15.4	15	14.8	14.0	48.23	61	41.4 (17.8)
Mn	86	100	71		43	30.1	5.9	182.86	218	135 (88.5)
Co	5.1	4.2	6.0	7.08	6.1	3.38	2.0	5.53	15.12	11.0 (6.5)
Ni	13	9.0	17	13.7	14	40.0	8.0	19.65	61.28	27.2 (17.4)
Pd	0.0074	0.013	0.001					0.018		
Ir	0.002	0.002	0.001							
Pt	0.035	0.065	0.005					0.02		

DPR Korea [9]	Czech [8]	Nigeria [19]	Brazil [11]	India [14]	Pakistan [5]	Bulgaria [12]	Australia [4]	Romania [4]	Russia [4]	Ukraine [4]	For most coals
2.4	21.38	9.4	29.2								1–20
0.3	1.34		5.21			0.28	0.27	0.82	0.41	0.32	
1.3	2.66		2.49			5.7	0.73	1.14	0.56	0.63	0.1–10
3.3	4.75		3.98			2.2	4.62	18.43	3.56	2.66	
1.4	2.46	2.2	6.30			3.6	1.47	3.67	1.37	1.48	0.5–10
44.7	32.98		26.8		132.69	71					5–400
	944.51				1891.2	120					10–3000
	205.14										20–500
400						2430	201	70	67	1127	50–2000
						26	3.6	1.0	4.1	11.1	
	88.08	13.3	19.8	20.65	27.61	28	68.8	69.3	62.7	69.0	0.5–50
0.3	0.17					0.16					0.02–2
						0.007					
81.0	67.43	20.5	117.1	40.4	52.26	56	25.2	56.2	25.9	7.4	5–300
0.5	0.34	0.2	<0.1	0.39							0.1–3
0.4							0.08	0.14		0.39	0.02–1
						0.019	0.02	0.03	0.03	0.02	
24.4	9.44	7.2	29.1	21.11		32					2–80
0.2	2.65		<0.1								
14.6	39.94	1.2	13.5	3.85	6.69	19	1.9	11.7	1.0	3.1	0.5–80
0.4	3.01		1.00	2.54		1.8	0.91	1.11	0.19	1.80	0.05–10
			4.09				2.4	5.6	3.1	12.8	0.2–4
17.8	57.43	41.5	52.9	70.00	51.71	16	65.3	118.7	18.2	20.2	0.5–60
		115.5	121.9	43.11	11.45	312	68.2	310.69	657.7	42.2	5–300
	8.28	32.7	7.59	36.14	4.73	9.3	10.0	17.7	6.2	7.6	0.5–30
	94.07	32.7	16.0	19.91	21.22	15	48.5	76.4	22.5	45.3	0.5–50
											2–80

Table 2.2 The average concentration of trace elements (TEs) in Chinese coals from different coal-forming periods (arithmetic mean, μg/g) [2].

TE	$C-P_1$	P_2	T_3	J_1-J_2	J_3-K_1	E–N
Hg	0.22 (0.29)	0.42 (0.39)	0.48 (0.50)	0.11 (0.24)	0.19 (0.27)	0.07
As	2.59 (3.71)	6.42 (6.99)	9.16 (10.37)	3.23 (3.94)	6.04 (6.49)	12.016
Be	1.92	2.42	2.67	2.41	1.88	0.93
Sb	0.68 (0.78)	1.67 (1.65)	3.62 (3.90)	0.61 (0.90)	1.37 (1.17)	1.03
Cd	0.30 (0.37)	0.62 (0.69)	0.99 (0.82)	0.13 (0.26)	0.13 (0.18)	0.44
F	149.40	200.90	84.10	84.46	134.20	352.80
Cr	15.81 (19.36)	29.02 (31.24)	50.40 (33.43)	11.35 (12.52)	11.89 (22.80)	43.33
Co	4.24	10.07	7.72	8.59	8.09	11.75
Pb	20.70 (22.34)	26.95 (24.07)	20.38 (21.81)	8.76 (15.00)	12.29 (15.80)	26.21
Mo	3.47	8.40	3.14	1.92	2.55	3.44
Ni	11.75 (15.66)	24.05 (26.21)	28.14 (20.30)	11.97 (10.80)	10.56 (13.03)	56.46
Se	4.84 (3.35)	4.71 (4.33)	1.47 (2.76)	0.43 (0.98)	0.63 (0.91)	1.01
Th	8.71	7.81	9.07	3.01	4.91	3.72
U	2.60	4.29	3.90	1.24	1.72	3.93

Weighted mean, from Dai and Ren [2].
() Bootstrap mean, from Tian et al. [1].

also shows that the contents of Co, Cr, Se, and Ni are higher than their averages for Chinese coals and world coals [23]. It should be pointed out that these results could be changed in some special region, for example, Wang et al. [24] reported a higher Hg concentration in anthracite than in bituminous coal. Thus these results can only be suitable for nationwide statistical levels.

According to the potential hazardous TE contents in Chinese coals of different forming periods, J_1-J_2 coal presents better quality because the contents of Hg and As are relatively low, and the average contents of Cr, Cd, and Pb are minimum in all the coal-forming periods. $C-P_1$ coal has minimal contents of As, Co, and Sb in the coals of all forming periods, and the contents of Be and Ni are also relatively low; however, Se content is the highest. J_3-K_1 coal shows minimal Ni and Be concentration, and the contents of Cd, Hg, and Se are relatively low, but the As content is much higher. These three coal-forming periods account for most of Chinese coal reserves, thus TE emission during exploitation and combustion should be given further attention.

Table 2.3 Contents of trace elements (TEs) in coals of different forming periods from Shanxi Province (μg/g) [21].

TE	N			J_2			P_1			C_3 (Late carboniferous)		
	Range	Average	Sample number	Range	Average	Sample number	Range	Average	Sample number	Range	Average	Sample number
As	7.6–18.4	13.5	4	0.1–82.4	12.3	17	bdl–17.5	1.7	52	<0.4–8.4	2.6	37
B							bdl–3.3	1.4	10	bdl–45	12	11
Ba	52.6–493.0	303.9	4	<15–200	71	17	31–427	103	52	<10–244	96	37
Cd	2.2–2.5	2.2	3	bdl–1.1	0.6	11	bdl–2.4	1.0	44	bdl–2.0	1.2	34
Cl	bdl–100	60	3	bdl–260	116	11	100–1865	366	45	bdl–940	557	34
Co	4.3–9.9	7.4	4	1.6–25.0	7.4	17	0.4–22.5	4.8	52	0.7–10.0	1.2	37
Cr	26.2	26.2	1	12.2–19.4	7.2	11	2.5–32.8	11.0	52	1.6–53.1	14.4	15
Cu	21–108	53	3	bdl–264	29.8	11	bdl–48	15.7	44	bdl–60.9	15.8	34
F	259–393	313	3	bdl–117.6	35.2	11	bdl–1900	157	43	bdl–3600	270	30
Hg	0.03–0.05	0.04	3	bdl–0.28	0.10	11	bdl–2.89	0.35	44	bdl–2.67	0.59	34
Mn						11	bdl–62.6	16.3	10	bdl–215	55.7	12
Mo	3.8–37.9	9.4	4	bdl–6.4	2.3	11	bdl–24.1	5.0	52	bdl–56.9	7.8	37
Ni				11.1–17.0	13.1	3	2.7–46.0	11.2	10	bdl–46.0	19.8	12
Pb	29.0–38.0	33.5	2	bdl–30.0	9.4	11	bdl–66	25.0	44	bdl–88.0	28.7	34
Sb	0.4–0.7	0.5	4	0.1–1.7	0.6	17	0.1–11.3	0.7	52	0.1–0.7	0.3	37
Se	bdl–12.6	4.5	4	0.1–1.9	0.6	17	0.1–12.2	5.4	52	bdl–12.1	6.1	37
Th	2.9–7.1	4.8	4	0.5–8.0	1.5	17	0.5–20.6	9.4	52	0.7–17.1	7.3	37
U	2.2–20.2	8.2	4	0.22–3.4	0.8	17	0.2–5.9	2.7	52	0.9–49.9	5.7	37
V							bdl–25.8	9.2	10	bdl–84.1	25.7	12
Zn	58.4–106	83.4	3	bdl–64.1	23.7	17	bdl–149	38.6	45	bdl–193.0	63.3	34

bdl, below detection limit.

Table 2.4 Contents of trace elements (TEs) in coals of different ranks from Shanxi Province (μg/g) [21].

	Anthracite			Bituminous coal			Lignite		
TE	Range	Average	Sample number	Range	Average	Sample number	Range	Average	Sample number
As	bdl–5.4	2.4	17	bdl–82.4	3.6	89	7.6–18.4	13.5	4
B	bdl–11	3.0	7	bdl–45	7.4	17	bdl	bdl	3
Ba	57.7–244	131	17	<10–200	88	89	52.6–493	3.4	4
Cd	bdl–1.4	0.5	14	bdl–2.5	1.1	75	2.2–2.5	2.2	3
Cl	bdl–522	217	14	bdl–1865	370	76	bdl–100	60	3
Co	0.4–13.7	3.9	17	0.6–25.0	4.7	89	4.3–9.9	7.4	4
Cr	4.5–53.1	13.6	17	1.6–32.8	10.3	24	26.2	26.20	1
Cu	bdl–26	13.5	14	bdl–246	17.9	75	21.0–108	53.0	3
F	bdl–400	73	11	bdl–3600	170	72	259–393	313	3
Hg	bdl–2.30	0.50	14	bdl–2.89	0.39	75	0.03–0.05	0.04	3
Mn	bdl–191	66.3	8	bdl–72.9	17.7	17			
Mo	bdl–56.9	8.9	17	bdl–24.1	5.1	83	3.8–37.9	9.4	4
Ni	6.69–46.0	26.8	8	1.8–29.4	10.1	17			
Pb	bdl–69.7	27.8	14	bdl-66.0	23.9	75	29.0–38.0	33.0	3
Sb	0.1–1.9	0.4	17	0.1–11.3	0.6	89	0.4–0.7	0.5	4
Se	1.7–11.0	5.5	17	bdl–13.5	4.7	89	bdl–12.6	4.5	4
Th	2.2–18.2	8.2	17	0.5–20.6	7.2	89	2.96–7.1	4.8	4
U	1.1–49.9	7.7	17	0.2–18.7	2.6	89	2.2–20.2	8.2	4
V	bdl–84.1	30.9	8	bdl–33	9.0	17			
Zn	bdl–193	26.8	14	bdl–173	45.2	82	58.4–106	83.4	3

bdl, below detection limit.

Table 2.5 Bootstrap mean concentration of trace elements in different coal ranks (μg/g) [1].

Coal ranks	Hg	As	Se	Pb	Cd	Cr	Ni	Sb
Bituminous	0.26	5.26	3.77	18.63	0.44	18.33	16.89	1.10
Anthracite	0.62	4.86	5.83	22.77	0.99	22.62	15.73	0.78
Lignite	0.24	11.79	1.98	14.90	0.49	17.94	21.86	1.14

P_2 coal has highest contents of Mo, Pb, and U; there are also relatively high concentrations Hg, Cd, and F, especially the contents of Hg and Cd, which are double the Chinese average value. E–N coal has highest contents of As, F, Co, and Ni, and the contents of Cr and U are relatively high. J_3–K_1 coal has relatively higher concentrations of As and Co, and the whole content of TEs is high. These three coals account for a minor part of the reserves in China, and whole TEs are relatively high and dispersed unevenly. When considering these characteristics, application of these coals could cause environmental hazards in limited locations. Exploitation of these coals should be considered thoroughly and restrained. The migration and transformation behaviors of TEs should be studied systematically.

Ketris et al. [15] summarized TE contents in coals of different ranks around the world, as shown in Table 2.1. The rank of coal relates to the forming period, because hard coals are Paleozoic and brown coals are Mesozoic and Cenozoic in general. Because hard coal lasts longer, it probably receives TEs through hydrothermal input. According to their data, more elements are enriched in hard coals compared with brown coals. Strong enrichments in hard coals include Rb, Be, Zn, and Ni, and weak ones include Co, Ge, V, Pb, and Se; brown coals are enriched in B, U, and Mn. These results reflect the overall variation tendency of TEs in coal with different forming periods and ranks. The depth of the coal horizon indicates the age of coals, because upper coal has a shorter forming period and lower rank. Eskenazy [12] reported the distribution of TEs in coal from the deepest coal deposit in Bulgaria, which is covered by Paleozoic up to Cenozoic sediment. The coal rank has exceeded highly volatile bituminous coal. The contents of Ba, Cl, Br, Mn, Ge, Pb, W, Mo, As, Zr, Li, Cs, Cu, Zn, Sb, Ti, V, Th, U, Ag, Be, and Y are enriched from 1.5 to 9.2 times as compared with the average world contents. Another study demonstrates a systematic increase in the concentration of Se, Be, and Sb with depth because these elements have a strong association with fixed carbon in coals with higher rank. Other TEs (As, Pb, Cd, Cu, Ni, and Zn) suddenly increase their concentrations at particular depths, but are unaffected by coal-forming period or rank. This phenomenon depends on the unexpected invasion of TE-combined minerals. On this occasion, there seems to be no obvious relation between coal rank and TE concentration [14,25].

However, with regard to specific coal samples or elements, the tendency could change somewhat. Some alkaline-earth elements (Mg, Ca, Sr, and Ba), Na and B in US coals, show a decrease in concentration with increasing coal rank. The phenomenon could be explained as elements combined with functional groups in lower rank coals. With the rank increases, the functional groups were destroyed and these elements were released. This tendency is also suitable for Pb, because it can associate with organic matter in low-rank coals [26,27]. Cl and Br concentrations also show a descendant tendency with increasing coal rank in UK coals, and a few TEs vary significantly [28,29]. Notably, some TE contents are not controlled by coal rank. Research on radium content in different coal basins in Bulgaria showed no remarkable relation between Ra content and coal ranks, because the Ra contents in different coals vary randomly. In addition, geological activity can also change the distribution of TEs. A study focused on Turkish coal showed that high concentrations of U and Th were related to blocks within the basin frames, which were enriched with elements and were related to volcanism during the coal-forming period.

To sum up, the distribution of TEs in coals with different forming periods and ranks can change significantly. However, some elements are found in specific coal ranks, and it is possible that these ranks are a potential source of these valuable elements. Lin et al. [30] evaluated the TEs in US coals and concluded that the most promising coal sample for Li, Ga, and Ge recovery was bituminous coal because it has a relatively higher content of these valuable metals. Ge is usually found with higher concentration in coals with low rank and low calorific value but with lower content in high-rank coals; this trait is useful for prospecting coal deposits enriched with Ge and exploiting the potential source of Ge from coal and subsequent by-products [28,31]. Statistical analyses on a high-ash, bituminous Iranian coal seam indicated enrichment of Ga and V, which could have potential for the extraction of these valuable elements [32]. Similar results were obtained in statistical assessment of TE values from selected African coals (from South Africa, Botswana, Egypt, Tanzania, Nigeria, and Zambia), because Y, V, Li, and Ga contents are higher than world average values [33].

2.4 Distribution of trace elements in Chinese coals from different coal-bearing regions

The geographic distribution of coal reserves in China is uneven. The main coal-bearing areas in China can be divided into five regions: northern China, northeastern China, northwestern China, southern China, and the Tibet–western Yunnan coal-bearing area, as shown in Fig. 2.1 [3]. More than 90% of the coal reserves are in the northern China, northeastern China, and northwestern China coal-bearing regions. Furthermore, coal reserves in Shanxi, Ningxia, Xinjiang, Shaanxi, Gansu, Guizhou, and Inner Mongolia Provinces account for almost 80% of national coal reserves. The concentration of TEs in Chinese coals from different provinces and coal-bearing regions also varies. Table 2.6 gives the arithmetic mean concentrations of TEs in Chinese coals from different coal-bearing regions.

The enrichment of TEs results from geologic processes, i.e., the emplacement of elements by hydrothermal solutions generated by tectonic activity. Different coal-bearing regions in China experience different geologic processes. Thus the concentrations of TEs in Chinese coals from different provinces and coal-bearing regions vary. The coal-bearing regions of northern China, northeastern China, and northwestern

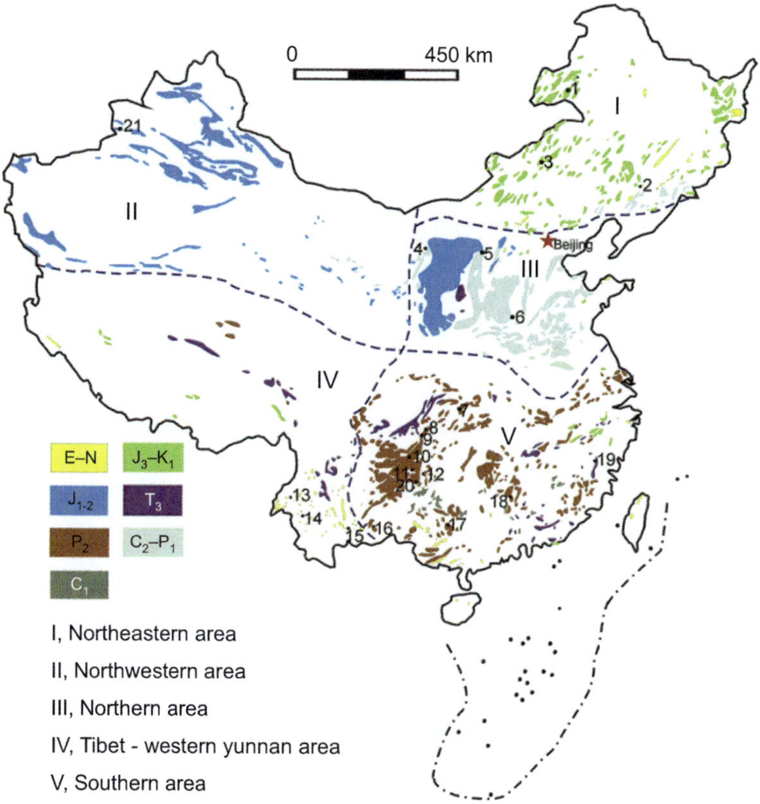

Figure 2.1 Coal distribution areas in China [3].

China usually have low Hg, As, Pb, Mn, Zn, Cd, and Cr concentrations, while coals distributed in southern China, especially southwestern China, are enriched in Hg, As, Pb, Cd, and Cr [1,3,38]. In the regional distribution of Chinese coals, northern China, northeastern China, northwestern China, and southern China are rich in Late Carboniferous to Early Permian coals, Late Jurassic to Early Cretaceous coals, Early to Middle Jurassic coals, and Late Permian coals, respectively [35]. As mentioned earlier, most TEs are enriched in P_2 and T_3 coals and are relatively low in J_1–J_2 and J_3–K_1 coals. Therefore the average concentration of Hg, As, Pb, Cd, and Cr in Chinese coals gradually increases from northern China to southern China.

Hg is the most volatile element around metals and could result in a serious environmental hazard, thus the distribution of mercury should be given special attention. Wand et al. [34] reported Hg contents in coals of the main coal production provinces in China, and the data are shown in Table 2.6. Mercury content in coals varies significantly in different regions and coal seams. Xingjiang, Heilongjiang, Shaanxi, and Jiangxi coals have relatively low mercury contents at <0.20 μg/g. Beijing, Jilin, and Henan coals show relatively higher mercury contents at >0.30 μg/g. Coals in other provinces range from 0.20 to 0.30 μg/g. High-Cr coals in China are found in the

Table 2.6 Average content of trace elements in coals by different provinces and coal-bearing regions (μg/g) [1].

Province and regions	Hg	As	Se	Pb	Cd	Cr	Ni	Sb	F
Anhui	0.43 (0.22[a])	2.89 (4.2[b])	7.54	13.24	0.11	31.25	19.57	0.25 (6.19[c])	
Chongqing	0.31	5.66 (3.2[b])	3.69	30.44	1.22	28.44	20.90	1.71	
Fujian	0.07	9.93 (11[b])	1.22	25.53	0.31	30.48	16.42	0.38 (0.38[c])	
Gansu	0.27	4.14 (2.9[b])	0.51	8.35	0.08	23.70	19.30	0.7 (0.70[c])	
Guangdong	0.07	8.30 (5.0[b])	0.60	24.40	0.25	74.00	24.90		
Guangxi	0.33	16.94 (19[b])	5.03	29.94	0.41	116.41	22.48	5.55 (3.67[c])	
Guizhou	0.39	6.68 (4.6[b])	3.82	23.81	0.79	28.47	22.87	6.01 (10.67[c])	
Hebei	0.15 (0.13[a])	4.88 (3.6[b])	2.31	29.30	0.23	32.52	14.61	0.41 (1.20[c])	
Heilongjiang	0.12 (0.12[a])	3.42 (2.8[b])	0.90	22.15	0.13	15.48	10.49	0.79 (0.90[c])	
Henan	0.20 (0.30[a])	2.20 (2.3[b])	4.86	16.78	0.54	24.94	11.84	0.37 (0.26[c])	
Hubei	0.20	5.30 (4.0[b])	8.76	47.39	0.36	40.52	18.61	1.17 (1.17[c])	
Hunan	0.12	10.59 (4.3[b])	3.72	26.29	0.64	37.03	13.25	1.54 (5.23[c])	
Inner Mongolia	0.22 (0.28[a])	5.77 (5.5[b])	1.10	26.27	0.10	13.02	6.35	0.70 (55.2[c])	
Jiangsu	0.69	2.74 (4.1[b])	6.11	20.98	0.06	19.82	15.48	0.55 (0.39[c])	
Jiangxi	0.16 (0.16[a])	7.41 (7.2[b])	8.39	19.33	0.56	39.75	22.66	1.83 (1.79[c])	
Jilin	0.40 (0.33[a])	11.57 (7.2[b])	4.06	29.00	0.15	23.09	15.34	1.02 (1.02[c])	
Liaoning	0.17 (0.20[a])	5.51 (4.9[b])	0.85	19.68	0.16	26.24	24.13	0.81 (4.44[c])	
Ningxia	0.22	3.65 (5.2[b])	4.27	14.05	1.10	10.63	10.95	0.27	

Qinghai	0.25	2.68	0.30	10.72	0.03	30.82	12.20	0.91 (0.35[c])	
Shaanxi	0.21 (0.16[a])	3.87 (2.2[b])	3.43	35.17	0.75	32.73	18.86	2.95 (0.69[c])	
Shandong	0.18 (0.17[a])	5.23 (3.9[b])	3.66	16.64	0.39	20.62	23.77	0.47 (0.53[c])	
Shanxi	0.17 (0.22[a])	3.84 (3.8[b])	3.85	26.23	0.75	21.57	15.41	1.13 (0.82[c])	
Shendong	0.41	1.22	2.63	14.84	0.15	27.04	9.82	0.47	
Sichuan	0.29 (0.13[a])	5.38 (4.2[b])	3.31	28.29	1.95	33.00	19.28	1.70 (1.25[c])	
Xinjiang	0.06 (0.03[a])	2.97 (2.7[b])	0.24	2.68	0.12	7.83	8.26	0.67 (0.82[c])	
Yunnan	0.36	8.82 (5.9[b])	1.48	42.54	0.80	73.62	24.32	0.97 (1.06[c])	
Zhejiang	0.65	12.04	12.02	17.25	0.47	24.20	9.95	0.73 (0.82[c])	
Northern China	0.23	3.92	3.42	20.54	0.67	19.91	13.55	0.74	
Northwestern China	0.18 (0.10[d])	3.86 (4.26[d])	3.84(0.36[d])	5.55 (7.47[d])	0.08 (0.46[d])	12.31 (19.24[d])	6.34 (16.30[d])	0.75 (0.78[d])	(135.14[d])
Northeastern China	0.21 (0.10[d])	6.62 (5.40[d])	1.61 (0.72[d])	19.94 (15.91[d])	0.15 (0.94[d])	22.68 (17.89[d])	18.52 (12.68[d])	0.65 (0.77[d])	(198.63[d])
Southern China	0.26 (0.24[d])	7.38 (7.39[d])	5.47 (5.32[d])	19.96 (19.70[d])	0.78 (1.15[d])	34.26 (32.52[d])	23.52 (21.71[d])	1.85 (2.06[d])	(323.26[d])
Northern China C_2–P_1	(0.18[d])	(1.75[d])	(5.30[d])	(23.40[d])	(1.01[d])	(13.67[d])	(13.45[d])	(0.44[d])	(144.13[d])
Northern China J_1–J_2	(0.17[d])	(5.54[d])	(0.60[d])	(14.28[d])	(0.87[d])	(15.87[d])	(15.00[d])	(0.51[d])	(128.23[d])

[a]From Wang et al. [34].
[b]From Wang et al. [35].
[c]From Qi et al. [36].
[d]From Bai [37].

Heshan [39–41], Guiding [42,43], Yanshan [44], Shenbei, Huaibei, Beipiao, and Weibei coalfields. Cr concentration in these coalfields ranges from 55.8 to 180 μg/g [42,43,45]. Ge-rich coals in China are mainly located in the Lincang germanium deposit, Yunnan Province, southwestern China; the average content of Ge is 850 μg/g. The Lincang germanium deposit is also notably rich in Nb, Li, Sb, W, Bi, and U and shows substantial enrichment of heavy rare earth elements (HREEs), which increases together with Ge [46]. Late Permian coal with significantly enriched rare earth elements (REEs) and yttrium (REY) (1423 μg/g) was reported from the Huayingshan coalfield in southwestern China; it is also enriched with Zr (695 μg/g), Nb (75.9 μg/g), Se (6.99 μg/g), and Hf (10.1 μg/g). Coalfields with high As concentrations (up to 35,000 ppm) are located in southern China, especially in southwestern China. It has been reported that the As concentration in coals in southwestern China is 267 μg/g [47]. In general, the content of arsenic in China is higher in northeastern and southern provinces, but lower in northwestern provinces, as shown in Table 2.6 [35].

Late Triassic coal samples in Tibet are enriched with As, Hg, Pb, and Se [48]. Dai et al. [49] also reported Jurassic coal in the Tibetan Plateau enriched with TEs of F (253 μg/g on average), P (2349 μg/g), Sr (526 μg/g), and Ba (790 μg/g). However, arsenic in a few coals from northern China (e.g., the Wulantuga coal mine in the Shengli coalfield in Inner Mongolia) is highly enriched, reaching up to 500 μg/g [50,51]. The enrichment of As could be influenced by minerals in coal, such as pyrite. Zhang et al. [52] reported higher As and Mo concentrations in Late Permian coal with pyrite veins; the content of TEs depends on the content of mosaic pyrite and its TE abundance. With the exception of Inner Mongolia (e.g., coals from the Wulantuga coal mine) [39,50], the concentration of Sb in coals from northern China is lower than that in coals from southern China [36]. Detailed data are shown in Table 2.6. In the regional distribution of Se in Chinese coals, coal from south-central China, which includes Anhui, Jiangsu, Henan, Chongqing, and Hubei Provinces, has a higher Se concentration compared to that of other provinces [53]. Enshi Prefecture in Hubei Province has a serious Se poisoning situation, and the Se concentration of carbonaceous–siliceous rock samples collected in Yutangba, Enshi Prefecture, reaches up to 5371–8290 μg/g [54–56]. The coal-bearing regions of northeastern China and southern China have high F concentrations, while the F content in northern and northwestern China coals is relatively low. Wu et al. [57] reported that the range of F content in Chinese coals varies from 25 to 1230 μg/g with a geometric mean of 136 μg/g, and the F content decreases gradually from sub-bituminous through bituminous coal to anthracite. Guizhou Province suffers from endemic fluorosis caused by coal burning [58]. However, the arithmetic mean concentration of fluorine in coals sampled from Guizhou Province is not high. The main fluorine source in these villages is not from coal but from clay mixed with coal to make coal–clay briquettes [59,60]. Highly elevated concentrations of F have been reported in the Late Permian coals from Heshan coalfield, reaching up to 3362 μg/g, and fluorine mainly occurs in fluorite [41].

Essentially, coals in southern China provinces, including Guangxi, Guizhou, and Yunnan, are usually enriched witth several kinds of TEs, as shown in Table 2.7. A typical example is the Late Permian coal from Yanshan coalfield, Yunnan Province, which is significantly enriched with B (268 μg/g), V (567 μg/g), Cr (329 μg/g),

Table 2.7 Concentrations of V, Cr, Mo, U, and Ni in coals from Guizhou, Guangxi, and Yunnan Provinces (μg/g) [44].

Coal-bearing region	V	Cr	Mo	U	Ni
Guiding, guizhou	264	59	141	106	50.3
Ziyun, guizhou	1290	943	153	200	85
Heshan, guangxi	326	192	124	72	9
Yanshan, yunnan	567	329	204	153	74
China average	35.0	15.4	3.19	2.4	13.7
World average	29	17	2	1.9	16

Ni (73.9 μg/g), Mo (204 μg/g), and U (153 μg/g), and slightly enriched with Se (25.2 μg/g), Zr (262 μg/g), Nb (20.1 μg/g), Cd (2.07 μg/g), and Tl (2.03 μg/g) [44]. Another representative sample is low-volatile bituminous coal in the Heshan coalfield, Guangxi Province. Most TEs in this coal are highly enriched with Mo, U, and W, more than 10 times the world mean. This is a result of TEs associated with either organic compounds or minerals [39]. Table 2.8 shows TE contents in Heshan coals and their comparison with world and Chinese averages. Tian et al. [61] reported average contents at approximately 0.61 μg/g for Cd, 30.37 μg/g for Cr, and 23.04 μg/g for Pb, but the arithmetic mean contents of these TEs in Chinese coals are higher in southern provinces than those in northern provinces. Higher values of TE contents occur in provinces such as Sichuan, Chongqing, Yunnan, Hubei, and Guangxi. The ranges of the 95% confidence interval for the arithmetic TE content means are relatively small, with relative uncertainties of −16.39% to +21.31% for Cd, −10.11% to +11.72% for Cr, and −8.55% to +8.64% for Pb.

According to the foregoing analysis, the overall performance of TE contents in Chinese coal is uneven in different coal-forming regions, but increase from the northeastern area to the southern area, especially As, Hg, F, Cd, and Sb. Although coal from the southern area only accounts for a limited part of the whole fuel consumption in China, it could result in endemic environmental pollution and a threat to health [58]. The main application for Chinese coal is combustion in power stations, so the emission and emigration behaviors of TEs in China, especially in southern provinces, should be given special attention. Because of the volatility of some TEs, such as Hg and F, flue gas derived from coal combustion will transfer toxic elements into the atmosphere and cause adverse impacts on human health. Thus TE contents and distribution in Chinese coal need to be investigated further.

2.5 Occurrence of trace elements in coals

Thorough and comprehensive investigation of the occurrence of TEs in coals is imperative for researching their provenance and migration behaviors. The occurrence of TEs

Table 2.8 Mean values of trace elements from Heshan coals and their comparison with world and Chinese averages (μg/g) [39].

	Heshan coals			Chinese coal range						Swaine's world ranges	
Trace element	**Min.**	**Max.**	**Mean**	**Min.**	**Max.**	**Mean**	**H/C[a]**	**Min.**	**Max.**	**Mean**	**H/W[a]**
As	2.1	70.1	19.2	0.21	32,000	276.61	0.07	0.5	80	29	1.92
Ba	27.4	321.9	79.2	4.1	1540	169.01	0.47	20	1000	120	0.40
Be	0.3	6.5	0.4					0.1	15	3	1.21
Bi	0.1	2.6	0.7					2	20		0.13
Cd	0.0	3.2	0.5	0.04	1.2	0.46	1.12	0.1	3	0.5	1.03
Co	3.4	23.0	10.2	0.03	39.6	6.72	1.51	0.5	30	4	2.03
Cr	5.3	321.3	76.2	0.46	942.7	34.87	2.19	0.5	60	8	3.81
Cs	0.5	18.0	6.3					0.3	5		3.59
Cu	4.3	30.2	12.6	4.28	133.7	28.22	0.45	0.5	50	7	0.84
Ga	3.4	33.9	14.6					1	20	2.3	1.69
Ge	0.2	1.7	0.6					0.5	50	7	0.12
Mn	4.0	270.9	82.0	6.02	8540	271.22	0.30	5	300	70	1.17
Mo	2.5	171.7	45.6	0.2	241	18.15	2.51	0.1	10	4	15.20
Nb	2.8	119.2	23.3					1	20		5.96
Ni	0.0	27.0	6.9	1.1	255	22.62	0.30	0.5	50	13	0.34
Pb	8.9	135.3	29.7	5.28	69.7	24.77	1.20	2	80	59	0.74
Rb	5.4	84.8	38.2	1.4	408	20.68	1.85	2	50	15	2.55
Sc	1.5	17.5	8.1					1	10	4	1.54
Sr	88.2	3388.7	551.8	4.9	894	175.96	3.14	15	500	240	2.76
Ta	0.4	29.4	2.9					0.1	1		29.42

Th	3.6	58.5	22.5	0.09	25.4	6.9	3.25	0.5	10	4	5.61
Tl	0.1	6.7	1.7					<0.2	1		6.72
U	6.9	175.8	48.3	0.16	199.3	7.52	6.42	0.5	10	2	24.14
V	11.0	444.2	138.3	3.4	1292	94.11	1.47	2	100	19	3.46
W	10.3	221.4	58.0					0.5	5		44.28
Y	5.5	99.4	37.4					2	50		1.99
Zn	4.1	98.5	40.5	0.56	193	43.24	0.94	5	300	20	0.81
Zr	40.9	881.9	246.7					5	200	28	4.41
La	5.0	88.0	28.2								
Ce	12.0	187.2	63.2								
Pr	1.6	21.2	7.7								
Nd	6.0	73.8	29.3								
Sm	1.5	17.4	7.0								
Eu	0.3	2.5	1.1								
Gd	1.5	19.8	7.6								
Tb	0.3	3.3	1.4								
Dy	1.7	20.2	8.8								
Ho	0.3	4.3	1.9								
Er	0.9	13.6	5.5								
Tm	0.1	2.1	0.8								
Yb	0.9	14.1	5.6								
Lu	0.1	2.0	0.8								

[a] *H/C*, ratio of concentrations in mean Heshan coal and mean Chinese coal; *H/W*, ratio of mean concentrations in Heshan coal and mean worldwide coal.

indicates how the elements enrich in coals. It also determines the toxicity and mobility of TEs during combustion and application procedures. In summary, understanding TE occurrence in coals requires evaluating atmospheric emissions pathways and potential values.

TEs in coal primarily associate with various inorganic minerals and organic matter. Each of these forms could be soluble in different and unique solvents, and because of this trait, a selective leaching method was developed for quantifying the occurrence of TEs in coals. Finkelman et al. [62] established a selective leaching protocol and chose ammonium acetate, hydrochloric acid, hydrofluoric acid, and nitric acid as solvents. TEs combined with various minerals and organic matter were continuously concentrated in different solvents and then quantitatively analyzed through inductively coupled plasma atomic emission spectroscopy and inductively coupled plasma mass spectroscopy. The extraction procedure is graphically described in Fig. 2.2 [62]. They analyzed TE occurrences in thousands of coals from a wide range of geographic locations, ranks, stratigraphic distributions, and compositions. It can be concluded that occurrence in the same coal rank is similar. However, several elements combine with different matter between bituminous and sub-bituminous coal. For example, about 75% As in bituminous coal is associated with pyrite, but only 40% in low-rank coals. All the detailed results are summarized in Table 2.9. This work is a representative study for recognizing how TEs occur in coals.

Occurrence of TEs in coals is influenced significantly by many factors, such as geological setting, coal rank, coal-forming period and processes, geological changes, and mineral content, so the occurrence condition could change notably. Because the

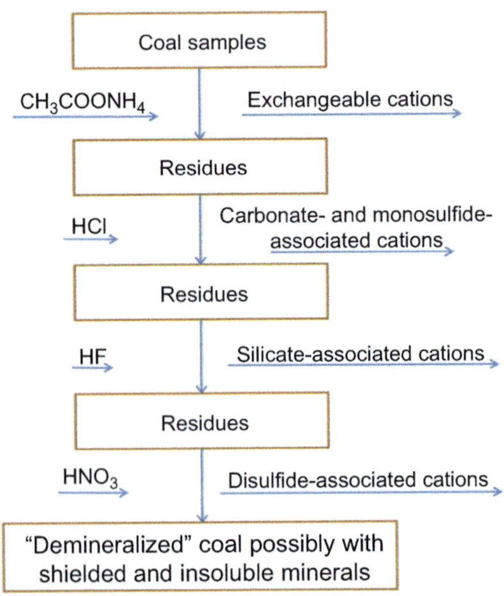

Figure 2.2 Diagram of the solvent extraction procedure [62].

Table 2.9 Modes of occurrence of elements in bituminous coals and low-rank coals [62].

Elements	Modes of occurrence	
	Bituminous coals	**Low-rank coals**
Al	90% silicate, 10% organic association	65% silicate, 35% organic association
Sb	90% silicate, 10% organic association 55% sulfide, 25% silicate, 10% monosulfide, 10% organic association and/or other phase	50% silicate, 20% sulfide, 10% monosulfide, 20% organic association, other
As	70% pyrite, 15% clays, 15% monosulfides, organic association and/or other phase	40% pyrite, 30% organic, 15% silicate, 15% monosulfides, carbonates
Ba	40% silicate, 15% carbonate, 45% barite, phosphate, organic association (US coal); 60% silicate, 15% carbonate, 25% barite, phosphate, organic association (non-US coal)	75% carbonate, 10% silicate, 15% organic association, sulfate, phosphate
Be	60% silicate, 40% organic, oxyhydroxide, or clay	30% silicate, 60% organic, oxyhydroxide, or clay
Br	70% organic, 10% silicate, 20%[a]	80% organic, 10% silicate, 10%[a]
Ca	70% carbonate, 20% organic, 5% silicate, 5% phosphate, etc.	60% carbonate, 35% organic association, 5% silicate
Cd	65% monosulfides (mainly sphalerite), 25% pyrite, 10% silicate	80% monosulfides (sphalerite), 10% pyrite, 10% silicate
Cs	100% silicate	80% silicate, 20% organic association
Cr	75% silicate, 10% sulfide, 10% organic association, 5% oxide	75% silicate, 10% sulfide, 10% organic, 5% oxide
Co	35% monosulfides, 20% silicate, 20% sulfide, 20% organic association	30% organic, 25% monosulfides, 25% clays, 15% pyrite
Cu	45% pyrite, 30% chalcopyrite, 20% clays, 5% organic association	30% pyrite, 30% chalcopyrite, 20% clays, 20% organic association
Hf	70% zircon, 25% clay, 5% other phase	70% zircon, 25% clay, 5% other phase

Continued

Table 2.9 Continued

	Modes of occurrence	
Elements	**Bituminous coals**	**Low-rank coals**
Fe	50% sulfides (pyrite), 25% carbonates, 20% silicates, 5% other; Australian coals have more carbonate iron and less sulfide iron.	60% carbonate and other, 20% sulfide, 15% silicate, 5% organic association
Pb	55% monosulfides (galena), 35% pyrite, 5% silicate, 5% other phase	50% monosulfides (galena), 10% pyrite, 25% silicates, 15% other phase
Li	90% silicate, 10% insoluble phases, organic association	70% silicate, 30% organic association, insoluble phases
Mg	30% silicate, 25% calcite, 25% ankerite/dolomite, 20% organic association	60% organic association, 30% calcite, 10% silicate
Mn	50% calcite, 25% clays, 15% siderite, 10% organic association	75% calcite, 10% silicate, 5% siderite, 10% organic association
Hg	90% sulfide (primarily pyrite), 10% all other forms	75% sulfide (primarily pyrite), 25% organic association and other phase
Mo	55% sulfide, 30% silicate, 15% carbonate, sulfate, organic association	65% silicate, 15% sulfide, 20% organic association, etc.
Ni	30% clays, 25% pyrite 30% linnaeite group and other monosulfides, 15% organic association, sulfates, carbonates	30% organic association, 30% monosulfides, sulfates, carbonates, 20% clays, 20% sulfides
P	95% phosphates, 5% organic association	85% phosphates, 15% organic association
K	95% silicate	75% silicates, 50% organic association
Light rare earths	70% phosphates, 20% clays, 10% carbonates, organic association	60% clays, 20% phosphates, 20% carbonates, organic association
Heavy rare earths	50% phosphates, 20% clays, 30% organic association, carbonates	50% clays, 25% phosphates, 25% organic association, carbonates
Rb	90% silicate, 10% other phase	85% silicate, 15% other phase
Sc	90% silicates, 5% organic, 5%[a]	95% silicate, 5% organic association

Table 2.9 Continued

Elements	Modes of occurrence	
	Bituminous coals	**Low-rank coals**
Se	70% sulfides and selenides, 30% organic association, shielded sulfides and selenides	75% organic, 20% sulfides and selenides, 5% silicate
Na	80% silicate (primarily clays), 20% organic association	65% organic association, 35% clay, salts, pore water
Sr	50% phosphates, carbonates, 25% silicate, 25% organic association	50% organic association, 50% phosphates, carbonates
Ta	75% oxides, 25% silicates	60% oxides, 40% silicates
Ti	65% clays, 35% Ti oxides	70% clays, 15% Ti oxides, 15% organic association
W	50% sulfide, 35% silicate, 15% organic association, etc.	60% silicate, 20% sulfide, 20% organic association, other phase
U	40% silicate, 30% insoluble minerals such as zircon, 15% phosphates, 5% carbonates, 5% organic association	55% organic association, 35% silicates, 5% phosphates, 5% insoluble phases such as zircon
V	65% silicates, probably clays, 35% organic association, etc.	50% silicate, 50% organic association, other phase
Zn	45% sphalerite, 30% pyrite, 25% silicates	55% sphalerite, 15% pyrite, 15% silicates, 15% organic association
Zr	70% zircon, 25% clay, 5% other phase	70% zircon, 25% clay, 5% other (maybe organic association)

[a]Indicates uncertainty in determining the mode of occurrence.

most toxic heavy metal derives from coal-fired flue gas, Hg occurrences in coal have an apparent influence on anthropogenic emission from coal-fired power plants. Hg in coal mainly combines with clay (Hg_{clay}), organic matter (Hg_{org}), and sulfides (Hg_{sulf}), and the latter two forms of Hg are dominant because their contents control Hg distribution in coal. Because of the affinity between Hg and sulfur, hydrothermal pyrites involved in coal may serve as Hg collectors, which enrich Hg during coal-forming procedures and geological processes. Because Hg_{org} and Hg_{sulf} provide most of the Hg concentration in coal, the contents of these two occurrences of Hg in coal have a negative correlation, which means a higher potential contribution of Hg_{org} in coal with lower sulfur content [63]. However, it should be pointed out that different testing methods could cause deviation among the proportions of occurrences of Hg. Luo et al. [64] researched mass distribution of different forms of Hg with two methods:

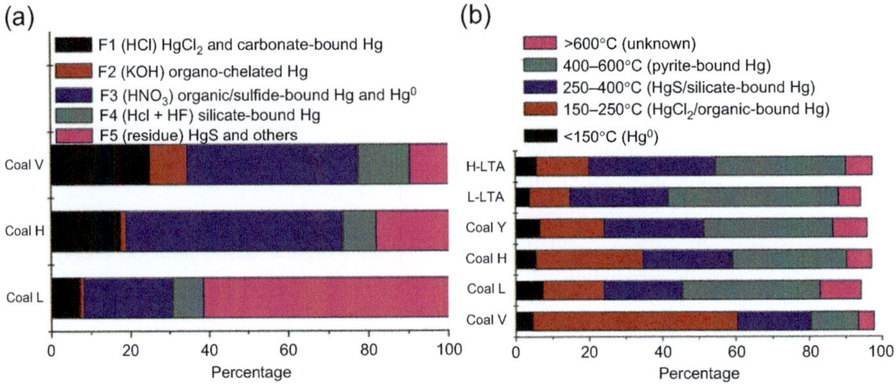

Figure 2.3 Mass distribution of different forms of Hg: (a) by sequential selective extraction and (b) by pyrolysis [64].

sequential selective extraction and pyrolysis. Although the results have shown discrepancies to some extent, Hg_{org} and Hg_{sulf} are still the dominant species, as shown in Fig. 2.3.

Temperature-programmed decomposition (TPD) technology is a practical method to identify mercury species and compounds in coal and coal deposits. Different modes of occurrence of mercury have unique decomposition and desorption temperatures. Occurrence of mercury species in coal could be separated and demonstrated through desorption peaks at various temperatures. Gao et al. [65] researched the modes of occurrence and thermal stability of mercury in raw coal, cleaned coal, middling, gangue, and slime samples collected from China. The results showed that the Hg contents of cleaned coal and middling are lower than that in raw coal, while the content of raw coal is less than that of gangue or slime. Mercury contents could be clarified by the intensity of mercury, with different mercury desorption peaks related to occurrences of mercury. The mercury TPD figure is shown in Fig. 2.4. Zhai et al. [66] researched four coal gangue samples in China. The mercury of these samples was characterized by a sequential extraction procedure coupled with a TPD method. Carbonate-bound Hg was released at a temperature range of 250–300°C, iron–manganese oxide-bound Hg was released at 200–600°C, and organic-bound Hg, sulfide-bound Hg, and silicate-bound Hg were released at 200–400, 400–600, and above 1200°C, respectively. The TPD curve is shown in Fig. 2.5. The TPD method has proved to be an effective approach in studying the modes of occurrence of mercury as well as thermal stability of coal gangues by clarifying the characteristic release temperatures of mercury species.

Occurrences of Se show a similar tendency with Hg. Sulfidic and organic matter are the dominant forms of Se. The contribution of Se in coal is significantly affected by sulfur content. In coal with abundant sulfides, an appreciable part of Se exists as sulfides and mixed with Cu, As, and Pb sulfides. As a result, in coal with lower sulfur content, Se combined with organic matter (Se_{org}) is the dominant form. It should be noted that during the coal-forming process, Se_{org} may be partly lost to higher-rank coals: this phenomenon indicates that a metamorphic change happens among different

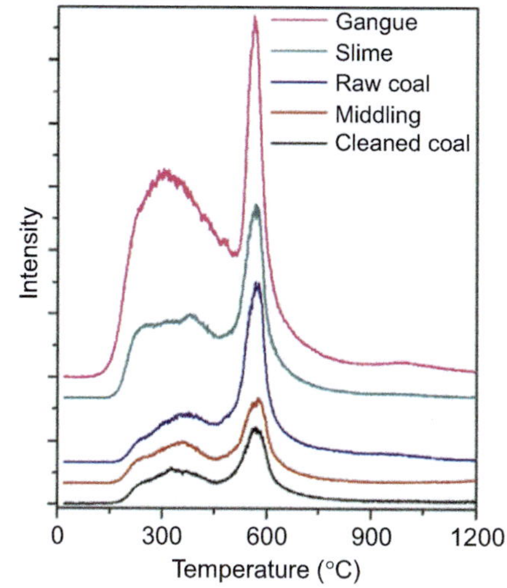

Figure 2.4 Temperature-programmed decomposition figure of gangue, slime, raw coal, middling, and cleaned coal [65].

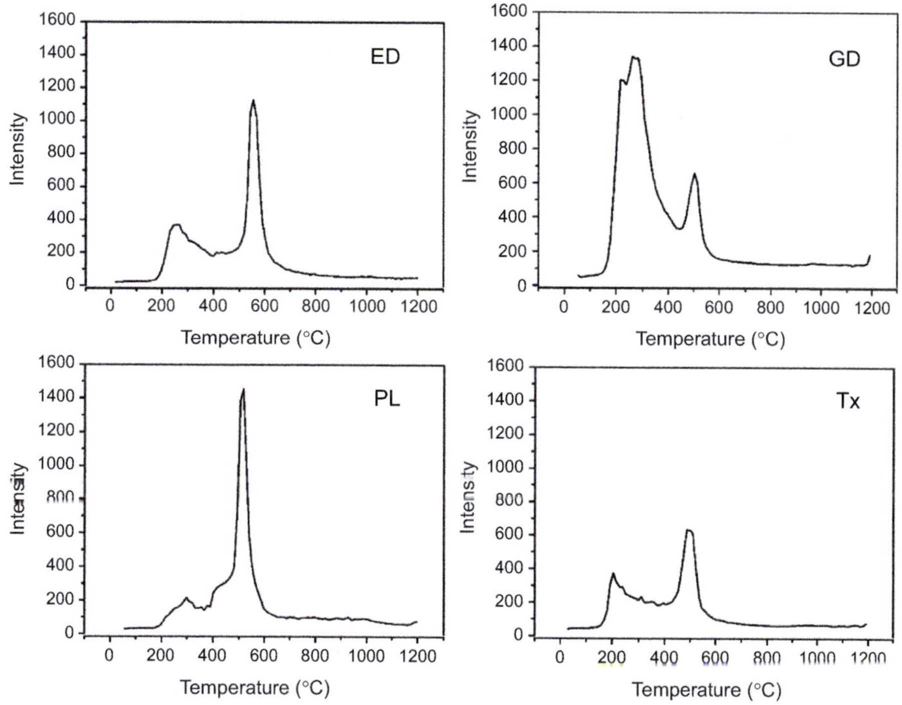

Figure 2.5 Hg release profiles during temperature-programmed decomposition of the coal gangue samples (ED, GD, PL and TX represent the gangue samples collected from Taiyuan power plant, Gundi coal mine, Pinglu coal preparation plant, and Taiyuan Coal washery) [66].

Se occurrences [67]. The dominant occurrences of As in coal are pyrite-bonded organic-bonded and arsenate-bound states. According to modern testing methods, As has been proved to occur in fine pyrite particles enclosed in microstructures [68]. The appearance of As in clay is also possible. Since pyrite is considered to be a main carrier of As, the content of pyrite and As in coal presents passive correlation. Higher-rank coals usually have lower As content, but coal rank is not the only impact; thermal water invasion and other coal chemistry activity could cause relatively high As contents in anthracite. The contact metamorphism of coal results in extreme high temperature and consumes organic matter in coal, so the organic As content decreases and influences the whole As content in coal [69]. The proportions of different As forms in coal samples obtained through a progressive release flotation test are shown in Fig. 2.6; these data indicate that sulfide-associated form and organic As are always the dominant occurrences [70]. Other studies also demonstrate this conclusion [71,72]. X-ray absorption near edge structure (XANES) spectra is also used to identify occurrences of As and Se. Huggins et al. found that there was a strong correspondence between Se and As XANES spectra for pyrite-rich coal fractions, indicating that these two elements share a similar chemical environment in the pyrite structure, most probably as As2-2 and Se2-2 for the S2-2 unit in the pyrite structure. Occurrences of Se also include organic-associated selenite, notably for the CLN (clean coal) fraction, as a result of oxidation. The spectra are shown in Fig. 2.7 [73]. Pb tends to occur in pyrite and other sulfide minerals. PbS is identified as the most predominant species of Pb in coal, which indicates that sulfide minerals are significant in hosting Pb. In addition, Pb apparently correlates with a number of elements, including Se, Cr, Cu, Al, and Si.

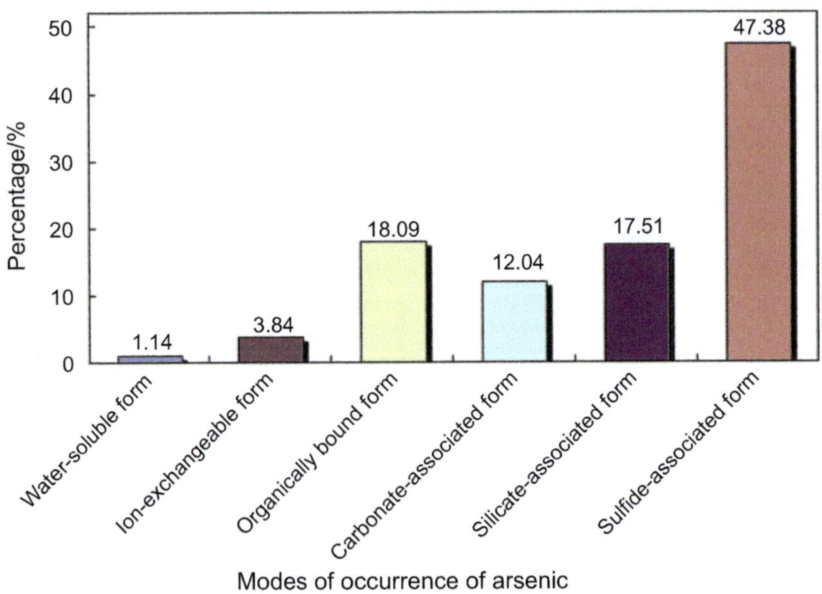

Figure 2.6 Modes of occurrence of arsenic in coal samples [70].

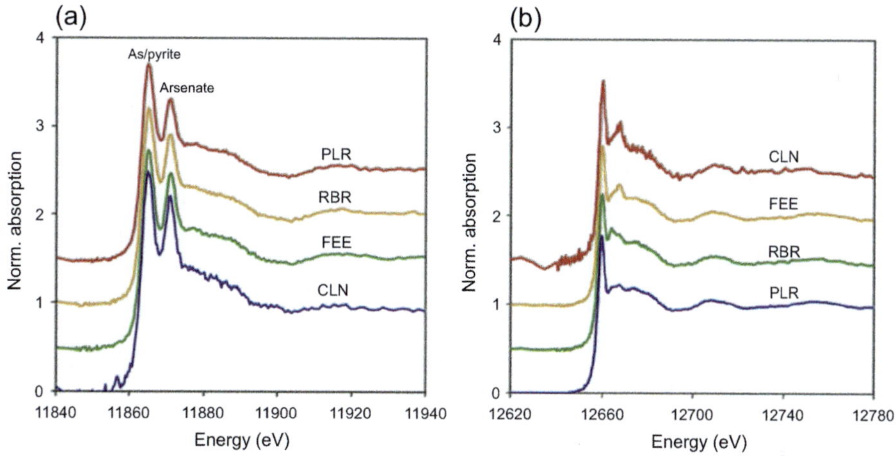

Figure 2.7 (a) Arsenic X-ray absorption near edge structure (XANES) spectra of Illinois #6 coal and fractions, (b) selenium XANES spectra of Illinois #6 coal and fractions (raw coal fed to the rotary breaker, the clean coal, the tailings fractions generated by the rotary breaker and the HM gravity separation at the plant are named as FEE, CLN, RBR and PLR) [73].

PbSe is a typical occurrence of Pb combined with a nonsulfide mineral. The remaining Pb is mainly present as an organically associated form, especially in low-rank coal [74–76]. Occurrences of Cd in coal are quite complex, because it can be combined with several carriers: sulfides, pyrite, silicates, carbonates, and clays. Among all the Cd species, sulfides and silicates are primarily the carriers of Cd, especially sphalerite. In low-rank coals, 80% of Cd is associated with monosulfides, and the rest is combined with pyrite and silicates. In some Chinese low-rank coal, up to 21% Cd is associated with organic matter, indicating that organic Cd primarily occurs in low-rank coals [62,77,78]. Some TEs show apparent affinity; this phenomenon is usually noticed on account of a similar occurrence of different elements. Li et al. [79] studied occurrences of TEs in Nantong coalfield and found that Pb appeared dominantly in sulfides and carbonates, and that Cd primarily occurred in silicate, followed by sulfides. Similar modes of occurrence resulted in their affinity. A positive correlation between Cr, V, and ash yield in coal indicates that these two TEs are primarily combined with mineral matter. According to the sequential extraction procedure, Cr and V are dominantly associated with silica and organic matter, which results in apparent correlation between Cr and V [80]. In some research, Cr combined with organic matter even accounted for more than half the Cr in coal, especially in bituminous coals [81]. There are diverse occurrences of Ni in coal. According to the analysis of coal samples from different coalfields, Ni was found in organic matter, sulfides, silicates, and carbonates [82]. Due to its complex occurrences, Ni has an affinity with many TEs, such as Cu, Cr, and Co [79]. The inorganically associated Sb predominantly appears in pyrite, followed by other minerals and metals, such as Nb, Bi, and Ca. However, Sb did not show an apparent association with organic matter [36]. Because the coal rank of lignite is low, it has a tendency to be enriched with several hazardous TEs and the occurrences

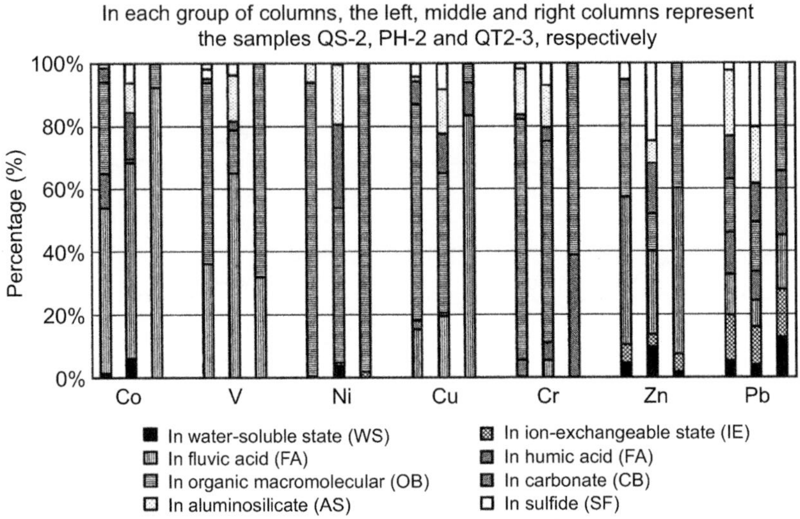

Figure 2.8 Percent leached data for Co, V, Ni, Cu, Cr, Zn, and Pb in the samples QS-2, PH-2, and QT-3 [83].

are complex. Ren et al. researched the occurrence of Co, V, Ni, Cu, Cr, Zn, and Pb in three lignite samples (called QS-2, PH-2, and QT-3) from the Shenbei coalfield, China, and the results are shown in Fig. 2.8 [83].

Except for the common and typical TEs discussed here, extremely toxic TEs should also be focused on. Uranium in coal has drawn attention because of its radiation and potential value, and confirmation of U occurrences in coal is the premise for its extraction, removal, and reuse. U is generally organically associated with low-rank coals, and with respect to Chinese coals, it is also combined with silicates, phosphate minerals, uranium minerals, or even physically adsorbed by pores. Higher-rank coals usually have lower contents of U. Chinese lignite and sub-bituminous coal are generally enriched in U; this tendency is a representation of the organic affinity of U [84]. Tl is a kind of rare element but with extreme toxicity. As a lithophile and chalcophile element, pyrite has been regarded as the main carrier of Tl in coal. This trait suggests a potential method of purifying Tl in coal before combustion; it is commonly purified with pyrite [85]. $BaSO_4$ is the predominant form of barium in coal. Although Ba has both organic and inorganic affinity, most Ba is associated with clay in much published research. Ba in coal samples located in a Yorkshire coalfield, UK, Guizhou Province, Malaysia, and Ordos Basin, northern China, showed apparent inorganic affinity, especially when combined with clay [86–89]. However, Ba has predominantly organic affinity in Turkish coal [90]. Cobalt is a TE with latent economic value. The occurrences of Co in coal are diverse. Co is widely combined with silicate, organic matter, carbonate, sulfide, and pyrite [91,92]. A coal sample from southwestern China was enriched with organic-bound Co, but the proportion of organic Co was not predominant as a result of complex occurrences of Co [92]. Cu is a valuable metal element. Occurrences of Cu include inorganic and organic associations, but it is primarily presented as co-occurring with other elements in minerals, such as Fe and Se [93,94]. It should be pointed out that mobility of Cu could be amplified in an acidic atmosphere and cause heavy metal pollution in the surrounding water [95]. Li is one of the most significant elements in batteries; coal and by-products of the coal combustion procedure enriched with Li could be latent sources for Li preparation. The dominant occurrences of Li are isomorphic chlorite and aluminosilicate, which show that inorganic affinity is primarily in occurrences of Li [96,97]. Fly ash after Li-enriched coal combustion had high concentration of Li. Most Li in fly ash is observed in the glass phase, and the rest was found in magnetic particles [98,99]. Identification of Li in coal and coal combustion by-products is beneficial for Li extraction. Mn is another useful metal found in coal. Mn in coal is closely related to minerals and has the potential for high removability, which provides the opportunity for extraction and recovery [100]. Mo and Re usually occur with other TEs. Seredin et al. [101] found that Mo mainly occurs as molybdenite in U-bearing coal deposits. Dai et al. [42] showed that Re occurs in sulfates derived from sulfide oxidation and carbonate minerals. The organic affinity of Mo and Re was also observed in coal from southern China, and a small part of these two elements was combined with illite or smectite [43].

Halogen elements (F, Cl, Br, I) are a series of elements in coal with relatively higher concentration. The most attractive element is F, which has caused serious environmental issues during the production and utilization of coal. Fluorine occurs in various forms in coal and presents mixed affinities in most coal samples. Inorganic affinity is the predominant character of F in coal, which means F combines with minerals. Fluorine minerals include fluorite, illite, kaolinite, fluorapatite, and montmorillonite [102,103]. The organic affinity of F was observed in specific coal samples. Organic and silicate associations of F accounted for more than 90% of total fluorine in a coal sample from Inner Mongolia, China [104]. The result of the sequential extraction/density separation procedure is shown in Fig. 2.9. F in Czech coal also presented with organic affinity [105]. Cl is relatively enriched in coal; most of the Cl in coal is combined with minerals. The most important mineral of chlorine is NaCl or KCl. Organic chlorine only accounted for 0.5%–25% of total Cl. A small part (5%–10%) of chlorine is even adsorbed on the surface of tiny pore walls in Cl^- form [106]. Cl could combine with other TEs. For example, Cl–Fe–Mn oxides were observed in coal gangue [107]. In high-Cl coals, Cl is associated with coal moisture, but other sources, such as silicate minerals, are negligible. This result indicates that Cl has a water-soluble occurrence as an ion. Br and Cl are closely related and their location within coal is thought to be common. These two elements have a similar occurrence in coal, and although Cl and Br are present in moisture, they are also adsorbed on the pore surface in coal. The proportion adsorbed in large pores is lower [29]. In summary, both Cl and Br have an organic occurrence and association with water-soluble (mainly ionic) and water-insoluble (ionic and covalent) organic combinations. The inorganic phases of Cl in coal are identified as clay mineral, mica, feldspars, polyhalite, gypsum, siderite, volcanic glass, and other carbonates and sulfates. A characteristic occurrence of Br is its association with illite and to a lesser extent with mica, kaolinite, and Fe-bearing minerals. Inorganic occurrences of both elements are water molecules, hydroxyl groups, and exchangeable cations in various minerals [108].

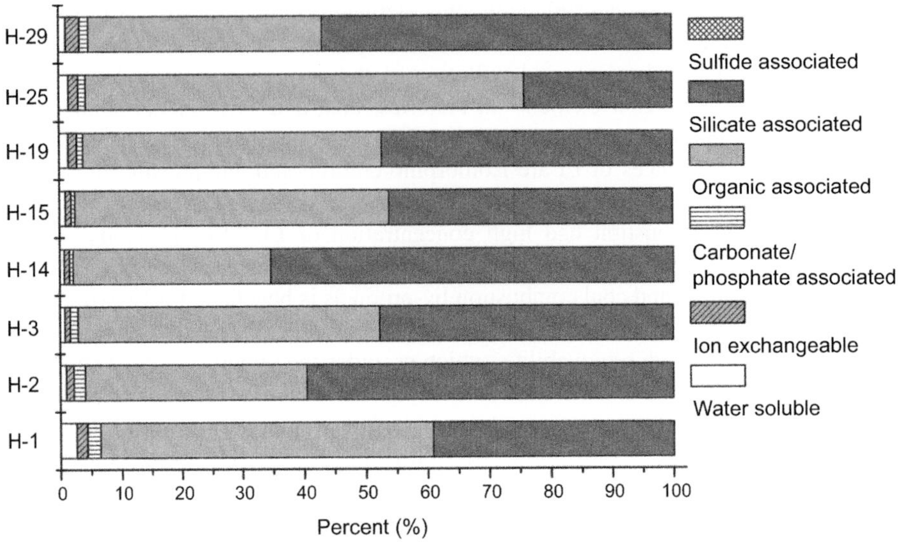

Figure 2.9 Percentage of six modes of occurrence of fluorine inferred from the sequential extraction/density separation procedure [104].

Peng et al. found that the iodine in both bituminous coal and anthracite occurs in a descending order of organic, bound with Fe–Mn oxide, water soluble, residue, ion exchangeable, and associated with carbonate. Iodine leached from coal can lead to an increase in iodine concentration in surrounding water and serious environmental contamination [109].

REEs are a series of TEs with extremely limited concentration in coal, including La, Ce, Pr, Nd, Sm, Eu, Gd, Tb, Dy, Ho, Er, Tm, Yb, and Lu. These elements can also be categorized as light REE (LREEs) and heavy REE (HREEs). All the REEs are either in the form of discrete minerals or associated with organic matter, but LREEs and HREEs show different extents of organic affinity. It was found that the whole REE concentration in coal and the LREE/HREE ratio increase as the ash yield increases; however, the REE concentration in ash shows a significant reduction. These results demonstrated that REEs in coal are associated with both organic and inorganic components. In addition, HREEs are relatively enriched with organic matter [110]. Although there are many studies on the occurrences of TEs in specific coal samples, common conclusions still need to be summarized. REY are strategically important materials in the aerospace, automation, defense, and electronic industries. Recovery of REY from coal, coal deposits, and coal combustion by-products is a promising method. Because the concentration of REY in coal is very low, fly ash obtained from coal combustion is a potential source of REY, because these elements migrate from coal to fly ash. A sequential chemical extraction study showed that silicate and aluminosilicate forms were the dominant occurrences of REY in coal fly ashes from Faer and Panbei power plants in China; these forms accounted for 80% of the whole REY [111]. Detailed data are shown in Fig. 2.10. In comparison with radioactive Y ores, fly ash has lower

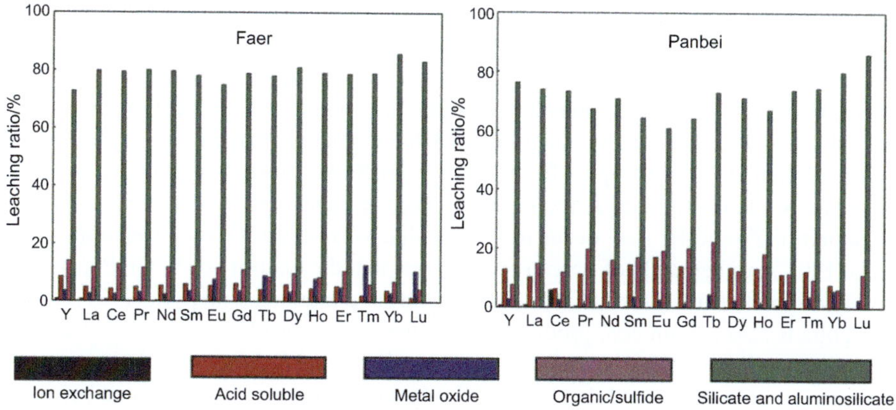

Figure 2.10 Modes of occurrence of individual rare earth elements and yttrium components in coal fly ash [111].

concentrations of Y but uses a simple extraction process. A dissolved fraction of Y corresponds with phosphate-associated Y; the insoluble part is related primarily to its association with poorly soluble Ti oxides/titanates [112]. Y is closely associated with lanthanides in nature, because both their ionic radii are very similar and their ionic charges are equal. Thus coal combustion fly ash enriched with REY is also a potential source for the extraction of La [113]. The efficient extraction technology of REY from coal, coal deposits, and fly ash should be investigated in the future.

To date, extensive studies on TE occurrence in coals from different countries have been conducted. Diehl et al. [114] studied the occurrence of mercury and other TEs in coals from northwestern Alabama, USA, and found that Hg, As, and Tl are mainly enriched and combined with pyrite veins and faults. Kolker et al. [115] also reported As mainly enriched with pyrite, but a small part of As is preliminary associated with organics as As^{3+} or arsenate. Significant (10%–40%) fractions of arsenate are derived from pyrite oxidation. Selenium in US coals occurs in several forms but is chiefly associated with organic fraction; other important occurrences include selenium-bearing pyrite, selenium-bearing galena, and lead selenide [116]. According to electron beam microanalysis of coal samples in the US Geological Survey, As is the most abundant TE in pyrite in coal, and the contents of Se, Ni, and other elements are less common at lower concentrations. Elements such as As, Se, and Sb substitute S in the pyrite structure, whereas the metals Hg and Pb are thought to substitute for Fe, so several TEs are enriched in coal samples with a high content of pyrite [117]. The affinity between pyrite and several TEs was found in coals from different locations. Spears et al. [28,118] researched TE contents in UK coals and found As, Ni, Pb, Cu, and Zn are enriched with pyrite, and that pyrite is the major source for Ni, Cu, and Zn. However, Ba and Sr are associated with phosphate minerals, and Ge and V appear to contain organic matter. Equeenuddin et al. [119]. found that both Cu and Cr are more closely associated with mineral matter in Indian coals, whereas Co is dominantly present in its organic form. Cd, Mn, Ni, Zn, As, and V occur in both organic and inorganic constituents.

Similar results were attained during the research of Turkish coal. Gürdal found that As, Cd, Hg, Se, Cu, Mo, Ni, and Zn show a possible association with pyrite; however, the elements Se, B, and Mo have both organic and inorganic associations [90].

In China, geological conditions for coalfields are complex, so the occurrence of TEs in coal could change notably. In northern China, Shanxi Province is an important source of coal. In coal samples from Ningwu coalfield, mostly Sr, Ba, Be, As, and Ga are associated with phosphates and carbonates or have an inorganic–organic affinity; a part of Zr and Hf may occur in anatase due to strong positive correlations with TiO_2 [120]. Another study of Shanxi coal showed that the majority of V, Cr, Pb, and Ni mainly originate from minerals, while Hg and Se are combined with pyrite [121]. Inner Mongolia is also a main coal producing province in northern China. An investigation of coal from Inner Mongolia reports that As, Cr, Hg, Li, Mn, and Pb are mainly associated with minerals, while Cd, Co, Ni, Sb, and Se are evenly distributed between minerals and organic matter. Be and Ba are mainly distributed in minerals with a minor proportion in organic matter [100]. These results indicate that elements could show affinity to both inorganic minerals and organic matter, which demonstrate the complexity of TE occurrences in coal. Guizhou Province is a typical coal production region in southern China, where the coal usually contains a higher concentration of TEs. Comprehensive studies were focused on TE occurrences in Guizhou coals. In summary, elements, including B, Ba, Be, Cs, F, Ga, Nb, Rb, Th, U, and Zr, are mainly associated with clay minerals, while As, Tl, Se, Sb, Cd, Hg, and Pb are mainly formed in pyrite and marcasite. Cr and Zn have affinity with both organic and inorganic matter, and Mo and Ni have an apparent strong organic affinity [52,122–124]. However, abnormal constituents in coal could change the occurrence of elements. A late Permian coal in southwest China, which is characterized by superhigh organic sulfur and high concentrations of TEs, was studied, and the results show that V, Cr, Se, Re, U, and Mo in this kind of coal are mainly associated with organic matter, while Cd is mainly distributed in sulfide minerals [43]. Thus an extremely high content of elements could impact its occurrence in coal. Zhao et al. [125] investigated the occurrence of Cr in high-chromium coals and found it mainly occurred as organic macromolecules, and the remaining Cr is associated with aluminosilicate and carbonate. Uranium is one of the most toxic and radioactive TEs in coal. Identification of its occurrence is the for avoiding radiation pollution during coal combustion. Yang et al. [126] studied a coal sample from Yunnan Province, southern China, through a sequential extraction method. It was concluded that most U in coal is bound to organic matter (58.48%) and Fe–Mn oxide fraction (31.62%); detailed data are shown in Table 2.10. In general, the occurrence of each element in coal is similar except for those samples with abnormal chemical constituents. The modes of occurrences for the main hazardous TEs in Chinese and world coals are summarized in Table 2.11.

In general, occurrences of TEs in coals around the world show apparent similarity, so the recognition of occurrences is practical to evaluate the insidious toxicity of TEs and provide information for coal preparation procedures. However, the results discussed previously are inaccurate for the given coal samples. Further investigation of the occurrence of TEs in coal is indispensable.

Table 2.10 Percentages of various forms of uranium in Yunnan coal [126].

Fractions	Content (mg/kg dry coal basis)	Percentage (%)
Exchangeable	0.17 ± 0.08	1.78 ± 0.87
Bound to carbonates	0.62 ± 0.11	6.48 ± 1.19
Bound to Fe and Mn oxides	3.06 ± 0.27	31.62 ± 2.77
Bound to organic matter	5.66 ± 0.39	58.48 ± 4.12
Residual	0.62 ± 0.20	6.39 ± 2.11
Total	10.15 ± 0.26	104.76 ± 2.72

Table 2.11 Summary of the modes of occurrence of eight key trace elements in Chinese and world coals [1].

Elements	Possible modes of occurrence
Hg	Sulfide bound (mainly pyrite), clay bound (correlated with Ca and Al), and organic bound
As	Sulfide affinity (pyrite and chalcopyrite), organic associations, carbonates, silicates
Se	Organic associations, pyrite, accessory selenides having aluminosilicate association (low sulfur coal), sulfide affinity (high sulfur coal)
Pb	Galena, organic association with Fe, sulfide (pyrite-rich coals), associated with aluminosilicate, selenides
Cd	Mainly organic affinity, sulfide, sphalerite
Cr	Organic association and clay (mainly illite), chromite ($FeCrO_4$) (ultramafic deposit origin)
Ni	Multiple associations: Sulfides, organics, aluminum-silicate, carbonates affinity
Sb	Organic and inorganic affinity: Organic association, pyrite and accessory sulfides, silicates

2.6 Summary

The average contents and occurrences of TEs in coal around the world with various forming periods, coal ranks, and coal-bearing regions are summarized. Apart from the abnormal enrichment of TEs in limited samples, most coals have similar TE contents. This conclusion indicates that homogeneity is the predominant tendency of TE distribution in coal. However, some high-concentration TEs in coal could result in harmful impacts on the environment surrounding coal combustion power plants.

Therefore TE contents and occurrences data will be useful for estimating the insidious hazard emissions. In addition, because coal is one of the most common fuels, it could be an alternative source of valuable elements. Thus the extraction process needs intensive study alongside coal, fly ash, or other coal-burning waste with high concentrations of valuable elements.

Acknowledgments

This project was supported by the National Key Technologies R&D Program: [Grant Number 2018YFB0605104].

References

[1] Tian HZ, Lu L, Hao JM, Gao JJ, Cheng K, Liu KY, et al. A review of key hazardous trace elements in Chinese coals: abundance, occurrence, behavior during coal combustion and their environmental impacts. Energy Fuels 2013;27:601−14.

[2] Ren D, Zhao F, Dai S, Zhang J, Luo K. Geochemistry of trace elements in coal. Beijing, China: Science Press; 2006.

[3] Dai SF, Ren DY, Chou CL, Finkelman RB, Seredin VV, Zhou YP. Geochemistry of trace elements in Chinese coals: a review of abundances, genetic types, impacts on human health, and industrial utilization. Int J Coal Geol 2012;94:3−21.

[4] Duliu OG, Culicov OA, Rădulescu I, Cristea C, Vasiu T. Major, trace, and natural radioactive elements in bituminous coal from Australia, Romania, Russia, South Africa and Ukraine: a comparative study. J Radioanal Nucl Chem 2005;264:525−34.

[5] Munir MAM, Liu G, Yousaf B, Ali MU, Abbas Q. Enrichment and distribution of trace elements in Padhrar, Thar and Kotli coals from Pakistan: comparison to coals from China with an emphasis on the elements distribution. J Geochem Explor 2018;185:153−69.

[6] Finkelman RB. Trace elements in coal. Biol Trace Elem Res 1999;67:197−204.

[7] Spears DA, Tewalt SJ. The geochemistry of environmentally important trace elements in UK coals, with special reference to the Parkgate coal in the Yorkshire-Nottinghamshire Coalfield, UK. Int J Coal Geol 2009;80:157−66.

[8] Bouška V, Pešek J. Quality parameters of lignite of the North Bohemian Basin in the Czech republic in comparison with the world average lignite. Int J Coal Geol 1999;40: 211−35.

[9] Hu J, Zheng B, Finkelman RB, Wang B, Wang M, Li S, et al. Concentration and distribution of sixty-one elements in coals from DPR Korea. Fuel 2006;85:679−88.

[10] Aftabi A, Shojaei SV, Nezhad RK. Geochemical and environmental baseline of major and trace elements in Zarand coals, southeastern Iran. Environ Earth Sci 2015;73:7457−76.

[11] Oliveira MLS, Ward CR, French D, Hower JC, Querol X, Silva LFO. Mineralogy and leaching characteristics of beneficiated coal products from Santa Catarina, Brazil. Int J Coal Geol 2012;94:314−25.

[12] Eskenazy GM. Trace elements geochemistry of the Dobrudza coal basin, Bulgaria. Int J Coal Geol 2009;78:192−200.

[13] Wagner NJ, Tlotleng MT. Distribution of selected trace elements in density fractionated Waterberg coals from South Africa. Int J Coal Geol 2012;94:225−37.

[14] Saha D, Chakravarty S, Shome D, Raviathul Basariya M, Kumari A, Kundu AK, et al. Distribution and affinity of trace elements in Samaleswari coal, Eastern India. Fuel 2016; 181:376–88.
[15] Ketris MP, Yudovich YE. Estimations of Clarkes for Carbonaceous biolithes: world averages for trace element contents in black shales and coals. Int J Coal Geol 2009;78: 135–48.
[16] Lin R, Soong Y, Granite EJ. Evaluation of trace elements in U.S. coals using the USGS COALQUAL database version 3.0. Part I: rare earth elements and yttrium (REY). Int J Coal Geol 2018;192:1–13.
[17] Goodarzi F, Grieve DA, Sanei H, Gentzis T, Goodarzi NN. Geochemistry of coals from the elk valley coalfield, British Columbia, Canada. Int J Coal Geol 2009;77:246–59.
[18] Sutcu EC, Karayigit AI. Mineral matter, major and trace element content of the Afşin-Elbistan coals, Kahramanmaraş, Turkey. Int J Coal Geol 2015;144–145:111–29.
[19] Sonibare OO, Jacob DE, Foley SF. Quantitative estimation of major and trace elements in coals of the Benue Trough, Nigeria. Energy Sources Part A Recovery Util Environ Eff 2013;35:753–61.
[20] Bureau CCG. Atlas of coal resources in major coal mines of China. 1996. Beijing, China.
[21] Zhang JY, Zheng CG, Ren DY, Chou CL, Liu J, Zeng RS, et al. Distribution of potentially hazardous trace elements in coals from Shanxi province, China. Fuel 2004;83: 129–35.
[22] Chen S, Wu D, Liu G, Sun R, Fan X. Geochemistry of major and trace elements in Permian coal: with an emphasis on No. 8 coal seam of Zhuji coal mine, Huainan Coalfield, China. Environ Earth Sci 2016;75.
[23] Chen J, Liu G, Jiang M, Chou CL, Li H, Wu B, et al. Geochemistry of environmentally sensitive trace elements in Permian coals from the Huainan coalfield, Anhui, China. Int J Coal Geol 2011;88:41–54.
[24] Wang SX, Zhang L, Li GH, Wu Y, Hao JM, Pirrone N, et al. Mercury emission and speciation of coal-fired power plants in China. Atmos Chem Phys 2010;10:1183–92.
[25] Mukherjee S, Srivastava SK. Trace elements in high-sulfur Assam coals from the Makum coalfield in the northeastern region of India. Energy Fuels 2005;19:882–91.
[26] Kortenski J, Sotirov A. Trace and major element content and distribution in Neogene lignite from the Sofia Basin, Bulgaria. Int J Coal Geol 2002;52:63–82.
[27] Hower JC, Robertson JD. Clausthalite in coal. Int J Coal Geol 2003;53:219–25.
[28] Spears DA, Zheng Y. Geochemistry and origin of elements in some UK coals. Int J Coal Geol 1999;38:161–79.
[29] Spears DA. A review of chlorine and bromine in some United Kingdom coals. Int J Coal Geol 2005;64:257–65.
[30] Lin R, Soong Y, Granite EJ. Evaluation of trace elements in U.S. coals using the USGS COALQUAL database version 3.0. Part II: non-REY critical elements. Int J Coal Geol 2018;192:39–50.
[31] Dai S, Finkelman RB. Coal as a promising source of critical elements: progress and future prospects. Int J Coal Geol 2018;186:155–64.
[32] Chelgani SC. Occurrences of valuable trace elements in Iranian coals as potential coal combustion byproducts. Int J Coal Prep Util 2018:1–13.
[33] Chelgani SC. Investigating the occurrences of valuable trace elements in African coals as potential byproducts of coal and coal combustion products. J Afr Earth Sci 2019;150: 131–5.
[34] Wang Q, Shen W, Ma Z. Estimation of mercury emission from coal combustion in China. Environ Sci Technol 2000;34:2711–3.

[35] Wang M, Zheng B, Wang B, Li S, Wu D, Hu J. Arsenic concentrations in Chinese coals. Sci Total Environ 2006;357:96−102.
[36] Qi C, Liu G, Chou C-L, Zheng L. Environmental geochemistry of antimony in Chinese coals. Sci Total Environ 2008;389:225−34.
[37] Bai X. The distributions, modes of occurrence and volatility of trace elements in coals of China, Beijing. 2003.
[38] Song D, Qin Y, Zhang J, Wang W, Zheng C. Concentration and distribution of trace elements in some coals from Northern China. Int J Coal Geol 2007;69:179−91.
[39] Shao L, Jones T, Gayer R, Dai S, Li S, Jiang Y, et al. Petrology and geochemistry of the high-sulphur coals from the Upper Permian carbonate coal measures in the Heshan Coalfield, southern China. Int J Coal Geol 2003;55:1−26.
[40] Zeng R, Zhuang X, Koukouzas N, Xu W. Characterization of trace elements in sulphur-rich late Permian coals in the Heshan coal field, Guangxi, South China. Int J Coal Geol 2005;61:87−95.
[41] Dai S, Zhang W, Seredin VV, Ward CR, Hower JC, Song W, et al. Factors controlling geochemical and mineralogical compositions of coals preserved within marine carbonate successions: a case study from the Heshan Coalfield, southern China. Int J Coal Geol 2013;109−110:77−100.
[42] Dai S, Seredin VV, Ward CR, Hower JC, Xing Y, Zhang W, et al. Enrichment of U−Se−Mo−Re−V in coals preserved within marine carbonate successions: geochemical and mineralogical data from the late Permian Guiding Coalfield, Guizhou, China. Miner Depos 2015;50:159−86.
[43] Liu J, Yang Z, Yan X, Ji D, Yang Y, Hu L. Modes of occurrence of highly-elevated trace elements in superhigh-organic-sulfur coals. Fuel 2015;156:190−7.
[44] Dai S, Ren D, Zhou Y, Chou C-L, Wang X, Zhao L, et al. Mineralogy and geochemistry of a superhigh-organic-sulfur coal, Yanshan Coalfield, Yunnan, China: evidence for a volcanic ash component and influence by submarine exhalation. Chem Geol 2008;255:182−94.
[45] Wu J, Yan J, Liu G, Zheng L. Advance of research on the distribution, mode of occurrence, and enrichment factors of chromium in Chinese coals. Bull Miner Petrol Geochem 2005;24:239−44.
[46] Hu RZ, Qi HW, Zhou MF, Su WC, Bi XW, Peng JT, et al. Geological and geochemical constraints on the origin of the giant Lincang coal seam-hosted germanium deposit, Yunnan, SW China: a review. Ore Geol Rev 2009;36:221−34.
[47] Tian C, Gupta R, Zhao Y, Zhang J. Release behaviors of arsenic in fine particles generated from a typical high-arsenic coal at a high temperature. Energy Fuels 2016;30:6201−9.
[48] Fu X, Wang J, Tan F, Feng X, Zeng S. Minerals and potentially hazardous trace elements in the late Triassic coals from the Qiangtang Basin, China. Int J Coal Geol 2013;116−117:93−105.
[49] Dai S, Hower JC, Ward CR, Guo W, Song H, O'Keefe JMK, et al. Elements and phosphorus minerals in the middle Jurassic inertinite-rich coals of the Muli Coalfield on the Tibetan Plateau. Int J Coal Geol 2015;144−145:23−47.
[50] Zhuang X, Querol X, Alastuey A, Juan R, Plana F, Lopez-Soler A, et al. Geochemistry and mineralogy of the Cretaceous Wulantuga high-germanium coal deposit in Shengli coal field, Inner Mongolia, Northeastern China. Int J Coal Geol 2006;66:119−36.
[51] Dai S, Wang X, Seredin VV, Hower JC, Ward CR, O'Keefe JMK, et al. Petrology, mineralogy, and geochemistry of the Ge-rich coal from the Wulantuga Ge ore deposit, Inner Mongolia, China: new data and genetic implications. Int J Coal Geol 2012;90−91:72−99.

[52] Zhang J, Ren D, Zheng C, Zeng R, Chou C-L, Liu J. Trace element abundances in major minerals of late Permian coals from southwestern Guizhou province, China. Int J Coal Geol 2002;53:55–64.
[53] Wang L, Ju Y, Liu G, Chou C-L, Zheng L, Qi C. Selenium in Chinese coals: distribution, occurrence, and health impact. Environ Earth Sci 2010;60:1641–51.
[54] Chengzu S. A brief description of the Yutangba sedimentary type selenium mineralized area in southwestern Hubei. Miner Deposits 1989;3:010.
[55] Zheng B, Yan L, Mao D, Thornton I. The Se resource in southwestern Hubei province, China, and its exploitation strategy. J Nat Resour 1993;8:204–12.
[56] Zheng B, Hong Y, Zhao W, Zhou H-Y, Xia W, Su H, et al. The Se-rich carbonaceous siliceous rock and endemic selenosis in southwest Hubei, China. Chin Sci Bull 1992;37: 1725–9.
[57] Wu D, Zheng B, Tang X, Li S, Wang B, Wang M. Fluorine in Chinese coals. Fluoride 2004;37:125–32.
[58] Chen J, Liu G, Kang Y, Wu B, Sun R, Zhou C, et al. Coal utilization in China: environmental impacts and human health. Environ Geochem Health 2014;36:735–53.
[59] Dai S, Ren D, Ma S. The cause of endemic fluorosis in western Guizhou Province, Southwest China. Fuel 2004;83:2095–8.
[60] Dai S, Li W, Tang Y, Zhang Y, Feng P. The sources, pathway, and preventive measures for fluorosis in Zhijin County, Guizhou, China. Appl Geochem 2007;22:1017–24.
[61] Tian H, Cheng K, Wang Y, Zhao D, Chai F, Xue Z, et al. Quantitative assessment of variability and uncertainty of hazardous trace element (Cd, Cr, and Pb) contents in Chinese coals by using bootstrap simulation. J Air Waste Manag Assoc 2011;61:755–63.
[62] Finkelman RB, Palmer CA, Wang P. Quantification of the modes of occurrence of 42 elements in coal. Int J Coal Geol 2018;185:138–60.
[63] Yudovich YE, Ketris MP. Mercury in coal: a review. Part 1. Geochemistry. Int J Coal Geol 2005;62:107–34.
[64] Luo GQ, Ma JJ, Han J, Yao H, Xu MH, Zhang C, et al. Hg occurrence in coal and its removal before coal utilization. Fuel 2013;104:70–6.
[65] Gao L, Wang Y, Huang Q, Guo S. Modes of occurrence and thermal stability of mercury in different samples from Guandi coal preparation plant. Fuel 2017;200:22–30.
[66] Zhai J, Guo S, Wei XX, Cao Y, Gao L. Characterization of the modes of occurrence of mercury and their thermal stability in coal gangues. Energy Fuels 2015;29:8239–45.
[67] Yudovich YE, Ketris MP. Selenium in coal: a review. Int J Coal Geol 2006;67:112–26.
[68] Kang Y, Liu G, Chou CL, Wong MH, Zheng L, Ding R. Arsenic in Chinese coals: distribution, modes of occurrence, and environmental effects. Sci Total Environ 2011; 412–413:1–13.
[69] Yudovich YE, Ketris MP. Arsenic in coal: a review. Int J Coal Geol 2005;61:141–96.
[70] Zhou C-C, Zhang N-N, Peng C B, Cong L F, Ouyang C-H, Han R. Arsenic in coal: modes of occurrence, distribution in different fractions, and partitioning behavior during coal separation-a case study. Energy Fuels 2016;30:3233–40.
[71] Pan J, Zhou C, Zhang N, Liu C, Tang M, Cao S. Arsenic in coal: modes of occurrence and reduction via coal preparation – a case study. Int J Coal Prep Util 2018:1–14.
[72] Zhang Y, Tian J, Feng S, Yang F, Lu X. The occurrence modes and geologic origins of arsenic in coal from Santanghu Coalfield, Xinjiang. J Geochem Explor 2018;186: 225–34.
[73] Huggins FE, Seidu LBA, Shah N, Huffman GP, Honaker RQ, Kyger JR, et al. Elemental modes of occurrence in an Illinois #6 coal and fractions prepared by physical separation techniques at a coal preparation plant. Int J Coal Geol 2009;78:65–76.

[74] Fang T, Liu G, Zhou C, Sun R, Chen J, Wu D. Lead in Chinese coals: distribution, modes of occurrence, and environmental effects. Environ Geochem Health 2014;36:563–81.
[75] Sia SG, Abdullah WH. Enrichment of arsenic, lead, and antimony in Balingian coal from Sarawak, Malaysia: modes of occurrence, origin, and partitioning behaviour during coal combustion. Int J Coal Geol 2012;101:1–15.
[76] Saikia BK, Goswamee RL, Baruah BP, Baruah RK. Occurrence of some hazardous metals in Indian coals. Coke Chem 2009;52:54–9.
[77] Shi J, Huang W, Chen P, Tang S, Chen X. Concentration and distribution of cadmium in coals of China. Minerals 2018;8.
[78] Song D, Wang M, Zhang J, Zheng C. Contents and occurrence of cadmium in the coals from Guizhou Province, China. Ann NY Acad Sci 2008;1140:274–81.
[79] Qingguang L, Chen P, Chen J, Hu Y. Modes of occurrence of Cr, Co, Ni, Cu, Cd, and Pb in the main coal seams of southwestern China's Nantong coalfield. Geochem Int 2018;56: 1220–32.
[80] Liu Y, Liu G, Qi C, Cheng S, Sun R. Chemical speciation and combustion behavior of chromium (Cr) and vanadium (V) in coals. Fuel 2016;184:42–9.
[81] Huggins FE, Shah N, Huffman GP, Kolker A, Crowley S, Palmer CA, et al. Mode of occurrence of chromium in four US coals. Fuel Process Technol 2000;63:79–92.
[82] Riley KW, French DH, Farrell OP, Wood RA, Huggins FE. Modes of occurrence of trace and minor elements in some Australian coals. Int J Coal Geol 2012;94:214–24.
[83] Ren D, Xu D, Zhao F. A preliminary study on the enrichment mechanism and occurrence of hazardous trace elements in the Tertiary lignite from the Shenbei coalfield, China. Int J Coal Geol 2004;57:187–96.
[84] Chen J, Chen P, Yao D, Huang W, Tang S, Wang K, et al. Abundance, distribution, and modes of occurrence of uranium in Chinese coals. Minerals 2017;7.
[85] Antonia López Antón M, Alan Spears D, Somoano MD, Rosa Martínez Tarazona M. Thallium in coal: analysis and environmental implications. Fuel 2013;105:13–8.
[86] Dai S, Ren D, Tang Y, Yue M, Hao L. Concentration and distribution of elements in late permian coals from western guizhou province, China. Int J Coal Geol 2005;61:119–37.
[87] Sia S-G, Abdullah WH. Concentration and association of minor and trace elements in Mukah coal from Sarawak, Malaysia, with emphasis on the potentially hazardous trace elements. Int J Coal Geol 2011;88:179–93.
[88] Spears DA, Borrego AG, Cox A, Martinez-Tarazona RM. Use of laser ablation ICP-MS to determine trace element distributions in coals, with special reference to V, Ge and Al. Int J Coal Geol 2007;72:165–76.
[89] Zhao CL, Sun YZ, Xiao L, Qin SJ, Wang JX, Duan DJ. The occurrence of barium in a jurassic coal in the huangling 2 mine, Ordos basin, northern China. Fuel 2014;128: 428–32.
[90] Gürdal G. Abundances and modes of occurrence of trace elements in the Çan coals (Miocene), Çanakkale-Turkey. Int J Coal Geol 2011;87:157–73.
[91] Singh PK, Singh VK, Rajak PK, Singh MP, Naik AS. Distribution and geochemistry of selected trace elements in the lignites of Cambay basin, Gujarat, western India. J Geol Soc India 2016;88:131–46.
[92] Rajabzadeh MA, Ghorbani Z, Keshavarzi B. Chemistry, mineralogy and distribution of selected trace-elements in the Parvadeh coals, Tabas, Iran. Fuel 2016;174:216–24.
[93] Silva LFO, DaBoit K, Sampaio CH, Jasper A, Andrade ML, Kostova IJ, et al. The occurrence of hazardous volatile elements and nanoparticles in Bulgarian coal fly ashes and the effect on human health exposure. Sci Total Environ 2012;416:513–26.

[94] Zhu JM, Johnson TM, Finkelman RB, Zheng BS, Sýkorová I, Pešek J. The occurrence and origin of selenium minerals in Se-rich stone coals, spoils and their adjacent soils in Yutangba, China. Chem Geol 2012;330–331:27–38.
[95] Black A, Craw D. Arsenic, copper and zinc occurrence at the Wangaloa coal mine, southeast Otago, New Zealand. Int J Coal Geol 2001;45:181–93.
[96] Dai S, Jiang Y, Ward CR, Gu L, Seredin VV, Liu H, et al. Mineralogical and geochemical compositions of the coal in the Guanbanwusu Mine, Inner Mongolia, China: further evidence for the existence of an Al (Ga and REE) ore deposit in the Jungar Coalfield. Int J Coal Geol 2012;98:10–40.
[97] Dai S, Li D, Chou C-L, Zhao L, Zhang Y, Ren D, et al. Mineralogy and geochemistry of boehmite-rich coals: new insights from the haerwusu surface mine, jungar coalfield, inner Mongolia, China. Int J Coal Geol 2008;74:185–202.
[98] Hu P, Hou X, Zhang J, Li S, Wu H, Damø AJ, et al. Distribution and occurrence of lithium in high-alumina-coal fly ash. Int J Coal Geol 2018;189:27–34.
[99] Dai S, Zhao L, Peng S, Chou C-L, Wang X, Zhang Y, et al. Abundances and distribution of minerals and elements in high-alumina coal fly ash from the Jungar Power Plant, Inner Mongolia, China. Int J Coal Geol 2010;81:320–32.
[100] Bai X, Wang Y, Li W. Distribution and occurrence of trace elements in the No.14 coal from the Huolinhe mine. Int J Coal Sci Technol 2017;4:199–213.
[101] Seredin VV, Finkelman RB. Metalliferous coals: a review of the main genetic and geochemical types. Int J Coal Geol 2008;76:253–89.
[102] Luo K, Ren D, Xu L, Dai S, Cao D, Feng F, et al. Fluorine content and distribution pattern in Chinese coals. Int J Coal Geol 2004;57:143–9.
[103] Dai S, Ren D. Fluorine concentration of coals in China—an estimation considering coal reserves. Fuel 2006;85:929–35.
[104] Wang X, Dai S, Sun Y, Li D, Zhang W, Zhang Y, et al. Modes of occurrence of fluorine in the late paleozoic No. 6 coal from the haerwusu surface mine, inner Mongolia, China. Fuel 2011;90:248–54.
[105] Yang N, Tang S, Zhang S, Huang W, Chen P, Chen Y, et al. Fluorine in Chinese coal: a review of distribution, abundance, modes of occurrence, genetic factors and environmental effects. Minerals 2017;7.
[106] Yudovich YE, Ketris MP. Chlorine in coal: a review. Int J Coal Geol 2006;67:127–44.
[107] Peng B, Li X. Release and transformation characteristics of modes of occurrence of chlorine in coal gangue during combustion. Energy Fuels 2018;32:9926–33.
[108] Vassilev SV, Eskenazy GM, Vassileva CG. Contents, modes of occurrence and origin of chlorine and bromine in coal. Fuel 2000;79:903–21.
[109] Peng B-x, Wu D-s. Modes of iodine occurrence in bituminous coal and anthracite and their environmental effects. J Fuel Chem Technol 2012;40:257–62.
[110] Lin R, Bank TL, Roth EA, Granite EJ, Soong Y. Organic and inorganic associations of rare earth elements in central Appalachian coal. Int J Coal Geol 2017;179:295–301.
[111] Pan J, Zhou C, Tang M, Cao S, Liu C, Zhang N, et al. Study on the modes of occurrence of rare earth elements in coal fly ash by statistics and a sequential chemical extraction procedure. Fuel 2019;237:555–65.
[112] Bartoňová L, Serenčíšová J, Čech B. Yttrium partitioning and associations in coal-combustion ashes prior to and after their leaching in HCl. Fuel Process Technol 2018; 173:205–15.
[113] Seredin VV, Dai S. Coal deposits as potential alternative sources for lanthanides and yttrium. Int J Coal Geol 2012;94:67–93.

[114] Diehl SF, Goldhaber MB, Hatch JR. Modes of occurrence of mercury and other trace elements in coals from the warrior field, Black Warrior Basin, Northwestern Alabama. Int J Coal Geol 2004;59:193–208.

[115] Kolker A, Huggins FE, Palmer CA, Shah N, Crowley SS, Huffman GP, et al. Mode of occurrence of arsenic in four US coals. Fuel Process Technol 2000;63:167–78.

[116] Coleman L, Bragg LJ, Finkelman RB. Distribution and mode of occurrence of selenium in US coals. Environ Geochem Health 1993;15:215–27.

[117] Kolker A. Minor element distribution in iron disulfides in coal: a geochemical review. Int J Coal Geol 2012;94:32–43.

[118] Spears DA, Manzanares-Papayanopoulos LI, Booth CA. The distribution and origin of trace elements in a UK coal; the importance of pyrite. Fuel 1999;78:1671–7.

[119] Equeenuddin SM, Tripathy S, Sahoo PK, Ranjan A. Geochemical characteristics and mode of occurrence of trace elements in coal at West Bokaro coalfield. Int J Coal Sci Technol 2016;3:399–406.

[120] Yang N, Tang S, Zhang S, Chen Y. Modes of occurrence and abundance of trace elements in pennsylvanian coals from the pingshuo mine, ningwu coalfield, Shanxi Province, China. Minerals 2016;6.

[121] Li C, Liang H, Wang S, Liu J. Study of harmful trace elements and rare earth elements in the Permian tectonically deformed coals from Lugou Mine, North China Coal Basin, China. J Geochem Explor 2018;190:10–25.

[122] Yang J. Concentrations and modes of occurrence of trace elements in the Late Permian coals from the Puan Coalfield, southwestern Guizhou, China. Environ Geochem Health 2006;28:567–76.

[123] Wang P, Yan X, Guo W, Zhang S, Wang Z, Xu Y, et al. Geochemistry of trace elements in coals from the yueliangtian mine, western guizhou, China: abundances, modes of occurrence, and potential industrial utilization. Energy Fuels 2016;30:10268–81.

[124] Cheng W, Zhang Q, Yang R, Tian Y. Occurrence modes and cleaning potential of sulfur and some trace elements in a high-sulfur coal from Pu'an coalfield, SW Guizhou, China. Environ Earth Sci 2014;72:35–46.

[125] Zhao Y, Zhang J, Zheng C. Release and removal using sorbents of chromium from a high-Cr lignite in Shenbei coalfield, China. Fuel 2013;109:86–93.

[126] Yang J, Liu D, Wang Y, Zhao Y, Zhang Y, Zhang J, et al. Release and the interaction mechanism of uranium and alkaline/alkaline-earth metals during coal combustion. Fuel 2016;186:405–13.

Trace element partition in coal combustion

3

Hailong Li[1], *Wenqi Qu*[1], *Zequn Yang*[2], *Jiexia Zhao*[1]
[1]School of Energy Science and Engineering, Central South University, Changsha, China;
[2]Department of Civil Engineering, The University of Hong Kong, Hong Kong, China

3.1 Introduction

Coal, responsible for 40% of electricity production around the world, will further solidify its dominant role in energy production in the future considering its relatively abundant reserves compared with crude oil and natural gas [1]. According to the newly released report by the World Coal Association, around 70% of coal had been used in combustion power generation by 2005 and the ratio will further climb to 80% in 2030 [2]. While the emissions of traditional pollutants like NO_x, SO_2, and CO_2 from coal combustion were strictly controlled, the emissions of trace elements (TEs) during combustion were a relatively new topic. The updated inventories on TE emissions from coal combustion sites and awareness of their toxicity on environmental health have urged limitations on their emissions. For example, the Minamata Convention came into force on August 16, 2018 to restrict mercury (Hg) emission from the anthropogenic sources with an emphasis on coal combustion power plants [3]. Selenium (Se) and arsenic (As) have also attracted attention due to their relatively high volatility and acute effect on human health [4,5]. To understand better the emissions of TEs from coal combustion, the primary mission is to fully understand their partition during the combustion process. Although TE partitions have been widely investigated for decades, special attention should also be paid to this area to provide fundamental aspects of future pollution control strategies. Thus, in this chapter, we have chosen five specific TEs, i.e., fluorine (F), Hg, Se, As, and chromium (Cr), to represent TEs with different levels of volatility (shown in Fig. 3.1) to generalize TE partitioning behavior during combustion. Their characteristics during coal combustion over laboratory-scale experiments are reviewed in detail. The effects of coal type, combustion temperature, and combustion mode on partitioning behavior and emission characteristics of TEs are discussed.

3.2 Reaction mechanism and chemical kinetics of trace elements during coal combustion

3.2.1 Reaction mechanisms

The mechanisms for TE decomposition and emission from the coal combustion process were never well defined. Most TEs coexist with iron sulfides (As, Hg, Se, F), organic

Emission and Control of Trace Elements from Coal-Derived Gas Streams
https://doi.org/10.1016/B978-0-08-102591-8.00003-9

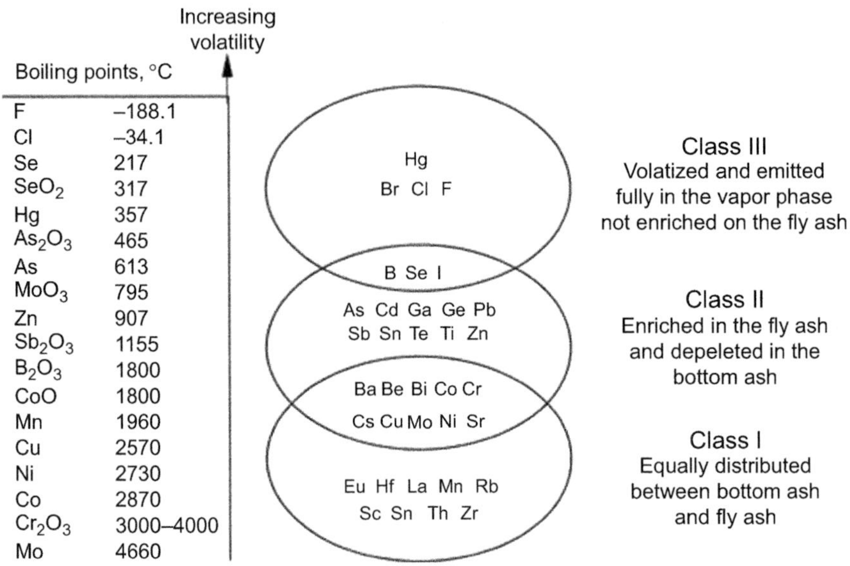

Figure 3.1 Categorization of trace elements based on volatility behavior.
From Xu M, et al. Status of trace element emission in a coal combustion process: a review. Fuel Process Technol 2003;85:215−37.

matter (Se, F), and clay minerals like kaolinite (Cr) in coal [6−8]. For TEs associated with iron sulfides, one plausible explanation is that the TEs vaporize during pyrite (FeS_2) to pyrrhotite (FeS) decomposition [9]. Zeng et al. proposed that under the combustion temperature, pyrite or pyrrhotite exist as melt or liquid phase to trap the TEs. Vaporizations of TEs were controlled by a series of complicated processes, including diffusion of TEs through the melt, diffusion of gaseous TEs through char pores, and decomposition reaction of TEs [10]. As for the TEs associated with organic matter like Se, decomposition of organics and the char matrix induce the TEs to vaporize as their elemental or reduced forms; in the case of Se it is Se(g) or H_2Se(g) and in the case of F it is HF(g) [7]. As for TEs in clay minerals like Cr, most of them are non-vaporized and will remain in bottom ash and fly ash during combustion [11].

3.2.2 Chemical kinetics

The chemical kinetics for TE partitions during coal combustion are classified into two categories, i.e., chemical kinetics of TE emission and chemical kinetics of TE transformation. Shen et al. used an online analysis system based on inductively coupled plasma optical emission spectroscopy and focused their research on the chemical kinetics of Se and As emissions. They found that the emission patterns of Se and As during coal combustion can both be expressed by the modified Arrhenius equations as shown in Eqs. (3.1)−(3.3)

$$r(x) = 0.94e^{-26.58/RT}\left(-0.56x^2 - 0.51x + 1.05\right) \tag{3.1}$$

$$r(x) = 290.3e^{-80.9/RT}\left(-5.86x^4 + 16.02x^3 - 17.43x^2 + 7.26x + 0.01\right) \quad (3.2)$$

$$x = \frac{q_0 - q}{q_0 - q_f} \quad (3.3)$$

In Eqs. (3.1) and (3.2), r (mg/kg s) is the release rate of Se and As from coal combustion, respectively, and x is the release degree of Se or As in coal. In Eq. (3.3), q, q_0, and q_f are the instantaneous concentration, initial concentration, and final concentration of Se or As, respectively. As shown in Fig. 3.2, the emission of Se (a and b) and As (c and d) from coal combustion was almost perfectly predicted by the Shen et al. theory. Combustion temperature has an obvious impact on Se and As vaporization rate because the increased combustion temperature accelerated the release of TEs. These give us a well-defined kinetic model to predict TE release from coal combustion gas.

On the other hand, Hg was classified as the most volatile part that was fully vaporized during the combustion process. In the past few decades, extensive studies have concentrated on Hg partition during coal combustion due to its long persistence in

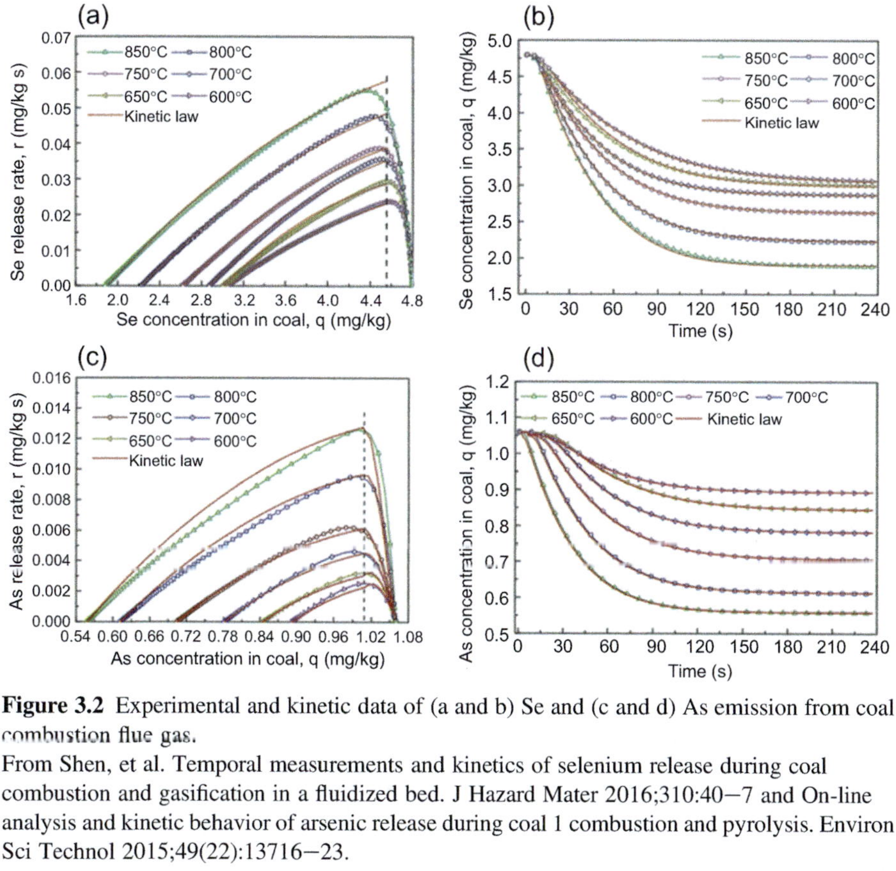

Figure 3.2 Experimental and kinetic data of (a and b) Se and (c and d) As emission from coal combustion flue gas.
From Shen, et al. Temporal measurements and kinetics of selenium release during coal combustion and gasification in a fluidized bed. J Hazard Mater 2016;310:40–7 and On-line analysis and kinetic behavior of arsenic release during coal 1 combustion and pyrolysis. Environ Sci Technol 2015;49(22):13716–23.

the environment and high toxicity to human health [12,13]. Hg exists in coal in the form of HgS or is accompanied by FeS_2 [14]. Due to its extremely high volatility, almost all of Hg decomposes into its gaseous elemental form (Hg^0) and discharges from the high-temperature combustion boiler irrespective of its original existence and the combustion temperature (800–1400°C) [14,15]. Cooling the flue gas temperature may result in the recombination of Hg^0 with other coal combustion by-products and the formation of oxidized mercury species, especially $HgCl_2$ [16].

Thus previous studies have focused on the reaction kinetics between Hg^0 with chlorine species emitted from coal combustion. Formulation of the reaction mechanism began with the kinetic framework of Widmer et al. [17]:

$$k = AT^n \exp(-E/RT) \tag{3.4}$$

in which k (cm^3/mol s) is the rate constant of the reaction. The typical reactions for Hg^0 oxidation by chlorine species are listed in Table 3.1.

Based on the foregoing kinetic model, Sliger et al. Hall et al., and Wilcox et al. predicted Hg^0 oxidation in the presence of HCl, the main form of chlorine species existing in combustion gas, as a function of initial Hg^0 and HCl concentration [18,19]. The scattered dots are the experimental results, while the lines are simulations by the kinetic models. As shown in Fig. 3.3(a) and (b), the kinetic models matched with the experimental results within the experimental uncertainties except when the concentration of HCl was 50 ppm. At a concentration of HCl below 300 ppm, the predicted and observed extent of oxidation seemed independent of the initial Hg^0 concentration. However, when the HCl concentration exceeded 4000-fold of the Hg^0 concentration, the extent of oxidation reached saturation regardless of the initial HCl concentration. However, on a common basis, the extent of oxidation of Hg^0 increased with the increased initial Hg^0 concentration and the HCl concentration. The kinetic models were a feasible method to predict Hg^0 homogeneous oxidation in combustion gas with relatively high HCl concentration.

Table 3.1 Hg^0 oxidation submechanism.

Reaction	**A**	**n**	**E (kcal/mol)**
1. $Hg^0 + Cl + M \leftrightarrow HgCl + M$	9.00×10^{15} cm^6/mol^2 s	0.5	0
2. $Hg^0 + Cl_2 \leftrightarrow HgCl + Cl$	1.39×10^{14} cm^3/mol s	0	34.0
3. $Hg^0 + HCl \leftrightarrow HgCl + H$	4.94×10^{14} cm^3/mol s	0	79.3
4. $Hg^0 + HOCl \leftrightarrow HgCl + OH$	4.27×10^{13} cm^3/mol s	0	19.0
5. $HgCl + Cl_2 \leftrightarrow HgCl2 + Cl$	1.39×10^{14} cm^3/mol s	0	1.0
6. $HgCl + Cl + M \leftrightarrow HgCl_2 + M$	1.16×10^{15} cm^6/mol^2 s	0.5	0
7. $HgCl + HCl \leftrightarrow HgCl_2 + H$	4.64×10^3 cm^3/mol s	2.5	19.1
8. $HgCl + HOCl \leftrightarrow HgCl_2 + OH$	4.27×10^{13} cm^3/mol s	0	1.0

From Niksa S, Kinetic modeling of homogeneous mercury oxidation: the importance of NO and H_2O in predicting oxidation in coal-derived systems. Environ Sci Technol 2001;35:3701–06.

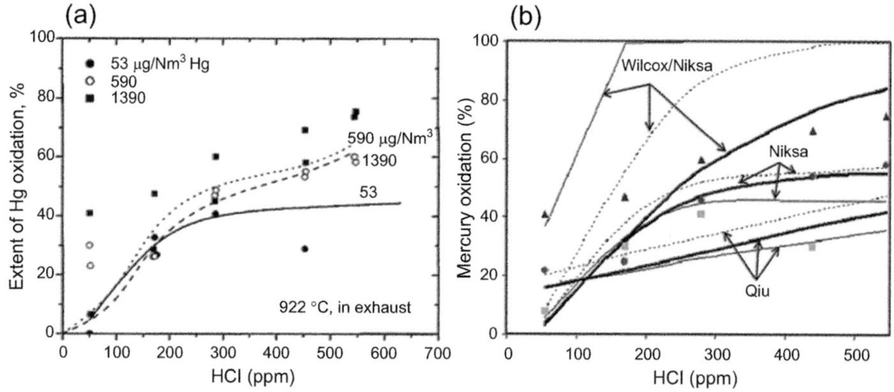

Figure 3.3 (a) Evaluation for Hg^0 oxidation with different Hg^0 and HCl concentrations and (b) comparison of Hg oxidation in Sliger et al. experiments with the predictions of Niksa, Qiu, and Wilcox/Niksa rate constants (*light solid*, *dotted*, and *dark solid curves* for I, II, and III, respectively). I, II, and III represent Hg levels of 53, 590, and 1390 μg/Nm^3, respectively.
(a) From Niksa S, Kinetic Modeling of Homogeneous Mercury Oxidation: The Importance of NO and H_2O in Predicting Oxidation in Coal-Derived Systems. Environ Sci Technol 2001;35: 3701−06. (b) From Krishnakumar, et al., Understanding Mercury Transformations in Coal-Fired Power Plants: Evaluation of Homogeneous Hg Oxidation Mechanisms. Environ Sci Technol 2007;41:7870−75.

Similarly, due to the high volatility, F existing as organic or inorganic fluoride in coal was almost emitted from the combustion flue gas in its reduced form (HF) [20]. However, cooling down the flue temperature will not induce the oxidation of HF. Thus few works have focused on the kinetic model of F; only the content of O will slightly influence the amount of fluorine emitted (shown in Fig. 3.4). The possible reason is that the presence of excess O_2 increased the furnace temperature and enhanced the release of F [20].

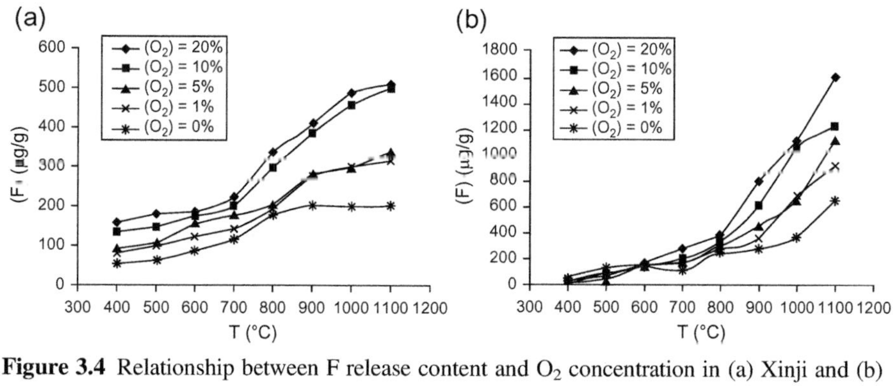

Figure 3.4 Relationship between F release content and O_2 concentration in (a) Xinji and (b) Liantang coals.
From Liu, et al. Behavior of fluorine in the combustion of chinese coal in small furnaces. Energy Fuels 2006;20:1406−10.

3.3 Trace element emissions during coal combustion

3.3.1 Trace element emissions from different coal types

Because the upper limits are set by TE concentration in coal for the emission concentration, coal type is the initial "controlling" factor that influences TE emissions during coal combustion. Raask et al. investigated the mode of occurrence of different TEs in six types of coals from the United States, the United Kingdom, and Australia [21]. Xu et al. reviewed the TE concentration in most coals based on previous studies [6]. As shown in Table 3.2, the TE concentration varies among different coal types, for example, Hg difference ranges from 0.02 to 1 ppm. Furthermore, for different TEs, their initial concentrations in coal are largely different from each other. In most coals, the concentration of F reaches 20–500 ppm; on the other hand, the concentration of Hg is only 0.02–1 ppm. The initial concentration of TEs will directly determine the upper limit of TE emissions during coal combustion. However, no significant correlation was observed between TE concentration and coal grade because sub-bituminous coal and bituminous coal have similar TE concentrations.

Yan et al. investigated TE emissions during the combustion of different types of coals under different temperatures and atmospheres [22]. As shown in Fig. 3.5(a), the volatile temperatures for Hg, Se, As, and Cr increased stepwise from less than 300K to more than 1400K. The order of volatile temperature is:

$$\text{Hg } (<300\text{K}) < \text{Se } (400-600\text{K}) < \text{As } (800-1000\text{K}) < \text{Cr (above } 1200-1500\text{K}).$$

Table 3.2 Concentrations of trace elements in US, UK, Australian, and most other coals (ppm).

Elements	Coals						For most coals
	United States		United Kingdom			Australia	
	a	b	c	d	e	f	
F	61	74	150	–	114	150	20–500
Hg	0.2	0.18	0.1	0.2	–	0.1	0.02–1
As	14	15	1.5	–	18	1.5	0.5–80
Se	2.1	4.1	0.8	–	2.8	0.8	0.2–4
Cr	14	15	6	–	34	6	0.5–60

[a]101 samples of chiefly Illinois bituminous coals [52].
[b]799 samples of bituminous and sub-bituminous coals [53].
[c]23 samples of bituminous coals [54].
[d]231 samples of bituminous coals [55].
[e]The number of samples not stated, bituminous coal (National Coal Board, reported by Lim, [56]).
[f]452 samples of New South Wales bituminous coals [57].
From Xu, et al. Status of trace element emission in a coal combustion process: a review. Fuel Process Technol 2003;85: 215–37.

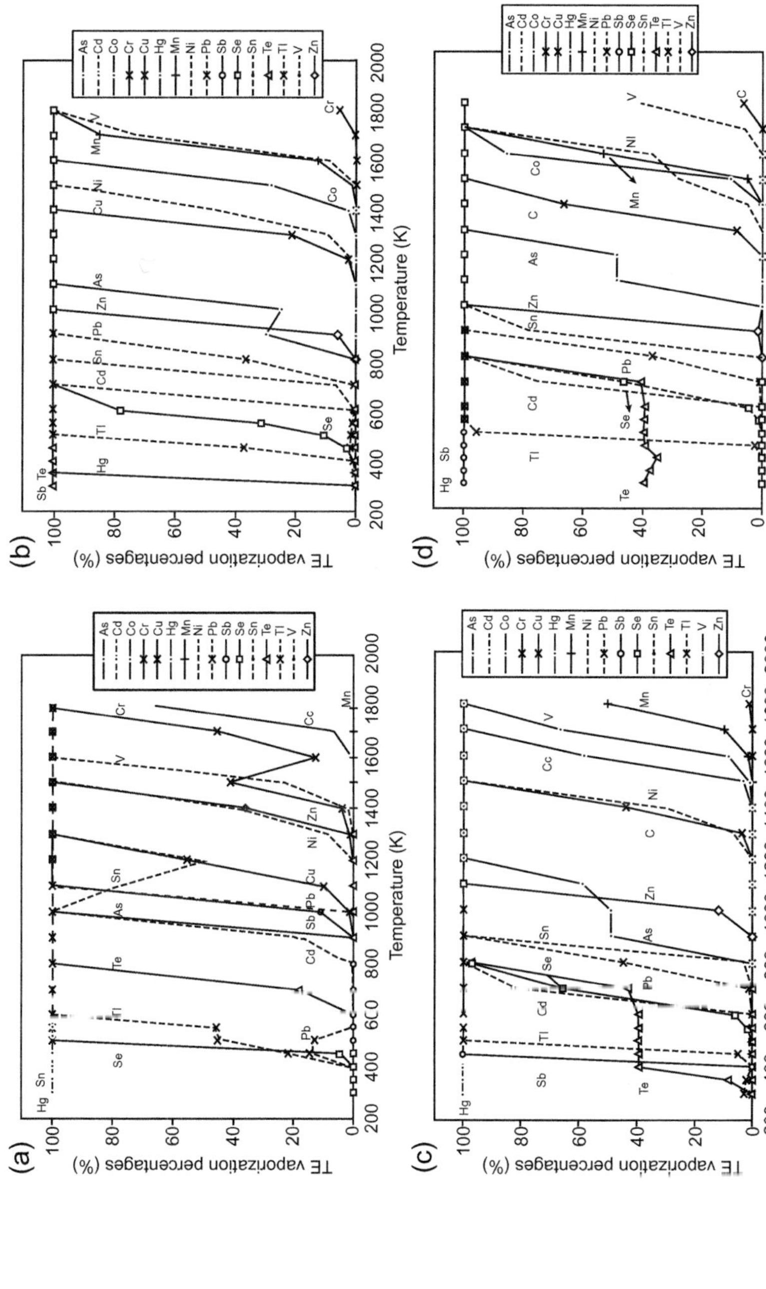

Figure 3.5 Trace element partition patterns of (a) low-ash coal under oxidizing conditions, (b) low-ash coal under reducing conditions, (c) high-ash coal under oxidizing conditions, and (d) high-ash coal under reducing conditions.
From Yan, et al. Volatility and chemistry of trace elements in a coal combustor. Fuel 2001;80:2217–26.

Therefore almost all Hg, Se, and As will evaporate during coal combustion except for Cr. As shown in Fig. 3.5(b), the reducing atmosphere slightly inhibits TE emissions during coal combustion. However, the extent is very limited when changing the volatility pattern of TE during coal combustion. In addition, some TE, such as Se and Cr, discharge temperatures are elevated in high-ash coal compared to low-ash coal.

Liu et al. investigated the As partition during the combustion of different ranks of coal (shown in Fig. 3.6) [23]. In their work, arsenic content from coal to coke and coal to ash was tested. For the first process (coal to coke), lignite was found to have the highest volatility from coal to coke, followed by bituminous coal, which has the least volatility. The main reason is that lignite has the highest volatile content, which promotes rapid coal combustion and oxygen diffusion into the pores of the carbon, thereby increasing the reaction rate and the volatilization of TEs. For the second process (coal to ash), they believe that the coal rank has little effect on the final arsenic volatilization over the entire range of fluctuations from coal to coke to ash. Anthracite and bituminous coal have an arsenic volatilization rate of about 60%, while lignite has a low volatilization rate. In the study by Liu et al. coal was not classified by coal grade but was divided into two groups: arsenic >4 μg/g and arsenic <4 μg/g. The results indicate that groups with higher arsenic concentrations tend to have greater levels of arsenic volatility at a given temperature. It can be seen that the volatilization of mercury, arsenic, and selenium in underground coal gasification (UCG) follows the order of Hg $>$ Se $>$ As. Mercury has the highest volatility and coal ash is above 98%. The total volatility of selenium from raw coal to ash is higher than 90% but lower than mercury. The volatility of arsenic is less than 60%, and arsenic enrichment is found in UCG ash.

Wang et al. studied the copartition patterns of As and SO_2 during the combustion of different ranks of coal under different temperatures (shown in Fig. 3.7) [24]. The results indicated that although arsenic levels were often associated with sulfur content, especially inorganic sulfur in coal, it was difficult to correlate the volatilization behaviors of arsenic and sulfur. Some researchers have found that the volatilization proportion of arsenic was reduced in coals with higher sulfur content. However, this conclusion is not confirmed by the necessary experimental studies. Contreras et al. [58] found the volatilization proportion of arsenic might be higher for coal with high calcium content, because sulfur was more likely to react with calcium in ash, thus reducing arsenic capture. In addition, sulfur and arsenic in coal may have competed with the active sites of CaO. However, the opposite conclusions were drawn about the reaction order of $CaO + As_2O_3(g)$ and $CaO + SO_2(g)$. To determine the effect of sulfur on arsenic volatilization, the online release curve of SO_2 was compared with the simultaneous release of arsenic during coal combustion (indirectly by changing the residence time of the raw coal in the tube furnace). To determine the effect of sulfur on arsenic volatilization, the online release curve of SO_2 and the simultaneous release curve of arsenic during coal combustion (indirectly obtained by changing the residence time of raw coal in the tube furnace) are compared, indicating that arsenic volatilization has a very strong correlation. In addition, these authors have found that coals with higher sulfur content often contain a larger proportion of sulfides combined with arsenic, which supports the positive effects of sulfur on arsenic

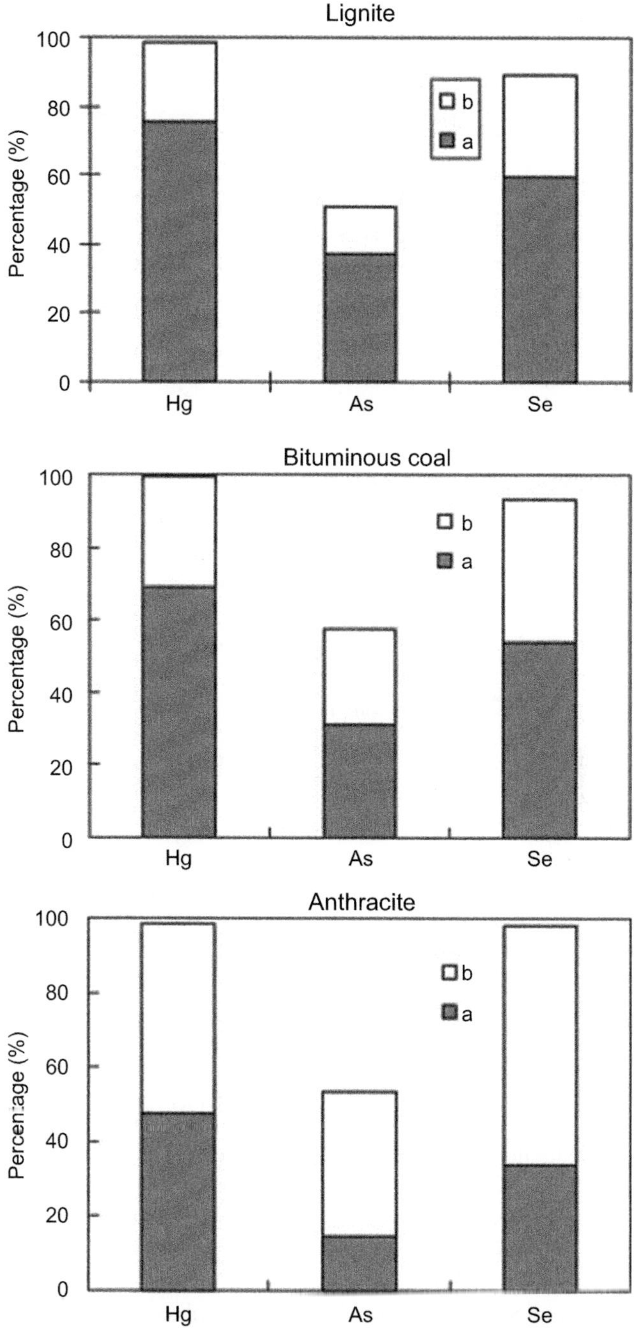

Figure 3.6 Comparison of the volatility of mercury, arsenic, and selenium during underground coal gasification of different coals (a, transformation of coal to char; b, transformation of char to ash). *TE*, trace element.
From Liu, et al. Volatilization of mercury, arsenic and selenium during underground coal gasification. Fuel 2006;85:1550–58.

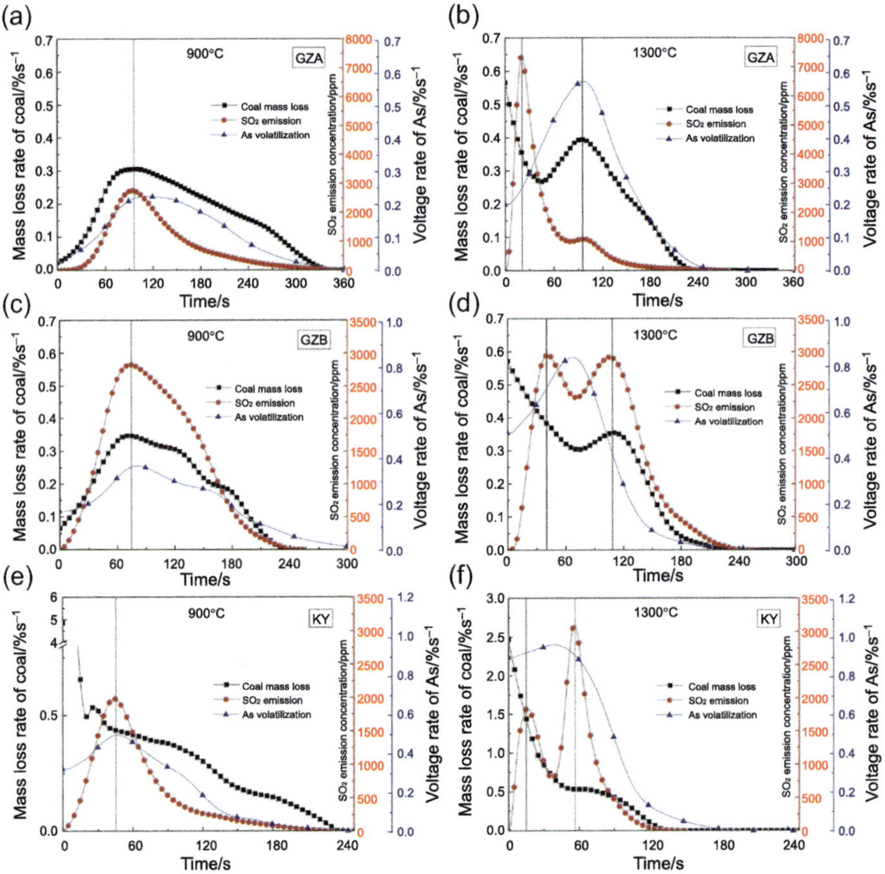

Figure 3.7 Simultaneous volatilization of arsenic and SO_2 during isothermal coal combustion: (a) GZA coal at 900°C; (b) GZA coal at 1300°C; (c) GZB coal at 900°C; (d) GZB coal at 1300°C; (e) KY coal at 900°C; and (f) KY coal at 1300°C.
From Wang, et al. Review of arsenic behavior during coal combustion: volatilization, transformation, emission and removal technologies. Prog Energy Combust Sci 2018;68:1–28.

volatilization. With the further development of online arsenic testing technology in the future, more quantitative conclusions can be expected. Although the decomposition mechanism of pyrite during coal combustion at different temperatures has been reported many times. The mechanism of decomposition of arsenopyrite is relatively less noticed. Although thermodynamic simulation software has been used to simulate possible chemical reactions for the decomposition of arsenopyrite, specific reaction temperatures and reaction kinetic constant *k* values have not been reported.

Furuzono et al. studied the partition of Se when combusting different coals (shown in Table 3.3) [25]. The partition patterns are shown in Fig. 3.8. The influence of flue gas temperature and kind of coal on the partitioning of Se from the gas to solid phase was examined. When five kinds of coals (A–E) were combusted under combustion

Table 3.3 Properties of feed coals.

Unit			Feed coal A	B	C	D	E
Inherent moisture		Wt% (air dried)	3.3	4.8	3.1	2.4	7.2
Ash			12.3	13.0	12.2	15.7	3.9
Volatile matter			34.4	32.7	31.4	26.0	43.0
Fixed carbon			50.0	50.0	53.3	55.9	46.0
Ash composition	SiO_2	Wt% of ash	60.4	47.8	56.8	46.6	38.1
	Al_2O_3		23.3	41.8	21.0	29.6	22.6
	Fe_2O_3		4.28	3.84	6.51	5.95	18.5
	CaO		3.20	2.11	3.50	6.95	7.37
	MgO		1.06	0.39	1.47	0.96	2.35
	Na_2O		0.32	0.23	0.60	0.36	1.14
	K_2O		1.40	0.51	2.04	0.60	0.95
Se		mg/kg (dry basis)	0.40	2.55	0.27	0.49	0.05

From Furuzono, et al. Behavior of selenium in the flue gas of pulverized coal combustion system: Influence of kind of coal and combustion conditions. Fuel Process Technol 2017;167:388–94.

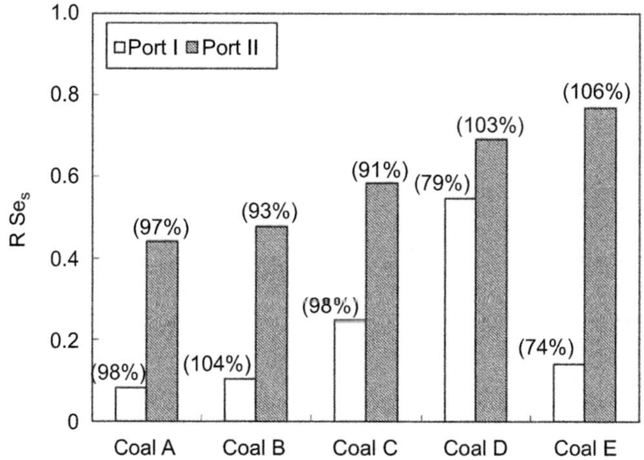

Figure 3.8 Influence of flue gas temperature and kind of feed coal on R Se$_g$. Mass balance of Se in the flue gas for each case is shown in parentheses.
From Furuzono, et al. Behavior of selenium in the flue gas of pulverized coal combustion system: influence of kind of coal and combustion conditions. Fuel Process Technol 2017;167: 388–94.

condition 1, the flue gas samples were collected at ports I and II. The fraction of Se in the solid phase (R Se_s) is given by the following equation.

$$R\,Se_s = Se_s/Se_s + Se_g$$

where Se_s and Se_g are the concentrations of Se in the solid phase and the gas phase, respectively, in the flue gas. Fig. 3.8 shows the effect of flue gas temperature and coal on R Se_s and the mass balance of Se in the flue gas. For port I, when coal with low Se levels (D and E) is used, an incomplete mass balance closure is obtained. For all five coals, R Se_s increases as the flue gas temperature decreases from 300° C (port I) to 100° C (port II). However, R Se_s differs between coal types, especially for samples collected at port I. In coal combustion flue gas, Se in the gas phase may react with specific inorganic components in the feed coal in a high-temperature zone above 300°C. In contrast, Se in the gas phase is expected to be physically adsorbed at a lower temperature (about 100° C) and condensed on the surface of the coal ash (solid phase). It is reasonable to promote chemical absorption in the higher-temperature region and physical absorption in the low-temperature region.

3.3.2 Trace element emissions from different coal combustion modes

In a typical conventional pulverized coal burner, combustion is carried out in a furnace having an operating temperature >1400°C. Coal is injected into the furnace and ignited upon suspension. As the particles are heated, the volatiles evaporate and burn. Minerals undergo thermal decomposition, melting, disintegration, and agglomeration. The final combustion products are typically spherical ash particles, which can then undergo other processes, such as coalescing with other particles or expanding due to internal gas release. The formation of molten ash droplets marks the highest temperature and a significant portion of the volatile elements will be present in the gas phase. Some of the larger mineral particles may only partially melt, and refractory minerals with a high melting point may escape and melt regardless of size. The main formation mechanism of coarse ash particles (>2 pm) is the carrying of a portion of the minerals in the feed coal. A portion of the incombustible material remains in the furnace as slag or bottom ash. The remaining inorganic material remains in the flue gas as fly ash and vapor.

As the combustion products leave the furnace and begin to cool, the volatile TEs condense. The particles can be formed by nucleation of the evaporating material and grown by coagulation and heterogeneous condensation. Condensation of most of the elements will be completed before the flue gas reaches the particle control unit and the gas temperature drops to 350°C. For some elements, condensation is delayed. For example, fly ash sampled from a high-temperature region prior to electrostatic precipitation may exhibit a completely different concentration of TEs than samples collected from cooler regions. This may be due to delayed condensation of volatile materials with low boiling points or slower cooling of ash particles compared to flue gas. The element forming the compound having a high vapor pressure can be discharged from the stack in the gas

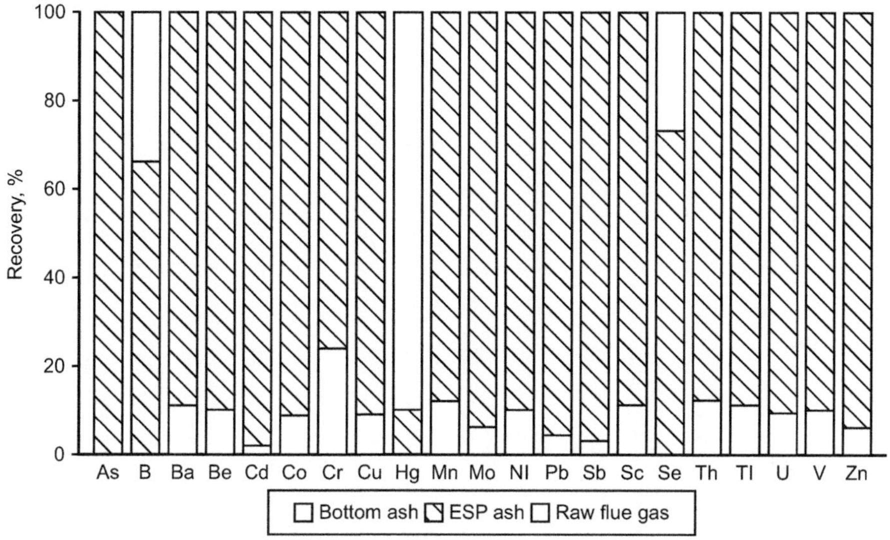

Figure 3.9 Relative distribution of trace elements between residues and flue gases during traditional pulverized coal combustion. *ESP*, electrostatic precipitation.
From Clarke, LB. The fate of trace elements during coal combustion and gasification: an overview. Fuel 1993;72:731–36.

phase (for example, Hg, Se). Fig. 3.9 shows the relative distribution of TEs between the residue and the flue gas. As shown in Fig. 3.9, >90% and 20%–30% of Hg and Se are released directly from coal combustion as a gas phase due to their high volatility. For As, although it will evaporate from the coal-fired boiler at high temperatures, 100% of As will accumulate on the fly ash surface after condensation and nucleation. In addition, Cr partially accumulates in the fly ash, and is partially nonvolatile and rich in bottom ash [26]. The volatile patterns of the TEs in a traditional pulverized combustor were in line with their categories as shown in Fig. 3.1.

To enhance the energy-using efficiency, the fluidized-bed combustion (FBC) of coal was designed for low-rank coals and applied increasingly widely worldwide [27]. The combustion characteristics of FBC systems are different from those of conventional pulverized coal-fired combustion. Emissions from atmospheric FBC systems, which operate at 750–900°C, are expected to be lower than from pulverized coal-fired combustion because of reduced volatilization at the lower operating temperature. This reduction could be partly offset by the longer residence time at a relatively high temperature in the fluidized bed, causing some increased volatilization.

Zhou et al. investigated TE partitions during FBC [28]. As shown in Fig. 3.10, arsenic is a major environmentally sensitive element and is classified as a chalcophile element. It is well known that arsenic is a highly volatile toxic element in which the volatilization rate of the feed coal is 23.5%. Arsenic is mainly associated with Fe–Mn oxide (52%), residue (34%), and carbonate (5%) in feed coal, indicating that As is mainly present in sulfide minerals. According to reports, arsenic in coal is mainly composed of arsenic pyrite, clay minerals, iron sulfide, etc. [29]. The positive

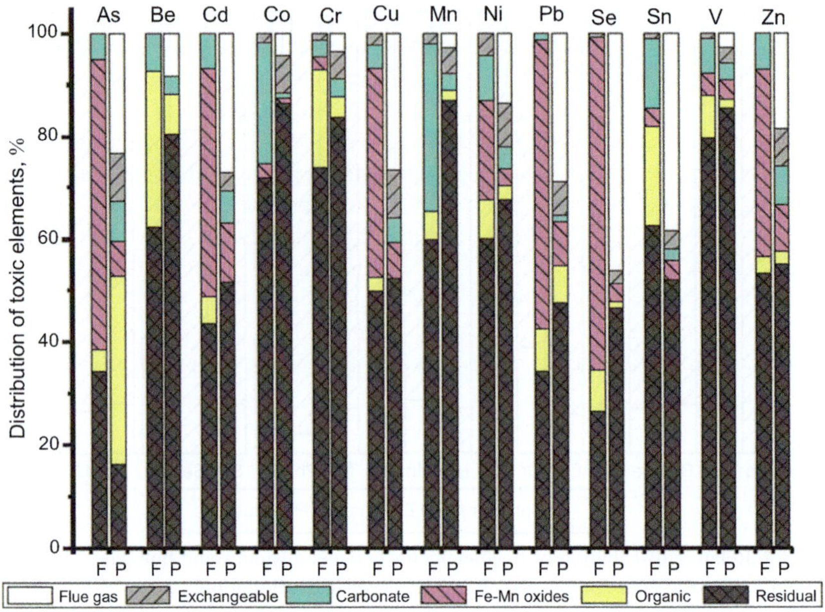

Figure 3.10 Trace element partition patterns during fluidized-bed coal combustion. *F*, feed coal; *P*, combustion products (bottom ash, fly ash, and flue gas).
From Zhou, et al. Partitioning and transformation behavior of toxic elements during circulated fluidized bed combustion of coal gangue. Fuel 2014;135:1–8.

correlation coefficient with Al ($r = 0.456$) suggests an inorganic affinity; furthermore, the correlation coefficient with Fe ($r= 0.532$) and Cu ($r = 0.918$) indicates the association of sulfide minerals. Arsenopyrite (FeAsS) is an example of arsenic sulfide in coal. In an oxidizing atmosphere, arsenopyrite could form ferric arsenate ($FeAsO_4$) and dissociate to hematite (Fe_2O_3) and arsenic oxides (As_2O_5); As_2O_5 could be captured by lime and form calcium arsenate ($Ca_3(AsO_4)_2$), which can be soluble in weak acid and associated in exchangeable fraction [29]. As a result, Fe–Mn oxide bonds and residues are reduced, while the exchangeable portion is increased from 1% to 9% during combustion (based on the original As). At the same time, As migration of particulate matter from the chimney can be absorbed or bound by the organic–organic complex and several specific functional groups, resulting in an increase in organic binding (from 4% to 36%). Huggins et al. proposed the appearance of more precise elements in organic matter [30].

Selenium is one of the most volatile TEs in coal that is released into atmosphere during combustion. Many studies have reported that Se in coal is mainly associated with sulfide minerals and/or organic matter [31]. How arsenic and selenium are related to coal composition is evident from similarities and significant differences between the two. Both elements can be present in FeS_2 where they directly replace the sulfur. FeS_2 is the most prominent form of arsenic in coal, and selenium can also be present in large quantities because of the formation of organic selenium (organic sulfides and thiophenes are dispersed in coal). The distribution of selenium in feed coal is bound for Fe—Mn oxides, with organic matter binding and residues of 64%, 9%, and 26%, respectively. After heat treatment, the exchangeable Fe—Mn oxides combine, the organic matter disappears, and the residual form increases. At the same time, 47% of Se (based on the original Se) is converted to a volatile phase.

Chromium is defined as a nonvolatile element (<5%) in feed coal since it mostly exists in residual form (71% and 73%, respectively). During thermal treatment, the increase in Cr in residual form (from 71% to 86%) probably results from the crystallization reaction of $Cr2O_3$ and the interaction of Cr with hematite [28]. In an oxidizing atmosphere, Co is integrated into iron oxide and enriched in the solid phase, bound Fe—Mn oxides disappear, while bound organic matter is transformed into an exchangeable state. For As, Se, or Cr, a number of other parts were transformed into the gas phase in the FBC compared with that in the traditional pulverized coal combustion system.

Current awareness of fossil energy shortage and environment pollution has resulted in an obvious shift to the exploitation of sustainable and renewable energy. To further reserve nonrenewable coal resources, cofiring of coal with biomass or sludge has been investigated in previous studies [32—34]. Zhou et al. investigated TE emission during the co-combustion of coal and peanut shells and the results are shown in Fig. 3.11(a) and (b) [32]. With the cofiring with biomass, the distribution pattern of As and Cr does not change significantly. The biomass itself shows a high proportion of As that is directly emitted as flue gas. In contrast, the evaporation rate of Cr in biomass is lower than the evaporation rate of coal. Coal was also cofired with coal and straw, and TE partition was studied during cofiring at 800 and 900°C of coal and straw, respectively [35]. Retention is defined as the percentage of an element introduced with the fuel, which is recovered with total combustion ash as shown in Fig. 3.12; no obvious difference was observed between the partition patterns of As, Se, and Hg during the co-combustion of coal/straw and the combustion of Polish coal. The difference

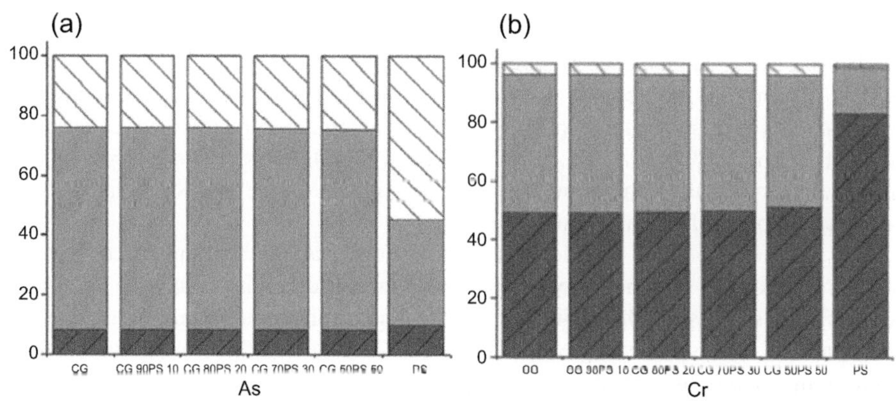

Figure 3.11 As and Cr partition patterns during the co-combustion of coal and peanut shells. From Zhou, et al. Thermochemical and trace element behavior of coal gangue, agricultural biomass and their blends during co-combustion. Bioresource Technol 2014;166:243—51.

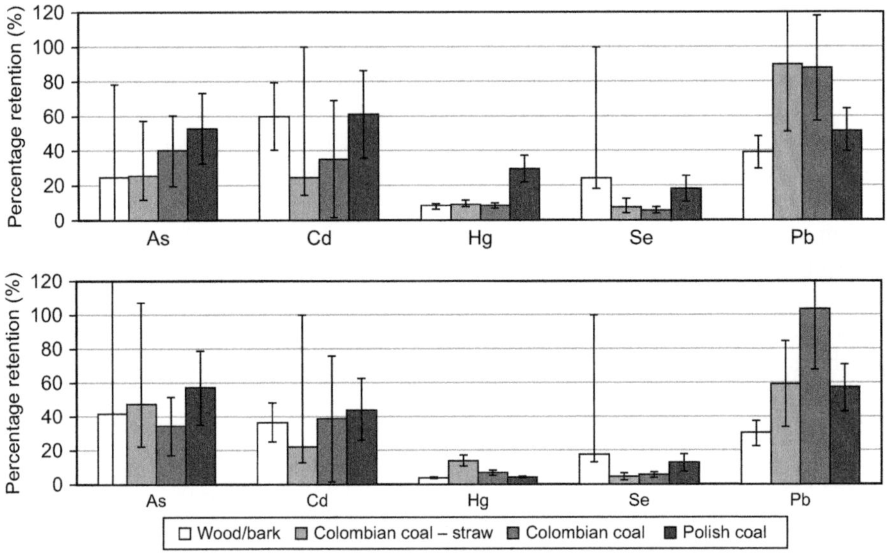

Figure 3.12 Percentage retention of trace elements during co-combustion of coal and straw. From Miller, et al. Trace element emissions from co-combustion of secondary fuels with coal: a comparison of bench-scale experimental data with predictions of a thermodynamic equilibrium model. Energy Fuels 2002;16:956–63.

between various combustion temperatures was within the range of experimental error. Thus the TE partition patterns will not be significantly influenced by co-combustion mode and co-combustion temperature in the Zhang et al. and Miller et al. studies.

Conventional coal-fired boilers use air for combustion where nitrogen from the air (about 79% by volume) dilutes the CO_2 concentration in the flue gas. The capture of CO_2 from such a dilute mixture using amine stripping is relatively expensive. Therefore oxy-fuel combustion was developed for the separation of inexpensive and advanced carbon dioxide from coal combustion flue gas (as shown in Fig. 3.13). During oxy-fuel combustion, a combination of oxygen and recycled flue gas, typically having greater than 95% purity, is used for combustion of the fuel. By recycling the flue gas, a gas consisting mainly of CO_2 and water is produced, ready to be sequestered without stripping CO_2 from the gas stream. Recycled flue gas is used to control the flame temperature and compensate for the missing N_2 volume to ensure that there is enough gas to transfer heat through the boiler.

Roy et al. explored TE partitions during oxy-fuel coal combustion. Since hexavalent chromium (Cr^{6+}) is more toxic than other forms of chromium, only the compositions of Cr^{6+} species are shown in Fig. 3.14(a). The results indicate that the amount of toxic gaseous Cr^{6+} species is greater for oxy-fuel combustion than for air combustion up to temperatures of 1200°C. However, at higher temperatures, these amounts are lower for oxy-fuel combustion. This is due to other chromium species that are stable at high temperature. It can also be seen that the amount of toxic gaseous chromium oxyhydroxide material ($CrO_2(OH)_2$) in oxy–fuel combustion is higher over the entire

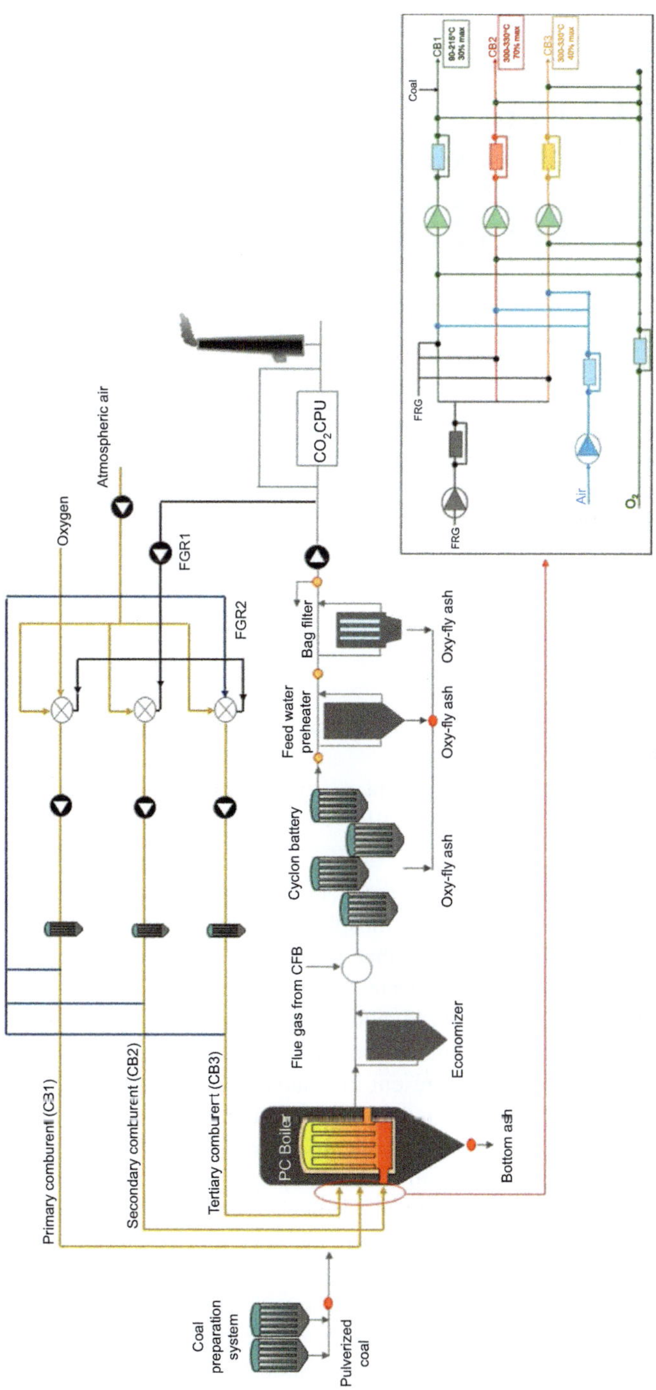

Figure 3.13 Diagram and process of the oxy-pulverized coal combustion demonstration plant. *CFB*, circulating fluidized bed; *CPU*, compression and purification unit; *FRG*, humid recirculation of flue gases.
From Cordoba, et al. Mass balance and partitioning of trace elements under oxy-coal combustion: first experiences. Fuel 2018;215:892–903.

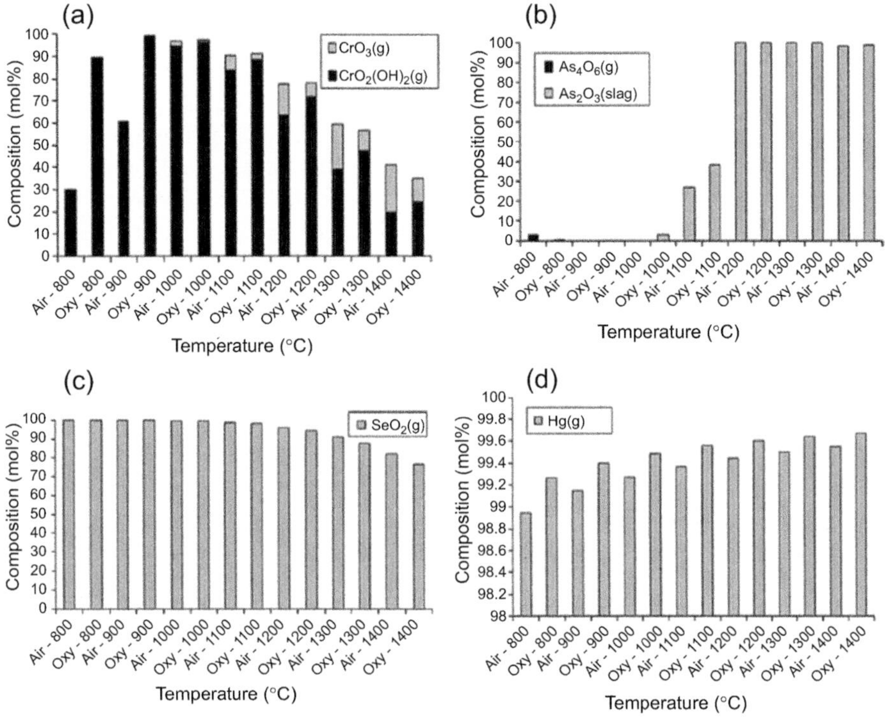

Figure 3.14 Partition patterns of (a) Cr, (b) As, (c) Se, and (d) Hg during oxy-fuel coal combustion.
From Roy, et al. Prediction of distribution of trace elements under oxy-fuel combustion condition using Victorian brown coals. Fuel 2013;114:135–42.

range of temperatures considered. The greater number of these species is due to the presence of water vapor in the oxy-fuel atmosphere, which is consistent with observations by Opila et al. [36], who concluded that water vapor in the atmosphere promoted the formation of volatile $CrO_2(OH)_2$ species.

From Fig. 3.14(b), it can be observed that at low temperature, toxic arsenic (As^{3+}) species are predicted in minor amounts for both air and oxy-fuel combustion. However, the As^{3+} species is found to be present in a substantial quantity from 1100°C in both air and oxy-fuel combustion. It is also noted that at higher temperatures (above 1200°C) the amounts of these species represent almost 100% of the total arsenic in both cases. It is thus clear that almost all of the As species will be confined in the solid slag phase at temperatures above 1100°C. However, it should also be noted that thermodynamic calculations do not consider the kinetics. Therefore experimental results are required to verify the predicted results.

In the case of selenium (as shown in Fig. 3.14(c)), $SeO_2(g)$ is considered to be one of the most toxic selenium (Se^{4+}) species. The results are shown in Fig. 3.14, indicating that the distribution of the compound is almost identical in air and oxy-fuel combustion, especially at low temperatures (up to 1000°C). It is also noted that in both cases the amount of the substance is almost 100% of total arsenic between 800 and 1000°C. However, at temperatures above 1000°C, these emissions are expected to decrease due to the corresponding increase in less toxic gaseous SeO compounds.

Since Hg^0 is more toxic than other forms of mercury, only compositions of gaseous Hg^0 are shown in Fig. 3.14(d). It can be seen that the predicted mercury emissions are relatively constant regardless of the atmosphere or temperature. In all cases, greater than 98.5% of the mercury leaves in the gas phase as Hg^0. Therefore according to the modeling, it can be concluded that mercury emissions from oxy–fuel combustion are slightly larger than mercury emissions from air combustion. In an earlier study, according to reports, due to the recirculation of flue gas containing mercury, the concentration of mercury in oxy-fuel combustion is higher than the concentration of mercury in air combustion [37]. It is expected that mercury concentration in an actual oxy-fuel unit might be higher than these predicted values. Since the mercury in the flue gas may result in corrosion in the CO_2 purification and compression unit, effluent gas treatment is necessary to eliminate this mercury during oxy-fuel combustion.

Roy et al. also investigated the emission patterns of TEs under different oxy-fuel combustion temperatures using Yallourn coal to realize a better control of TEs in real-world applications. As shown in Fig. 3.15(a), under oxy-fuel combustion

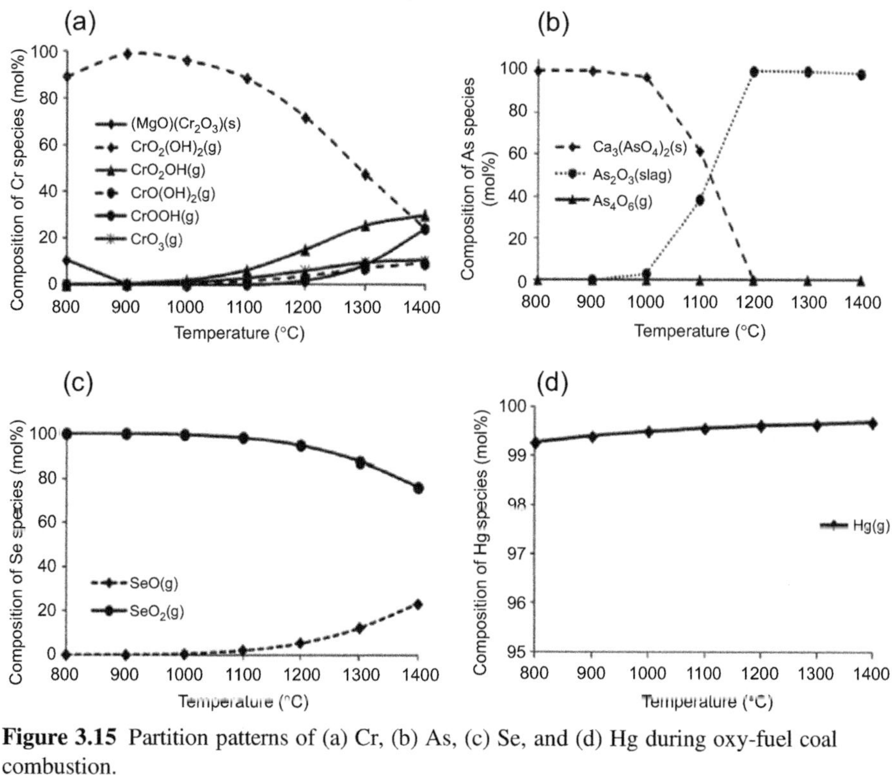

Figure 3.15 Partition patterns of (a) Cr, (b) As, (c) Se, and (d) Hg during oxy-fuel coal combustion.

From Roy, et al. Prediction of distribution of trace elements under oxy-fuel combustion condition using Victorian brown coals. Fuel 2013;114:135–42.

conditions, most Cr will be released as gas-phase $CrO_2(OH)_2$ at combustion temperatures below 1000°C. When the temperature was further raised to beyond 1000°C, the species of Cr emitted gradually transformed into $CrO_2(OH)(g)$, $CrO(OH)_2(g)$, CrOOH(g), and $CrO_3(g)$. In particular, when the combustion temperature is as high as 1300°C, CrO_2 (OH) becomes the main phase of Cr emission. Compared to conventional coal combustion and fluidized combustion modes, Cr appears to be more easily discharged from coal as a gas phase.

The equilibrium distribution of As during oxy-fuel combustion of three Victorian brown coals at different temperatures is shown in Fig. 3.15(b). The results indicate that almost all of the As was found to be present as As_2O_3(slag), $Ca_3(AsO_4)_2(s)$, and $As_4O_6(g)$ under oxy-fuel combustion. As shown in Fig. 3.15(c), it can be observed that almost all of the arsenic is predicted to be released as solid $Ca_3(AsO_4)_2$ and As_2O_3(slag) over the entire temperature range. This behavior is supported by Contreras et al. [38], who reported that the presence of silicon has an effect of increasing arsenic volatility. Due to low silica content in Yallourn coal, a lesser amount of volatile As_4O_6 is predicted compared to that in other coals.

For Se, the results indicate that total selenium in the system will be present in the gas phase as $SeO_2(g)$ and SeO(g) during the oxy-fuel combustion of all considered Victoria brown coals. As shown in Fig. 3.15(c), the dominant species is $SeO_2(g)$ over the entire temperature range studied. At higher temperatures, a large amount of SeO(g) is expected to form, which results in a small decrease in the $SeO_2(g)$ compound. However, gaseous selenium dioxide is known to condense below 100°C. Therefore in a commercial system, this will condense on the ash particles and be captured by the fabric filter. Its presence may require downstream gas treatment. Fig. 3.15(d) shows the equilibrium distribution of mercury as a function of temperature during the oxy-fuel combustion of Yallourn coal. The results indicate that elemental mercury is the dominant mercury species over the entire temperature range with a minor contribution of HgO(g) at low temperature. It is also noted that the mercury emissions are predicted to increase slightly with the increase in temperature. This could have important implications for downstream processes due to the corrosive nature of mercury.

Fu et al. explored the partitions of various TEs during coal combustion in a circulating fluidized-bed boiler with emphasis on As, Se, and Cr (shown in Fig. 3.16) [39]. Selenium and arsenic in coal are mainly related to pyrite, organic matter, and selenate/arsenate. In this study, selenium and arsenic were mainly associated with oxidizable fractions, accounting for 75% and 64.4%, respectively. The results of the continuous extraction experiments show that the two elements are mainly combined with sulfide (pyrite) and organic matter. Inorganic-related Se and As, such as pyrite, selenium, and arsenic, can evaporate with the conversion of pyrite; selenium/arsenic associated with organic matter would be released with particle fragmentation and char decomposition during combustion and can then form a certain amount of gaseous arsenic phases such as Se^0, SeO_2, As_2O_3, $AsCl_3$, and As_2O_5 according to thermodynamic equilibrium calculations. In this study, heterogeneous chemical reactions between gaseous Se/As and ash surface components as described previously are critical for arsenic enrichment in size-separated particles. Gas-phase selenium and arsenic may be captured by

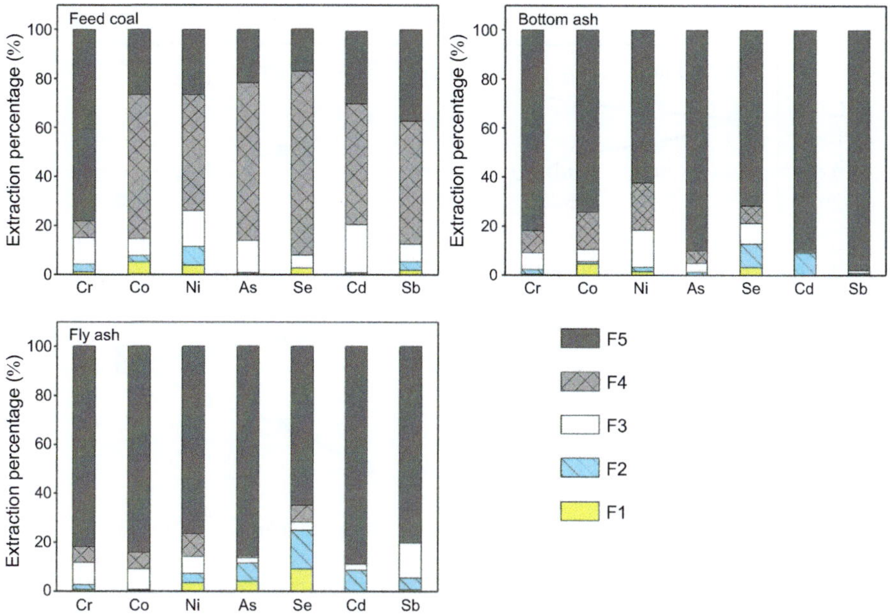

Figure 3.16 Distribution of hazardous trace elements extracted by sequential chemical extraction for feed coal, fly ash, and bottom ash: ion-exchangeable (F1), carbonate-bound (F2), Fe–Mn oxides (F3), organic and sulfide bound (F4), and residual (F5).
From Fu, et al. Emission and transformation behavior of minerals and hazardous trace elements (HTEs) during coal combustion in a circulating fluidized bed boiler. Fuel Process Technol 2018; 242:1950–60.

calcium and unburned carbon in specific materials. Some of the As and Se (mainly silicate) bound in the residue may be less volatile and will remain in the glassy material of the residual ash. Therefore Se is mainly converted into the easily soluble component F1 + F2 (up to 25.07%) in the fly ash as the oxidizable component F3 in coal. For arsenic, the proportion of F3 drops considerably, while F1 + F2 increases in fly ash.

Chromium occurs as Cr^{3+} in illite, and can also be associated with sulfide minerals in coal. The proportion of residual Cr in feed coal was 64.5%, which accounted for the low volatility of Cr during coal combustion since residual Cr would be retained in the glassy materials of ash. Typical flue gas temperatures in the circulating fluidized bed (CFB) furnace are around 850–900°C. Organically bound Cr would partially evaporate into gaseous CrO_3(g) and CrO_3(g), which could be stabilized with calcium oxide or other alkaline elemental oxides forming $CaCrO_4$ or (Mg, K, Na, Fe) CrO_4(s) on the ash surface. Formation of Cr(VI) increased the easily soluble fractions (F1 + F2) from 4.13% in coal to 6.37% in the CFB fly ash.

The redistribution of arsenic in a combustion system was not only highly dependent on volatility and association of arsenic during combustion, but was also determined by its interactions with gas-phase species and ash compositions. Zhou et al. investigated the partition of As in coal fly ash (shown in Fig. 3.17) [40]. Fe_2O_3 has a great influence on the redistribution of arsenic in the combustion system. With the increase in Fe

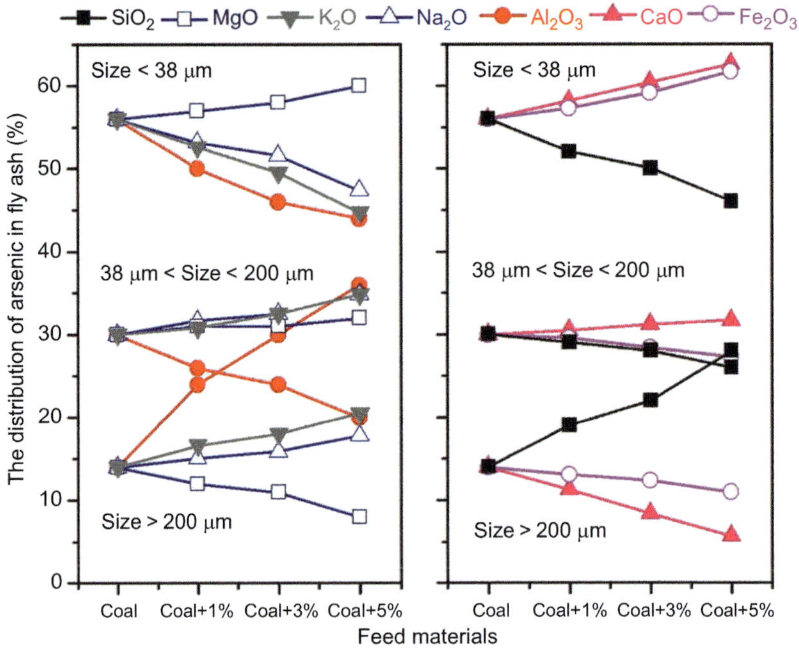

Figure 3.17 The effect of the addition of ash composition on arsenic distribution in fly ash during combustion at 1000°C.
From Zhou, et al. Effect of ash composition on the partitioning of arsenic during fluidized bed combustion. Fuel 2017;204:91–7.

content in the combustion process, retention efficiency is greatly improved. At the same time, laser ablation inductively coupled plasma mass spectrometry and scanning electron microscopy with energy dispersive X-ray spectroscopywere used to determine the possibility of fly ash capturing arsenic and selenium. It was found that arsenic in the particles was well correlated with Fe. The results show that the proportion of arsenic in the fine particles (<38 μm) increases with the addition of Fe_2O_3. These results indicate that Fe_2O_3 reacts easily with gaseous arsenic during combustion and forms fine particles as the temperature of the flue gas decreases. Thermodynamic equilibrium calculations show that $FeAsO_4$ is the main stable species in the Fe–As interaction process. Since the surface area of the selected Fe_2O_3 (14.27 m^2/g) is small, physical adsorption is considered to be negligible. At the same time, Fe_2O_3 is known to be catalytically oxidized to promote the formation of Fe arsenate. Therefore Fe_2O_3 can be considered as a suitable additive for the purpose of arsenic in the combustion system. The comprehensive reaction mechanism between arsenic species and Fe_2O_3 remains to be further studied.

Ca-based materials were extensively applied for the removal of sulfur during coal combustion. The ratio of arsenic in small particles (<200 μm) increased gradually by adding CaO, indicating that calcium arsenate mainly existed in small particles. The arsenic retention efficiency of Mg oxides augmented to a maximum and then decreased with the rising combustion temperature, which was similar to that of Fe_2O_3 and CaO. However, the retention efficiency of MgO was much lower than that of Fe_2O_3 and CaO. Like As–Ca interaction, $Mg_3(AsO_4)_2$ most probably resulted from As–Mg interaction. Meanwhile, the proportion of arsenic in fine particles increased by adding MgO. The presence of Al_2O_3 may result in the formation of $AlAsO_4$, which is a very stable compound even at 1400°C. Therefore the adsorption efficiency of Al_2O_3 increased with increasing combustion temperature and adding ratio. Meanwhile, physisorption may also be suggested for Al_2O_3 due to the great surface area (85.32 m^2/g) and increasing ratio of arsenic in coarse particles. However, the existence of Na and K may cause agglomeration, and result in an increase in particle size. It could be found that arsenic in large particles (>38 μm) increases with the increase in Na and K.

Arsenic and alkali metal (K and Na) oxides interact to form new species, including K_3AsO_4, KH_2AsO_4, KAs_3O_8, Na_3AsO_4, NaH_2AsO_4, and $NaAs_3O_8$. Among them, K_3AsO_4 and Na_3AsO_4 were predicted to be the most probable compounds. The sizes of most particles decreased during combustion as a result of particle collisions causing attrition.

As for F, a few studies have focused on F partitions during coal combustion. Liu et al. used two kinds of coals, i.e., Liantang and Xinji coals, to investigate F partition behavior during coal combustion [20]. As shown in Fig. 3.18(a) and (b), in each of the Chinese coals analyzed here, the figures indicate that, as the temperature of coal combustion increases, the amount of fluorine in the emitted gases and particles increases, while fluorine in the combustion residue decreases. It is thought that when the temperature increases, the combustion rate of organic fluorides and the pyrolysis rates of apatite and muscovite likewise increase.

Cordoba et al. also studied F partition during the oxy-fuel combustion system (shown in Fig. 3.19) [41]. Cordoba et al. classified F as moderately volatile elements with high condensation potential, which can be ascribed to the CO_2-rich flue gas continuously recirculated to the boiler with relatively high content of F (61 mg).

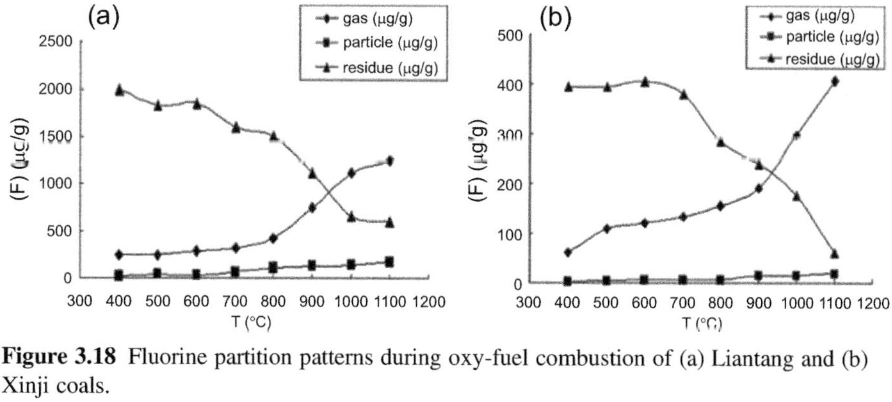

Figure 3.18 Fluorine partition patterns during oxy-fuel combustion of (a) Liantang and (b) Xinji coals.
From Liu, et al. Behavior of fluorine in the combustion of Chinese coal in small furnaces. Energy Fuels 2006;20:1406–10.

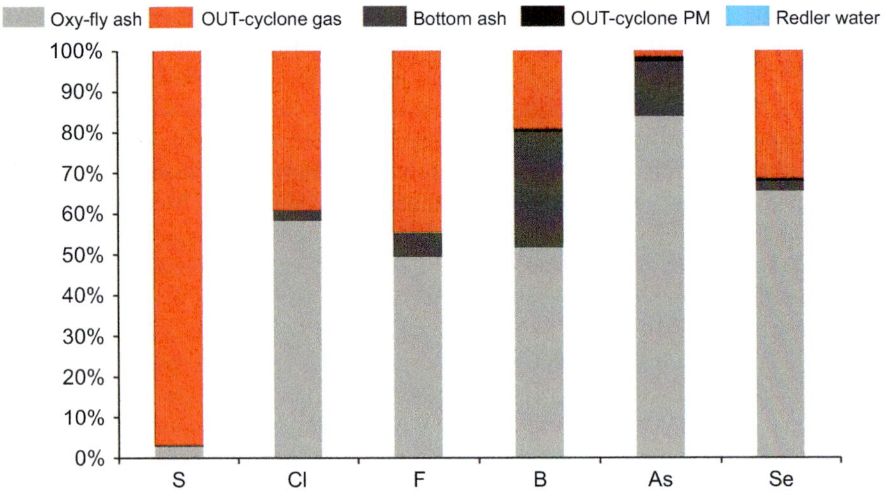

Figure 3.19 Partitioning of highly and moderately volatile elements after oxy-pulverized coal combustion.
From Cordoba, et al. Mass balance and partitioning of trace elements under oxy-coal combustion: first experiences. Fuel 2018;215:892–903.

However, over 29% of gaseous F is recirculated to the boiler with CO_2-rich flue gas. The behavior of As and Se were also investigated by Cordoba et al. for which both were classified as moderately volatile elements of F in Cordoba's study.

3.4 Trace element transformation during coal combustion

Once combustion gases are away from the combustion zone of a coal-fired boiler, the key factor influencing final TE transformation/partitioning behavior is the conversion of vaporized components into various solid and/or liquid forms. It is determined, basically, by three complex and interrelated processes: adsorption, condensation, and chemical transformation. Fig. 3.20 shows possible ways along which TEs are partitioned into vapor, submicron aerosol, and supermicron aerosol proposed by Linak et al. [42]. Several paths that pertain specifically to coal are distinguished by thick bold arrows and lines.

The involved physicochemical phenomena that control TE partitioning and their final physical form are as follows:

1. Heterogeneous condensation on the existing fly ash particles and heat exchange surfaces;
2. Physical/chemical adsorption on fly ash particles;
3. Homogeneous condensation (nucleation) and coalescence as submicron aerosols if local supersaturation condition exists;

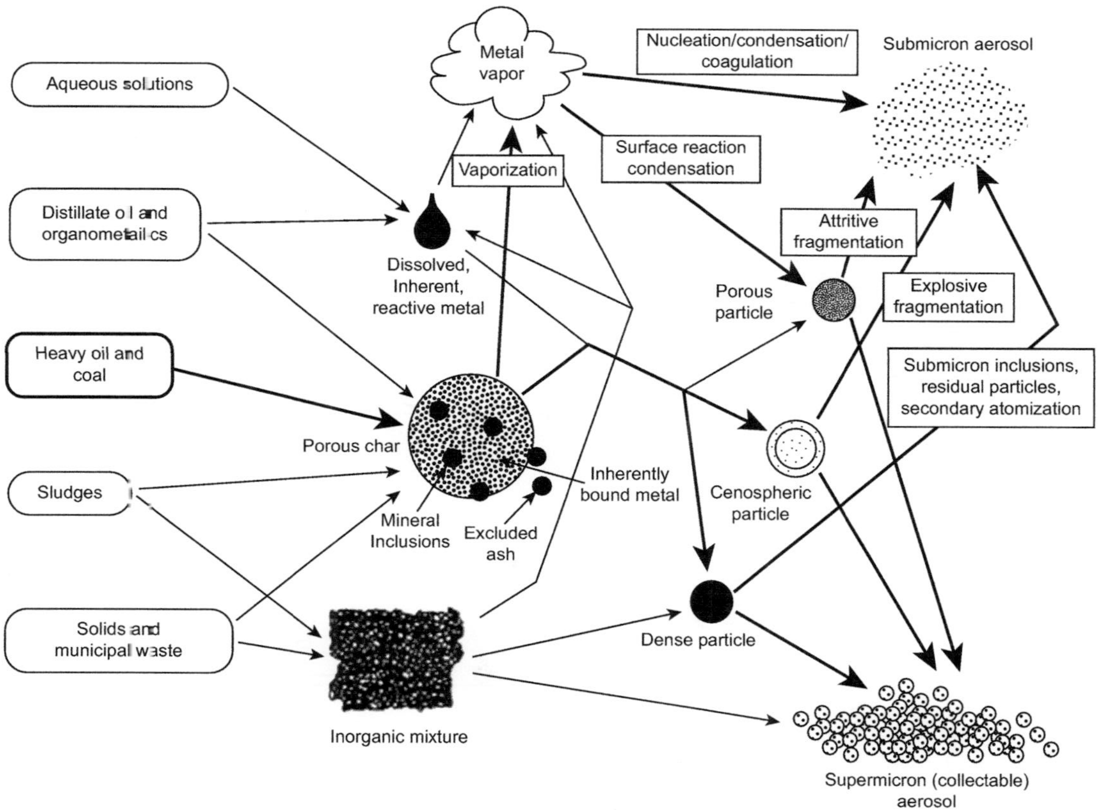

Figure 3.20 Mechanisms for particle formation in a combustion system.
From Linak, et al. Trace metal transformation mechanisms during coal combustion. Fuel Process Technol 1994;39(2):173–98.

4. Homogeneous and heterogeneous chemical reaction among TEs, fly ash, and flue gas constituents;
5. Remaining in the vapor phase for species with high vapor pressure at typical boiler exit temperature.

In addition, metal aerosol dynamics is the theoretical base deciding fly ash formation and particle size distribution.

Senior et al. investigated the transformation of Hg in traditional coal combustion flue gas containing CO, CO_2, Cl_2, H_2, HCl, H_2O, Hg, $HgCl_2$, HgO, N_2, NO, NO_2, O_2, and SO_3 with a thermal dynamic model. The predicted results of Senior et al. are shown in Fig. 3.21. As shown, below 725K (450°C), all Hg is expected to be present as $HgCl_2$. Above about 975K (700°C), 99% of Hg is expected to be present in gaseous Hg^0. The remaining 1% is expected to be gaseous HgO. Between 725 and 975K, the partition between $HgCl_2$ and Hg is determined by the chlorine content of the coal. The mercury content of coal has no effect on the equilibrium distribution of mercury species. For the coals studied herein, the equilibrium HCl concentration in the predicted gas is in the range of 24−111 ppm. Even at these low concentrations, the reaction between Hg and HCl dominates the equilibrium chemistry. In addition, the effect of different chlorine concentrations in the flue gas on the degree of mercury oxidation was discussed (as shown in Fig. 3.21). The relatively high concentration

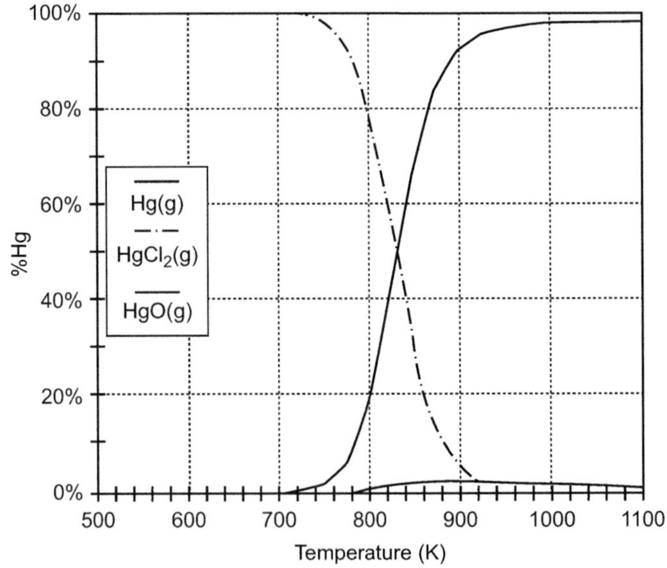

Figure 3.21 Equilibrium mercury speciation in flue gas as a function of temperature (Pittsburgh coal).
From Senior, et al. Gas-phase transformations of mercury in coal-fired power plants. Fuel Process Technol 2000;63:197−213.

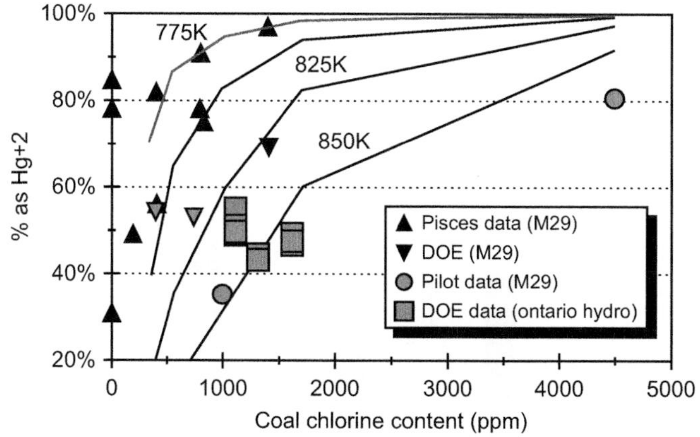

Figure 3.22 The content of oxidized Hg in flue gas as a function of chlorine concentration and temperature.
From Senior, et al. Gas-phase transformations of mercury in coal-fired power plants. Fuel Process Technol 2000;63:197–213.

and low temperature promoted the transformation of Hg between Hg^0 and $HgCl_2$. The kinetic model was also applied to predict the transformation of Hg species in the Senior et al. study.

Suriyawong et al. investigated Hg speciation during both traditional and oxy-fuel combustion modes and the results are shown in Fig. 3.22 [43]. For conventional air combustion, an average of 0.1 $\mu g/m^3$ is oxidized mercury and 0.4 $\mu g/m^3$ is elemental mercury. The range of species formation is consistent with the values reported in the literature. Reynolds reported that about 10%–20% of mercury emissions from western bituminous coal are oxidized and the rest are in the elemental state. Lee and Srivastava reported the results of gas phase similarity formation during Powder River Basin coal combustion. For 25% O_2/75% CO_2 combustion, an average of 0.08 $\mu g/m^3$ is oxidized mercury and 0.38 $\mu g/m^3$ is elemental mercury. For 20% O_2/80% CO_2 combustion, an average of 0.1 $\mu g/m^3$ is oxidized mercury and 0.6 $\mu g/m^3$ is elemental mercury. The experimental results were averaged from four replicates under the same conditions. The average total gaseous mercury concentration in the combustion chamber under the experimental conditions ranged from 0.65 to 0.8 $\mu g/m^3$, which is consistent with the mercury content measured in coal (ranging from 0.069 to 0.108 ppm). The total gaseous mercury concentration under O_2–CO_2 conditions is comparable to that of air, indicating that the O_2–CO_2 system has no effect on the evaporation of mercury. Mercury is released by direct evaporation rather than by chemical reduction–oxidation (Fig. 3.23). In air and O_2–CO_2 systems, the combustion temperature of the coal particle surface is high enough to evaporate mercury from the coal matrix (mercury has a boiling point of only 629K at 1 atm). The ratio of oxidized mercury to elemental mercury in O_2–CO_2 combustion conditions is also close to that of air, indicating that

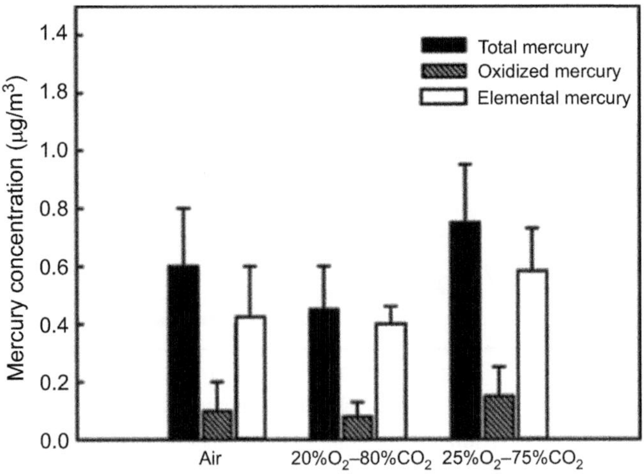

Figure 3.23 Hg concentration measured at the combustor exit under traditional and oxy-fuel combustion conditions.
From Suriyawong, et al. Submicrometer particle formation and mercury speciation under O_2-CO_2 coal combustion. Energy Fuels 2006;20:2357–63.

the O_2-CO_2 system has no significant effect on the form of mercury at the exit of the combustion chamber. The experimental results here are in good agreement with previous studies showing that mercury is predominantly elemental at high temperatures.

However, in the Senior et al. and Suriyawong et al. studies, only the oxidized and elemental forms of Hg were taken into consideration but the existence of the presence of particulate-bound mercury (Hg^P) was omitted. The particulate-bound mercury was defined as mercury in the flue gas attached to the fly ash and bound to it.

Otero-Rey et al. took Hg^P into consideration using three different coal samples and the results are shown in Table 3.4 [44]. Besides, Otero-Rey et al. also investigated the partition patterns of As and Se. Hg is more volatile than As and Se, and therefore the proportion of this element in the gas phase is larger (76.5%, 1.2%, and 3.7%, respectively). As and Se are mainly condensed in fly ash (97.7% and 90.3%, respectively). There is no obvious difference in the partition patterns observed for the three different samples of coals, which indicated that Hg will be mostly vaporized during the coal combustion process.

Moreover, by calculating the enrichment factor (ΔE_f) of different TEs, Seames et al. explored the possible enrichment of As and Se on submicron and supermicron particles [45]. As shown in Fig. 3.24(a), the negative value of the ΔE_f of As is calculated. Therefore the volatilization and subsequent heterogeneous transformation to the surface of submicron particles does not appear to be an important mechanism for the postcombustion phase distribution of arsenic during the combustion of Pittsburgh coal. These results indicate that almost all of the arsenic in the submicron fly ash enters the postsolids combustion zone. For Se (as shown in Fig. 3.24(b)), the differential

Table 3.4 Partition patterns of As, Hg, and Se during the traditional coal combustion process.

		Bottom ash total (%)	Fly ash total (%)	Flue gas total (%)
As	Sample 1	0.5	97.6	1.9
	Sample 2	1.4	97.6	1.0
	Sample 3	1.4	97.9	0.7
Hg	Sample 1	0.1	21.8	78.1
	Sample 2	0.3	19.4	80.3
	Sample 3	0.4	28.4	71.2
Hg	Sample 1	9.8	87.2	3.0
	Sample 2	4.1	91.8	4.1
	Sample 3	4.2	91.9	3.9

From Otero-Rey, et al. As, Hg, and Se flue gas sampling in a coal-fired power plant and their fate during coal combustion. Environ Sci Technol 2003;37:5262–67.

enrichment factor for Se in Illinois a6 fly ash submicron particles is greater than the calculated value for arsenic. Therefore the volatilization and subsequent heterogeneous transition to the submicron particle surface appears to be an important mechanism for the postcombustion phase distribution of selenium during combustion of the Illinois a6 coal, rather than the Pittsburgh coal.

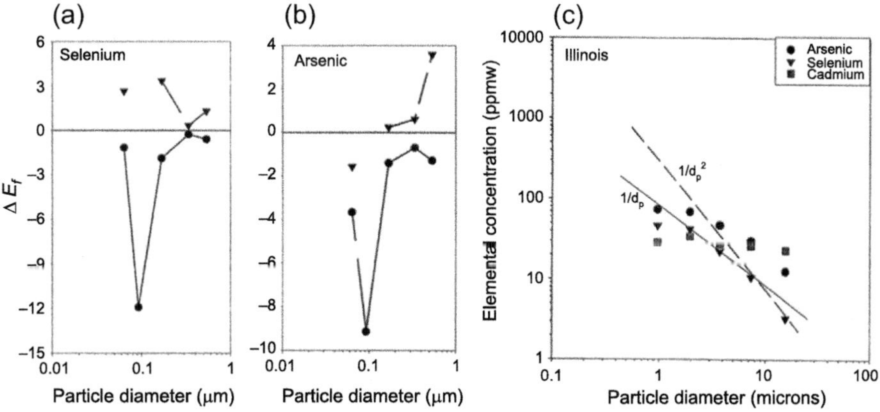

Figure 3.24 Differential enrichment factor comparison of (a) Se, (b) As, and (c) Illinois a6 coal particle size dependence of trace elements in the supermicron region.
From Seames, et al. Partitioning of arsenic, selenium, and cadmium during the combustion of Pittsburgh and Illinois a6 coals in a self-sustained combustor. Fuel Process Technol 2000;63: 179–96.

The heterogeneous partitioning of trace metals onto supermicron size particles can occur by condensation of metal vapor upon existing particles and reaction of metal vapor with other chemical species on the surface or within the pores of a particle. The rate of transformation by reaction may be limited by diffusion of the metal vapor through the gas film surrounding the particle, diffusion into the pores of the particle, the reaction rate on the exterior surface of a nonporous particle, or the reaction rate on the internal surface within a porous particle [42]. The logarithmic-log plot of As and Se concentrations and particle sizes in the sample provides some insight into these heterogeneous distribution mechanisms. The $1/d_p$ particle size dependence suggests that surface reactions controlled by external surface reactions or pore diffusion may be the primary mechanism of formation in the continuum region. The first dependence P indicates that the film transfer is a rate-controlling process; this is true for condensation or infinitely fast surface reactions in a continuous system. The result is shown in Fig. 3.24(c). The data in Fig. 3.24(c) show that during the combustion of six coals in Illinois, surface reaction is the most likely rate-limiting step for converting gas-phase As and Se into a solid phase. This conclusion is consistent with the findings of Senior et al. about the accumulation of As and Se on the surface of submicron and ultrafine particles [46].

Yan et al. explored the transformation of Se during coal combustion under oxidizing and reducing atmospheres [47]. As shown in Fig. 3.25, under oxidizing conditions (Fig. 3.25(a)), the dominant gaseous species is SeO_2, which is present at temperatures as low as about 500K (227°C), whereas solid SeO_2 is dominant when $T < 500K$. Under reducing conditions (Fig. 3.25(b)), gaseous H_2Se is dominant

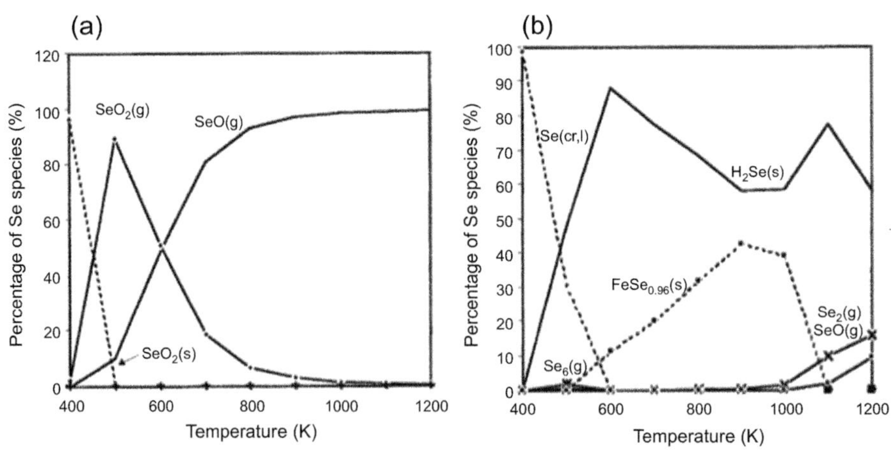

Figure 3.25 Thermodynamics of Se equilibrium distribution in the flue gas from Se-added coke combustion (a) under oxidizing conditions and (b) under reducing conditions.
From Yan, et al. Fate of selenium in coal combustion: volatilization and speciation in the flue gas. Environ Sci Technol 2001;35:1406–10.

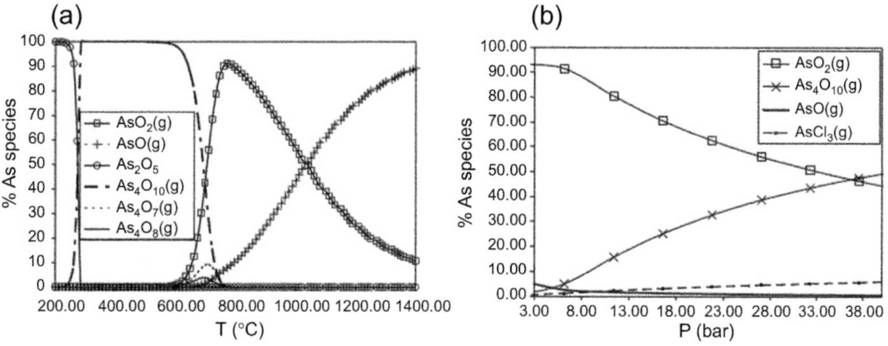

Figure 3.26 (a) Equilibrium compositions for As in a combustion atmosphere and (b) equilibrium compositions for arsenic in different pressures at 800°C.
From Contreras, et al. Arsenic interactions during co-combustion processes based on thermodynamic equilibrium calculations. Fuel 2009;88:539–46.

beyond 500K. Below this temperature, atomic Se, i.e., Se (cr,l), is almost the only existing species. Contreras et al. investigated the transformation of As during coal combustion under the oxidizing atmosphere [38]. As shown in Fig. 3.26(a), arsenic is predicted to form a mixture of gaseous oxides at temperatures higher than 300°C; in this sense, arsenic may be totally volatilized at co-combustion temperatures, with AsO_2(g) and AsO(g) the most probable species at temperatures higher than 800°C. At temperatures below 300°C, As_2O_5 is predicted as the dominant species. The effect of increased pressure on the equilibrium composition of arsenic was evaluated (Fig. 3.26(b)). Results show that, at 800°C, variations in pressure modify the proportion of the gaseous species of arsenic, with a decrement in AsO_2(g) concentration and an increment in As_4O_{10}(g), as pressure increases.

Goodarzi et al. and Huffman et al. investigated the partition patterns of Cr during the coal combustion process using the X-ray absorption near edge structure method [48,49]. Due to its low volatility, almost all Cr present in coal was collected in the bottom ash and fly ash. As shown in Figs. 3.27 and 3.28, Cr^{3+} dominated the Cr species in coal combustion bottom ash and fly ash. According to Goodarzi et al. research, 19% of Cr in fly ash is water soluble, which indicated the coexistence of Cr^{6+}. These results indicate that Cr may be mobilized from the fly ash under basic conditions: Cr_2O_3 becomes soluble at a pH slightly below 13 to form CrO_4. Moreover, the existence of metallic Cr in the fly ash was proven by Goodarzi et al. to be from the contamination of the gas carrier tubes. In conclusion, Cr in bottom ash and fly ash dominantly existed as the Cr^{3+} form during the traditional coal combustion process, while some of the Cr^{3+} was oxidized into Cr^{6+} with higher toxicity.

Moreover, co-combustion of coal with other biomass was an economic way to reserve nonrenewable coal resources. Contreras et al. investigated the interaction of As during the co-combustion of coal with possible minor and trace elements [38]. A representative picture is shown in Fig. 3.29(a)–(d). The behavior of arsenic in a

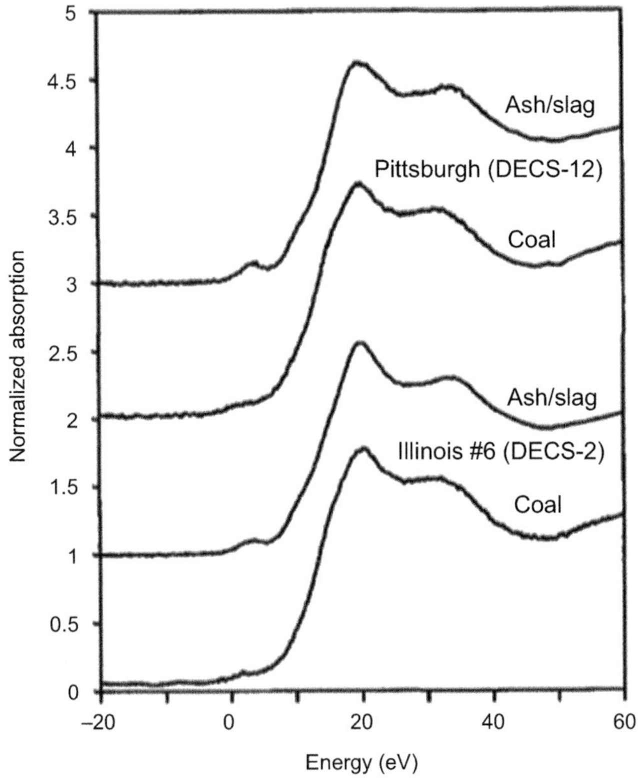

Figure 3.27 Comparison of Cr K-edge X-ray absorption near edge structure spectra of chromium in coal and ash.
From Huffman, et al. Speciation of arsenic and chromium in coal and combustion ash by XAFS spectroscopy. Fuel Process Technol 1994;39:47–62.

combustion system depends on the amount of arsenic vaporized during combustion and its interaction with gaseous and ash particles. Due to the formation of thermally stable materials, the interaction of arsenic with mineral components results in a small loss of evaporation of arsenic. If we introduce all possible chemical forms that can present a particular element in the combustion atmosphere, the resulting equilibrium composition can represent the phase in which the reaction can take place. Although some arsenic interactions with fly ash components have been previously described, few of the available patterns show the resulting compounds from these interactions. As–fly ash component interaction favors the formation of arsenates. $FeAsO_4$ is the most stable species resulting from As–Fe interaction (Fig. 3.20(a)); this species vaporizes even at 1100°C. Because arsenic in coal is well documented to have a strong affinity to concentrate in sulfide minerals, and its association with FeS_2, arsenic is

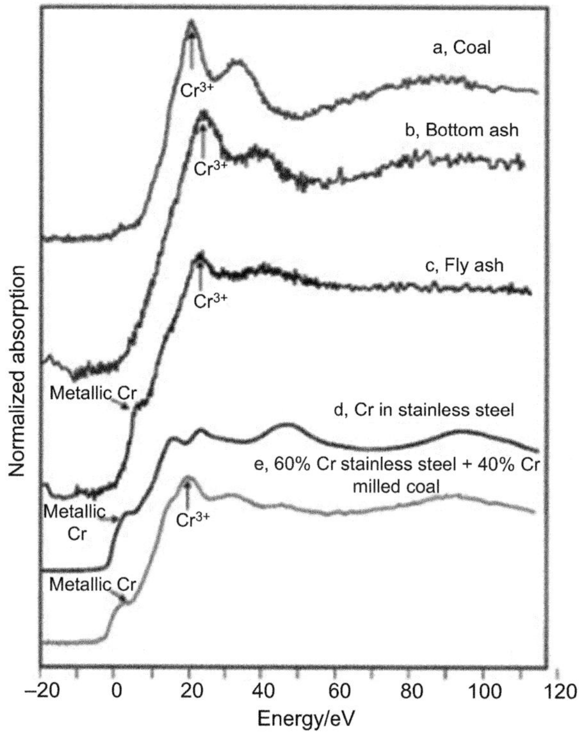

Figure 3.28 Comparison of simulated chromium X-ray absorption near edge structure spectra for: (a) milled coal; (b) bottom ash; (c) electrostatic precipitation of fly ash; (d) Cr in stainless steel; and (e) a mixture of 60% Cr in stainless steel and 40% Cr in milled coal.
From Goodarzi, et al. Monitoring the species of arsenic, chromium and nickel in milled coal, bottom ash and fly ash from a pulverized coal-red power plant in western Canada. J Environ Monit 2001;3:1–6.

predicted to form mainly condensed species, which would be captured by control devices. K–As interactions (Fig. 3.20(b)) originate new species (K_3AsO_4 and KAs_3O_8), stable at temperatures below 900°C. $NaAs_3O_8$ (Fig. 3.20(c)) is the most probable compound resulting from As–Na reaction ($T < 800$°C). The presence of aluminum favors the formation of $AlAsO_4$ (Fig. 3.20(d)), a very stable species that sustains an extremely low equilibrium vapor pressure even at 1400°C. Mg–As interacts to form $Mg_3(AsO_4)_2$ (Fig. 3.20(e)), stable in the range of temperatures between 400 and 1100°C, while Ca–As forms $Ca(AsO_2)_2$ at 650–750°C and $Ca_3(AsO_4)_2$, which is stable at temperatures over 700°C (Fig. 3.20(f)). Calcium arsenate is widely known as an arsenic species in fly ash. These results suggest that most arsenic may be condensed at temperatures lower than 400°C. However, if these polyatomic arsenates are omitted from the model, arsenic is predicted to form a mixture of gaseous oxides.

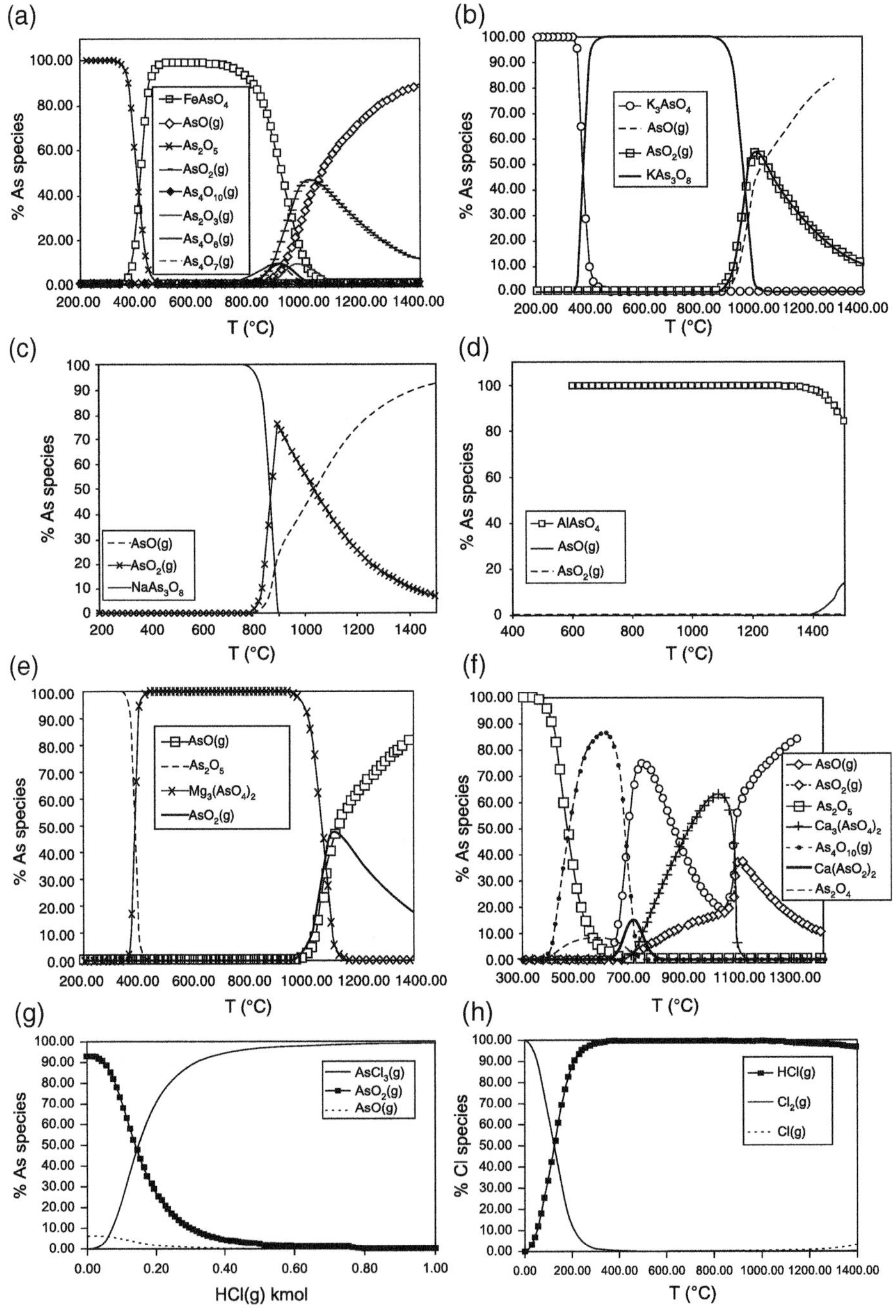

Figure 3.29 Interactions of Hg with (a) Fe, (b) K, (c) Na, (d) Al, (e) Mg, (f) Ca, and (g and h) Cl in the co-combustion process.
From Contreras, et al. Arsenic interactions during co-combustion processes based on thermodynamic equilibrium calculation. Fuel 2009;88:539–46.

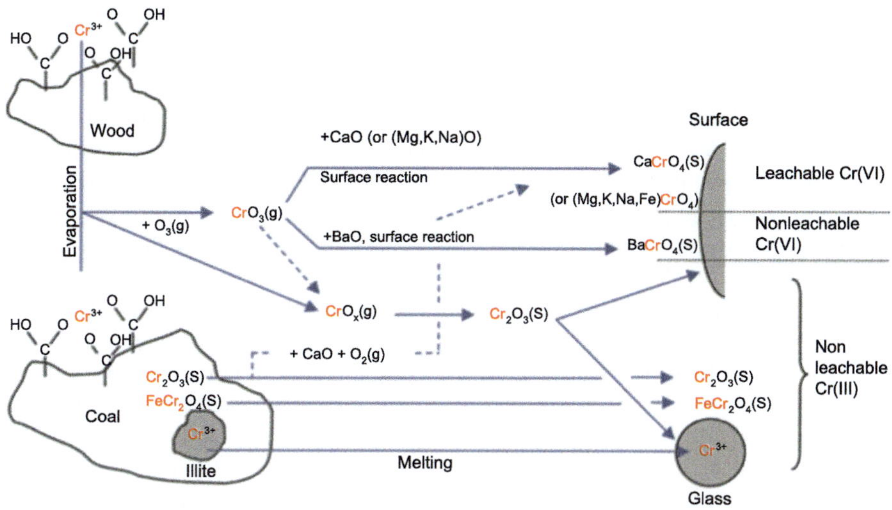

Figure 3.30 Possible reaction mechanisms for the formation of leachable and nonleachable Cr(VI) and nonleachable Cr(III).
From Stam, et al. Chromium speciation in coal and biomass co-combustion products. Environ Sci Technol 2011;45:2450–56.

It has been reported that the presence of other elements in the fuel, especially chlorine, has an important effect on the emission of TEs. Chlorine has a significant effect on arsenic behavior by forming volatile chlorides, increasing the evaporation of arsenic. In this sense, it is important to understand the chlorine content of biofuels to predict the type of arsenic present in the gas phase and then to select the most appropriate arsenic control technology. Fig. 3.20(g) shows that $AsCl_3(g)$ is the dominant arsenic species at 800°C for high chlorine concentrations, and $AsO_2(g)$ is the most likely species for low chlorine fuels. On the other hand, Fig. 3.20(h) shows that the atomic chlorine concentration decreases with decreasing temperature because the atomic Cl reacts to form HCl and a lower amount of molecular chlorine Cl_2. Thermodynamic calculations indicate that arsenic chloride is formed at low temperatures (400–800°C) when chlorine is considered to be in the form of HCl.

Stam et al. proposed the Cr emission model during the co-combustion of coal [50]. As shown in Fig. 3.30, Cr is ultimately Cr(III) or nonleaching Cr(VI). Our equilibrium calculations indicate partial volatilization of Cr, which is more loosely organically bonded to Cr (as opposed to illite-related Cr or as an oxide or spinel, which is less loosely bound). Depending on the local oxygen content, $CrO_3(g)$ may be formed, or gaseous Cr in a lower oxidation state may be formed. The latter may form chromium oxide, which condensed at high temperatures (>1500°C). A possible reaction scheme is that $CrO_3(g)$ is stabilized with free lime by $CrO_3(g) + CaO(s) \rightarrow CaCrO_4(s)$. This hypothesis is supported by our measurements, although not extensively, correlating the Cr(VI) percentage with the free lime content in the ash (Fig. 3.30). It is stressed that the reaction is assumed to occur with free lime. This also explains the reason why fewer higher Cr(VI) percentages have been found with meat and bone meal cofiring, although Ca levels are high. Calcium is not available because it is bound to phosphate, most likely hydroxyapatite or bioapatite. Also, cofiring tests have been performed in our research with chicken manure in a 1.5 MWth test boiler. Cofiring 10%, 15%, and 20% chicken manure with Colombian coal resulted in Cr(VI) percentages of total chromium of 9%, 14%, and 11%, as determined with X-ray absorption fine structure (XAFS). Such values are elevated compared to pure coal firing. It is questionable how far test boiler conditions reflect those of large-scale plants and therefore they have not been taken into account.

In Ref. [26], the authors proposed another reaction mechanism involving free lime. They studied the leachability of Cr (to evaporate TEs) in waste incinerator ash after heating at 990 ± 10°C. Formation of $CaCrO_4$ was probable and the synthesis reaction $2Cr_2O_3 \mid 4CaO \mid 3O_2 \rightarrow 4\ CaCrO_4$ was proposed. It must be noted, however, that the latter takes a long residence time (1 h) and because the experiment takes place in the cup, strong contact between the particles allows the surface to react. These conditions are different from those in pulverized coal boilers. However, the reaction pathway cannot be excluded.

Aside from the reaction with CaO, CrO_3(g) may react with other components to form chromate: BaO, FeO, MgO, K_2O, or Na_2O are possibilities. That different chromates can be formed is supported by the leaching results, indicating forms of Cr(VI) that are leachable and not leachable. Values of 25% and 31% from $PM_{2.5}$ and $PM_{2.5+}$ leaching, respectively, lie within this region. This means that a combination of no, less, and good soluble chromates is likely to be present. $BaCrO_4$ is insoluble, whereas the solubilities of other chromates of K, Na, Ca, and Mg vary from 4.9 to 238 g/L or are indicated as very soluble. The very stable, nonsoluble $BaCrO_4$ can account for the low liquid/solid (L/S) = 100 values, whereas the higher L/S = 100 values are accounted for by other chromates.

In theory, Cr(VI) can even be present in the glass phase (for ash, this is not confirmed by any measurement data), which can account for low L/S = 100 values. The formation of Cr(VI) is a high-temperature process that far exceeds the theory of 1000°C. It can be argued that high Cr(VI) values (29% and 41%) are found in FBC ash, while temperature limits in fluid gas FBC are about 900°C. However, burned coke particles have their own microclimate, which is quite different from gas temperature conditions, and the temperature can be as high as 2000°C.

Jano-Ito et al. compared the emission patterns of Hg under air-combustion and oxy-combustion conditions (shown in Fig. 3.31) [51]. In the oxy-fuel combustion system, gaseous $HgCl_2$(g) and Hg^0(g) are the main species existing in coal combustion flue gas and the dominant role was determined by the flue gas temperature. When the flue gas temperature was below 600°C, Hg^0 would be oxidized by the chlorine species in flue gas into $HgCl_2$, which indicated that $HgCl_2$ was the dominant Hg species when the flue gas temperature was relatively low. HgO(g) theoretically existed in the flue gas, for which the concentration was one to two magnitudes lower than that of Hg^0(g) and $HgCl_2$(g). Thus its presence was negligible. The partition patterns of Hg exhibited no obvious difference during oxy-fuel and air-fired conditions.

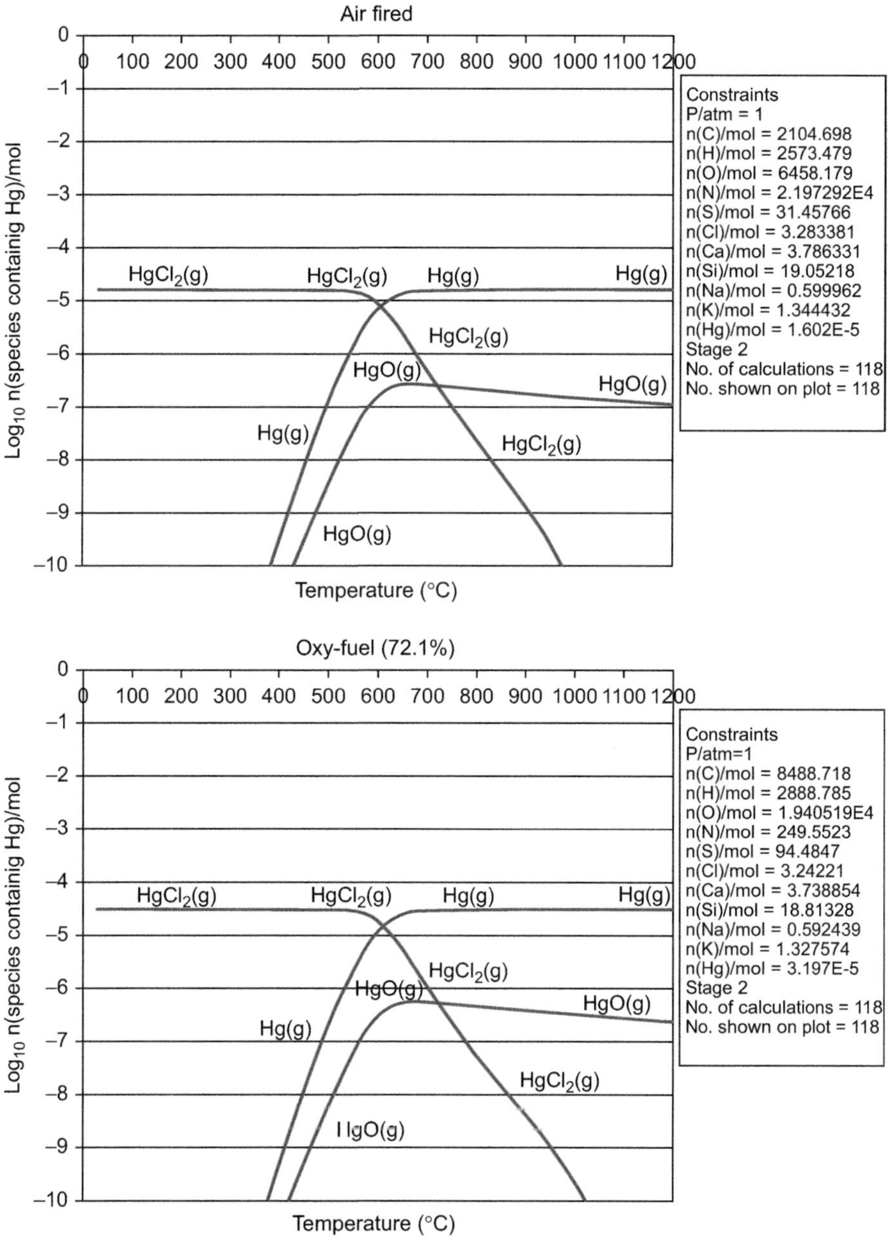

Figure 3.31 Hg species transformation during air-fired and oxy-fired conditions.
From Jano-Ito, et al. Comparison of thermodynamic equilibrium predictions on trace element speciation in oxy-fuel and conventional coal combustion power plants. Energy Fuels 2014;28: 4666–83.

3.5 Conclusions

Different TEs establish different patterns of partition during the coal combustion process. In this chapter, five representative elements (Hg, Se, As, Cr, and F) were chosen to investigate the partition patterns of TEs with various volatilities during coal combustion. Hg will be emitted from the coal mainly as Hg^0(g), due to its extremely low boiling point, from the boiler. Cooling the flue gas temperature in the downstream will lead to Hg^0(g) transforming into $HgCl_2$(g) considering the presence of HCl(g). Se will be discharged as SeO(g) from the high-temperature boiler and be further oxidized into SeO_2 when the flue gas temperature is relatively low. As shared similar partition patterns with Se and the content of O_2 in flue gas; vapor pressure and temperature significantly influence their presence in the postcombustion sessions. Cr is a nonvolatile element in coal, which will be accumulated in the bottom ash and fly ash after combustion. F will be mostly emitted as HF, no obvious transformation of which was observed in the flue gas due to its stability. Different combustion modes on TE partitions were reviewed and no significant differences were discovered between them. When the coal was cofired with other fuels, e.g., biomass, the impurities in biomass were found to influence the partition patterns of the TEs. The introduction of more substrates and their high concentrations in the co-combustion process make the partition patterns of the TEs more complicated. Moreover, the interactions between different TEs, like the interaction between HF and Hg^0(g), will result in a more intricate situation for the researcher to explore TE partitions. The mechanisms for TE emission and transformation were proposed by previous studies. However, they are as complex as the occurrence of TEs in coal. Future studies are needed to make them clear although it is very hard considering the complicated composition of coal combustion flue gas and the interactions between them.

Acknowledgments

This project was supported by the National Natural Science Foundation of China (51776227), Natural Science Foundation of Hunan Province, China (2018JJ1039, 2018JJ3675).

References

[1] Vejahati F, Xu Z, Gupta R. Trace elements in coal: associations with coal and minerals and their behavior during coal utilization—a review. Fuel 2010;89:904—11.
[2] WC Institute, editor. Coal resource overview of coal report; 2009.
[3] Minamata convention on mercury. 2017.
[4] Spallholz JE. On the nature of selenium toxicity and carcinostatic activity. Free Radical Biol Med 1994;17:45—64.
[5] Hughes MF. Arsenic toxicity and potential mechanisms of action. Toxicol Lett 2002;133: 1—16.

[6] Xu M, Yan R, Zheng C, Qiao Y, Han J, Sheng C. Status of trace element emission in a coal combustion process: a review. Fuel Process Technol 2004;85:215–37.

[7] Bool LE, Helble JJ. A laboratory study of the partitioning of trace elements during pulverized coal combustion. Energy Fuels 1995;37:880–7.

[8] Zheng B, Ding Z, Huang R, Zhu J, Yu X, Wang A, Zhou D, Mao D, Su H. Issues of health and disease relating to coal use in southwestern China. Int J Coal Geol 1999;40:119–32.

[9] Bool LE, Peterson TW, Wendt JOL. The partitioning of iron during the combustion of pulverized coal. Combust Flame 1995;100:262–70.

[10] Zeng T, Sarofim AF, Senior CL. Vaporization of arsenic, selenium and antimony during coal combustion. Combust Flame 2001;126:1714–24.

[11] Querol X, Fernández-Turiel J, López-Soler A. Trace elements in coal and their behaviour during combustion in a large power station. Fuel 1995;74:331–43.

[12] Zahir F, Rizwi SJ, Haq SK, Khan RH. Low dose mercury toxicity and human health. Environ Toxicol Pharmacol 2005;20:351–60.

[13] Carocci A, Rovito N, Sinicropi MS, Genchi G. Mercury toxicity and neurodegenerative effects. In: Whitacre DM, editor. Reviews of environmental contamination and toxicology. Cham: Springer International Publishing; 2014. p. 1–18.

[14] Pavlish JH, Sondreal EA, Mann MD, Olson ES, Galbreath KC, Laudal DL, Benson SA. Status review of mercury control options for coal-fired power plants. Fuel Process Technol 2003;82:89–165.

[15] Zheng C, Liu J, Liu Z, Xu M, Liu Y. Kinetic mechanism studies on reactions of mercury and oxidizing species in coal combustion. Fuel 2005;84:1215–20.

[16] Rizeq RG, Hansell DW, Seeker WR. Predictions of metals emissions and partitioning in coal-fired combustion systems. Fuel Process Technol 1994;39:219–36.

[17] Niksa S, Helble JJ, Fujiwara N. Kinetic modeling of homogeneous mercury oxidation: the importance of NO and H_2O in predicting oxidation in coal-derived systems. Environ Sci Technol 2001;35:3701–6.

[18] Sliger RN, Kramlich JC, Marinov NM. Towards the development of a chemical kinetic model for the homogeneous oxidation of mercury by chlorine species. Fuel Process Technol 2000;65–66:423–38.

[19] Hall B, Schager P, Lindqvist O. Chemical reactions of mercury in combustion flue gases. Water Air Soil Pollut 1991;56:3–14.

[20] Liu D, Sakai Y, Yamamoto M, Sadakata M. Behavior of fluorine in the combustion of Chinese coal in small furnaces. Energy Fuels 2006;20:1406–10.

[21] Raask E. The mode of occurrence and concentration of trace elements in coal. Prog Energy Combust Sci 1985;11:97–118.

[22] Yan R, Gauthier D, Flamant G. Volatility and chemistry of trace elements in a coal combustor. Fuel 2001;80:2217–26.

[23] Liu S, Wang Y, Yu L, Oakey J. Volatilization of mercury, arsenic and selenium during underground coal gasification. Fuel 2006;85:1550–8.

[24] Wang C, Liu H, Zhang Y, Zou C, Anthony EJ. Review of arsenic behavior during coal combustion: volatilization, transformation, emission and removal technologies. Prog Energy Combust Sci 2018;68:1–28.

[25] Furuzono T, Nakajima T, Fujishima H, Takanashi H, Ohki A. Behavior of selenium in the flue gas of pulverized coal combustion system: influence of kind of coal and combustion conditions. Fuel Process Technol 2017;167:388–94.

[26] Clarke LB. The fate of trace elements during coal combustion and gasification: an overview. Fuel 1993;72:731–6.

[27] Basu P. Combustion of coal in circulating fluidized-bed boilers: a review. Chem Eng Sci 1999;54:5547–57.
[28] Zhou C, Liu G, Fang T, Wu D, Lam PKS. Partitioning and transformation behavior of toxic elements during circulated fluidized bed combustion of coal gangue. Fuel 2014;135:1–8.
[29] Langner P, Mikutta C, Suess E, Marcus MA, Kretzschmar R. Spatial distribution and speciation of arsenic in peat studied with microfocused X-ray fluorescence spectrometry and X-ray absorption spectroscopy. Environ Sci Technol 2013;47:9706–14.
[30] Huggins FE, Huffman GP. Modes of occurrence of trace elements in coal from XAFS spectroscopy. Int J Coal Geol 1996;32:31–53.
[31] Zhao Y, Zhang J, Sun J, Bai X, Zheng C. Mineralogy, chemical composition, and microstructure of ferrospheres in fly ashes from coal combustion. Energy Fuels 2006;20:1490–7.
[32] Zhou C, Liu G, Cheng S, Fang T, Lam PK. Thermochemical and trace element behavior of coal gangue, agricultural biomass and their blends during co-combustion. Bioresour Technol 2014;166:243–51.
[33] Åmand LE, Leckner B. Metal emissions from co-combustion of sewage sludge and coal/wood in fluidized bed. Fuel 2004;83:1803–21.
[34] Gogebakan Z, Selçuk N. Trace elements partitioning during co-firing biomass with lignite in a pilot-scale fluidized bed combustor. J Hazard Mater 2009;162:1129–34.
[35] Miller BB, Kandiyoti R, Dugwell DR. Trace element emissions from Co-combustion of secondary fuels with Coal: a comparison of bench-scale experimental data with predictions of a thermodynamic equilibrium model. Energy Fuels 2002;16:956–63.
[36] Opila EJ, Myers DL, Jacobson NS, Nielsen IMB, Johnson DF, Olminsky JK, Allendorf MD. Theoretical and experimental investigation of the thermochemistry of $CrO_2(OH)_2(g)$. J Phys Chem A 2007;111:1971–80.
[37] Mitsui Y, Imada N, Kikkawa H, Katagawa A. Study of Hg and SO_3 behavior in flue gas of oxy-fuel combustion system. Inter J Greenhouse Gas Control 2011;5:S143–50.
[38] Contreras ML, Arostegui JM, Armesto L. Arsenic interactions during co-combustion processes based on thermodynamic equilibrium calculations. Fuel 2009;88:539–46.
[39] Fu B, Liu G, Sun M, Hower JC, Mian MM, Wu D, Wang R, Hu G. Emission and transformation behavior of minerals and hazardous trace elements (HTEs) during coal combustion in a circulating fluidized bed boiler. Environ Pollut 2018;242:1950–60.
[40] Zhou C, Liu G, Xu Z, Sun H, Lam PKS. Effect of ash composition on the partitioning of arsenic during fluidized bed combustion. Fuel 2017;204:91–7.
[41] Córdoba P, Diego R. Mass balance and partitioning of trace elements under oxy-coal combustion: first experiences. Fuel 2018;215:892–903.
[42] Linak WP, Wendt JOL. Trace metal transformation mechanisms during coal combustion. Fuel Process Technol 1994;39:173–98.
[43] Suriyawong A, Gamble M, Lee M-H, Axelbaum R, Biswas P. Submicrometer particle formation and mercury speciation under O_2-CO_2 coal combustion. Energy Fuels 2006;20:2357–63.
[44] Otero-Rey JR, López-Vilariño JM, Moreda-Piñeiro J, Alonso-Rodríguez E, Muniategui-Lorenzo S, López-Mahía P, Prada-Rodríguez D. As, Hg, and Se flue gas sampling in a coal-fired power plant and their fate during coal combustion. Environ Sci Technol 2003;37:5262–7.
[45] Seames WS, Wendt JOL. Partitioning of arsenic, selenium, and cadmium during the combustion of Pittsburgh and Illinois #6 coals in a self-sustained combustor. Fuel Process Technol 2000;63:179–96.

[46] Senior CL, Bool LE, Srinivasachar S, Pease BR, Porle K. Pilot scale study of trace element vaporization and condensation during combustion of a pulverized sub-bituminous coal. Fuel Process Technol 2000;63:149–65.
[47] Yan R, Gauthier D, Flamant G, Peraudeau G, Lu J, Zheng C. Fate of selenium in coal combustion: volatilization and speciation in the flue gas. Environ Sci Technol 2001;35: 1406–10.
[48] Goodarzi F, Huggins FE. Monitoring the species of arsenic, chromium and nickel in milled coal, bottom ash and fly ash from a pulverized coal-fired power plant in western Canada. J Environ Monit J 2001;3:1–6.
[49] Huffman GP, Huggins FE, Shah N, Zhao J. Speciation of arsenic and chromium in coal and combustion ash by XAFS spectroscopy. Fuel Process Technol 1994;39:47–62.
[50] Stam AF, Meij R, te Winkel H, Eijk RJ, Huggins FE, Brem G. Chromium speciation in coal and biomass co-combustion products. Environ Sci Technol 2011;45:2450–6.
[51] Janoito MA, Reed GP, Millan M. Comparison of thermodynamic equilibrium predictions on trace element speciation in oxy-fuel and conventional coal combustion power plants. Energy Fuels 2014;28:4666–83.
[52] Ruch RR, Gluskorer HJ, Smm NF. Occurrence and distribution of potentially volatile trace elements in coal, Illinois St Geol Sun Urbana, U.S.A: Envir Geol Notes No 72; 1974.
[53] Swanson VE, Medlin JH, Hatch JR, Coleman SL, Wood GH, Woodruff SD, Hilderbrand RT. Collection. chemical an/dysis and evaluation of coal samples. Washington. U.S.A: U.S. Dept Int Geol Surv Open File Report 76-468; 1976.
[54] Hislop JS, Fisher EMR, Morton AG, Salmon L, Wood DA. Multi-element analysis of British coals, 22 CortJ Anal Chem in Energy Technol, Gatlinburgh, U.S.A. 1978.
[55] Taylor CJ. Mercury and other potentially toxic trace elements in British coals. U.K: National Coal Board Yorkshire Regional Laboratory; 1973.
[56] LIM MY. Trace elements from coal combustion. London, U.K.: IEA Coal Research Report ICTIS/'rR05; 1979.
[57] SWAINE DJ. Trace elements in coal, II. A. Conf on Trace Substances in Environmental Health. Columbia, U.S.A.: University of Missouri; 1977. p. 107.
[58] Contreras ML, Armesto Arostegui JM. Arsenic interactions during co-combustion processes based on thermodynamic equilibrium calculations. Fuel 2009;88(3):539–46. ISSN 0016-2361, https://doi.org/10.1016/j.fuel.2008.09.028.

Trace element partition in coal fires

4

Hailong Li [1], *Zequn Yang* [2], *Jianping Yang* [1], *Yingchao Hu* [1]
[1]School of Energy Science and Engineering, Central South University, Changsha, China;
[2]Department of Civil Engineering, The University of Hong Kong, Hong Kong, China

4.1 Introduction

One of the consequences of coal mining is the exposure of coal to air and moisture resulting in the ignition of coal through the processes of chemisorption, oxidation, and spontaneous combustion. The ignition of coal is a global concern and burning coal may cause significant environmental problems [1–7]. In China alone, several hundred million people commonly burn raw coal in unvented stoves that permeate their homes with high levels of toxic metals and organic compounds. At least 3000 people in Guizhou Province in southwest China are suffering from severe arsenic poisoning. The primary source of arsenic appears to be the consumption of chili peppers dried over fires fueled with high-arsenic coal. Coal samples in the region were found to contain up to 35,000 ppm arsenic. Chili peppers dried over high-arsenic coal fires adsorb 500 ppm of arsenic on average. More than 10 million people in Guizhou Province and surrounding areas suffer from dental and skeletal fluorosis [8]. Excess fluorine is caused by eating corn dried over burning briquettes made from high-fluorine coals and high-fluorine clay binders. Polycyclic aromatic hydrocarbons formed during coal combustion are believed to cause or contribute to the high incidence of esophageal and lung cancers in parts of China [9]. Domestic coal combustion has also caused selenium poisoning and possibly mercury poisoning [10,11]. Better knowledge of trace element (TE) compositions in coal mining and partition patterns during spontaneous coal firing may help to reduce some of these health problems [12]. Thus, in this chapter, we investigate the partition patterns of some typical TEs during coal fires. The effects of coal type and fire temperature on the partitioning behavior and emission characteristics of TEs are discussed.

4.2 Locations of typical active coal fires

As shown in Fig. 4.1, coal fires occur most frequently in exposed or underground coal seams [13]. They are often triggered via spontaneous combustion of coal, which is an exothermic oxidation reaction between coal and air. Coal temperature will increase if heat released by coal oxidation is not sufficiently dissipated. When the temperature reaches a critical level (80–130°C), the coal will start to smolder and burn [14].

Emission and Control of Trace Elements from Coal-Derived Gas Streams
https://doi.org/10.1016/B978-0-08-102591-8.00004-0

Figure 4.1 Coal fires and their hazards in the Wuda coalfield, Inner Mongolia Autonomous Region. (a) Burning coal; (b) a coal dust explosion triggered by a coal fire when the bulldozer moves the burning coal; (c) pungent gas and smoke emitting from an underground coal fire; (d) a crack induced by a coal fire. Its width and temperature are approximately 40 cm and 177.8°C, respectively.
From Song, et al. Coal fires in China over the last decade: a comprehensive review. Inter J Coal Geol 2014;133:72–99.

A large number of factors influence the spontaneous combustion of coal, which can be categorized into three groups: coal properties, coal seam properties, and external conditions. Specifically, each group involves a couple of subfactors [15]. Apart from factors influencing the spontaneous combustion of coal, forest fires, lightning, and human activities such as mining activities, discarding of burning cigarettes, burning trash piles, and ignition of coal for heating, cooking, and generating “red dog” for ash used for traction on ice also trigger coal fires [13]. Among these factors, mining activities exposing coal to air or creating ventilation pathways are the most common factor triggering coal fires. According to their origin, age, location, or burning stage, coal fires can be categorized into various groups [13].

Coal fires occur in numerous countries around the globe, typically in China, India, the United States, Australia, and South Africa [16,17]. If underground coal seams are burned and reduced to ash, most of the landscape associated with coal fires comes from underground volume loss. Different forms of bedrock surface fracturing and ground subsidence endanger adjacent mines and related industries. Coal fires often cause collapse of mining wells and may completely stop other prosperous mining activities.

China. China is the world's largest coal producer and consumer. Annual output exceeds 2 billion tons. China's coal fires were mentioned in the historical literature of 1000 AD, for example, during the "Northern Song Dynasty" (960–1280) to explore Li Daoyuan's travel report in western China. Marco Polo (1254–1324 BC) mentioned in his travel documents that "burning the mountain along the Silk Road," the ancient coal fire in Xinjiang has passed the age of the Pleistocene [18]. Currently, the main coal-fire areas stretch along the coal mining belt in China, which extends for 5000 km from east to west along the northern half of the country. According to Van Genderen and Guan, more than 50 coalfields in this belt are affected by fires. These are located in the provinces of Xinjiang, Inner Mongolia, Ningxia, Shanxi, and Shaanxi. The largest and best-researched coal-fire area in China is the Wuda area in Inner Mongolia [19]. Wuda is located at ~39°51′N and 106°60′E. The city and its mining operations are located on the west side of the Yellow River north of Helan Mountain. The north–south direction of the western part of the city is inclined to coal mining with an area of 35 km^2 and a geological reserve of 630 million tons of coal. Currently, 16 of the 20 fires are still burning in the syncline west of Udda. In addition, Ruqigou in Ningxia, Keerjian, Tielike in Xinjiang, and mining areas in northeastern Urumqi, Xinjiang, are experiencing coal fires. China burns up to 20 million tons of coal per year, which is equivalent to the annual production of hard coal in Germany. This amount of resources is lost 10-fold because coal adjacent to coal fires becomes inaccessible [20]. Song et al. summarized the typical locations for coal firing over the last few decades in China (shown in Fig. 4.2) [21].

India. The largest known coal-fire area in India is in the Jharia coal mining belt. This strongly industrialized area hosts a well-known coalfield, whose coal fires have been extensively studied by Bhattacharya et al. and Kuenzer et al. [22,23]. Jharia is the largest coalfield in India. It is located about 250 km northwest of Kolkata and 1150 km southeast of Delhi. It extends from ~23°38′N and 86°07′E to 2350′N and 86°30′E with a coverage area >450 km^2. Coal mining can be traced back more than a century; both open-air and underground mining exist. More than 90 surface and underground coal fires were reported in the area, the first of which was discovered in 1916. According to Agarwal et al., ~70 level surface and underground fires are still burning. Fire causes severe land subsidence and overall environmental degradation [24]. The Jharia fires are of special concern, because the region is densely settled, and the local population is endangered by sudden bedrock collapse. Other coal fires are burning in the Ray coal mining area, the Ranigani area, and the Singareni coal belt.

United States. Numerous coal fields in the United States are affected by fires. The federal Office of Surface Mining manages a database (Abandoned Mine Land Inventory System) that contained 150 fires in 1999. Coal fires occur in Alaska, Kentucky, Pennsylvania, and West Virginia east of the Appalachian coalfields, as well as in Wyoming, Montana, and Colorado in the Rocky Mountains [13]. The Colorado Division of Mines and Geology regularly publishes reports on the status of coal fires in abandoned underground mines in the state. Of the 1736 abandoned coal mines, 32 are currently burning, and 50 fires are considered dormant [25]. However, the most

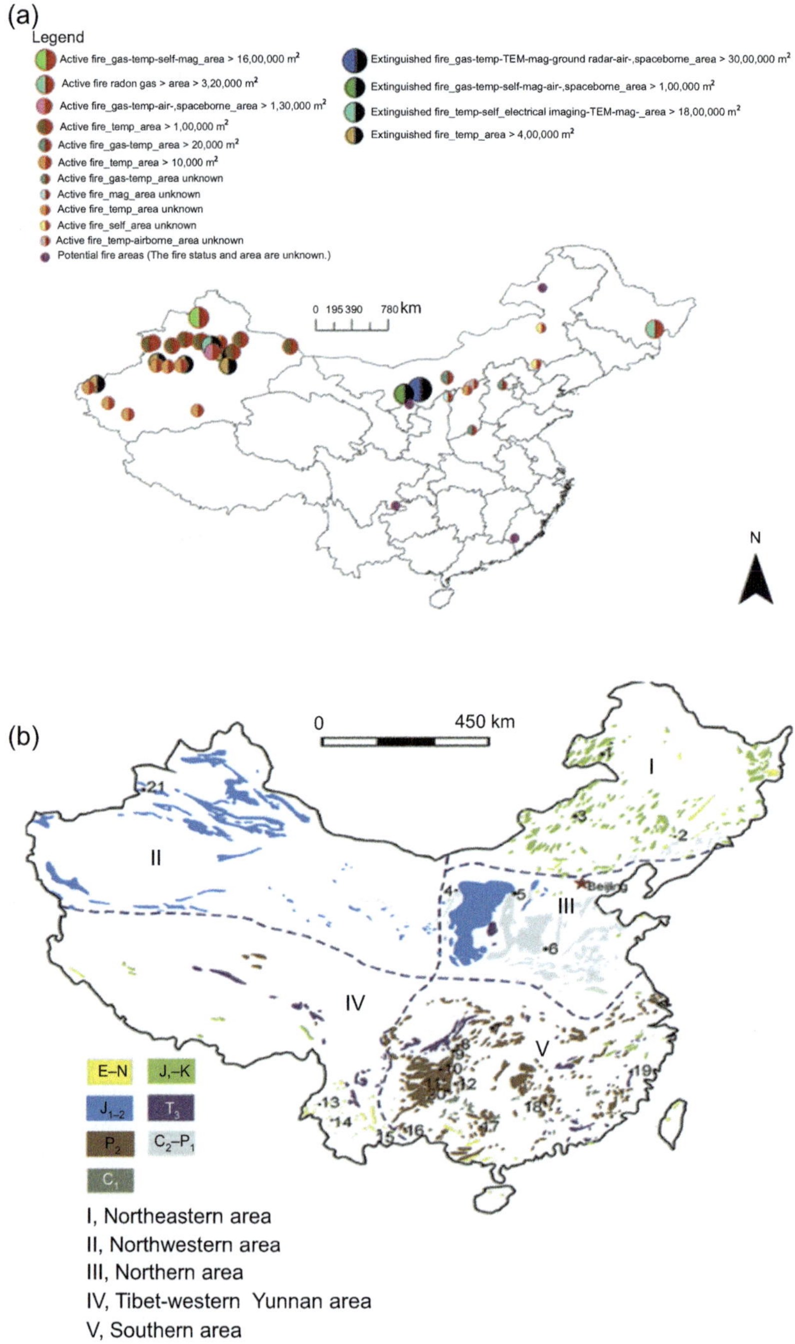

Figure 4.2 (a) Current coal-fire distribution and development in China over the course of the last decade according to publications and reports. The legend is comprised of three parts: fire status, detection techniques used in these areas, and fire area. In the part on detection

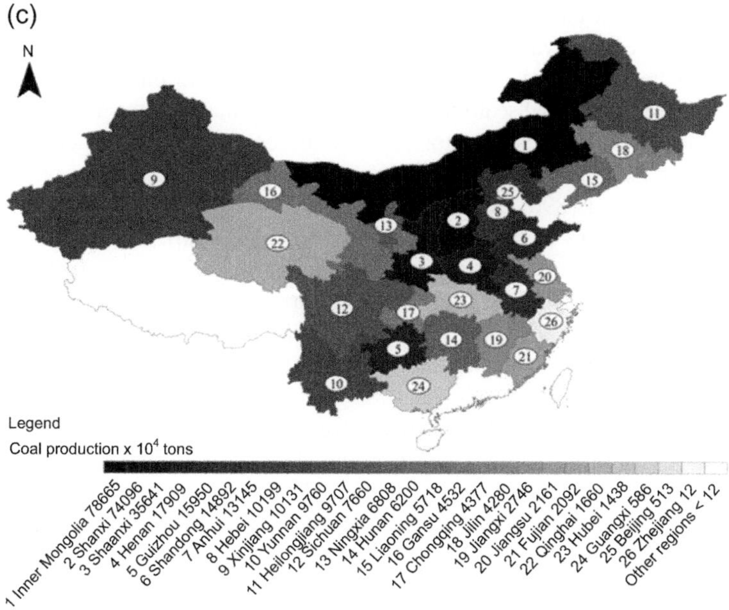

Figure 4.2 cont'd.

famous coal fire in the United States is located in Pennsylvania, beneath the city of Centralia. The Centralia mine fire started in 1962 in the city's garbage dump, when a small trash fire ignited a nearby coal vein [26]. The fire spread steadily underground, and fire fighters could not control it. The slowly advancing coal fire endangered homes and infrastructure. Hence the entire city had to be evacuated and relocated between 1980 and 1998. Most houses were demolished, and the US postal service revoked the town's zip code. Even today, the fire continues to lead to the collapse of formerly inhabited ground and infrastructure.

techniques, Temp stands for temperature measurement; Self, self-potential; Electrical imaging, 2D electrical imaging method; Mag, magnetic technique; Airborne, airborne remote sensing; and Spaceborne, spaceborne remote sensing. This map does not cover all coal fires because not all data are available. (b) Coal distribution areas in China [51]. 1, Yimin coalfield; 2, Shenbei coalfield; 3, Wulantuga; 4, Wuda coalfield; 5, Jungar coalfield; 6, Fengfeng–Hangdan coalfield; 7, Enshi; 8, Nanchuan; 9, Songzao coalfield; 10, Dafang coalfield; 11, Zhijin coalfield; 12, Guiding; 13, Tengchong; 14, Lincang; 15, Jianshui; 16, Yanshan; 17, Heshan; 18, Meitian coalfield; 19, Jianou; 20, Ziyun; 21, Yili. There are six major coal-forming periods: Late Carboniferous and Early Permian (C_2–P_1), Late Permian (P_2), Late Triassic (T_3), Early and Middle Jurassic (J_1–J_2), Late Jurassic and Early Cretaceous (J_3–K_1), and Paleogene and Neogene (E–N). (c) Coal production of the different administrative regions in 2010.
From Song, et al. Coal fires in China over the last decade: a comprehensive review. Inter J Coal Geol 2014;133:72–99.

South Africa. On the African continent, the largest coal deposits occur in South Africa, Zimbabwe, Botswana, Mozambique, and Zambia [1,27]. Coal mining in the world market has produced a large number of coal gangue piles, from which fires are common. However, coal seam fires also exist, particularly in the Witbank coalfield in South Africa, which accounts for 53% of the country's coal production and is responsible for generating 41% of the country's energy [28]. Mining operations in the field began around 1890. Due to the use of slab-column mining technology, a large amount of coal remains in the ground (pillars and roof coal). As early as 1947, the collapse of this structure and increased ventilation through surface fracturing led to spontaneous combustion. These fires are still burning. Open mining began in 1996 in areas that had been mined underground, igniting additional coal fires. The threat of fracturing and subsidence jeopardizes many mining operations and local settlements in Witbank town [13].

Other countries. Additional coal-fire areas exist in Australia, Russia [3], Indonesia [5], and Europe [29]. In Sumatra, for example, coal fires are limited to the southern region where coal is mined in larger quantities. Coal seams in the area are not covered by soil erosion and the conditions are conducive to spontaneous combustion. In dry years, coal fires ignite forest fires, which in turn ignite additional coal and peat fires. In Europe, coal fires occurred near Zagreb, Croatia. Small-scale controlled fires also exist in the Ruhr coal mine area of Germany [30]. Smaller fires are reported in France, Poland, Czech Republic, and Ukraine. The Ukrainian coal industry has deposited coal remains in >2100 heap piles, with a capacity of 2000 million tons [27]. More than 120 of these heap piles are burning; they produce 15,000 tons of CO_2 and 5000 tons of CO annually. In 1998, 74 coal fires were burning in Russia; updated data are not available to the authors. The regions affected are Kuzbass, the Pechora Basin, and the Donetsk Basin (Fig. 4.3).

In summary, coal fires could happen anywhere around the world. Gürdal et al. studied the intrinsic reasons accounting for the occurrence of coal fires [31]. According to their results, the low grade and high pyrite content of Çan coal enhances the spontaneous combustion process. As a result of the lowering of the grade, joint moisture content, oxygen content, internal surface area, and air permeability tend to increase. An increase in the natural moisture content of coal can release heat. Greater surface area and gas permeability have similar effects. Among the samples analyzed, sulfur minerals were observed to form fibrites in the fragile and secondary pyrite (cubic crystals) in the edge of the fracture and organic matter, which are known to accelerate the self-heating of the coal. Surface wetting and sulfur flowering are important in this area. They proposed a coal-fired temperature of about 1100°C. Onifade et al. suggested that both materials indicate that carbon, moisture, hydrogen, volatiles, nitrogen, and ash content can complement the tendency of spontaneous combustion [32]. When the major and trace elemental concentrations of the samples analyzed were compared to those of basin coals, significant concentration increases were observed. The TEs that generally increased included As, Ba, Be, Bi, Cd, Cu, Co, Cs, F, Ni, Nb, Se, Sb, Tl, Ta, Pb, U, V, and W. When compared to world coals it was observed that As, Cu, Co, Cs, F, Mo, Ni, Nb, Pb, Se, Tl, Ta, Zr, Zn, U, V, and W displayed higher values.

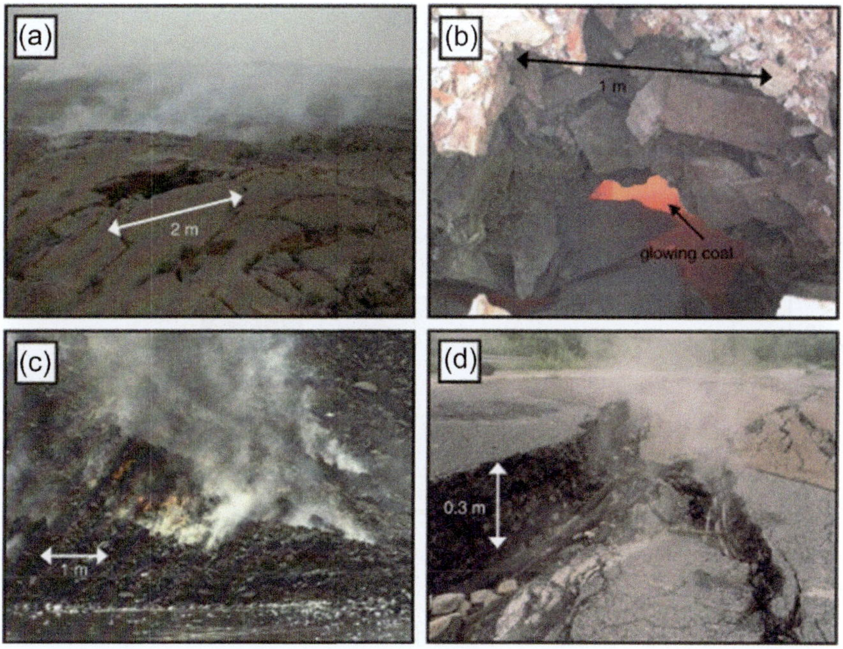

Figure 4.3 Coal fires worldwide. (a and b) Underground coal fires in Wuda, Inner Mongolia, China. (c) Coal heap fire in the Witbank coalfield in South Africa. (d) Underground coal fire under a completely cracked road in Centralia, Pennsylvania, USA.
From Kuenzer, et al. Geomorphology of coal seam fires. Geomorphology 2012;138:209–22.

4.3 Trace element emissions during coal fires in different coal deposits

TE emissions from coal fires in different coal deposits were rarely studied. Goodarzi et al. became aware of TE emissions from different coal deposits in the 1990s in a self-burning coal seam in Coalspur, Alberta, Canada [33]. Goodarzi divides the coal seam into three distinct zones: the oxidation zone (sample 1), the combustion zone (sample 2), and the distillation zone (samples 3–11) (as shown in Fig. 4.4). The oxidized zone consists of the uppermost part of the burning coal seam and contains vents for conducting volatile materials. Coal fragments in this area do not burn because they are oxidized; however, they show oxidized edges and cracks. Oxidation cracks are usually filled with fluorescent hydrocarbons (coal tar). The alteration temperature in this region is estimated to be 400°C. The burning zone is an intensive part of burning coal seams. The coke in this area has the form of burning coal. The combustion of coal in this area is due to the narrow high-temperature areas and coal fragments entering the air. The alteration temperature of this region is estimated to be about 600°C. The distillation zone consists of a majority of the burning coal seam and comprises three subzones, including two zones marked as focal. One is near the top of the seam just below the burning zone and the other is at the bottom. Sandwiched

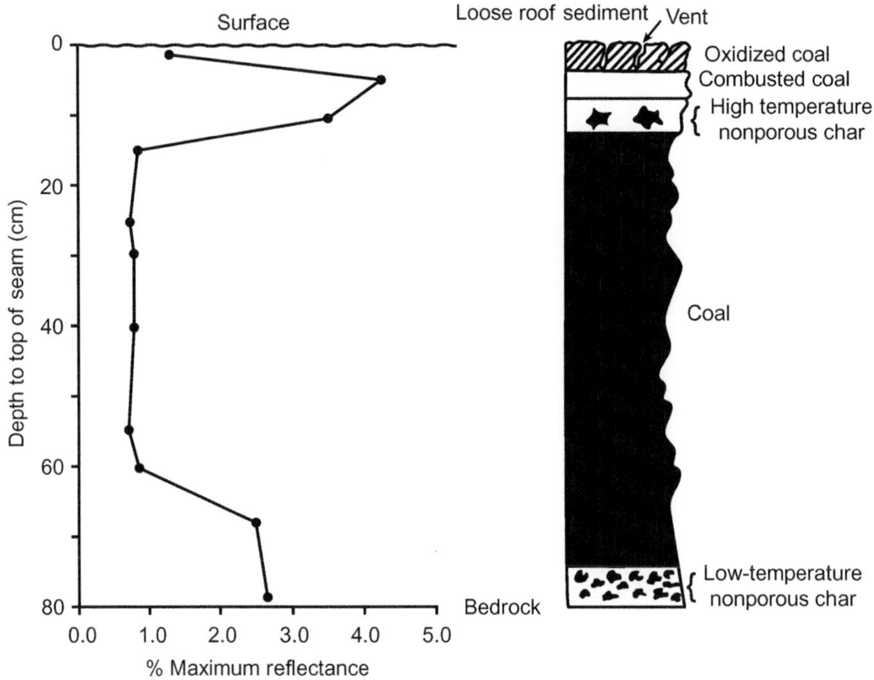

Figure 4.4 Reflectance of the burning coal seam.
From Goodarzi. Variation of elements in self-burning coal seam from Coalspur, Alberta, Canada. Energy Sources 1990;12:345–61.

between the two is a distillation (precarbonization) zone. Carbonization at the top of the coal seam is due to the heat generated in the combustion zone. The reason for carbonization near the seam floor is the heat transferred by the bedrock (baked mudstone) because the coal seam below the coal seam burns.

Variation of elements associated with carbonate. Ba, Ca, Fe, Mg, and Mn are generally associated with carbonate minerals in coal. Calcite is the only carbonate mineral detected in this set of samples. In the absence of pyrite from a mineral combination of these samples, iron is expected to be associated with carbonate, i.e., siderite. However, it was not detected. Changes in Ba, Ca, Fe, Mg, and Mn on the weld profile are almost similar, especially in the upper part of the weld (as shown in Fig. 4.5(a)), indicating that the concentrations and depletion levels of these elements are similar. Since coal is devolatilized at 550°C, Ba, Ca, Fe, Mg, and Mn are concentrated in carbonized subzone 2. The concentration of Ca, Fe, Mg, and Mn was continuously reduced to a minimum at the top of the zone by distillation (precarbonization zone) at a temperature of 300°C. Ba follows a similar trend except that it has a minimum peak at the top of carbonized subzone 2. All of these elements show the maximum concentration in the intensive area of the seam (burning area 650°C). Concentration of these elements upon combustion indicates the concentration of devolatilization of aliphatic and alicyclic carbons and aromatic carbon, which is evident by the sharp

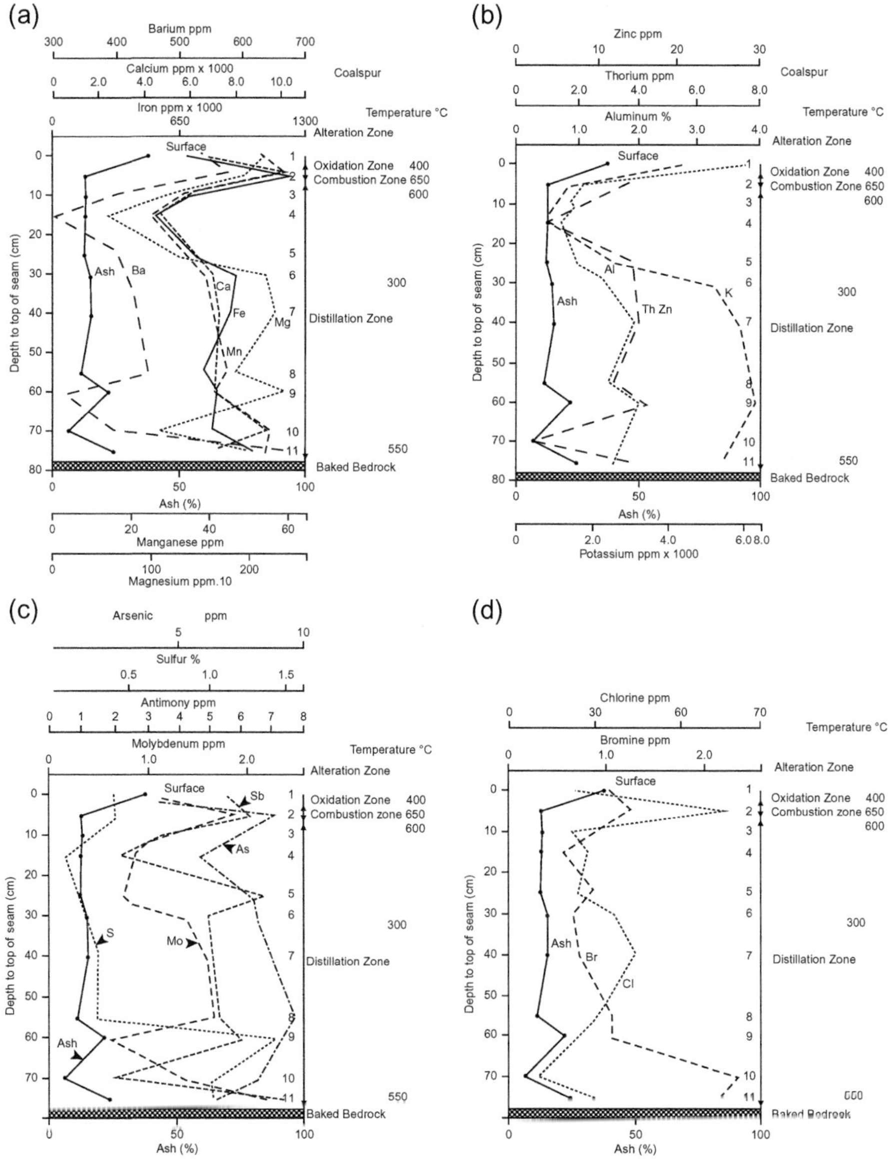

Figure 4.5 Variations of trace elements associated with (a) carbonate; (b) clay minerals; (c) sulfides and sulfates; and (d) halogens.
From Goodarzi. Variation of elements in self-burning coal seam from Coalspur, Alberta, Canada. Energy Sources 1990;12:345–61.

increase in carbon content in this region. These elements do not devolatilize during combustion of the coal and have the same distribution in the fly ash and slag. Therefore during coal seam combustion, they are concentrated in the coal ash of the combustion zone and are not removed from the system.

Variation of elements associated with clay minerals. The elements normally associated with clay minerals, namely Al, Si, K, Cs, Rb, and rare earth elements (REEs), are associated with minerals in coal. Therefore their changes are also related to the ash content (Fig. 4.5(b)). Compounds of alkali metal elements (Na, K, Rb, and Cs) are very stable substances at high temperatures. These elements are the main components of the deposits (soils) formed on the convective heat transfer surface of 600−1000°C, and if they are in a water-soluble form (for example, Na, Cl), they may become devolatilized. However, there is no evidence that these elements are devolatilized and mobilized from the burning coal seam, probably because these elements are present in the clay minerals and are only dehydrated in the high-temperature zone. Other elements associated with aluminosilicates (e.g., Th and Zn) and other unrelated elements follow a similar trend to Al, indicating that their concentrations are more controlled by the ash content of the coal rather than the hot zone of the burning coal seam.

Variation of elements associated with sulfides and sulfates. The elements As, Mo, S, Sb, and Se are usually associated with sulfides or sulfates in coal. These elements have an intermediate behavior between elements that volatilize and emit in the gas phase, such as halogens, and preferentially concentrate to smaller particles, such as Ti, during coal combustion. These elements are potentially hazardous in the development of coal resources because they are released into the atmosphere during coal use. There is no sulfur-containing mineral other than gypsum in the burning coal seam. This indicates that sulfur is mainly related to the microstructure of coal. Further evidence of this organic association usually comes from changes in sulfur. Sulfur concentration does not increase with increasing coal ash in different regions (Fig. 4.5(c)), especially in the upper part of the coal seam, which indicates the flow of S in the high-temperature region, i.e., samples 1−3. The appearance of orthogonal sulfur crystals is the main mineral (48%) in the sample collected from the top of the coal seam and the vent to the oxidation zone, indicating that the volatile S is released as HS_2 and it decomposes into a surface.

Variation of halogens. During the combustion of coal, halogens are volatilized and discharged in the gas phase. Cl and Br are the only halogen elements whose distribution is detected in this set of samples. Bromine has the highest concentration in sample 10, probably due to the devolatilization of Br in sample 1, which is in contact with the thermal bedrock and carbonized (Fig. 4.5(d)). The devolatilized Br was captured in the cooler material (sample 10) above the carbonized coal (sample 11). Chlorine showed that the coal in the distillation zone (sample 4) gradually concentrated toward the combustion zone (sample 2) and the oxidation zone (sample 1) was suddenly depleted. This behavior of C1 is unexpected because C1 is unstable at high temperatures. However, this behavior of C1 may be related to the nature of the altered residue in the burning coal seam. Unlike char, coke is porous and can be exposed to volatile materials. Coke formed by changes in coal in Coalspur is nonporous, dense, and massive; therefore volatile materials cannot easily escape from high-temperature regions. The oxidized zone is present in the first relatively porous layer at the top of the coal seam. The coal in this area is broken and loosely packed. Therefore it allows volatile substances to pass through the surface. Organic petrological studies of burning coal seams have indeed shown that changes in coal at temperatures of 600−650°C result in saturation of oxidized coal fragments by fluorescent hydrocarbons (coal tar).

The coal seam is prone to spontaneous combustion due to external ignition sources such as forest fires or lightning strikes. Sparse information is available about element mobilization during the spontaneous combustion of coal seams. The elements Fe, Mg, Ca, and Mn are concentrated in the lower half of the distillation zone, wherein the temperature reaches 500−550°C. The upper portion of the carbonization zone (300°C) was significantly reduced before concentration in the combustion zone (600−650°C) increased sharply. This behavior is expected based on a model of a coal-fired power plant; the foregoing elements do not volatilize during coal combustion but are concentrated in bottom ash and fly ash [34].

Elements associated with clay minerals in coal, such as Al, Cs, K, Rb, Ti, and REEs, exhibit changes associated with ash changes. These elements are concentrated in the lower half of the carbonization zone, but are significantly depleted above the carbonization zone directly below the combustion zone. As the clay in the hottest part of the coal seam dehydrates, its concentration in the combustion and oxidation zones increases again. The concentration of halogens Br and Cl is steadily decreasing throughout the carbonization zone and then slightly increased in the combustion zone. These two elements are highly volatile and are present in the gas phase during combustion of the coal. Goodarzi attributed the high concentration of Cl at 650°C to the nonporous and dense nature of this noncoking coal-burning coke. The chlorine is again depleted in the oxidation zone. Unlike porous coke, which allows volatiles to escape, the bulk and dense coke formed by such low-grade coal captures volatiles (and halogens) and prevents them from escaping. Changes in other elements (As, Mo, and Sb) are unclear. These elements indicate a gradual increase in concentration in the combustion zone after depletion of the upper portion of the distillation zone.

Querol et al. explored the characterization of burnt coal gangue banks at Yangquan, Shanxi Province, China. The representative sample of coal gangue is characterized by low moisture (1.1%, ad) and high-ash content (79%, db). The gangue has relatively low coal content (10% C, 1.3% H, 0.2% N, db) and calorific value (3.75 MJ/kg, db). X-ray diffraction (XRD) and scanning electron microscopy (SEM) analyses indicated that mineral matter is characterized by high proportions of subangular to subrounded kaolinite ($Al_2Si_2O_5(OH)_4$, 50%−60%), and minor proportions of quartz (SiO_2, 10%−20%), palygorskite ($(Mg,Al)_5(Si,Al)_8O_{20}(OH)_2 \cdot 8H_2O$, <10%), clinochlore ($(Mg,Fe)_5Al(Si_3Al)O_{10}(OH)_8$, <5%), illite ($(K_{0.5}(Al,Fe,Mg)_3(Si,Al)_4O_{10}(OH)_2$, <5%), siderite ($FeCO_3$, <5%), and angular to subangular pyrite (FeS_2, <5%) (as shown in Figs. 4.6 and 4.7). The occurrence of Pb and Ba attached to clays was evidenced by SEM-energy-dispersive X-ray spectroscopy. Chemical composition is characterized by a relatively high content of Al (15.2%), Fe (2.5%), K (0.9%), and S (0.8%). TEs are not especially enriched with this material when compared with the kaolinite-rich lithologies analyzed in this study. All the TE contents

Figure 4.6 Sampled coal gangue gob showing details of the spontaneous combustion in the gob slope, the fired coal gangue, gas vents, and the fresh coal gangue.
From Querol, et al. Environmental characterization of burnt coal gangue banks at Yangquan, Shanxi Province, China. Int J Coal Geol 2008;75:93–104.

measured in the coal gangue fall within the concentration range reported by Zhao et al. (2008) for the Shanxi coal gangue [35]. Only Cd and Co showed slightly higher concentrations in the present study (1.1 vs. <DL (detectable level) −0.7 mg Cd/kg, and 9.1 vs. 3.4−6.3 mg Co/kg). Therefore the sample can be regarded as representative of the Shanxi coal gangue.

The coal in the study area is a low-moisture (1.5%, ad), high-S (4.4%, db), and high-ash (33%, db) coal. The C content is 82.1% daf and the calorific value reached 32.1 MJ/kg (maf). The N, H, and O contents are 1.4, 3.7, and 2.7% db, respectively. The volatile matter content (15.5% daf) suggests a low volatile bituminous rank. The high-sulfur content is typical of the coal of the local Pennsylvanian Taiyuan formation. The coal is characterized by a high vitrinite content and is rich in molten stone, giving the coal high microporosity. Inorganic impurities in coal particles include subangled to subcircular kaolinite (10%−15%) and palygorskite (<10%) and pyrite (4%), both of which are discrete to subangled or microclip (as shown in Fig. 4.8) ankerite ($Ca(Fe,Mg)(CO_3)_2$ <5%) and quartz (<2%). Chemical composition is marked by high concentrations of Al (5.1%), Fe (4.0%), and S (4.4%). These data and the XRD analysis suggest that a high proportion of S occurs as pyritic S. This coal sample is characterized by markedly high concentrations of Hg (1.8 mg/kg), Se (23 mg/kg), As (37 mg/kg), and Pb (94 mg/kg), and relatively high contents of Th (12 mg/kg), Tl (1.8 mg/kg), Hf (4.4 mg/kg), and Zr (174 mg/kg). All the TE contents measured

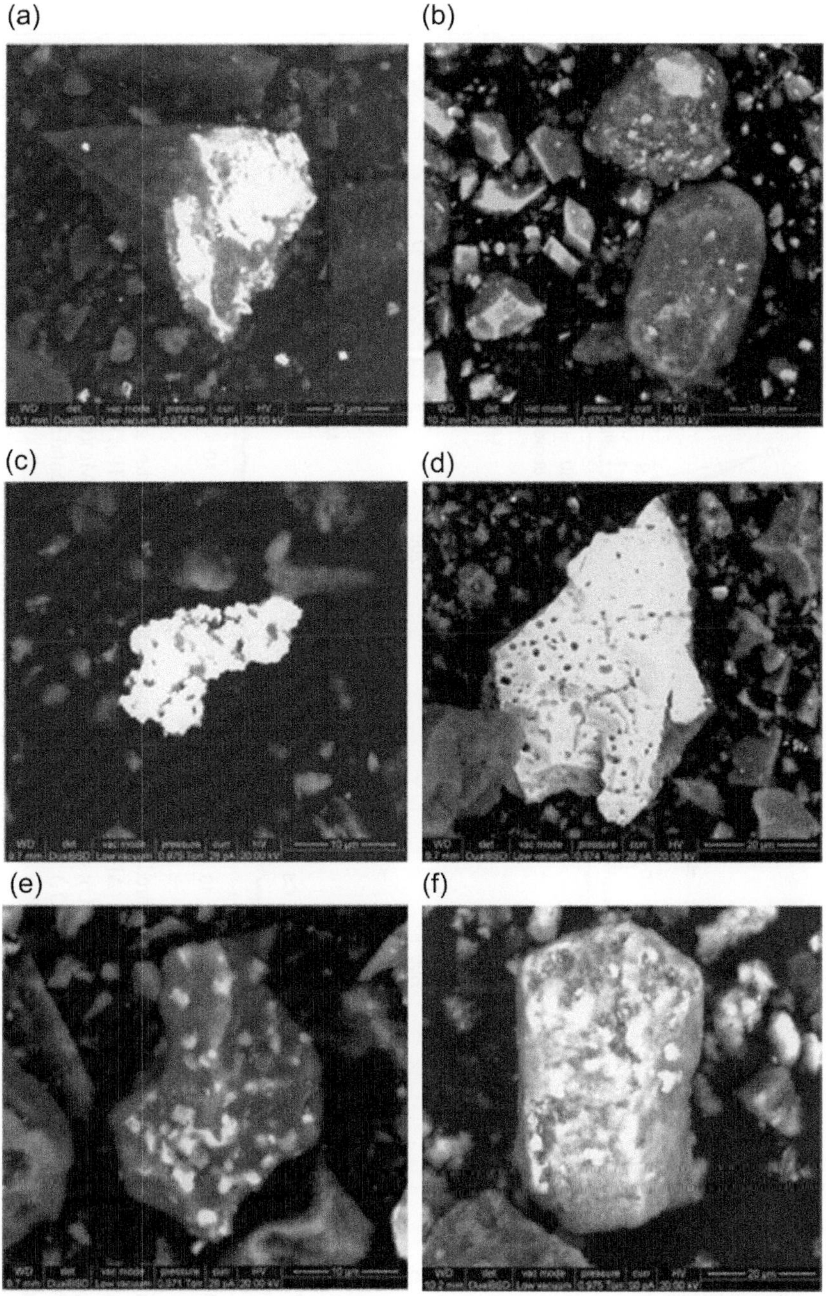

Figure 4.7 Scanning electron images (back-scattered images) of (a) pyrite (bright) in micro-cleats in coal, (b) clay particles coated with smaller bright pyrites, (c) irregular Cu oxide identified in the quartzite sample, (d) fired clay with a typical porous anthropogenic texture, (e) Fe oxides (bright hematites) attached to a larger anhedral silicate particle (mullite), (f) large clay particle (kaolinite) covered with smaller sulfur (bright) particles in the coal-fire gas mineral sample.
From Querol, et al. Environmental characterization of burnt coal gangue banks at Yangquan, Shanxi Province, China. Int J Coal Geol 2008;75:93–104.

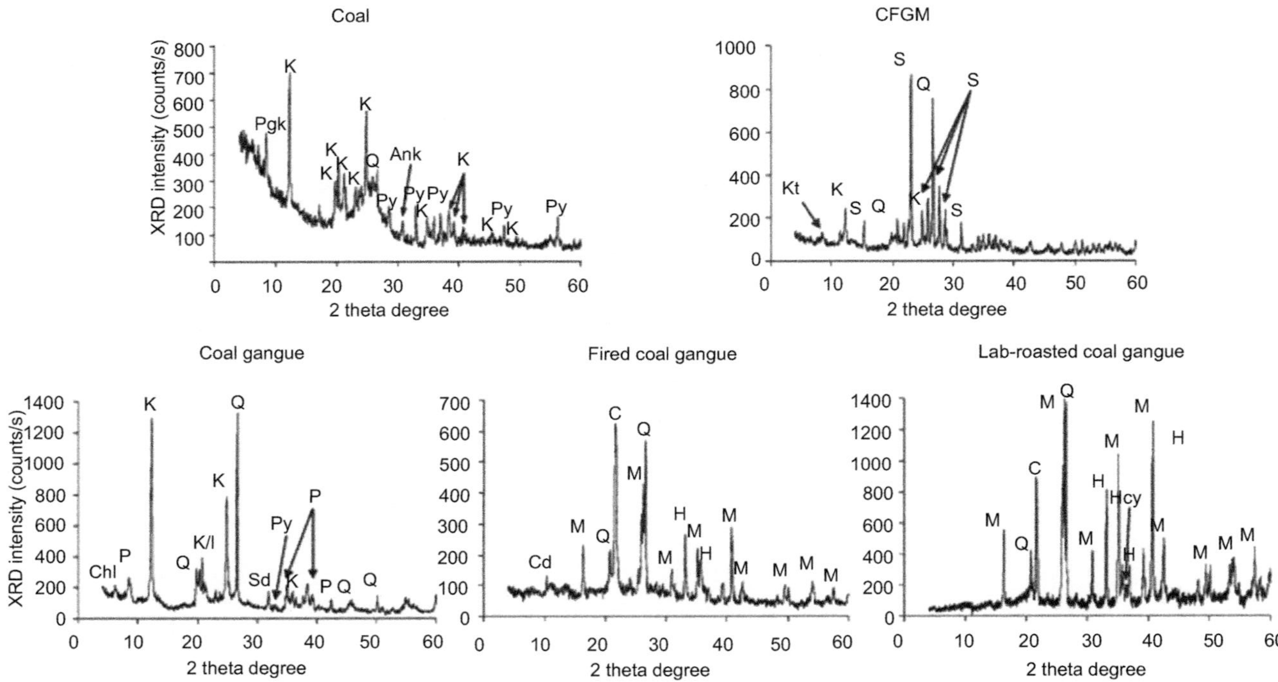

Figure 4.8 X-ray diffraction patterns for coal, coal gangue, fired coal gangue, condensate material (coal-fire gas minerals [CFGM]) and for the laboratory-roasted coal gangue. Q: quartz (SiO_2); K: kaolinite ($Al_2Si_2O_5(OH)_4$); I: illite ($KAl_2Si_3AlO_{10}(OH)_2$); Chl: clinochlore ($(Mg,Fe)_5Al(Si_3Al)O_{10}(OH)_8$); P: palygorskite ($(Mg,Al)_5(Si, Al)_8O_{20}(OH)_2 \cdot 8H_2O$); py: pyrite ($FeS_2$); Ank: ankerite ($Ca(Fe,Mg)(CO_3)_2$); Sd: siderite ($FeCO_3$); C: cristobalite ($SiO_2$); M: mullite ($Al_6Si_2O_{13}$); H: hematite ($Fe_2O_3$); Cd: cordierite ($Mg_2Al_4Si_5O_{18}$); S: elemental sulfur ($S_0$-alpha); Kt: koktaite ($(NH_4)_2Ca(SO_4)_2H_2O$); Hcy: hercynite ($FeAl_2O_4$).
From Querol, et al. Environmental characterization of burnt coal gangue banks at Yangquan, Shanxi Province, China. Int J Coal Geol 2008;75:93–104.

in the coal sample fall in similar concentration ranges of Shanxi coal with the exception of Pb (94 vs. <DL−88 mg Pb/kg) and Se (23 vs. <DL−13 mg Se/kg). Therefore the sample can be regarded as representative of the Shanxi coal, and is probably enriched with clausthatite (PbSe), galena, or any other Pb- and Se-bearing species.

The volatilization potential of major pollutants and trace contaminants during spontaneous combustion comes from the on-site coal gangue/gangue concentration ratio. The very low ratios obtained for Cd, Hg, Pb, Sn (<0.1), Ge (0.6), and Se (0.7) provide evidence that these elements partially volatilize during autoignition. Hg, Pb, Se, and Ge condense on the outside of the chimney of the fire vent. Attention should be paid to the similar content of S (0.8%) and As (10 mg/kg) in coal gangue and fired gangue. This can be attributed to the fact that after combustion, S and As from the vents of the deeper self-igniting flame may condense in the burning coal gangue. As described by Querol et al. [36], the appearance of free CaO in coal ash is decisively beneficial to the formation of calcium sulfate and arsenide. This may explain the presence of anhydrite ($CaSO_4$) in the fired gangue but not in the coal gangue, in which S may be present in different modes of occurrence such as pyrite (identified by SEM) and probably minor amounts of gypsum.

The low-coal gangue/laboratory-calcined coal gangue ratios (<0.5) revealed volatilization of S, As, Ga, Ge, Pb, Sn, and Zn. These results are in line with those obtained by Zhao et al. [35]. It should be pointed out that the volatile potential of S and As in calcined samples is higher than that obtained using field-fired coal gangue because of the postcombustion reactions between fired coal gangue and combustion gases in field conditions.

Coal gangue and most of its constituents gave rise to neutral (pH values 6.8−7.4) and low-conductive leachates (mostly <250 μs/cm), which indicates a low overall solubility with respect to fired gangue or the coal-fire gas minerals (CFGMs) (>2300 μs/cm, Table 4.1). Elements occurring in salts such as Ca, Cl^-, Na, and SO_4^{2-} showed the highest releases (most values in the range of 40−500 mg/kg), followed by minor amounts of K, Mg, F^-, NO_3^-, Al, Si, Sr, and NH_4^+, whereas the remaining elements showed low releases (<0.5 mg/kg). The leachable concentrations of a number of elements of environmental concern were close to or below the detection limit, which suggests that these are highly immobile, at least under neutral conditions. Therefore no environmental consequences from the leaching of coal gangue are expected.

Sulfide mineralization produces a slightly acidic leachate (pH 5.7). S and Ca may act as secondary sulfates (possibly gypsum) associated with pyrite weathering, showing the highest release, followed by Mg and Fe. Some potentially harmful elements, such as Hg, Pb, As, Ni, or Se, are very unstable, although they are abundant in content, due to their relatively low leaching concentration. Their presence in the sulfide lattice may be responsible for this low mobility. Release of the remaining trace contaminants is very low, indicating that these contaminants are also highly retained.

In contrast to coal gangue, results showed that calcium sulfate (presumably gypsum) is the prevalent soluble species, whereas halite content is markedly lower as evidenced by the lower releases of Na and Cl^-. The relatively high releases of NO_3^- and the different leaching behavior of very few trace pollutants in coal,

Table 4.1 Leachable concentrations according to EN 12457-2 of major and trace elements (mg/kg) and rare earth elements (REE) in coal gangue and different lithologies sampled.

	Coal	Coal gangue	Fired gangue	CFGMs	Sulfide mineralizations	Clay	Quartzite	Sandstone
pH	6.8	7.3	6.7	3.9	5.7	6.9	7.4	6.8
Conductivity (μs/cm)	796	201	2305	5348	1610	70	241	67
mg/kg								
Al	<2	6	17	176	<10	12	27	6
As	0.01	0.1	0.1	0.1	0.03	0.04	0.3	<0.01
B	0.1	0.3	0.3	0.3	0.5	0.4	0.2	0.4
Ba	0.7	0.06	0.8	0.4	0.4	0.1	0.5	0.05
Be	<0.01	<0.01	<0.01	0.1	<0.01	<0.01	<0.01	<0.01
Bi	<0.01	<0.01	<0.01	<0.01	<0.01	<0.01	<0.01	<0.01
Ca	1084	3	5096	3014	2553	5	5	36
Cd	<0.01	<0.01	<0.01	0.05	0.02	<0.01	<0.01	<0.01
Cl^-	44	39	31	1654	66	31	<1	51
Co	0.1	<0.01	<0.01	3	0.01	<0.01	<0.01	<0.01
Cr	<0.01	<0.01	<0.01	0.02	<0.01	<0.01	<0.01	<0.01
Cs	<0.01	<0.01	<0.01	0.05	<0.01	<0.01	<0.01	<0.01
Cu	0.02	0.01	0.07	0.1	0.05	0.01	0.05	0.01
F^-	<1	11	<1	579	<1	8	<1	6
Fe	<0.2	0.2	<1	283	502	1	3	4
Ga	<0.01	0.01	0.05	0.01	<0.01	<0.01	0.02	<0.01
Ge	<0.01	<0.01	<0.01	0.01	0.01	<0.01	<0.01	<0.01

Hf	<0.01	<0.01	<0.01	0.01	<0.01	<0.01	<0.01	<0.01
Hg	0.002	0.002	0.001	0.001	<0.001	0.001	0.001	0.001
K	12	13	243	118	<25	<10	38	5
Li	0.3	0.04	19	3	2	0.01	0.01	0.02
Mg	309	1	658	183	678	3	3	9
Mn	17	<0.01	5	39	41	0.01	0.1	0.02
Mo	<0.01	0.3	0.8	<0.01	<0.01	0.1	0.1	0.04
Na	139	474	241	147	110	112	477	137
Nb	<0.01	<0.01	<0.01	<0.01	<0.01	<0.01	<0.01	<0.01
NH^{4+}	10	4	8	5829	48	4	3	<1
Ni	0.3	<0.01	0.1	3	1	<0.01	<0.01	<0.01
NO_3	40	<1	3	<1	23	1	<1	17
P	<1	<1	<5	<5	<5	<2	<2	<2
Pb	<0.01	<0.01	<0.01	0.01	<0.01	<0.01	0.01	<0.01
Rb	0.01	0.01	0.3	1	0.1	<0.01	0.02	<0.01
SO_4^{2-}	3521	290	13,832	23,511	9402	119	87	70
Sb	<0.01	0.02	<0.01	<0.01	0.03	<0.01	<0.01	<0.01
Sc	<0.01	0.02	<0.01	0.1	0.04	0.02	0.03	0.02
Se	0.2	0.2	0.04	0.1	0.1	0.04	<0.01	<0.01
Si	21	41	8	185	109	57	58	59
Sn	<0.01	<0.01	<0.01	<0.01	<0.01	<0.01	<0.01	<0.01
Sr	18	0.1	44	14	65	0.03	0.3	0.2
Ta	<0.01	<0.01	<0.01	<0.01	<0.01	<0.01	<0.01	<0.01

Continued

Table 4.1 Continued

	Coal	Coal gangue	Fired gangue	CFGMs	Sulfide mineralizations	Clay	Quartzite	Sandstone
Th	<0.01	<0.01	<0.01	<0.01	<0.01	<0.01	<0.01	<0.01
Ti	<0.01	0.02	0.03	0.1	0.04	0.02	0.02	0.02
Tl	<0.01	<0.01	<0.01	<0.01	<0.01	<0.01	<0.01	<0.01
U	<0.01	<0.01	<0.01	0.02	<0.01	<0.01	<0.01	<0.01
V	<0.01	0.1	0.1	0.1	0.02	0.1	0.1	0.01
W	<0.01	<0.01	0.05	<0.01	<0.01	<0.01	<0.01	<0.01
Y	<0.01	<0.01	<0.01	1	0.02	<0.01	<0.01	<0.01
Zn	0.2	0.01	<0.01	13	0.2	<0.01	0.02	0.02
Zr	<0.01	<0.01	<0.01	<0.01	<0.01	<0.01	<0.01	<0.01
Y	<0.01	<0.01	<0.01	1	0.02	<0.01	<0.01	<0.01
REE								
La	<0.01	<0.01	<0.01	0.5	0.01	<0.01	<0.01	<0.01
Ce	<0.01	<0.01	<0.01	2	0.02	<0.01	<0.01	<0.01
Pr	<0.01	<0.01	<0.01	0.2	<0.01	<0.01	<0.01	<0.01
Nd	<0.01	<0.01	<0.01	1	<0.01	<0.01	<0.01	<0.01
Sm	<0.01	<0.01	<0.01	0.3	<0.01	<0.01	<0.01	<0.01
Gd	<0.01	<0.01	<0.01	0.3	<0.01	<0.01	<0.01	<0.01
Dy	<0.01	<0.01	<0.01	0.2	<0.01	<0.01	<0.01	<0.01
Er	<0.01	<0.01	<0.01	0.1	<0.01	<0.01	<0.01	<0.01
Yb	<0.01	<0.01	<0.01	0.1	<0.01	<0.01	<0.01	<0.01

CFGMs, coal-fire gas minerals.
From Querol, et al. Environmental characterization of burnt coal gangue banks at Yangquan, Shanxi Province, China. Int J Coal Geol 2008;75:93−104.

i.e., Co, Ni, and Zn, should be pointed out. This behavior entails slightly higher leachable concentrations than in the coal gangue despite lower bulk contents. This increased mobility may be attributed to a weak binding to the organic phase of the coal with the result that these elements are more available for leaching. Nevertheless, values are not very significant, especially when compared with those observed in CFMG condensates.

Although the S content in coal gangue and burning coal gangue is very similar, the manner of occurrence is significantly different due to the large release of S. In coal gangue, the leaching concentration of Ca is also significantly higher than that of coal gangue. These observations are consistent with the appearance of anhydrite in the fired coal gangue sample. The higher leachable concentrations of Ba, K, Li, Mg, and Sr, combined with higher overall levels, may be related to the formation of other sulfates during spontaneous combustion. In contrast, the release of Cl and F in the burned coal gangue is reduced, indicating that these elements are lost during the combustion process. The release of most trace contaminants from coal gangue did not change significantly after spontaneous combustion and in many cases remained at a low level, approaching the detection limit.

The CFGM condensate produces a significantly acidic leachate with high electrical conductivity (>5 ms/cm), which may be primarily due to the large release of readily leached sulfate. In addition to the natural sulfur and trace amounts of koktaite determined in the XRD spectrum, ion balance closure provides us with clear evidence of the presence of ammonium and calcium sulfate. This results in highly leachable concentrations of SO_4^{2-} (23,511 mg/kg), NH_4^+ (5829 mg/kg), and Ca (3014 mg/kg), which should be regarded as a matter of environmental concern since these mobile species could cause soil, surface, and groundwater pollution, especially the high levels of sulfate. Ammonium originates from the transformation of the N occurring during gasification of coal under the high temperature reached in combustion/gasification with low O_2 fugacity. Even if the N levels in coal are low, the NH_3 formed will condensate as ammonium sulfate by interaction with SO_2. Other elements occurring in soluble salts, i.e., Cl^- (1654 mg/kg), F^- (579 mg/kg), Na, Mg, K, and Fe are largely released. Environmental consequences may result from very few trace pollutants such as Co (3 mg/kg), Ni (3 mg/kg), and particularly Zn (12 mg/kg). Although Be, Cd, Cs, Cu, REEs, Sc, Se, U, and Y showed relatively low releases, they could be potentially harmful. It is worth noting that the these elements are highly immobile in coal, coal gangue, fired coal gangue, and typically in conventional coal combustion by-products and natural rocks. This unusual leaching behavior can be attributed to (1) a more soluble form and mode of occurrence of these elements in the condensate and (2) a distinct acidic pH of the leaching material, which enhances the solubility of some trace contaminants. Given the environmental significance of these factors, it is necessary to further investigate the possible reasons for investigating increased liquidity. In contrast, despite the low pH, other potentially harmful elements such as Hg, Mo, Pb, Sb, Sn, or Tl remain highly retained.

The column leaching test allowed Querol et al. to study the release patterns of elements in CFGM condensates (shown in Table 4.2). Important releases of some major elements were produced along the percolated water volume, which is equivalent

Table 4.2 Leached concentrations (mg/kg) of major and trace elements from coal-fire gas mineral condensates as derived from column leaching studies.

	Leached fractions						
mg/kg	**1**	**2**	**3**	**4**	**5**	**6**	**Total leached amount**
Al	20	5	2	<1.5	<1.5	<1.5	<32
As	0.01	<0.01	<0.01	<0.01	<0.01	<0.01	0.01
B	<0.01	<0.01	<0.01	<0.01	<0.01	<0.01	<0.01
Ba	0.03	0.04	0.05	0.05	0.06	0.06	0.3
Be	0.02	0.01	<0.01	<0.01	<0.01	<0.01	0.03
Bi	<0.01	<0.01	<0.01	<0.01	<0.01	<0.01	<0.01
Ca	549	261	155	114	86	70	1235
Cd	<0.01	<0.01	<0.01	<0.01	<0.01	<0.01	<0.01
Co	0.4	0.1	0.04	0.03	0.02	0.02	0.6
Cr	0.06	0.02	0.01	<0.01	<0.01	<0.01	0.1
Cs	0.01	<0.01	<0.01	<0.01	<0.01	<0.01	0.01
Cu	0.2	0.05	0.02	0.01	0.01	0.01	0.3
Fe	91	25	10	7	5	4	140
Ga	<0.01	<0.01	<0.01	<0.01	<0.01	<0.01	<0.01
Ge	<0.01	<0.01	<0.01	<0.01	<0.01	<0.01	<0.01
Hf	<0.01	<0.01	<0.01	<0.01	<0.01	<0.01	<0.01
K	4	<3	<3	<3	<3	<3	<19
Li	0.3	0.1	0.05	0.04	0.03	0.02	0.5
Mg	28	8	4	2	2	1	45
Mn	9	3	1	0.8	0.5	0.4	14
Mo	<0.01	<0.01	<0.01	<0.01	<0.01	<0.01	<0.01
Na	16	5	2	1	1	1	27
Nb	<0.01	<0.01	<0.01	<0.01	<0.01	<0.01	<0.01
Ni	0.4	0.1	0.05	0.03	0.02	0.02	0.7
P	<1	<0.75	<0.75	<0.75	<0.75	<0.75	<5
Pb	0.01	<0.01	<0.01	<0.01	<0.01	<0.01	0.01
Rb	0.06	0.02	0.01	0.01	0.01	<0.01	0.1
SO_4^{2-}	5787	2016	1020	711	519	411	10,470
Sb	<0.01	<0.01	<0.01	<0.01	<0.01	<0.01	<0.01

Table 4.2 Continued

	Leached fractions						
mg/kg	**1**	**2**	**3**	**4**	**5**	**6**	**Total leached amount**
Sc	0.09	0.02	0.01	<0.01	<0.01	<0.01	0.1
Se	0.02	0.01	<0.01	<0.01	<0.01	<0.01	0.03
Si	1	0.7	<0.75	<0.75	<0.75	<0.75	<5
Sn	<0.01	<0.01	<0.01	<0.01	<0.01	<0.01	<0.01
Sr	2	1	0.5	0.4	0.3	0.3	4
Ta	<0.01	<0.01	<0.01	<0.01	<0.01	<0.01	<0.01
Th	0.07	<0.01	<0.01	<0.01	<0.01	<0.01	0.07
Ti	<0.01	<0.01	<0.01	<0.01	<0.01	<0.01	<0.01
Tl	<0.01	<0.01	<0.01	<0.01	<0.01	<0.01	<0.01
Tm	<0.01	<0.01	<0.01	<0.01	<0.01	<0.01	<0.01
U	0.03	<0.01	<0.01	<0.01	<0.01	<0.01	0.03
V	0.01	<0.01	<0.01	<0.01	<0.01	<0.01	0.01
W	<0.01	<0.01	<0.01	<0.01	<0.01	<0.01	<0.01
Y	0.2	0.05	0.02	0.01	<0.01	<0.01	0.3
Yb	0.02	<0.01	<0.01	<0.01	<0.01	<0.01	0.02
Zn	2	0.5	0.2	0.1	0.1	0.1	3
Zr	<0.01	<0.01	<0.01	<0.01	<0.01	<0.01	<0.01
La	0.07	0.02	0.01	<0.01	<0.01	<0.01	0.09
Ce	0.4	0.08	0.03	0.01	0.01	0.01	0.5
Pr	0.05	0.01	<0.01	<0.01	<0.01	<0.01	0.07
Dy	0.06	0.01	<0.01	<0.01	<0.01	<0.01	0.07
Nd	0.3	0.06	0.02	0.01	0.01	<0.01	0.4
Sm	0.08	0.02	0.01	<0.01	<0.01	<0.01	0.1
Gd	0.08	0.02	0.01	<0.01	<0.01	<0.01	0.1
Dy	0.06	0.01	<0.01	<0.01	<0.01	<0.01	0.07
Er	0.02	<0.01	<0.01	<0.01	<0.01	<0.01	0.02
Yb	0.02	<0.01	<0.01	<0.01	<0.01	<0.01	0.02

From Querol, et al. Environmental characterization of burnt coal gangue banks at Yangquan, Shanxi Province, China. Int J Coal Geol 2008;75:93−104.

to an annual rainfall of 600 mm. These high releases corresponded to SO_4^{2-} (10,470 mg/kg), Ca (1235 mg/kg), Fe (140 mg/kg), Mg (45 mg/kg), Al (<32 mg/kg), Na (27 mg/kg), and K (<19 mg/kg). During the first 100 mm of rainfall, most of these quantities (about 45%–60%) were released during the first stage of leaching. Afterward, leaching is gradually reduced to approximately 80% of the annual leachable amount (equivalent to 300 mm of rainfall). Attenuated element release from the CFGM condensate is then produced. Regarding environmental factors, Zn (3 mg/kg), Ni (0.7 mg/kg), Co (0.6 mg/kg), and Cu (0.3 mg/kg) are those in which higher amounts are leached. The leaching tendency of TEs is similar to the leaching tendency of major elements. Therefore during the first 100 mm of rainfall, 60%–70% of the total leaching amount is discharged, and in the first half of the rainy season, about 90% is discharged. Other TEs, As and Se, although leached at lower levels (0.01 and 0.03 mg/kg, respectively), should also be noted because their toxicity thresholds are lower for humans and other organisms. Based on these results, it can be said that once pollutants are produced, or at least before the start of the rainy season, measures should be taken to prevent the dispersion of pollutants from the CFGM condensate to other parts of the ecosystem, because some of the current pollutants show very high, mainly It is leached from the early stages of water penetration.

With regard to leaching potential and subsequent soil and water pollution, CFGM condensate constitutes a major environmental problem for spontaneous combustion to be addressed. This will require measures to prevent spontaneous combustion and protect the environment. Therefore it is recommended to quickly cover the soil on coal gangue to avoid oxidation of the organic matrix and sulfide phase to reduce the risk of spontaneous combustion. Recommended remedies include the addition of alkaline substances such as limestone, magnesite, or coal fly ash. This addition neutralizes the pH and reduces leaching emissions. However, further research is needed to optimize the dose and test for possible leaching of elements from fly ash.

In addition, it has been noted that most fires occur on the slopes of coal gangue gobs rather than on flat terraces. The first reason is that the slope is exposed to the wind and the air is forced to seep into the oxygen in the goaf, thereby enhancing the oxidation of the outer layer of the goaf. Therefore avoiding the construction of goafs in windy areas may reduce the possibility of spontaneous combustion. Moreover, the stratification of the coal gangue with the low permeability material in the goaf and/or the compaction of the coal gangue (to reduce the porosity) will minimize the penetration of air and water and thus minimize combustion. It has also been observed that most fires begin at gob slopes at depths of 1–2 m and in most cases gas vents are built into the outer unburned coal gangue layer. The fact that the outside of the goaf slope is not ignited may be due to the cooling effect of the wind, which prevents heat build-up through various cycles of oxidation of coal and sulfide.

Moreover, Kribek et al. compared the TE distributions in the burnt and unburnt zones at the Novator mine, Czech Republic (shown in Fig. 4.9) [37]. The results of Al or Ti standardization are very similar and demonstrate that the combustion portion of the waste heap is rich in many elements, especially total sulfur, Cu, As, Se, Mo, Ag, Cd, Sn, Sb, Tl, Pb, Bi, and U. Nonparametric Kruskal–Wallis

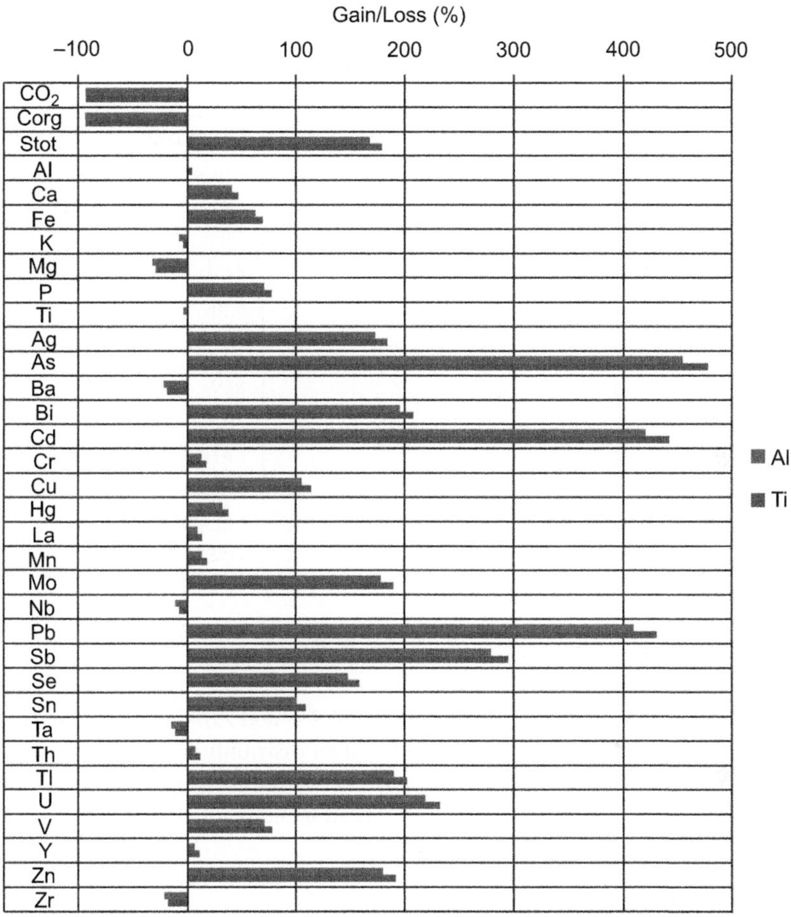

Figure 4.9 Changes (in rel. %) in the contents of elements (median values) in the burnt zone in relation to their median values in the unburnt zone. The calculation is based on Al and Ti as immobile elements.
From Kribek, et al. Trace element geochemistry of self-burning and weathering of a mineralized coal waste dump: the Novátor mine, Czech Republic. Int J Coal Geol 2017;173:158−75.

one-way analysis of variance (KW ANOVA) and nonparametric Mann−Whitney (MW) U tests study significant differences between chemical elements and Al ratios. The results of MW U and KW ANOVA testing organic carbon, Fe, P, Ag, As, Bi, Cd, Mo, Nb, Pb, Sb, Ta, Te, and Zn/Al ratios were very significant ($P < .0001$). These chemical element/Al ratios in the unburned portion and the burned portion of the stack are significantly different from each other. For other variables, the element/Al ratio results are slightly different in the two tests. The MW test is a more powerful (or sensitive) nonparametric alternative to the KW test and, in some cases, can provide a greater ability to reject the null hypothesis than the KW test. Therefore the MW P value is usually higher than the KW P value (Table 4.3). Therefore the

Table 4.3 Emissions estimates of sulfur, arsenic, mercury, and fluoride from combustion of coal gangue and coal in China.

	Coal		
Coal gangue	**Raw coal**	**Clean coal**	
Production in 2013 (million tons)	750	1,472	2,208
Sulfur content (wt%)	1.01	1.34	0.5−0.8
Sulfur emission (million tons/a)	1.13	4.93	2.76−4.42
Sulfur dioxide emission (million tons/a)	2.26	9.86	5.52−8.84
Arsenic emission (ton/a)	246	195 236	
Mercury emission (ton/a)	45	202 257 172	
Fluoride emission (ton/a)	63,298	66,398 162,161	

a, annually.
From Wang, et al. Estimate of sulfur, arsenic, mercury, fluorine emissions due to spontaneous combustion of coal gangue: an important part of Chinese emission inventories. Environ Pollut 2016;209:107−13.

MW *P* values of the total sulfur, K, Mg, Ba, U, V, Cr, and Zr/Al ratios in the range of 0.001−0.01 may reflect the difference in the data distribution of the unburned and combusted portions of the waste heap. In other words, these differences come not only from combustion but also from the inhomogeneity of the original unburned matrix.

Wang et al. estimated the emission of As, Hg, and F based on the following equation and the TE content in the raw coals in Chinese coal gangue [38]. The atmospheric emission of sulfur and potentially toxic elements from coal gangue combustion (E, ton) can be calculated as follows:

$$E = A_{coalgangue} \times [C_{coalgangue} - C_{fired} \times C_{ash}]$$

Among them, coal gangue and C_{fired} are the concentration of sulfur and potentially toxic elements in coal gangue and cinder (gangue), and C_{ash} is the ash content of coal gangue. Meteorite represents the total amount of coal gangue. In 2013, the emissions of sulfur, arsenic, mercury, and fluorine from coal gangue spontaneous combustion in China were about 1.13 million tons, 246 tons, 45 tons, and 63,298 tons (complete combustion conditions) (Table 4.3). The estimate is based on an autoignition ratio of one-third and an ash content of 70% (based on the average of our results), with a coal gangue production of 750 million tons in 2013. To verify the credibility of this estimate, they chose emissions factors of SO_2 (9.84 mg/g), hydrogen sulfide (0.61 mg/g), As (1.08 mg/kg), Hg (0.40 mg/kg), and F (271 mg/kg) during spontaneous combustion of coal gangue. The emission factor-based verification results are close to the results obtained, showing S (131 million tons), As (270 tons), and F (67, 750 tons), but the mercury emissions from this work are lower than the verification results (100 tons).

Hong et al. studied the emission of one of the most volatile TEs (mercury) during self-firing in the Wuda coalfield, Inner Mongolia, China [39]. In detail, Fig. 4.10(a) clearly shows the concentration of gaseous mercury in the boreholes of the five fire zones. The red circle represents the high mercury content (1000–10,000 ng/m^3) in the borehole. The temperature of the gas in these holes at a distance of 3 m from the orifice (T_a) can reach about 100°C, and it is observed that water vapor rises from the openings of some of the holes. This may indicate a relatively strong coal seam smoldering below the area. The red diamond shows the highest mercury concentration (>10,000 ng/m^3), whose value T_a is about 180°C. The walls and lids of these holes are severely corroded, indicating the release of corrosive materials during combustion of the coal seam. Blue diamonds mean mercury concentrations of 200–1000 ng/m^3, and T_a is about 50°C. Near-surface concentration of mercury in the atmosphere surrounding a borehole with a concentration represented by a blue circle (<200 ng/m^3) of the bore (92 ng/m^3) is, however, similar or slightly higher, indicating a possible fire. In general, the range of T_a is 26.3–192°C, and if the logarithm of mercury concentration is used, then a significant positive correlation between the logarithm of mercury concentration and T_a is observed (Fig. 4.10(b), $R = 0.95$). Combined with the mercury concentration in most boreholes, high temperatures may also indicate to some extent that the underlying coal seam is smoldering.

4.4 Trace element distribution in coal fire products

Pone et al. investigated TE distribution in coal-fire by-products in Witbank and Sasolburg coalfields [40]. As shown in Fig. 4.11, XRD and SEM analyses established that condensation by-products were mostly sulfur and sulfur-bearing minerals. Salammoniac (NH_4Cl) was the dominant mineral phase found as CFGMs. Heavy metals such as arsenic were detected. A grain of pure mercury was also found and an aggregate of zinc, copper, lead, iron, and germanium was detected. XRD diffractograms confirmed the presence of mascagnite ($(NH_4)_2SO_4$), illite ($(Al,Si)_4O_{10}[(OH)_2,H_2O]$), letovicite ($(NH_4)_3H(SO_4)_2$), and phlogopite ($KMg_3(AlSi_3)O_{10}(F,OH)_2$). The compounds titanium dioxide (TiO_2), barite ($BaSO_4$), iron sulfate ($FeSO_4$), gypsum ($CaSO_4\cdot 2H_2O$), and silicate were also identified. An unknown and unclassified CFGM composed of sulfur, nitrogen, and chlorine is reported. Other minerals detected during SEM analysis were composed of chlorine and bromine, chlorine and iodine, sulfur and nitrogen, sulfur, nitrogen, and oxygen, and sulfur and chlorine. The minerals detected at the Witbank and Sasolburg coalfields are similar to minerals and compounds derived from burning anthracite seams in Pennsylvania. The minerals are interpreted as having formed during gas exhalation and condensation, and by the interaction of gases and CFGMs with surface water, atmospheric gases, and surrounding rocks.

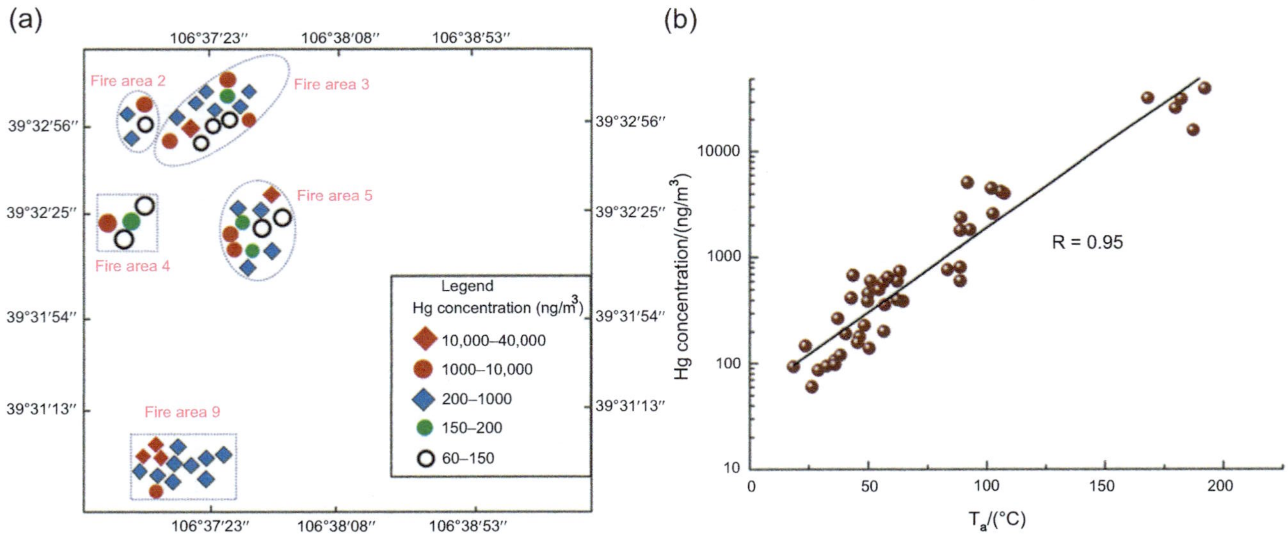

Figure 4.10 (a) Distribution of gaseous mercury in the exhaust gas of the drill holes in different fire areas and (b) relationship between drill hole mercury concentration and T_a.
From Hong, et al. Mercury emissions from dynamic monitoring holes of underground coal fires in the Wuda Coalfield, Inner Mongolia, China. Int J Coal Geol 2017;181:78–86.

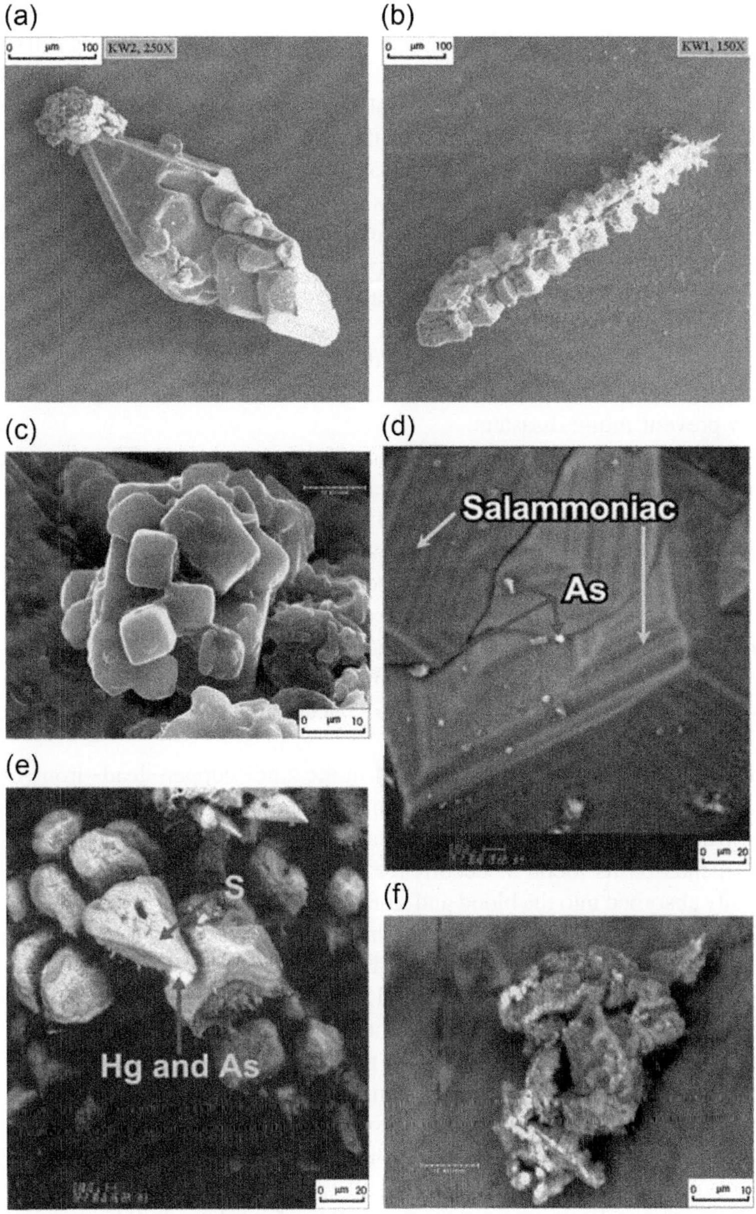

Figure 4.11 Scanning electron photomicrographs of coal fire minerals. (a) Hexagonal sulfur crystal from Witbank coalfield. (b) Skeletal to dendritic sulfur crystal from Witbank coalfield. (c) Cubic salammoniac crystal from Sasolburg coalfield. (d) Salammoniac-bearing arsenic compound from Witbank coalfield. (e) Sulfur bearing mercury–arsenic compound from Witbank coalfield. (f) Skeletal aggregate of sulfur, lead, zinc, copper, iron, and germanium from Sasolburg coalfield.
From Pone, et al. The spontaneous combustion of coal and its by-products in the Witbank and Sasolburg coalfields of South Africa. Int J Coal Geol 2007;72:124–40.

Arsenic was detected in both coalfields. Arsenic and related volatile compounds are mobilized during coal combustion and enter the environment as gas or as solid phases depending on the burning conditions. Thermodynamic modeling by Bel'kova et al. showed that during the combustion of coal at low temperatures, arsenic is released as solid phases. Yan et al. (2001) stated that the dominant solid arsenate released during low-temperature combustion of coal is thallium arsenate ($TlAsO_4$) [41,42]. When the combustion temperature exceeds ~600°C (873K), arsenic enters the gas phase in the form of As_2, As_3, and/or As_s. Since the temperature measured in this study is generally higher than 600°C, it can be explained that most of the arsenic in the coal-fire vents of the Witbank and Sasolburg coalfields are in the gas phase, possibly AsO and As_2O_3, because they are the dominant gas species at around 1000°C. Although workers in Witbank and Sasolburg coalfields have not reported symptoms of arsenic poisoning, an assessment of the environmental impact of arsenic compounds in these areas may prevent future disasters.

Mercury, detected as a CFGM by-product, is the most volatile TE released during the combustion of coal. Furimsky stated that between ~1527 and 927°C, more than 99% of mercury present in coal is vaporized in the form of elemental mercury [43]. He stated that mercury compounds, which appeared in the solid phase at about 327°C, consisted of $HgCl_2$ and $HgSO_4$. The detection of a pure grain of mercury during CFGM analyses provides limited evidence that mercury is also emitted into the atmosphere as a solid-phase pollutant. Although there is no direct evidence of health problems caused by mercury from coal combustion, the proportion of anthropogenic mercury in the environment is a matter of concern for the Witbank and Sasolburg area, given its toxicity even at low doses.

Lead detected as a by-product of CFGM in the zinc−copper−lead−iron−bismuth combination is usually associated with minerals such as galena (PbS), lead sulfate (PbSe), pyrite, aluminosilicates, and carbonates. The release of lead during coal combustion depends on its mode of occurrence and combustion conditions. In any case, lead is easily absorbed into the blood and stored in the bone where it returns to the blood during aging. If exposed enough, lead can have serious and chronic health effects.

Zhao et al. explored the emissions of TEs in spontaneous coal mine fires in coal mines in Shanxi Province, China (Fig. 4.12). The burning part is light yellow and red. A gob heap shows a fresh, unburned portion on the left, dark gray to black. There are many burning sites (yellow) on the gob pile on the right side of the figure. The air around the goaf is dusty and full of the pungent odor of sulfur compounds.

Ranges and arithmetic means of 20 TEs in solid waste samples of gob piles in Shanxi are listed in Table 4.4. The concentrations of TEs show broad variations. As compared with coals from Shanxi [44], solid waste is highly enriched with V, Cr, Mn, Cu, and Th, with enrichment factors of 1.5−5, and is slightly enriched in F, Co, Ni, Sb, Zn, and U. It should be noted that solid waste is lower in As, Se, Mo, Cd, Hg, and Pb than in the average Shanxi coals.

As compared with the crustal abundances [45], the solid waste is highly enriched with Se (by a factor of 20), Tl (up to 12×), and F, As, Mo, Sn, Sb, Hg, Th, and U (2×−5×), and is slightly enriched with Be, Cd, and Pb. In addition, the solid waste has lower contents of V, Cr, Mn, Co, Ni, Cu, and Zn than the earth's crust.

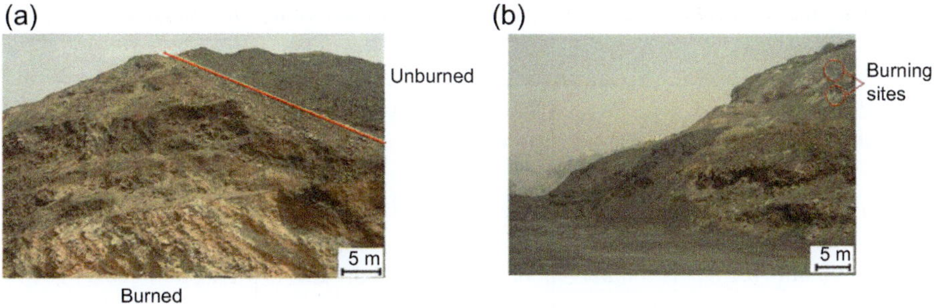

Figure 4.12 (a) The combusted and unburned portions of a gob pile, Shanxi, China, and (b) the burning areas on a gob pile, Shanxi, China.
From Zhao, et al. Trace element emissions from spontaneous combustion of gob piles in coal mines, Shanxi, China. Int J Coal Geol 2008;73:52–62.

The relationship between TEs and chemical composition of the solid waste in gob piles is shown in Fig. 4.13. The ash yield of the solid waste varies from 27.7% to 95.4%, mostly within 65%–90%. The concentration of Pb is positively, while F is negatively, correlated with the ash. Other TEs do not have a clear relationship with the ash. The sulfur content in the solid waste is from 0% to 6.5%, mostly less than 0.5%. Arsenic is positively correlated with sulfur with a correlation coefficient of 0.69 ($n = 70$, $r = 0.69$), indicating that arsenic is associated mainly with sulfide [46]. The Fe_2O_3 content in the ash of gob pile samples varies from 0.2% to 10.5%, and mostly within 0.5%–3.0%. The relationship between four elements (Mn, Zn, As, and Cd) and Fe_2O_3 is shown in Fig. 4.13. Most Fe_2O_3 in the solid waste is derived from pyrite. Arsenic is positively correlated with Fe_2O_3 with a correlation coefficient of 0.51 ($n = 70$, $r = 0.51$), indicating that As is associated with pyrite. The correlation coefficients of Fe_2O_3 versus Mn, Zn, and Cd are 0.63, 0.40, and 0.36, respectively. It appears that pyrite contains significant amounts of these TEs. The Mn content increases with the combined (CaO + MgO) content increasing in the ash. The Mn content is correlated with the (CaO + MgO) content ($r = 0.49$, $n = 70$), indicating an appreciable amount of Mn is sited in carbonate minerals.

A mass balance approach may be used to estimate TE emissions. The emission factor of TEs during spontaneous combustion may be calculated as follows:

$$E = CSW - CASH \times (1 - LOI) \tag{4.1}$$

where E is the emission factor of a TE (mg/kg), C_{SW} is the content of a TE in solid waste (mg/kg), C_{ASH} is the TE content in clinker (ash) (mg/kg), and LOI is the weight loss during spontaneous combustion (%).

Using this equation, emission factors of TEs during spontaneous combustion of solid waste are calculated. The TE emission factors vary widely, as listed in Table 4.5. The amount of solid waste in Yangquan is bigger than in the other eight coal mines, and the emission factor of Hg and Pb is also higher than others, up to 0.67 mg/kg (Hg) and 9.9 mg/kg (Pb). The maximum emission factors of other TEs are located at

Table 4.4 Trace element contents in samples of solid waste and coal from Shanxi, China (ppm).

Element	Solid waste			Shanxi coal		Clarke values
	Range	Average	Number of samples	Range	Average	
As	bdl−27.5	3.6	90	bdl−82.4	4.1	1.8
Be	2.2−2.8	2.5	3			2.1c
Cd	bdl−0.7	0.21	90	bdl−2.5	1.1	0.2
Co	3.4−6.3	5.3	3	0.41−25.0	4.1	25
Cr	bdl−87.3	26.8	90	1.63−53.1	11.3	100
Cu	3.2−93.0	30.0	90	bdl−264.0	18.6	55
F	20.6−912.5	223.0	90	bdl−3600	186	62.5
Hg	bdl−3.4	0.3	90	bdl−2.89	0.40	0.08
Mn	0.8−821.0	175.6	90	bdl−215	38	950
Mo	3.0−3.9	3.3	3	bdl−56.9	5.9	1.5
Ni	7.1−21.5	16.0	8	1.8−46.0	15.6	75
Pb	0.5−108.0	18.2	90	bdl−88.0	24.7	12.5
Sb	0.5−0.6	0.6	8	0.1−11.3	0.5	0.2
Se	bdl−2.9	1.0	90	bdl−12.6	4.9	0.05
Sn	2.8−5.0	3.8	3			2.1c
Th	14.3−22.5	19.2	3	0.5−20.6	7.2	4.2
Tl	0.6−0.7	0.6	3			0.05c
U	4.6−5.9	5.2	3	0.2−49.9	3.6	1.8
V	43.1−96.5	67.0	3	bdl−84.1	18.2	135
Zn	2.0−122.0	57.2	90	bdl−193.0	45.9	70

bdl, below detectable limit.
From Zhao, et al. Trace element emissions from spontaneous combustion of gob piles in coal mines, Shanxi, China. Int J Coal Geol 2008;73:52−62.

Wangzhuang for As, Fenghuangshan for Cr and Se, and Zhenchengdi for F and Cd, respectively. Based on the amount of solid waste spontaneous combustion in different coal mines, the average TE emission factors are calculated as shown in Table 4.5. Spontaneous combustion of gob piles shows higher TE emission factors than coal combustion.

The emission percentages of TEs can also be obtained from simulation tests of spontaneous combustion (shown in Fig. 4.14). The percentages normally depend on the volatility of the elements. Emission percentages of volatile elements Se and Hg are higher than 90%. The emission percentage for Hg during spontaneous combustion is higher than combustion in coal-fired boilers in China [47]. Fluorine emission percentage is about 80% and other TEs are less than 30%.

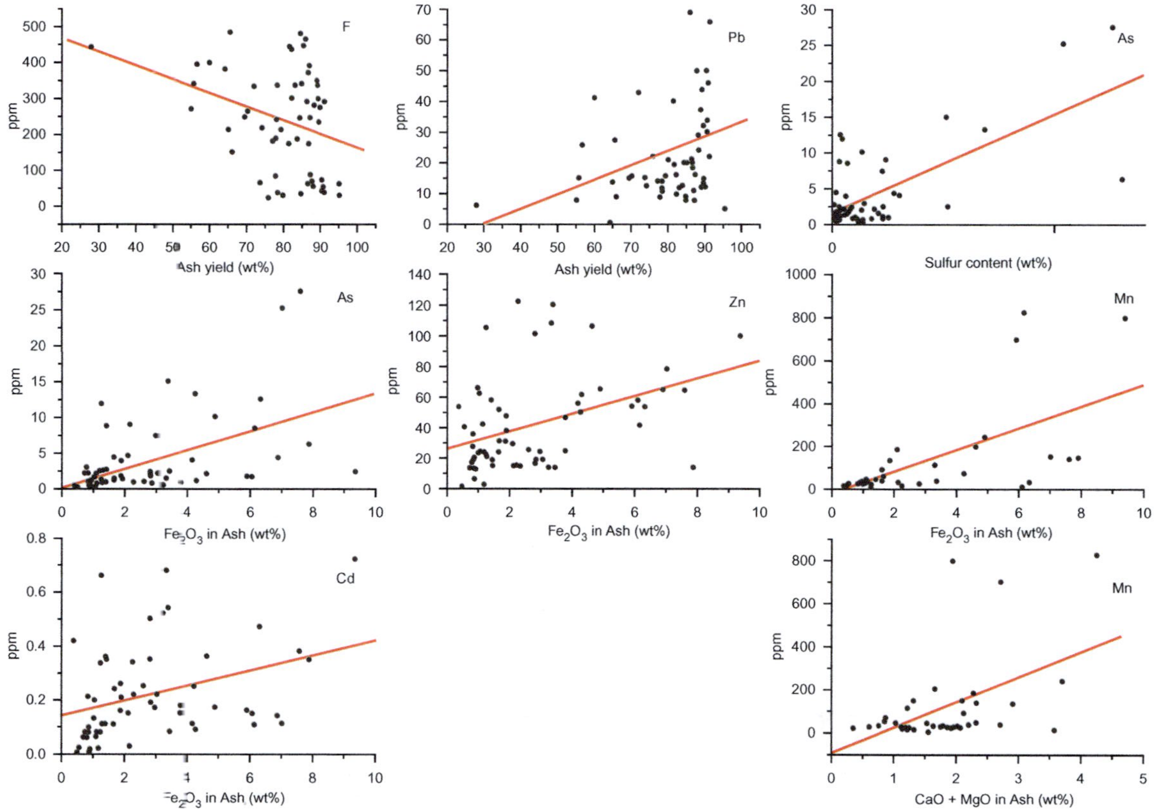

Figure 4.13 Correlation of trace elements with chemical composition of solid waste samples of gob piles.
From Zhao, et al. Trace element emissions from spontaneous combustion of gob piles in coal mines, Shanxi, China. Int J Coal Geol 2008;73:52–62.

Table 4.5 Emission factors of trace elements during simulated spontaneous combustion of solid waste (in mg/kg).

Location	Amount of solid wastes (10^4 t)	F	Hg	Pb	As	Cd	Cr	Se
Yangquan	16.1	30.1	0.67	9.9	0.05	0.01	6.5	0.09
Wangzhuang	9.8	299	0.32	3.0	0.68	0.01	15.4	0.27
Fenghuangshan	6.6	225	0.27	4.4	0.14	0.02	18.2	0.86
Gaoyang	4.6	107	0.03	3.2	0.10	0.01	13.4	0.29
Jiaojiazai	1.1	102	0.21	4.1	0.26	0.01	6.2	0.03
Guandi	13.8	116	0.17	2.2	0.14	0.02	2.4	0.80
Duerping	13.7	166	0.04	4.4	0.05	0.01	8.5	
Xiqu	15	287	0.45	3.5	0.42	0.06	17.4	
Zhenchengdi	7.6	319	0.13	2.9	0.11	0.07	10.7	
Average (based on the solid waste amount)		182	0.3	4.6	0.2	0.02	10.6	0.25

From Zhao, et al. Trace element emissions from spontaneous combustion of gob piles in coal mines, Shanxi, China. Int J Coal Geol 2008;73:52–62.

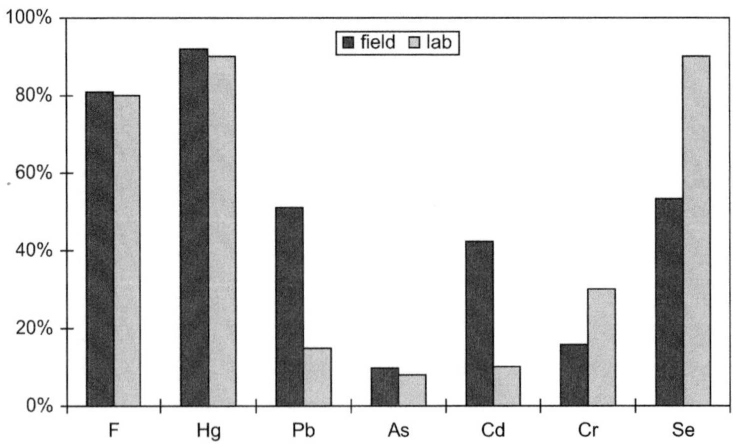

Figure 4.14 Emission factors of trace elements (in %) during spontaneous combustion of gob piles in field and laboratory simulation tests.
From Zhao, et al. Trace element emissions from spontaneous combustion of gob piles in coal mines, Shanxi, China. Int J Coal Geol 2008;73:52–62.

The mass balance approach is used to estimate TE emissions during spontaneous combustion of gob piles in the field. The TEs in gob piles and clinkers were determined. Most of the clinker is not entirely composed of ash because of incomplete combustion. For field samples, the LOI can be determined in the laboratory. Assuming the ash weight is not changed during spontaneous combustion, the weight loss can then be calculated as follows:

$$B_1/A_1 = (B_2 + B_{Loss})/A_2 \tag{4.2}$$

$$LOI = B_{Loss}/(A_2 + B_2 + B_{Loss}) \tag{4.3}$$

where A_1, A_2 is the ash content of solid waste of the gob pile and clinker, respectively (%), B_1, B_2 is the sum of another fraction (M + V + FC) in proximate analysis of the solid waste from the gob pile and clinker, respectively (%), and B_{Loss} is the weight loss during spontaneous combustion of solid waste (%).

During spontaneous combustion of solid waste in the gob pile, many mineral transformation processes (shown in Table 4.6) take place, for example, clays are dehydrated. So, the ash content of clinker, A_2, is relatively small. To ensure the accuracy of the emission factor, A_2 is modified as follows:

$$A_2' = 1.25 \times A_2 \tag{4.4}$$

Thus LOI may be expressed as follows:

$$LOI = B_{Loss}/(1.25 \times A_2 + B_2 + B_{Loss}) \tag{4.5}$$

Table 4.6 Emission factors of trace elements during spontaneous combustion for samples from gob piles in the field (in mg/kg).

Element	Emission factor
As	1.08
Cd	0.08
Cr	6.89
F	271
Hg	0.40
Pb	17.9
Se	0.24

From Zhao, et al. Trace element emissions from spontaneous combustion of gob piles in coal mines, Shanxi, China. Int J Coal Geol 2008;73:52–62.

The samples of gob piles have ash contents between 47.4% and 68.3%, and those from clinker have ash contents in the range between 89.6% and 95.7%. The emission factors of TEs during spontaneous combustion are listed in Table 4.6. The emission factors of TE emissions in percentages during spontaneous combustion are shown in Fig. 4.14. More than half of F, Se, Hg, and Pb are emitted to the atmosphere as a result of spontaneous combustion at a high temperature. However, emission factors of As and Cr are significantly lower than other elements possibly because of their lower volatility and modes of occurrence.

Because of the long duration of reaction in the field, most of the TEs show higher emission percentages as compared with simulation tests, especially for the volatile elements Pb and Cd. The emission percentages of Cr and Se are lower in the field samples. Field samples contain calcite, which may decompose to CaO and react with selenium compound in flue gas, thus restraining selenium emission during combustion.

To quantify the impact of TE emissions on the environment, TE abundances in flue gas are calculated as follows [48]:

$$V_0 = Q/4.145 + 0.455 \tag{4.6}$$

$$V = 1.04Q/4.187 + 0.54 + 1.0161 \times (\alpha - 1) \times V_0 \tag{4.7}$$

$$C_{Gas} = E/V \tag{4.8}$$

where V_0 is the theoretical atmospheric value needed by spontaneous combustion (m^3/kg), Q is the lower heat value of the solid waste application group (kJ/kg), α is the surplus air coefficient (1.10), V is the exhaust gas value from spontaneous combustion of each unit (m^3/kg), and C_{Gas} is the TE concentration in flue gas (mg/m^3). As shown in Table 4.7, TE concentrations in flue gas are much higher than the values allowed by the Resource Conservation and Recovery Act of American. Fluorine concentration in flue

Table 4.7 Concentrations of trace elements (TEs) in flue gas during spontaneous combustion of gob piles.

Element	TEs in flue gas (mg/m^3)
As	0.32
Cd	0.02
Cr	1.84
F	89.1
Hg	0.16
Pb	5.98
Se	0.1

From Zhao, et al. Trace element emissions from spontaneous combustion of gob piles in coal mines, Shanxi, China. Int J Coal Geol 2008;73:52–62.

gas is the highest among the seven TEs (89.1 mg/m^3). The Hg concentration in flue gas is 0.16 mg/m^3, which exceeds the limit of 0.05 mg/m^3 of municipal solid waste incineration in some countries [49,50]. The Cr, As, Cd, and Pb concentrations are much higher than the limits of 0.5 mg/m^3. There are no emission standards for other elements.

4.5 Conclusions

Coal fires can be classified according to their origin, age, location, and burning stage. Coal fires occur in numerous countries around the globe, typically in China, India, the United States, Australia, and South Africa. Most coal-fire-related landforms occur from volume loss underground if an underground coal seam burns and is reduced to ash. Different forms of bedrock surface fracturing and land subsidence endanger adjacent mines and related industry. Coal fires frequently lead to the collapse of mining shafts and can put a complete halt to otherwise prosperous mining activities.

Elemental mobilization in a self-burning coal seam is related to the nature of thermally altered residues and the temperature of alteration zones. There are two types of thermally altered residues (chars): oxidized char, which is cracked, fractured, and loosely packed, and combusted and carbonized char, which is nonporous and compact. Three alteration zones resulting from this burning are present in the coal from the top to the bottom of the seam. In the first, the oxidation zone, the temperature is about 400°C. Char fragments are often impregnated with volatile matter and elements characteristically abundant in such volatile matter (e.g., H). In the second and third zones, the combustion and carbonization zones, temperatures are about 650 and 600°C, respectively. The chars in these zones are enriched with elements that typically volatile during heat treatment as these temperature levels, such as As, Br, CI, Mo, and Sb. The concentration of these elements in the combustion and carbonization zones is unexpected and is due to the nonporous and compact nature of chars that trap the elements.

Volatilization and mobilization of elements in the burning coal seam are limited to C, H, N, halogens, and elements typically associated with sulfides and sulfates, such as As, S, Sb, and Mo. These elements are mobilized in the upper part of the seam. They are often trapped in the nonporous, compact combustion, and carbonization zones. However, they are eventually mobilized and migrate toward the surface; this is evidenced by the occurrence of sulfur crystals on the surface and also in vents leading to the oxidation zone. Elements associated with clay minerals, e.g., Al, Ti, K, Cs, Rb, and REEs, show very little change in the high-temperature zones.

As for the emissions of TEs from coal-fire by-products, S and Hg are concentrated at the top of the coal bed, suggesting that this portion of the coal was absent in much of the mine. Indeed, the S content of the delivered coal increases through the delivery period, indicating changing quality of the mined coal. A number of elements are increasingly concentrated in the cooler fractions of the ESP fly ash. Among them, S, As, Bi, Cd, Ga, Ge, Mo, Pb, Sb, Te, Tl, W, and Zn show significant increases in concentration in the cooler fractions. Se is highly concentrated in first-row ESP fly ash, but the concentration diminishes in cooler fractions, complicating attempts at understanding its behavior.

Acknowledgments

This project was supported by the National Natural Science Foundation of China (51776227) and Natural Science Foundation of Hunan Province, China (2018JJ1039, 2018JJ3675).

References

[1] Bell FG, Bullock SET, Hälbich TFJ, Lindsay P. Environmental impacts associated with an abandoned mine in the Witbank Coalfield, South Africa. Int J Coal Geol 2001;45: 195–216.
[2] Heffern EL, Coates DA. Geologic history of natural coal-bed fires, Powder River basin, USA. Int J Coal Geol 2004;59:25–47.
[3] Stracher GB. Coal fires burning around the world: a Global Catastrophe. Int J Coal Geol 2004;59:1–6.
[4] Stracher GB, Taylor TP. Coal fires burning out of control around the world: thermodynamic recipe for environmental catastrophe. Int J Coal Geol 2004;59:7–17.
[5] Whitehouse AE, Mulyana AAS. Coal fires in Indonesia. Int J Coal Geol 2004;59:91–7.
[6] Sheail J. 'Burning bings': a study of pollution management in mid-twentieth century Britain. J Hist Geogr 2005;31:134–48.
[7] Chatterjee RS. Coal fire mapping from satellite thermal IR data – a case example in Jharia Coalfield, Jharkhand, India, ISPRS J. Photo. Rem Sens 2006;60:113–28.
[8] Zheng B, Yu X, Zhand J, Zhou D. Environmental geochemistry of coal and endemic arsenism in southwest Guizhou, PR China. In: 30th international geologic congress abstracts; 1996. p. 410.
[9] Ando M, Tadano M, Asanuma S, Tamura K, Matsushima S, Watanabe T, Kondo T, Sakurai S, Ji R, Liang C. Health effects of indoor fluoride pollution from coal burning in China. Environ Health Perspect 1998;106:239.
[10] Finkelman RB, Belkin HE, Zheng B. Health impacts of domestic coal use in China. Proc Natl Acad Sci USA 1999;96:3427–31.
[11] Zheng B, Hong Y, Zhao W, Zhou H-Y, Xia W, Su H, Mao D, Yan L, Thornton I. The Se-rich carbonaceous siliceous rock and endemic selenosis in southwest Hubei, China. Chin Sci Bull 1992;37:1725–9.
[12] Hussain R, Luo K, Chao Z, Xiaofeng Z. Trace elements concentration and distributions in coal and coal mining wastes and their environmental and health impacts in Shaanxi, China. Environ Sci Pollut Res 2018;25:19566–84.
[13] Kuenzer C, Stracher GB. Geomorphology of coal seam fires. Geomorphology 2012;138: 209–22.
[14] Taraba B, Michalec Z. Effect of longwall face advance rate on spontaneous heating process in the gob area – CFD modelling. Fuel 2011;90:2790–7.
[15] Song Z, Zhu H, Jia G, He C. Comprehensive evaluation on self-ignition risks of coal stockpiles using fuzzy AHP approaches. J Loss Prev Process Ind 2014;32:78–94.
[16] Kuenzer C, Zhang J, Tetzlaff A, van Dijk P, Voigt S, Mehl H, Wagner W. Uncontrolled coal fires and their environmental impacts: investigating two arid mining regions in north-central China. Appl Geogr 2007;27:42–62.
[17] Ellyett CD, Fleming AW. Thermal infrared imagery of the Burning Mountain coal fire. Remote Sens Environ 1974;3:79–86.

[18] Gielisch H. Loeschung von Kohlebraenden–Statusbericht zum GTZ Projekt, Internal unpublished report of the DMT (German Montan Technology) on a project on coal fire extinction in China within the framework of a German GTZ project. 2002. p. 52.
[19] Van Genderen J, Haiyan G. Environmental monitoring of spontaneous combustion in the North China coalfields: final report to European Commission. ITC; 1997.
[20] van Dijk P, Zhang J, Jun W, Kuenzer C, Wolf K-H. Assessment of the contribution of in-situ combustion of coal to greenhouse gas emission; based on a comparison of Chinese mining information to previous remote sensing estimates. Int J Coal Geol 2011;86: 108–19.
[21] Song Z, Kuenzer C. Coal fires in China over the last decade: a comprehensive review. Int J Coal Geol 2014;133:72–99.
[22] Bhattacharya A, Reddy S, Mukherjee T. Multi-tier remote sensing data analysis for coal fire mapping in Jharia coalfield of Bihar, India. In: Proceedings of the Twelfth Asian conference on remote sensing. Singapore: National University of Singapore; 1991. p. 21–2.
[23] Kuenzer C, Hecker C, Zhang J, Wessling S, Wagner W. The potential of multidiurnal MODIS thermal band data for coal fire detection. Int J Remote Sens 2008;29:923–44.
[24] Agarwal R, Singh D, Chauhan D, Singh K. Detection of coal mine fires in the Jharia coal field using NOAA/AVHRR data. J Geophys Eng 2006;3:212.
[25] Renner S. Report on the status of fires at abandoned underground coal mines in Colorado. Unpublished report of the Colorado Division of Mineral and Geology. 2005.
[26] Elick JM. Mapping the coal fire at Centralia, Pa using thermal infrared imagery. Int J Coal Geol 2011;87:197–203.
[27] Pennig L. Vergleich zweier Kohlebrandregionen in Nordchina mit Hilfe von Fernerkundungs-daten und GIS [Master thesis at the University of Jena]. 2003.
[28] Uludag S, PhMips H, Eroglu HN. Assessing spontaneous combustion risk in South African coal mines using a GIS tool [Unpublished MSc dissertation]. Johannesburg: University of the Witwatersrand; 2001. 154 p.
[29] Klika Z, Kozubek T, Martinec P, Kliková C, Dostál Z. Mathematical modeling of bituminous coal seams burning contemporaneously with the formation of a variegated beds body. Int J Coal Geol 2004;59:137–51.
[30] Mrasek V. Die gluehenden Berge der Kohle. Spiegel Online; March 22, 2005.
[31] Gürdal G, Hoşgörmez H, Özcan D, Li X, Liu H, Song W. The properties of Çan Basin coals (Çanakkale—Turkey): spontaneous combustion and combustion by-products. Int J Coal Geol 2015;138:1–15.
[32] Onifade M, Genc B. Spontaneous combustion of coals and coal-shales. Int J Min Sci Technol 2018;28:933–40.
[33] Goodarzi F. Variation of elements in self-burning coal seam from Coalspur, Alberta, Canada. Energy Sources 2007;12:345–61.
[34] Thomas G, Fariborz G. Effect of geological processes on coal quality and utilization potential: review with examples from western Canada. J Hazard Mater 2000;74:109–24.
[35] Zhao Y, Zhang J, Chou C-L, Li Y, Wang Z, Ge Y, Zheng C. Trace element emissions from spontaneous combustion of gob piles in coal mines, Shanxi, China. Int J Coal Geol 2008; 73:52–62.
[36] Querol X, Fernández-Turiel J, López-Soler A. Trace elements in coal and their behaviour during combustion in a large power station. Fuel 1995;74:331–43.
[37] Křibek B, Sýkorová I, Veselovský F, Laufek F, Malec J, Knésl I, Majer V. Trace element geochemistry of self-burning and weathering of a mineralized coal waste dump: the Novátor mine, Czech Republic. Int J Coal Geol 2017;173:158–75.

[38] Wang S, Luo K, Wang X, Sun Y. Estimate of sulfur, arsenic, mercury, fluorine emissions due to spontaneous combustion of coal gangue: an important part of Chinese emission inventories. Environ Pollut 2016;209:107−13.
[39] Hong X, Liang H, Lv S, Jia Y, Zhao T, Liang W. Mercury emissions from dynamic monitoring holes of underground coal fires in the Wuda Coalfield, Inner Mongolia, China. Int J Coal Geol 2017;181:78−86.
[40] Pone JDN, Hein KAA, Stracher GB, Annegarn HJ, Finkleman RB, Blake DR, McCormack JK, Schroeder P. The spontaneous combustion of coal and its by-products in the Witbank and Sasolburg coalfields of South Africa. Int J Coal Geol 2007;72:124−40.
[41] Bel'kova V, Bychinsky V, Tshadov I, Kotel'nikov N. The behaviour of trace elements at the Cheremkhovo coal combustion. Ugol Coal 2000;5:43−70.
[42] Yan R, Gauthier D, Flamant G. Partitioning of trace elements in the flue gas from coal combustion. Combust Flame 2001;125:942−54.
[43] Furimsky E. Characterization of trace element emissions from coal combustion by equilibrium calculations. Fuel Process Technol 2000;63:29−44.
[44] Zhang J, Zheng C, Ren D, Chou C-L, Liu J, Zeng R, Wang Z, Zhao F, Ge Y. Distribution of potentially hazardous trace elements in coals from Shanxi province, China. Fuel 2004; 83:129−35.
[45] Rudnick RL, Gao S. Composition of the continental crust. Treatise Geochem 2003;3:659.
[46] Dai S, Ren D, Tang Y, Yue M, Hao L. Concentration and distribution of elements in late Permian coals from western Guizhou province, China. Int J Coal Geol 2005;61:119−37.
[47] Wang Q, Wenguo Shen A, Ma Z. Estimation of mercury emission from coal combustion in China. China Environ 1999;34:2711−3.
[48] Ge Y. Environmental impact of trace elements in coal gangue from Shanxi Province. Coal Geol China 1996;4:58−62.
[49] Linak WP, Wendt JOL. Toxic metal emissions from incineration: mechanisms and control. Prog Energy Combust Sci 1993;19:145−85.
[50] Zheng C, Xu M, Zhang J, Liu J. Emissions and control of trace elements from coal combustion. Wuhan: Hubei Scientific and Technical Press; 2002.
[51] Dai S, Ren D, Chou C-L, Finkelman RB, Seredin VV, Zhou Y. Geochemistry of trace elements in Chinese coals: a review of abundances, genetic types, impacts on human health, and industrial utilization. Int. J. Coal Geol 2012;94:3−21.

Trace element partition in coal gasification

5

Hailong Li [1], Jiexia Zhao [1], Yingchao Hu [1], Zequn Yang [2]
[1]School of Energy Science and Engineering, Central South University, Changsha, China;
[2]Department of Civil Engineering, The University of Hong Kong, Hong Kong, China

5.1 Introduction

Gasification, which is a means of converting fossil fuels, biomass, and wastes into either a combustible gas or a synthesis gas for subsequent utilization, offers the potential for both clean power and chemical production [1]. On a worldwide basis, there are some 160 modern gasification plants in operation and a further 35 at the planning stage. The primary products that can be produced in such plants include electricity, ammonia, oxy-chemicals, syngas, methanol, and hydrogen. Although trace element (TE) classification was unchanged during the coal gasification process [2], the behavior of TEs is different during coal gasification under reducing atmospheres from that of coal combustion under oxidizing conditions [3]. For instance, TEs, which are able to form hydrides such as B_2H_4, SeH_2, and AsH_3, can be expected to increase volatility in reducing conditions [4]. For an in-depth insight into TE partitions during the coal gasification process, TE partitioning among ashes, particles, and gases in different coal gasification processes, the distribution of elements depending on the gasification conditions, and the mode of occurrence of TEs in coal have been calculated [5−7]. In this chapter, similar to the previous two chapters, the partition of five specific TEs (Hg, As, F, Se, and Cr) during coal gasification over laboratory-scale experiments is reviewed in detail. Effects of coal type, gasification temperature, and gasification mode on the partitioning behavior and emission characteristics of TEs are discussed.

5.2 Chemical thermodynamics and kinetics studies of trace elements during coal gasification

Modern gasification of coal is operated typically under a pressure of 1−2 MPa (10−20 atm), which makes experimental measurements of TE emissions and partitions more difficult than that under coal combustion conditions [8]. Thermodynamics provides a powerful calculation method for studying the effects of different process variables on equilibrium chemical systems. However, the calculation does not give a reaction time, which is necessary to reach the theoretical equilibrium state. Without complete general theory or basic data, thermodynamics, nonideality, and dynamics can

Emission and Control of Trace Elements from Coal-Derived Gas Streams
https://doi.org/10.1016/B978-0-08-102591-8.00005-2

be successfully combined for all chemical processes. The use of thermodynamic models to predict TEs is a viable method for studying TE distribution in coal gasification processes.

Besides thermodynamics, the kinetics of TE vaporization is another critical parameter to evaluate TE emission speed during coal gasification. However, related research in this area is very rare. Shen et al. investigated selenium emission kinetics and thermodynamics during coal gasification [9]. The kinetic law dominating Se emission during the coal gasification process can be expressed as:

$$r(x) = 11.96e^{-45.03/RT}\left(-0.53x^2 - 0.56x + 1.09\right)$$

The mathematical equations of the proposed dynamic law are similar to the mathematical equations for the coal combustion section. The effect of temperature is taken into account, which facilitates its use under complex gasification conditions. Fig. 5.1(a) shows a comparison between experimental kinetics and the laws of kinetics mathematically determined. The red line is calculated by the laws of dynamics. For all

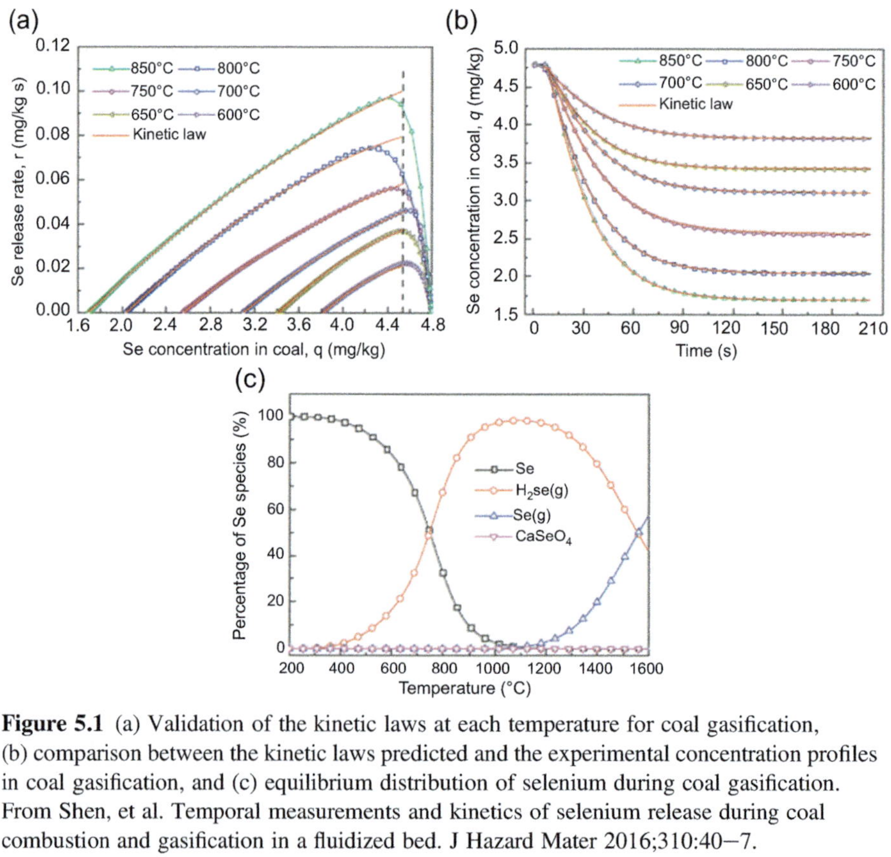

Figure 5.1 (a) Validation of the kinetic laws at each temperature for coal gasification, (b) comparison between the kinetic laws predicted and the experimental concentration profiles in coal gasification, and (c) equilibrium distribution of selenium during coal gasification. From Shen, et al. Temporal measurements and kinetics of selenium release during coal combustion and gasification in a fluidized bed. J Hazard Mater 2016;310:40–7.

temperatures, good agreement between the laws of kinetics and experimental data can be seen. This demonstrates that the kinetic method successfully predicted the kinetics of selenium release during coal gasification. The dynamic characteristics of selenium release during coal combustion are different from the release of selenium during coal gasification. This illustrates the effect of operating conditions on the kinetics of selenium release. With the increase in temperature, the release rate of selenium in the coal gasification process is more obvious than that in the coal combustion process. This indicates that the release kinetics of selenium during coal gasification is more sensitive to temperature than during combustion. When the release rate of selenium reaches a peak, the concentration of selenium remaining in the coal particles during coal gasification is similar to that during coal combustion. This indicates that the residual selenium concentration in coal has little effect on the release kinetics of selenium, which is most likely related to the form of selenium in coal.

Fig. 5.8(b) shows the total amount of selenium remaining in coal particles at each temperature. The red line is calculated by the laws of dynamics. In all cases, the amount of residual selenium in coal particles gradually decreased and became constant after a period of time. Residual selenium is reduced faster at low temperatures than at high temperatures. This corresponds to an increase in the rate of selenium release as the temperature increases. The release of selenium during coal gasification lasts for approximately 120 s. As the temperature increases, the percentage of selenium released during coal gasification increases significantly. As the coal gasification temperature rose from 600 to 850°C, the percentage of selenium released increased from 20.3% to 64.3%. It can be concluded that the kinetic method accurately predicts the temporal evolution of selenium during coal gasification.

During coal gasification (Fig. 5.8(c)), gaseous $H_2Se(g)$ possesses a high proportion at temperatures from 750 to 1550°C. The transformation of $H_2Se(g)$ to Se(g) increases with a rise of temperature from 1150°C. When gas temperature decreases to below 750°C, the selenium presents dominantly in the form of atomic selenium. In the Shen et al. study, the presence of CaO was taken into consideration and they found that CaO hardly reacted with Se species in the fly ash. This is because CaO is more likely to react with CO_2 to form $CaCO_3$.

5.3 Trace element emission during coal gasification

5.3.1 *Influence of coal types*

Liu et al. investigated the influence of coal types on TE emission during the underground coal gasification (UCG) process [10]. UCG involves the conversion of coal in the seam into a combustible fuel gas, thus avoiding the costs of mining and transportation. It is potentially a clean method of coal conversion into a high-energy fuel gas and may be adopted to replace small coal-fired power plants and to maximize coal resources [11]; this is partly achieved through its implementation with unminable coal seams or those that have geological problems. It has been widely tested in China and used at a commercial scale; both low- and high-pressure variants are being considered. UCG fuel gas can be used for power generation, hydrogen production, and as a chemical feedstock.

The volatility of mercury, arsenic, and selenium and the volatility of carbon conversion to ash during the conversion of coal to coke are calculated as shown in Fig. 5.2. It can be seen that the volatilization of mercury, arsenic, and selenium in UCG follows the order of Hg > Se > As. Mercury has the highest volatility and coal ash is above 98%. The total volatility of selenium from raw coal to ash is higher than 90% but lower than mercury. The volatility of arsenic is less than 60%, and arsenic enrichment is found in UCG ash. It has also been found that the volatility of arsenic during the conversion of char to ash at elevated temperatures corresponds to an increase in the volatilization of coal with temperature. In contrast, arsenic volatilization during the combustion of crushed coal does not always increase with increasing temperature [12].

It is clear that coal type has an effect on the volatilization behavior of the elements. With regard to the UCG of lignite, the majority of mercury, arsenic, and selenium evaporates during transformation to char, while for anthracite the main evaporation occurs in the conversion of char to ash during the high-temperature gasification stage. Therefore it can be stated that the volatilization of mercury, arsenic, and selenium depend largely on the properties of the specific element, the mode of its occurrence, coal reactivity, etc.

Regarding the influence of coal types, it is obvious that in the process of coal conversion to carbon, the element has the highest volatility to lignite, followed by bituminous coal and anthracite. This is consistent with the order of gasification reactivity, which is lignite > bituminous > anthracite. For the underground gasification of coal in coal seams, its reactivity affects the fracture of coal, which in turn affects the degree of reaction and the decomposition of sulfides and other minerals, which is closely related to the volatilization of elements. Lignite has a high moisture and volatile content and is easily released upon heating. This in turn helps the coal seam to decompose to form a loose structure with an increased reaction area. On the other hand, the coarse pore structure in lignite is favorable for the coal seam to crack at high temperature, allowing air and oxygen to diffuse to the surface of fresh coal, increasing the reaction rate and the volatilization of TEs. In contrast, anthracite has a compact structure, a low reaction area, and low reactivity, and is considered to be unsuitable for gasification. However, UCG field trials with anthracite in Xiyang County showed that this was feasible. Therefore the UCG of anthracite can be explained as follows: the coal seam is difficult to crack during the pyrolysis stage, but when the temperature rises above 900°C, the internal and external temperature difference causes the coal in the coal seam to expand and crack, expanding the reaction area and making gasification possible. Therefore for UCG charcoal, mercury, arsenic, and selenium are particularly low in volatility and rise sharply during the conversion of coke to ash at high temperatures. Due to its high volatile content, bituminous coal exhibits high reactivity at the beginning of gasification. After the pyrolysis stage, its strong miscibility will affect its gasification reactivity. However, during the UCG process, as the fire front moves forward along the gasification tunnel and gas flows through most of the gasification tunnel, the coal remains in contact with the gasifying agent to maintain the UCG process. On the other hand, the higher stability of bituminous coal is detrimental to coal

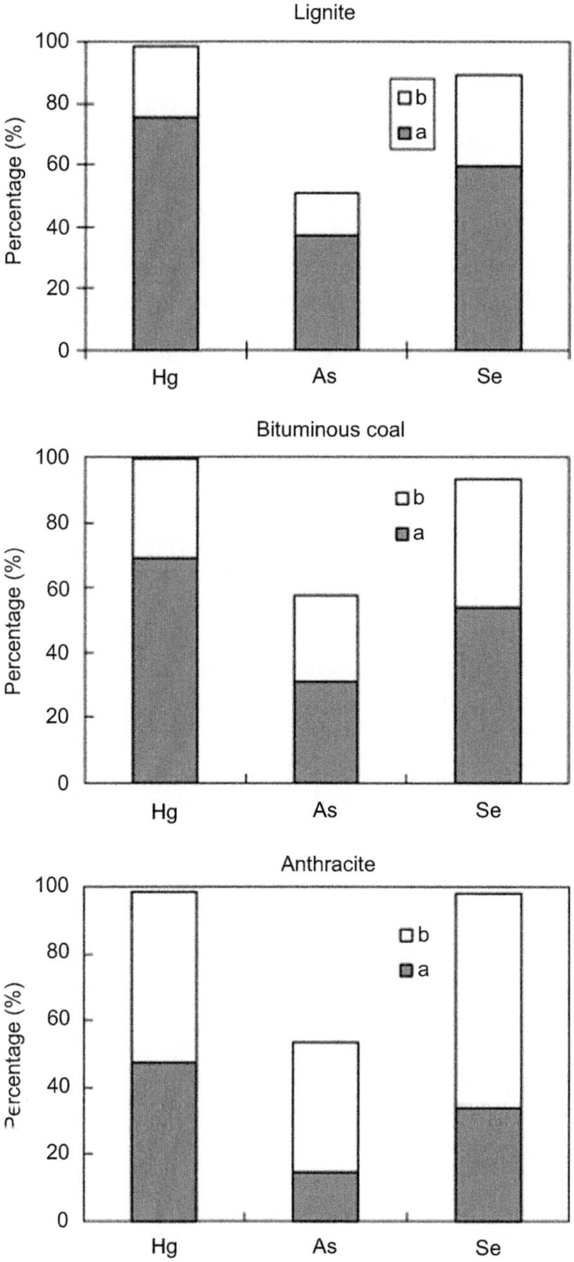

Figure 5.2 Comparison of the volatility of mercury, arsenic, and selenium during underground coal gasification of different coals: (a) transformation of coal to char and (b) transformation of char to ash.
From Liu, et al. Volatilization of mercury, arsenic and selenium during underground coal gasification. Fuel 2006;85:1550–58.

cracking, which leads to a reduction in the reaction surface area, which in turn reduces the gasification rate in the final stage. Therefore higher volatility of TEs was observed during the conversion of coal to coke.

The foregoing results only indicate the tendencies for volatilization of mercury, arsenic, and selenium during UCG. The volatile elements do not always remain in the gas phase when they are cooled, adsorbed onto surfaces, and redistributed into different phases as the gas flows through the long gasification tunnel and production well. This influence on the distribution of mercury, arsenic, and selenium in the final UCG products needs further investigation, depending largely on the length of the passage before gas reaches the surface. It has been found that gas-phase mercury is about 20% of the total mercury in UCG products, lower than from surface gasification and coal burning.

5.3.2 Trace element partitions during the gasification process equipped with a hot gas filter

The economics of power generation systems employing gasification may be greatly improved by substituting waste with a significant heating value and low or negative costs, such as sewage sludge, for part of the solid fuel. Gasification schemes for such applications are usually modest in scale (c. 10 MW_t) because of the limited amount of waste available in a single location. Air-blown gasification is often selected, because of the lower infrastructure requirements and advantages for fuel with a high or refractory ash content and moderate sulfur content [13]. Reed et al. explored the TE emissions during the co-gasification of coal with 25% of sewage sludge addition [14]. As shown in Fig. 5.3, Hg and Se were fully vaporized during the co-gasification process. On the contrary, As and Cr will be accumulated on the dust fines. However, Hg was not quantifiable in the tests with coal alone, but was found in most of the tests with 25% sewage sludge pellet addition. However, the measured values are believed to be an underestimate, as discussed elsewhere [15]. Se was only quantifiable in the fuel gas, never in any of the solid output streams. This was due to a high limit of detection for Se (10 μg/g) in the solids. Se was detected in the fuel gas in all of the tests with 25% sewage sludge pellet addition to the feed.

In addition, Hg partitions in coal gasification using two coal types were studied, namely Illinois #8 (asphalt, R8) and Highvale (sub-bituminous coal, R5) [14]. Based on the assumption that the output stream measures all Hg, Fig. 5.4(a) gives a comparison of the output distribution data for the two test measurements. The data from the two tests indicated that the mercury content in the hot gas filter was very high; a smaller proportion of the total production was found in the filtered fine powder of test R8 compared to test R5. Also shown is correction data for test R8, which is based on the assumption that all input Hg not counted by the hot gas filter fines is carried out in the fuel gas. According to the revised data of test R8, the proportion of Hg in the hot gas filter is much lower than the Hg in test R5.

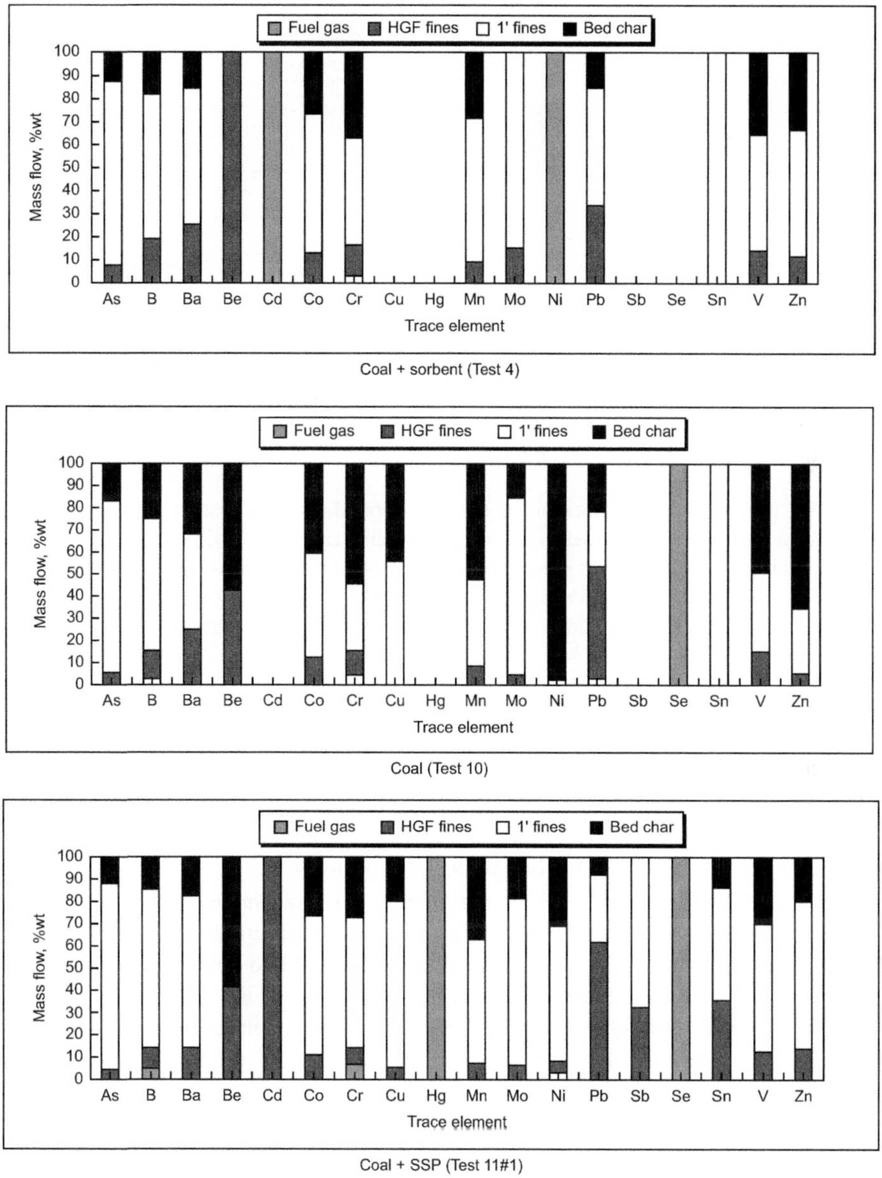

Figure 5.3 Typical distribution of trace elements in the co-gasification of coal and sewage sludge output streams. *HGF*, hot gas filter; *SSP*, sewage sludge pellets.
From Reed, et al. Control of trace elements in gasification: distribution to the output streams of a pilot scale gasifier. Energy Fuels 2001;15(4):794–800.

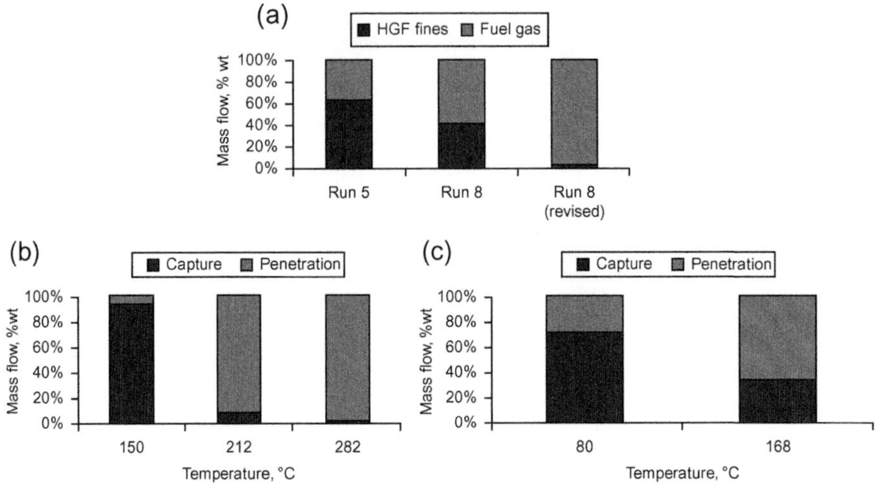

Figure 5.4 (a) Distribution of mercury in the underground coal gasification gasifier output streams with different types of coal, and effect of temperature on the distribution of mercury in the hot gas cleanup reactor output streams for a bed depth of (b) 5 mm and mean gas residence time of 0.5 s and (c) 0.5 mm and mean gas residence time of 0.05 s. *HGF*, hot gas filter. From Reed, et al. Control of gasifier mercury emissions in a hot gas filter: the effect of temperature. Fuel 2001;80:623–34.

The limited amount of data obtained from the pilot plant suggested that, in contrast to the thermodynamic equilibrium predictions, significant removal of Hg had taken place in the hot gas filter. The hypothesis that the hot gas filter fines were responsible for Hg removal needed to be investigated in a system where this could be studied in isolation from other process variables. The next stage of the work was consequently carried out using a bench-scale hot gas cleanup reactor.

Data from laboratory-scale hot gas purification reactors can be examined using the same assumptions applied to pilot-scale gasifiers, i.e., capture and permeation measurements account for all Hg. The effect of the adsorbent bed temperature on Hg distribution is shown on the 5 mm deep bed in Fig. 5.4(b) for the 0.5 mm deep bed in Fig. 5.4(c). Fig. 5.4(b) shows that the Hg ratio in the "captured" bed (i.e., the hot gas filter fines) rises sharply as the temperature decreases. For the 0.5 mm deep bed in Fig. 5.4(c), the same trend of increased capture as temperature decreases is observed. However, it is likely that in the case of such a shallow bed, turbulence may be important. However, the effect of temperature on the Hg-capturing ability of the filtered fine particles was clearly observed.

In the hot gas cleanup reactor, hot gas filter fines were shown to reach Hg loadings in excess of 30 μg/g (initial Hg concentration of sorbent: 0.01 μg/g). However, in this reactor, the Hg is vaporized as a single batch; it is not therefore possible to work with a steady Hg concentration in the gas stream. A low estimate may still be arrived at for the peak value of the concentration as follows; if the Hg were vaporized at the same rate throughout the duration of the test, the concentration would be about 300 $\mu g/m^3$. The rate of vaporization must reach a peak value as the temperature of the first stage increases, so the peak concentration must exceed 300 μg/m^3. Vidic et al. showed that the loading achieved by a commercial activated carbon at 140°C is approximately directly proportional to the Hg concentration in the gas, in the range 110−1080 μg/m^3 [16]. Although direct comparisons are not possible because of the lack of Hg concentration data for the present hot gas cleanup reactor, it can at least be stated that the Hg loadings achieved by the hot gas filter fines in the hot gas cleanup reactor are of a similar order of magnitude to literature values for commercial activated carbons.

In the hot gas filter of the UBC gasifier, the Hg concentration in the fuel gas would have been about 2.4 μg/m^3 in test R5 and >100 μg/m^3 in test R8. At 140°C, Vidic et al. reported a commercial activated carbon as having achieved an equilibrium loading of 10 μg/g at a concentration of 110 μg/m^3, whereas the hot gas filter fines from the gasifier only achieved a loading of about 0.07 μg/g in both tests [16]. This suggests that the equilibrium Hg loading capacity of the hot gas filter fines from the UBC gasifier was not as high as the commercial activated carbon, and/or that these fines were not fully utilized for reasons associated with the UBC gasifier, such as insufficient residence time or inadequate mixing.

Bunt et al. also explored TE partitions according to their volatility with Fact-Sage software [2,17,18]. Bunt et al. separated the gasifier into different zones and used the software to predict TE behavior in the separated zones (shown in Fig. 5.5). For modeling purposes the gasifier was treated in 5.5 distinct zones:

1. The drying and devolatilization zone;
2. The reduction (gasification) zone; and
3. The combustion and ash zone.

Each area is considered a separate area for Fact-Sage modeling purposes. The model works by coal (inorganic and organic components) flowing down from the top and downward in the gasifier, while gas from subsequent areas flows up into the area being modeled, i.e., into the dry coal. The devolatilization zone contacts

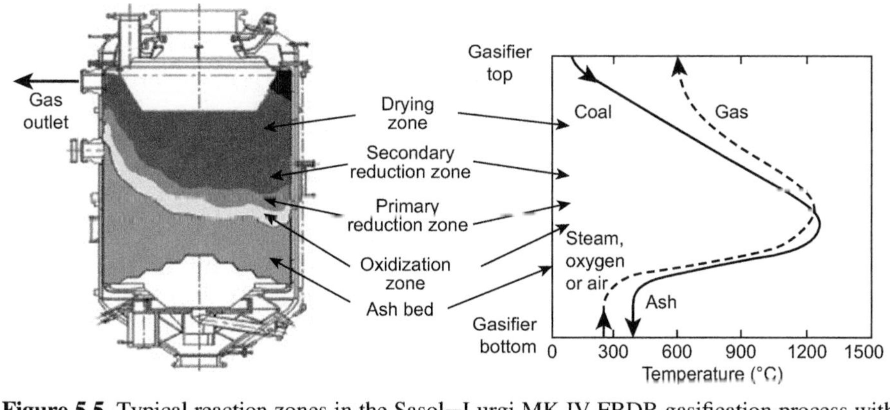

Figure 5.5 Typical reaction zones in the Sasol−Lurgi MK IV FBDB gasification process with respect to the temperature profile.
From Bunt, et al. Trace element behavior in the Sasol−Lurgi MK IV FBDB gasifier. Part 1 − the volatile elements: Hg, As, Se, Cd and Pb. Fuel 2008;87:2374−87.

and reacts with the gas flowing out of the gasification zone, thus creating a countercurrent effect. Illustratively, this means that the upward flowing gas from the gasification zone reacts with the downward flowing coal and minerals into the gasifier. The drying and devolatization zones only consider a temperature range of 25–725°C, while the reduction zone temperature ranges from 725 to 1316°C. The gas flow rate into the bottom of the gasifier consists of oxygen (11,856 kg/h) and steam (50,900 kg/h) at a pressure of 28 bar. In this case, the feed coal and gas flow rates and composition input characteristics used in the model are the same as those performed by Van Dyk, except for TE composition and concentration [19].

From the results of the Sasol–Lurgi MK IV FBDB gasifier profile, it can be seen that, as shown in Fig. 5.6(a), the normalized mercury content (%) from the dry to the combustion (gray bed) area is reduced, and almost all of the Hg is in heat. The end of the solution zone becomes a gas phase. No further significant changes in Hg

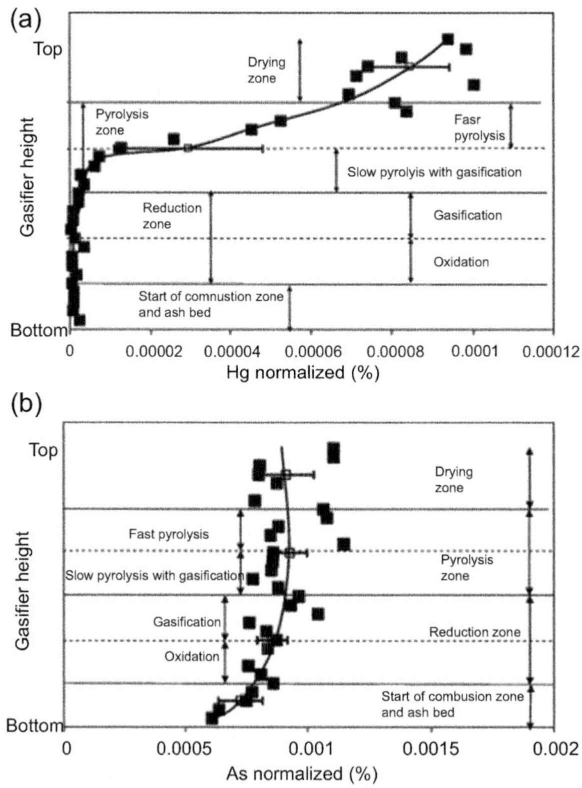

Figure 5.6 Line graph showing the normalized mercury, arsenic, selenium, and chromium percentage profile occurring within the Sasol–Lurgi MK IV FBDB gasifier.
From Bunt, et al. Trace element behavior in the Sasol–Lurgi MK IV FBDB gasifier. Part 1 – the volatile elements: Hg, As, Se, Cd and Pb. Fuel 2008;87:2374–87; Trace element behaviour in the Sasol-Lurgi fixed-bed dry-bottom gasifier. Part 3 – the non-volatile elements: Ba, Co, Cr, Mn, and V. Fuel 2010;89:537–48.

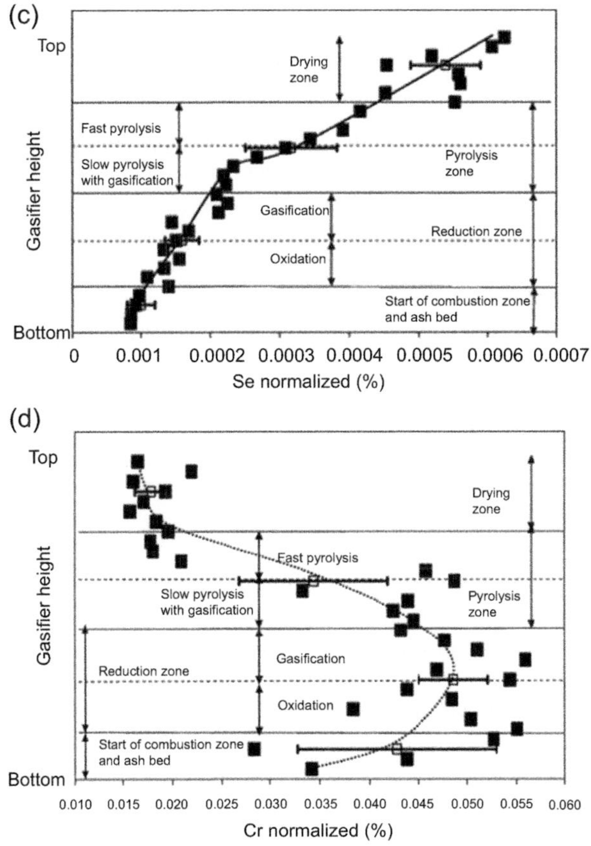

Figure 5.6 cont'd.

concentration occurred in the gray bed area, which means that the reduction and gray bed areas were not statistically significantly different in terms of Hg behavior. Thus 44% of the Hg is rapidly volatilized in the drying zone and then further distributed to the gas phase by 42% in the rapid pyrolysis zone of the gasifier. At the end of pyrolysis, 98% of the coal-bound Hg has been devolatilized, clearly indicating significant volatility under the reducing conditions prevalent in fixed-bed gasifiers. The 1% Hg content present in the ash discharged may be due to the possibility of a carbon/coke adsorption mechanism occurring in the lower half of the fixed-bed gasifier.

From the Sasol–Lurgi MK IV FBDB gasifier measured results, given in Fig. 5.6(b), it can be seen that the normalized As content (%) is not significantly different between the drying, pyrolysis, and reduction zones. There is, however, a statistically significant difference in As behavior evident between the reduction and combustion (ash-bed) reaction zones of the gasifier, suggesting that As speciation is different under reducing and oxidizing conditions.

As can be seen from Fig. 5.6(b), <15% of As is volatilized in the drying zone. At the end of pyrolysis, a slight overbalance of As was found from the measurements, indicating that this may be due to sampling and/or analytical errors, or recondensation of volatile As in this region of the gasifier. In the reduction zone, the As content is further reduced by 5% compared to the pyrolysis zone, and a 33% change in the As trend is clearly seen in the gas bed of the gasifier. In summary, based on the measured distribution results, it was observed that 45% of the As present in the raw coal was distributed into the gas phase.

Linking the As partition and morphological analysis data (Fig. 5.6(b)) to the measured normalized As distribution map (Fig. 5.6(b)), it is clear that As is solid in the pyrolysis process (first Ni, then Cu) and combines the area of the fixed-bed gasifier. The modeling and measurement results of As are not very consistent, indicating that approximately half and all of the As are still present in the solid state, respectively. This divergence may be due to heat transfer and mass transfer limitations that occur between lump coal. Taking into account the high temperatures present in the combustion zone (1316°C), all remaining As is distributed to the gas phase, supporting a statistically significant difference in As behavior observed in the measurements. Condensation of gaseous As species is expected to be responsible for the nonstatistical differences that occur in the reduction and pyrolysis regions of the fixed-bed gasifier. Future Simu-Sage iterative simulation of As species condensation during the countercurrent process is expected to improve the understanding of the observed variability for this particular species.

From the Sasol–Lurgi MK IV FBDB gasifier measured profile results, shown in Fig. 5.6(c), it can be seen that the normalized Se content (%) decreased from 0.0006% to 0.0001% from the drying to the combustion (ash-bed) zone. The average normalized Se content is thus significantly different between all the reaction zones.

From Fig. 5.6(c) it can also be observed that ca. 30% Se had volatilized in the drying zone. At the end of pyrolysis about two-thirds of the Se has partitioned to the raw gas phase. In the reduction zone, a further reduction in Se content is evident, and less than 15% of bound Se remains in the discarded ash of the gasifier.

The volatilization of Hg, As, and Se in the case of this study follows the sequence Hg > Se > As, the same as reported earlier for the underground gasification of bituminous coal [20]. In both studies, Hg had the highest volatility, which for coal to ash reaches more than 98%. The gross volatility of Se from raw coal to ash was around 90%, but less than that of Hg. The volatility of As is lower than 60% in both cases and As enrichment has been found in the ash fraction.

Fig. 5.6(d) shows the normalized chromium content (%) of the measured Sasol–Lurgi FBDB gasifier sample. It is apparent that the average normalized Cr content between the drying, pyrolysis, and reduction zones is significantly different, indicating a significant increase in Cr content with the depth of the gasification bed. It can also be observed that the Cr associated with coal in the drying zone is overbalanced by 111% and this overbalance is increased to 262% at the end of pyrolysis. In the reduction zone, Cr superbalance was further increased to 334% and finally to 208% in the gray bed. The shape of the trend curve also indicates that the Cr content increases steadily until the beginning of the assumed oxidation zone, and after the shape is reversed, the Cr may change the oxidation state, causing some volatilization.

The final closure overbalance of 208% in this case is higher when compared with the findings of Wilde and Halbrook [21] where a closure of 114% was reported for Cr. In their study a TE mass balance of coal, ash, tar, gas-liquor, and limited raw gas samples was taken during commercial-scale tests conducted at a Sasol–Lurgi FBDB gasification plant in South Africa. Samples were taken from a single isolated gasifier, as well as from a bank of gasifiers. It was found that Hg, As, Se, Cd, and Pb partitioned to the raw gas, while the other TEs partitioned to the ash fraction. Large overbalances of Ni, Cr, and Mo (in excess of the coal input) were found in the gasifier ash and were attributed to probable minor erosion of the gasifier internals.

Gasifier ash plows are known to comprise a composite Ni/Cr-rich alloy having a continuous operating life of approximately 3 years. In calculating the Cr superbalance reported here and correlating it with the Cr content present in the ash plow, it is clear that plow erosion contributes only 1.3% to the superbalance data reported here. Therefore, either external contamination, or chemical accumulation during sampling of the gasifier, or both, are likely contributors to the over-balance and should be studied further.

The chrome equilibrium compositional profiles calculated for the drying, pyrolysis, and reduction zones of the Sasol–Lurgi FBDB gasifier show that Cr (assumed to be in the form of chromite [$FeCr_2O_4$] in the feed coal) indeed remains in that form throughout the drying, pyrolysis, and reduction zones, up to a temperature of 1316°C. This stable chromite species is also predicted to be stable within the combustion and ash-bed zone of the fixed-bed gasifier, showing no volatilization or melting behavior. These findings are therefore in good agreement with the literature [21,22].

For fluorine, Li et al. investigated the behavior of fluorine during the Zhungeer and Yima coal CO_2-gasification process performed at 800–950°C with a residence time of 30 min [23]. Fig. 5.7(a) illustrates the relationship between the residue and the reaction temperature. As the temperature increases, the residue production of Yima coal gradually decreases, while the residue yield of Zhungeer coal decreases linearly. The residue production of Yima coal is much lower than that of Zhungeer coal, especially at low temperatures, indicating that the CO_2 gasification reactivity of Yima coal is higher. The fluorine content in the gasification residue is shown in Fig. 5.7(b), where the value of 25°C represents the fluorine content in the raw coal. The fluorine content in Yima slag is higher than that of raw coal, but as the temperature increases, the fluorine content decreases significantly. For Zhungeer coal, the fluorine content in the residue is about half that of the raw coal. This means that the volatility or reactivity of fluorine in Yima coal is lower than the volatility or reactivity of carbon in the carbon dioxide gasification process, but the situation of Zhungeer coal is just the opposite. Fig. 5.7(c) shows that the fluorine volatility of Zhungeer coal increases with increasing temperature and is higher than that of Yima coal below 900°C. The fluorine volatility of Yima coal did not change much at low temperatures, and increased significantly above 850°C. The different volatilities of fluorine in Yima and Zhungeer coal indicate that the release of fluorine depends not only on its mode of occurrence, but also on the gasification reactivity of the raw coal.

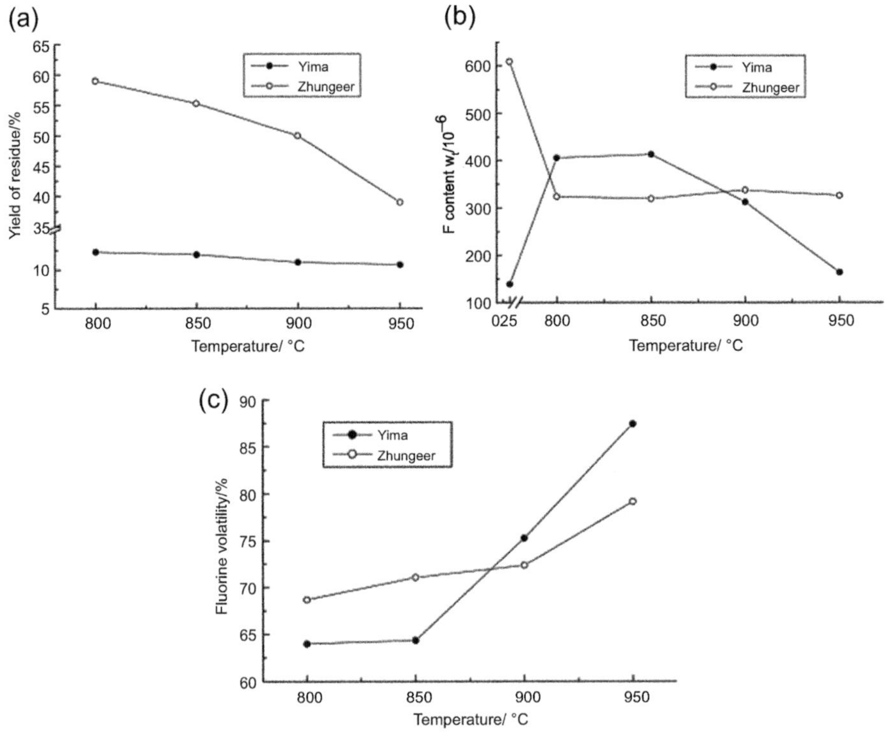

Figure 5.7 Effect of gasification temperature on (a) the yield of residue; (b) fluorine contents in the residue; and (c) fluorine volatility.
From Li, et al. Volatilization behavior of fluorine in coal during fluidized-bed pyrolysis and CO_2-gasification. Fuel 2005;84:353–57.

Guo et al. explored the release of fluorine during CO_2 gasification with Datong (DT) and Jincheng (JC) coals in a temperature range of 800–1100°C for 30 min under a flow of CO_2 or H_2O [24]. Fig. 5.8(a) shows the RR (release ratio) and GY (gas yield) of F and Cl for the two coals under carbon dioxide. A significant coal–carbon dioxide reaction occurs only at temperatures above 900°C, as indicated by the high GY at 1000 and 1100°C. However, for DT coal, the RR of F is 47% and for JC coal at 800°C it is 12%. They are higher than GY and are significantly higher than the RR in the temperature programmed decomposition (TPD) at the same temperature. Since the heating rate and residence time have little effect on the coke yield at 800°C (GY in gasification is similar to volatile yield [VY] at 800°C in TPD), although further comparisons are below 200°C, CO may promote F release. Further verification is needed in the future. The RR of F increases with increasing temperature in the study temperature range, DT coal reaches 90%, and JC coal reaches 85% at 1100°C. The Cl of both coals has a high RR and increases with increasing temperature; DT coal is 86%–96% and the temperature range is 82%–88% for JC coal. Similar to F release, the RR value of JC coal at

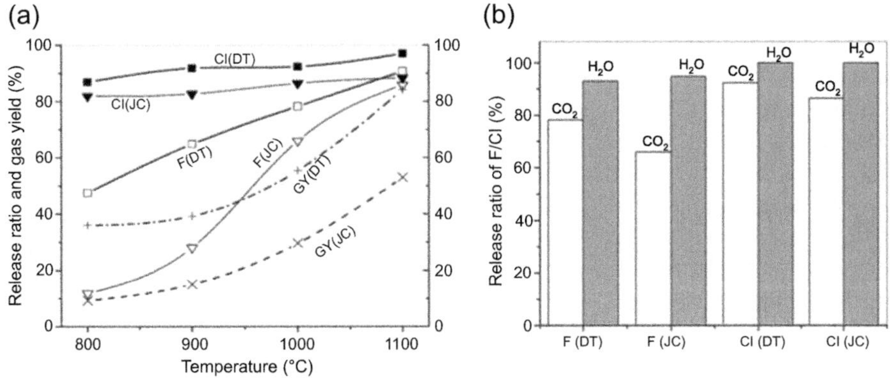

Figure 5.8 (a) Release ratios of F/Cl and gas yields in CO_2 gasification of Datong (DT) and Jincheng (JC) coals at a residence time of 30 min and (b) effect of H_2O on F and Cl releases in the gasification of DT and JC coals at 1000°C and a residence time of 30 min.
From Guo, et al. The fate of fluorine and chlorine during thermal treatment of coals. Environ Sci Technol 2006;40:7886–89.

800°C (82%) is also much higher than GY and RR (43%) in TPD at the same temperature. Since CO_2 has no significant effect on the GY of JC coal at 800°C and lower, the higher RR of Cl at 800°C in gasification indicates the promoting effect of CO_2 on Cl release.

Fig. 5.8(b) shows that, similar to the TPD results, the RRs of Cl are always higher than that of F and the RRs of F and Cl for DT coal are always higher than that for JC coal in CO_2 gasification. Furthermore, the RRs of F and Cl are higher than the corresponding GYs in CO_2 gasification. Fig. 5.8(b) compares the RRs of F and Cl at 1000°C under H_2O and CO_2 flows at a residence time of 30 min. Clearly, in most cases, H_2O promotes F and Cl releases from the coals and diminishes the difference in RR between F and Cl. Compared to the data in the CO_2 atmosphere the H_2O atmosphere results in additional 28.6 and 14.6 percentage points in RR of F and 13.5 and 7.7 percentage points of Cl for JC and DT coals, respectively. The promoting effect of H_2O on F and Cl releases can be attributed to reactions of inorganic F salts with H_2O at high temperatures, such as the following:

$$CaF_2 + H_2O \rightarrow CaO + 2HF$$

$$MgF_2 + H_2O \rightarrow MgO + 2HF$$

$$CaF_2 + SiO_2 + H_2O \rightarrow CaSiO_3 + 2HF$$

If this is the case, it also indicates that JC coal contains more inorganic F and Cl than DT coal does. Clearly, CO_2 and H_2O have promoting effects on the release of F and Cl, which results in chars with low Cl and F contents. The promoting effect of H_2O on F and Cl releases is more significant than that of CO_2, which diminishes the difference in releases of F and Cl at higher temperatures. A significant H_2O promotion on release of F and Cl may be an indication of the presence of large amounts of coexisted mineral salts of F or Cl.

Pinto et al. investigated fluorine emission during co-gasification of coal and feedstock [25]. Although the fluorine concentration in the feedstock was very low, below the detection limit of the method used, the fluoride content in the condensate was analyzed using capillary ion electrophoresis as described in the experimental section. A low level of fluorine was detected in the condensate, probably due to the gradual accumulation of fluorine during sampling. The amount of fluorine retained increases with the length of the sampling time. In Fig. 5.9(a), the concentration of fluorine released into the gas phase is shown in parts per million by volume (ppmv). It is not possible to calculate the percentage of fluorine released into the gas phase because the corresponding value in the feedstock is below the detection limit of the method used.

Due to the very low fluorine contents in both coal and wastes, fluorine concentration in the condensation system was usually very low, varying between 22 and 43 ppmv, depending on fuel fluorine content. As experienced with chlorine, most fluorine was also released to the gas phase, hence fluorine capture in the solid phase was usually very low, probably almost zero.

Fig. 5.9(a) shows the fluorine content obtained using different coal and waste mixtures. Although all raw materials have a fluorine content below the detection limit, a closer look at Fig. 5.9 shows that a higher fluorine content is detected when a higher level of coal or pine waste is used, especially the last one. This suggests that coal and pine may contain large amounts of fluorine. For the mixture with 10% of pine, 10% of polyethylene, and the sentence was corrected (PE) and 5% (w/w) of petcoke it was also detected an increase in fluorine release, this would suggest that petcoke also had a high fluorine content, which also agrees with the results obtained when the blend with 20% of pine, 20% of PE and 5% of petcoke was used.

The results presented in Fig. 5.9(b)–(d) show that for all the feedstocks studied, fluorine was never detected in the condensation system when dolomite or Ni–dolomite were added to the gasification medium, probably due to HF reacting with CaO through reaction (12) and (13), leading to the formation of CaF_2, which was probably retained in the char:

$$CaO + 2HF \rightarrow CaF_2\ H_2O$$

$$Ca(OH)_2 + 2HF \rightarrow CaF_2 + 2H_2O$$

These results agree with those reported by Weinell et al. [26], who stated that HF, HBr, and HI could participate in the same kind of reactions as HCl, being CaO ($Ca(OH)_2$), together with limestone ($CaCO_3$), nahcolite ($NaHCO_3$), soda (Na_2CO_3), or trona ($Na_2CO_3NaHCO_3$), the most suitable sorbents for the control and retention of halogens.

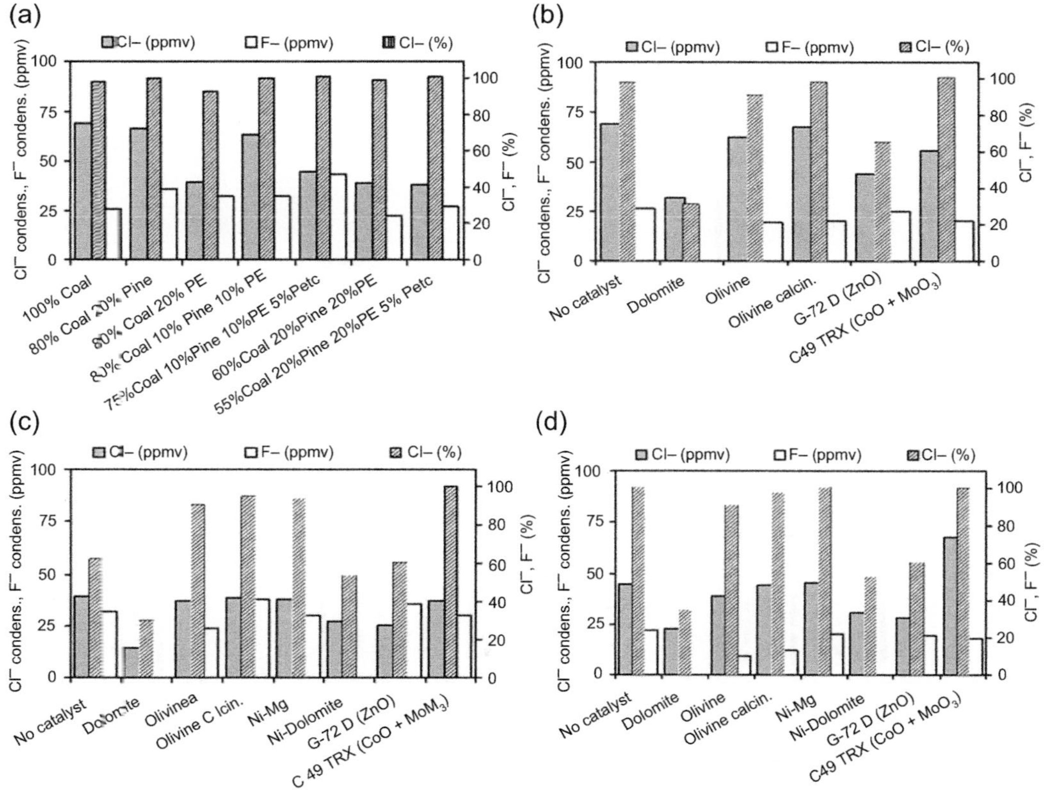

Figure 5.9 (a) Effect of waste type and content on Cl and F formation during co-gasification of coal mixed with wastes at 850°C, and the effect of catalyst or sorbent type on Cl and F formation during gasification of coal at (b) 850°C, (c) 850°C of 80% of coal mixed with 20% of PE, and (d) 850°C of 75% of coal mixed with 10% of pine, 10% of PE, and 5% (w/w) of petcoke.

From Pinto, et al. Effect of catalysts in the quality of syngas and by-products obtained by co-gasification of coal and wastes. 2: heavy metals, sulfur and halogen compounds abatement. Fuel 2008;87:1050–62.

5.4 Trace element transformation during coal gasification

Diaz-Somoano et al. explored the TE transformation patterns under different gasification temperatures with the HSC software [3], as can be observed in Fig. 5.10.

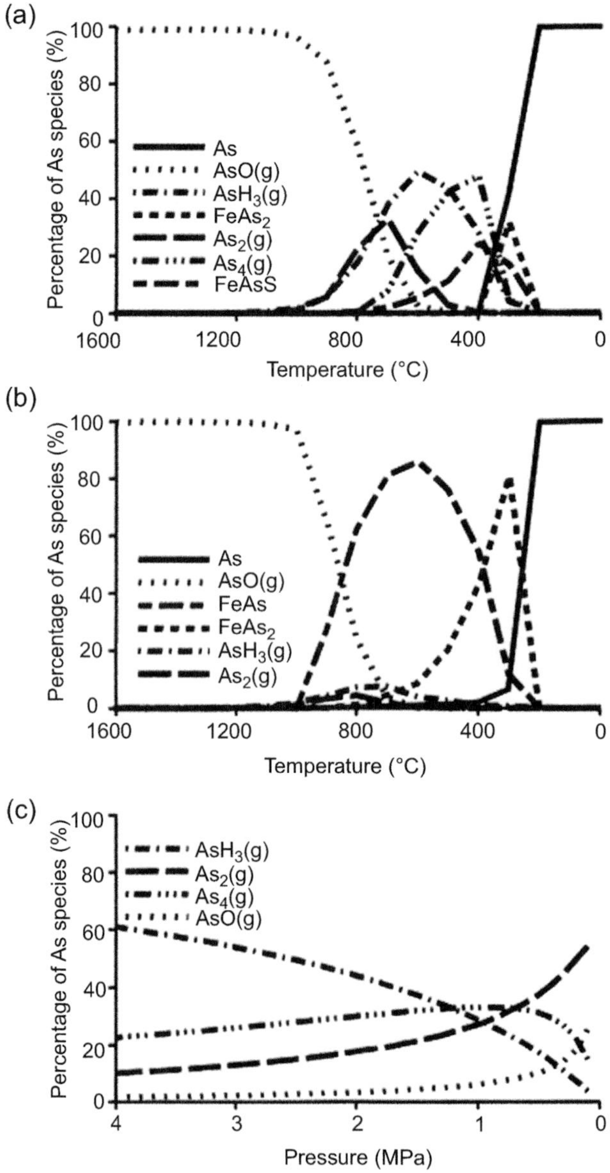

Figure 5.10 Equilibrium compositions for arsenic in (a) mixture III at 2.5 MPa; (b) mixture II at 2.5 MPa; and (c) different pressures at 600°C.
From Diaz-Somoano, et al. Trace element evaporation during coal gasification based on a thermodynamic equilibrium calculation approach. Fuel 2003;82:137–45.

At temperatures above 1000°C, all arsenic is present as AsO(g), at which temperature, changes in pressure and gas atmosphere do not change the equilibrium composition. When the temperature is lowered, that is, at those temperatures at which gas cleaning is possible, different condensation and gas-phase compounds may be present. Between 500 and 800°C, AsO(g), AsH_3(g), As_4(g), and As_2(g) are the species formed in the gas phase. These are the dominant species in a gas atmosphere containing H_2S(g) (Fig. 5.10(a)), but if the gas atmospheres do not contain H_2S(g) (Fig. 5.10(b)) the dominant species are condensed FeAs and $FeAs_2$. Between 300 and 500°C, As_2(g), As_4(g), AsH_3(g), FeAsS, FeAs, and $FeAs_2$ are the most stable species. When pressure increases, the concentration of AsH_3(g) increases significantly, while AsO(g), As_2(g), and As_4(g) decreases slightly (Fig. 5.10(c)). In an atmosphere free of H_2S(g), As_4(g) is not formed and the FeAsS is reduced to FeAs and $FeAs_2$. In this case, the condensed matter is most stable between 200 and 700°C at atmospheric pressure, while only AsO(g) is present at temperatures above 700°C. At temperatures below 200°C at atmospheric pressure, only condensed As is formed. Based on these theoretical results, it is expected that all of the arsenic present in the coal evaporates in the gas above 1000°C and also partially vaporizes in the gas phase under gas-cleaning conditions. However, at the low temperature of the emitted gas, only the condensed phase can be expected.

Mercury species at equilibrium are all in the gas phase (Fig. 5.11(a)). Above 400°C the dominant species is Hg(g) but at lower temperatures, $Hg(CH_3)_2$(g) may also be formed, their relative proportions depending on pressure (Fig. 5.11(b)). This composition does not vary when the gas atmosphere contains HCl(g) and H_2S(g). Mercury is present in the gas phase and is mostly emitted in flue gases.

Minimal variations in chromium species at equilibrium were observed with temperature and pressure. In all the conditions studied, at temperatures between 1400 and 300°C, chromium remains in the form of condensed Cr_2O_3 (Fig. 5.12(a) and (b)). At temperatures higher than 1500°C, at pressures ranging between 0.1 and 0.5 MPa, the formation of Cr(g) takes place, its concentration depending on gas in the atmosphere. At temperatures lower than 60°C and 1 atm, some $Cr(CO)_6$(g) may form in

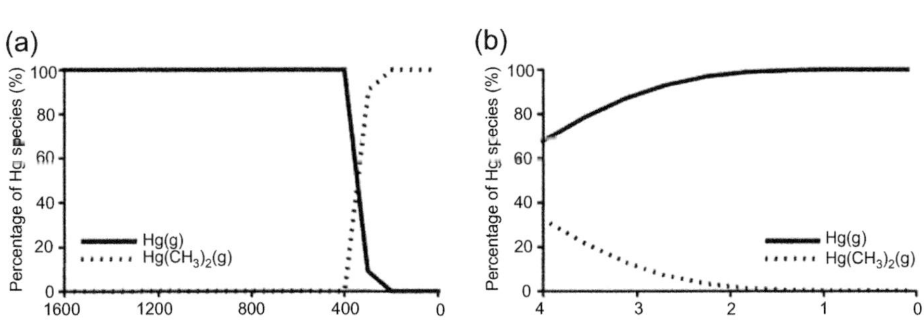

Figure 5.11 Equilibrium compositions for mercury (a) at different atmospheres and temperatures and (b) at different pressures and at 350°C.
From Diaz-Somoano, et al. Trace element evaporation during coal gasification based on a thermodynamic equilibrium calculation approach. Fuel 2003;82:137–45.

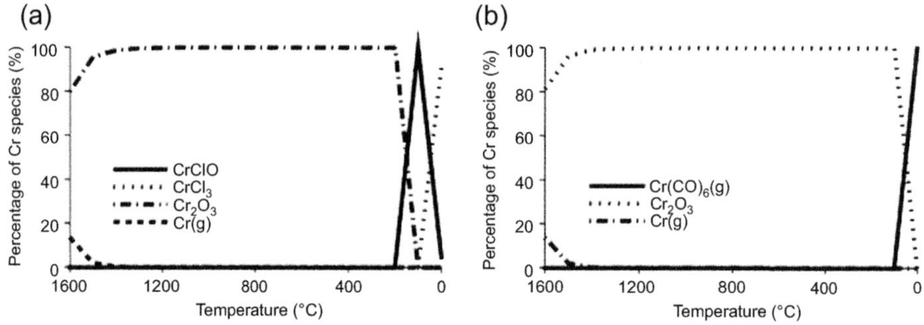

Figure 5.12 Equilibrium compositions for chromium between 0.1 and 2.5 MPa in (a) a gasification atmosphere containing HCl (mixtures II and IV) and in (b) a gasification atmosphere without HCl (mixtures I and III).
From Diaz-Somoano, et al. Trace element evaporation during coal gasification based on a thermodynamic equilibrium calculation approach. Fuel 2003;82:137–45.

the absence of HCl(g). In short, chromium species in the gas phase may form in the gasifier, but they cannot be expected to do so in gas-cleaning systems. However, a minor proportion of chromium may be emitted into the atmosphere in particular cases.

Selenium exists in the gas phase over the entire temperature range (Fig. 5.13(a) and (b)). The composition at equilibrium is not affected by changes in the gas atmosphere. However, temperature and pressure can change the relative proportions of selenium compounds, but mainly at low pressures and temperatures. Therefore these factors have no significant effect on the gas composition in the gas, gas-cleaning system, or gas emissions. H_2Se(g) is the main species present under these conditions. Se(g) can be formed at high temperatures, the concentration of which depends on the pressure.

Reed et al. investigated the transformation of Hg in different contents of sulfur from coal [27], as shown in Fig. 5.14(a). Gas phase Hg^0(g) is the predominant species until

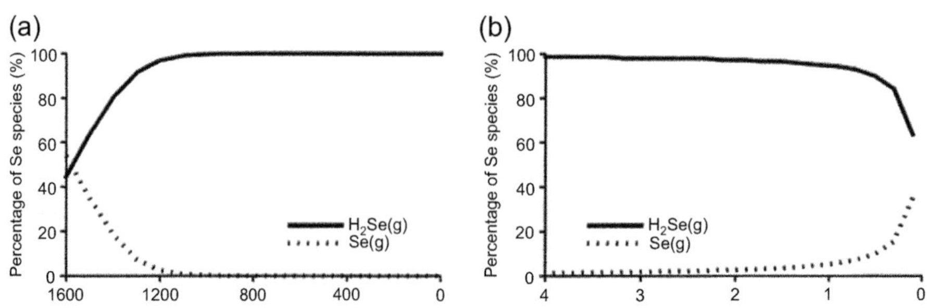

Figure 5.13 Equilibrium compositions for selenium in different gas atmospheres (a) at 1 atm and (b) at different pressures and at 1500°C.
From Diaz-Somoano, et al. Trace element evaporation during coal gasification based on a thermodynamic equilibrium calculation approach. Fuel 2003;82:137–45.

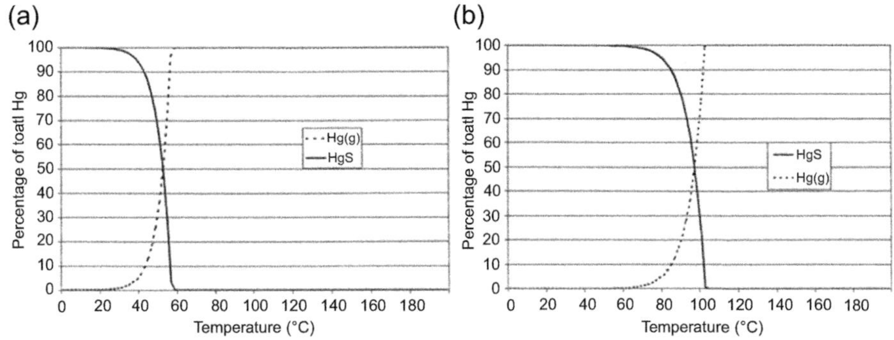

Figure 5.14 Effect of hot gas filter temperature on predicted thermodynamic equilibrium composition of Hg species in (a) low sulfur content and in (b) high sulfur content. From Reed, et al. Control of gasifier mercury emissions in a hot gas filter: the effect of temperature. Fuel 2001;80:623–34.

the hot gas filter temperature falls to below a point at which HgS can condense; in this system condensation was calculated to occur below about 508°C. Increasing amounts of S and Hg in the system should increase the temperature at which HgS may condense, as shown in Fig. 5.14(b) for a system composition containing more S, where an increase of 508°C is predicted in the temperature at which HgS condensation may occur. In actual operation, the gas filter temperature is normally kept higher than this to avoid condensation of water from the fuel gas; condensed HgS should not therefore be expected to form. Thus in view of the volatility of Hg and the high upstream temperatures, it seems likely that all of the Hg that enters the hot gas filter occurs in the form of Hg^0(g), which is not removable by a particle separator.

Helbel et al. took H_2S and HCl vaporized from coal gasification into consideration and predicted the transformation behavior of TEs in their presence [7]. The possible formation of compounds containing more than one TE was not considered; vapor/condensed partitioning was considered for each TE in isolation. For input conditions, measured coal composition and experimental conditions were used. The experimentally determined fractional vaporization of each TE was also used to define the initial vapor concentrations.

For arsenic, as shown in Fig. 5.15(a), AsO(g) is predicted to be the dominant vapor-phase species. AsO(g) has been detected in molecular beam studies of arsenic vapor evolved upon heating of solid As_2O_3 and an As_2O_3-CaO mixture; however, it has not been experimentally observed in a combustion or gasification system. If equilibrium is restricted to prevent the formation of AsO(g), AsS(g) and AsH_3(g) are high-temperature vapor-phase materials expected under postgasification conditions, and As_2(g) and As_4(g) become significant in the middle temperature range (600–1000K). Under postcombustion conditions, studies have shown that for Pittsburgh bitumen and Black Thunder sub-bituminous coal, concentrated arsenic oxide or calcium arsenate species dominate the calculated equilibrium distribution of arsenic at concentrations <1000K. Here, when oxygen is not present in the postgasification

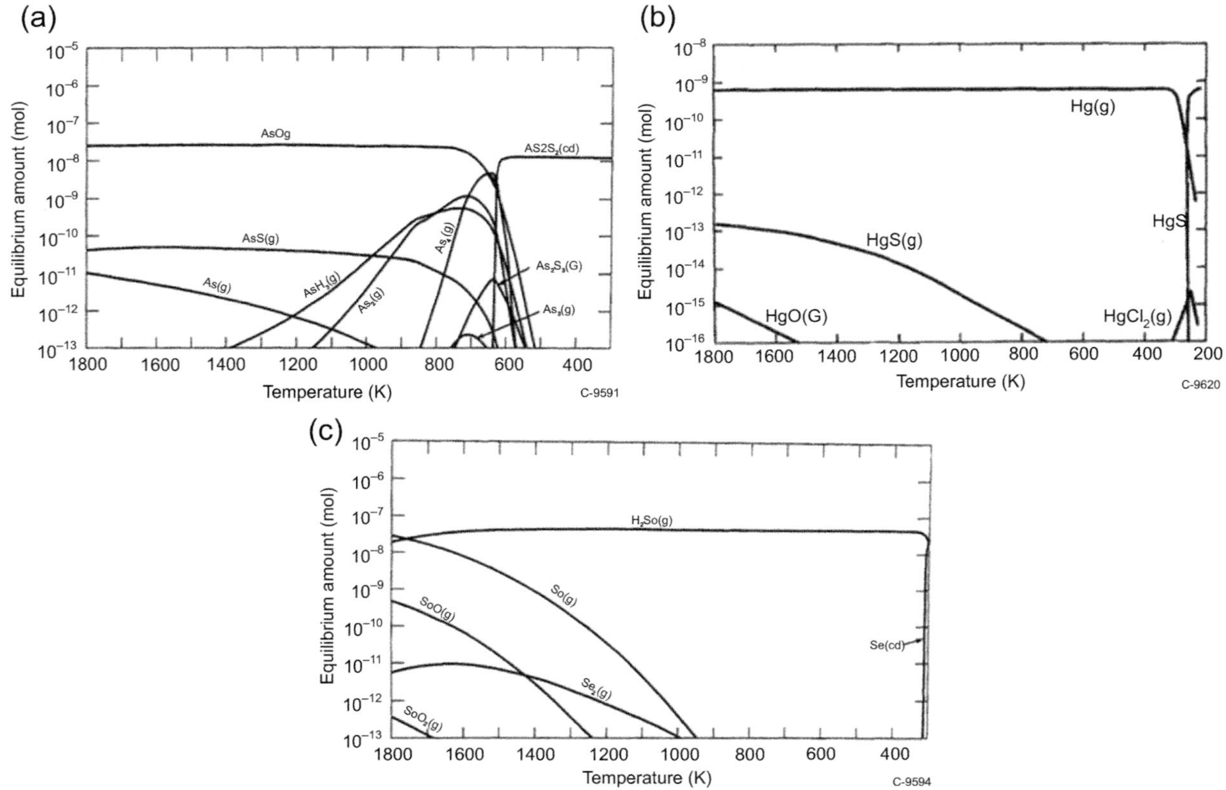

Figure 5.15 Partition patterns of (a) As, (b) Hg, and (c) Se during coal gasification in the presence of H_2S and HCl. From Helbel, et al. Trace element partitioning during coal gasification. Fuel 1996;75:931–39.

condition, the concentrated-phase arsenic species does not appear until the temperature falls below 650K.

For mercury, Hg(g) is dominant at all temperatures examined (Fig. 5.15(b)). This is identical to results reported previously for combustion of a sub-bituminous coal under reducing conditions. For selenium, the reduced gas-phase species $H_2Se(g)$ is dominant under postgasification conditions. In contrast, $SeO_2(g)$ is favored under postcombustion conditions at temperatures >500−600K. As shown in Fig. 5.15(c), the higher volatility of H_2Se relative to SeO_2 indicates that vapor-phase selenium will persist to lower temperatures under postgasification conditions than it will under postcombustion conditions. It may therefore be more difficult to remove selenium from gasifier gases than from postcombustion gases.

Liu et al. investigated TE transformations during UCG, i.e., conversion of coal in the seam into combustible gas without mining and transportation [20], under both oxygen−steam and air-blown conditions with MDTATA software. Under oxygen−steam conditions, the proportion of oxygen in the gas is slightly increased to promote gasification of the coal. The mole fraction of arsenic species in the underground gasification of the oxygen-vapor large section (1500 Pa) is shown in Fig. 5.16(a). It was observed that $As_2(g)$ was formed at 250°C and became a main gaseous substance at a temperature of 300−900°C. $As_s(g)$ appears above 500°C and becomes the dominant species between 900 and 1300°C. The formation of AsO(g) and As(g) occurs at 800°C and dominates at temperatures above 1300°C. At temperatures below 300°C, arsenic is present in the solid phase in the form of As and As_2S_3. Based on these results, it is apparent that elemental arsenic evaporates into the gas phase during the early gasification stage in the large cross-section UCG process. When an underground thermal cleaning system is applied, it still exists in the form of a gas phase in which $As_2(g)$ is the main gaseous substance. As the gas flows through the production well and reaches the surface, the gas temperature drops below 200°C and the arsenic-containing material will condense to form a solid compound. This indicates that arsenic is redistributed as a solid phase under vapor-phase cooling and involves a complex process of converting and condensing arsenic on solid particles by physical adsorption and chemical reaction. Therefore, at this stage, arsenic will be enriched with ash particles and entrained in the UCG gas, which facilitates the removal of arsenic from the fuel gas.

The results in the presence of mineral elements show that only potassium has a significant effect on arsenic species, while effects of Ca, Fe, and Na on species partitioning in equilibrium are not observed. The species of $As_2Ca_3O_8$, $AsFeO_4$, $As_2Fe_3O_8$, K_3AsO_4, Na_3AsO_4, and Na_3As are considered in Scientific Group Thermodata Europe (SGTE) in the calculated system, while K_3As is excluded. It is found that Na_3As and Na_3AsO_4 are not formed and arsenic sulfide is still the dominant condensed form in the presence of sodium. When the amount of potassium is 20 times higher than that of arsenic, arsenic becomes less volatile and K_3AsO_4 becomes the dominant condensed phase, as shown in Fig. 5.16(b). The condensation temperature rises with the increase in the amount of potassium in the calculated system. However, since potassium in coal is tightly combined with aluminosilicate and is not easily released, only a small quantity will enter the gasification system and react with gaseous species of other elements. Thus arsenic volatilization and species distribution are not markedly affected.

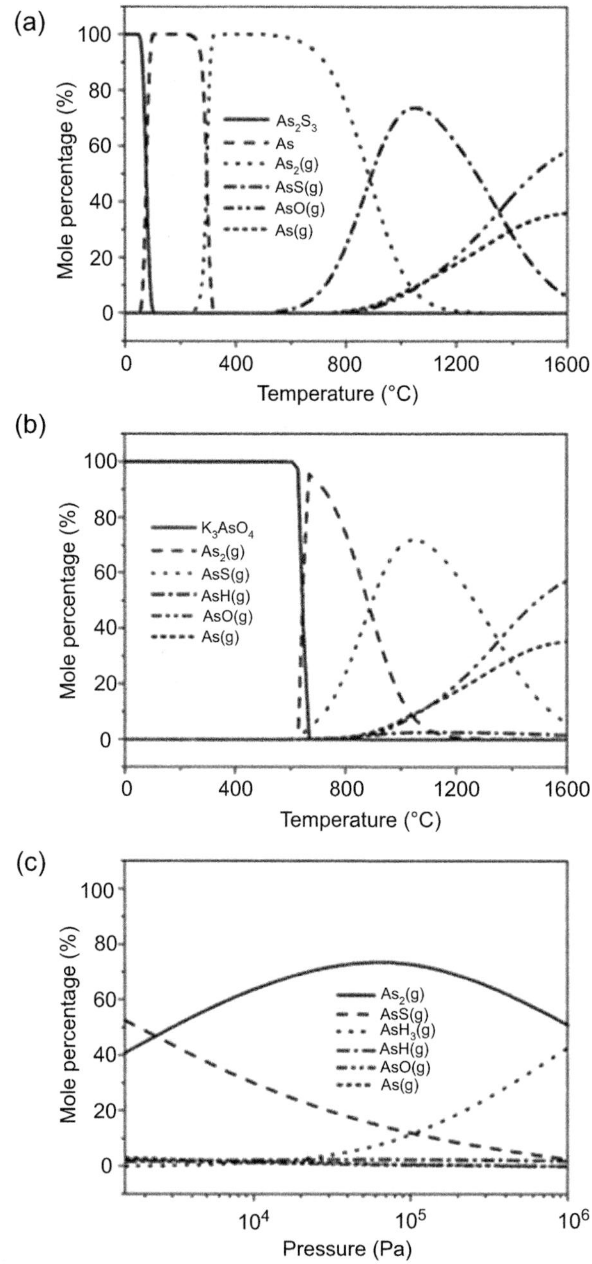

Figure 5.16 Equilibrium partitioning of arsenic for (a) oxygen–steam gasification at 1500 Pa, (b) in the presence of potassium (K/As = 20) for oxygen–steam gasification at 1500 Pa, and (c) with different pressures.

From Liu, et al. Thermodynamic equilibrium study of trace element transformation during underground coal gasification. Fuel Process Technol 2006;87:209–15.

$As_2Ca_3O_8$, $AsFeO_4$, and $As_2Fe_3O_8$ are not found in the equilibrium composition of arsenic.

Regarding the distribution of arsenic under air-blown gasification under the same pressure, no significant change was observed at temperatures higher than 400°C. However, at a temperature of 200−400°C, As_4(g) is produced and remains in the flue gas from coal combustion in a reducing atmosphere, consistent with the reducing atmosphere in oxygen−steam gasification. In addition, it was found that the condensed point of arsenic was lower when air was blown into the gasification system, indicating that arsenic became more volatile in this case.

The effect of elevated pressure on equilibrium composition of arsenic was evaluated at pressures from 1500 Pa to 10 MPa. The results indicate that pressure variation remarkably changes the distribution of the gaseous species, as shown in Fig. 5.16(c). With the increase in pressure, the mole fraction of AsS(g) decreases, while that of the reduced species, AsH_3(g), remarkably increases. When pressure attains 10 MPa in the deep UCG process, AsH_3(g) and As_2(g) coexist at nearly the same fractions, indicating that reducing atmosphere is significantly enhanced at higher pressures. Meanwhile, condensation of arsenic species arises at lower temperatures with increasing pressure. For instance, the condensed phase of arsenic occurs at 300°C at 1500 Pa, being significantly lower than 420°C needed at 10 MPa. Therefore arsenic will not exist in the gas phase below 420°C at 10 MPa and it always happens in UCG of a deep coal seam.

Selenium is in the gas phase almost over the entire temperature range (Fig. 5.17(a)). However, the relative distribution of gaseous species can vary with temperature. H_2Se(g) is the main species with a temperature below 1250°C. The formation of Se(g) increases sharply with increasing temperature and dominates at temperatures above 1200°C. HSe(g) and SeS(g) appear as secondary species between 600 and 1250°C. When the gas temperature drops below 50°C, the condensed phase of selenium exists in the form of Se. Therefore it seems that all of the above gaseous substances coexist in the gasification stage. Even when the temperature of the production well is below 200°C, selenium can remain in the gas phase. H_2Se(g) is the only gaseous substance in the hot and cold gas-cleaning system, which creates a burden on selenium discharge. Na_2Se and K_2Se are species considered in the presence of Na and K, but they are not easily formed and have no effect on the distribution of selenium species and selenium condensation. The effect of Ca and Fe on selenium partitioning has not been observed.

Selenium equilibrium composition under air-blown gasification at 1500 Pa is shown in Fig. 5.17(b). It is evident that SeS(g), instead of the reduced species Se(g), possesses a high proportion at temperatures from 950 to 1250°C. It indicates that air gasification is favorable for the emission of sulfur and the formation of sulfide species for those thiophil elements present in coal. With the increase in pressure, H_2Se(g) extends to hold a broader range. When the pressure attains 1 atm, H_2Se(g) becomes the only dominant gaseous species. Moreover, the condensation point of selenium increases by 100°C at a pressure of 10 MPa.

For oxygen−steam gasification at a pressure of 1500 Pa, the gaseous species of chromium are only formed at temperatures higher than 1000°C (Fig. 5.18). Cr(g) is the dominant gaseous species at temperatures above 1200°C, CrO(g) and Cr_2O(g)

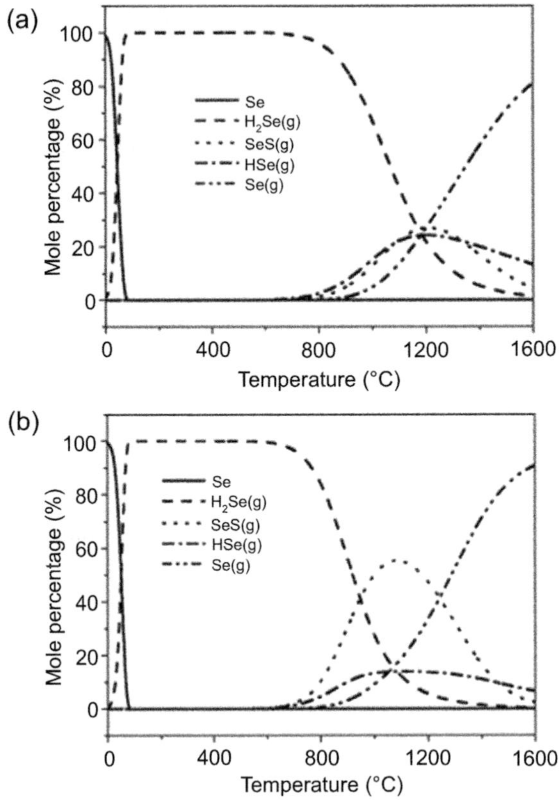

Figure 5.17 Equilibrium partitioning of selenium for (a) oxygen–steam gasification and (b) air-blown gasification at 1500 Pa.
From Liu, et al. Thermodynamic equilibrium study of trace element transformation during underground coal gasification. Fuel Process Technol 2006;87:209–15.

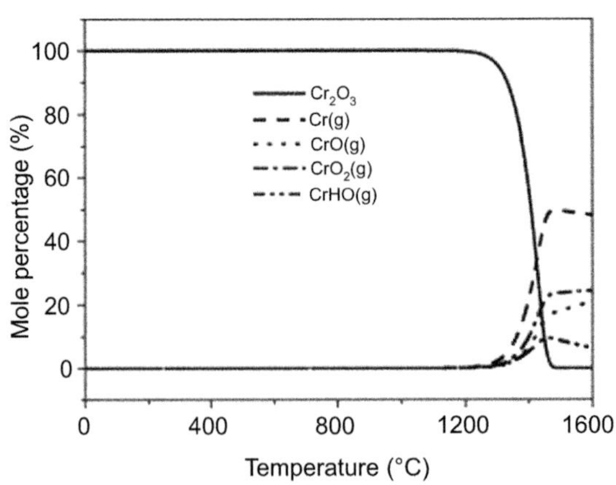

Figure 5.18 Equilibrium partitioning of chromium for oxygen–steam gasification at 1500 Pa.
From Liu, et al. Thermodynamic equilibrium study of trace element transformation during underground coal gasification. Fuel Process Technol 2006;87:209–15.

only occurring at trace quantities. Below 1000°C, chromium is in the form of Cr_2O_3. It is evident that chromium becomes volatile only at high gasification temperatures, e.g., 1200°C. Once gas leaves the gasification zone and cools, it is condensed and transformed into the solid phase. Therefore chromium cannot occur in UCG gas in a gas-cleaning system and the majority of chromium is enriched in bottom ash left underground. Moreover, the occurrence of chromium species is not influenced by sulfur and chlorine contained in the system. With the improvement of pressure, the condensation point of chromium increases and the mole fraction of gaseous species remarkably decreases. At a pressure of 10 MPa, no gaseous species occurs even at the gasification stage. Similar results are obtained for air gasification.

5.5 Conclusion

Under the reducing conditions associated with the gasifier product gas, related information was quite limited compared to that from conventional combustion modes. Thus relying on thermodynamics was the mainstream when studying TE partitions during the coal gasification process. As, Hg, F, and Se would remain in the vapor phase at the gasification temperature for most cases, while Cr with low volatility will remain in bottom ash and fly ash in different forms. The lower temperatures associated with fluidized and fixed beds should result in lower fractions vaporized for all elements except mercury. In addition, for nearly all the elements considered, the volatile fraction decreased as the partial pressure of oxygen was increased to a stoichiometric ratio of 1.0. Although not conclusive, this result is suggestive of increased reaction between the TEs and the coal minerals (e.g., aluminosilicates) at the higher coal particle temperatures associated with higher oxygen partial pressures of stoichiometric combustion. Moreover, the operation of many commercial gasifiers at elevated pressure will slightly reduce the fraction of an element vaporized for all elements except mercury. When the gas temperature was relatively low in the tail gases, the transformation of TEs possibly occurred to induce more complicated compositions of TEs.

Acknowledgments

This project was supported by the National Natural Science Foundation of China (51776227) and the Natural Science Foundation of Hunan Province, China (2018JJ1039, 2018JJ3675).

References

[1] Minchener AJ. Coal gasification for advanced power generation. Fuel 2005;84:2222–35.

[2] Bunt JR, Waanders FB. Trace element behaviour in the Sasol–Lurgi MK IV FBDB gasifier. Part 1 – the volatile elements: Hg, As, Se, Cd and Pb. Fuel 2008;87:2374–87.

[3] Dıaz-Somoano M, Martınez-Tarazona MR. Trace element evaporation during coal gasification based on a thermodynamic equilibrium calculation approach. Fuel 2003;82:137–45.

[4] Beishon D, Hood J, Vierrath H. The fate of trace elements in the BGL gasifier. In: Proceedings of the 6th annual international Pittsburgh coal conference; 1989. p. 539–47.
[5] Clarke LB. The fate of trace elements during coal combustion and gasification: an overview. Fuel 1993;72:731–6.
[6] Mojtahedi W. Trace metals volatilisation in fluidised-bed combustion and gasification of coal. Combust Sci Technol 1989;63:209–27.
[7] Helble JJ, Mojtahedi W, Lyyränen J, Jokiniemi J, Kauppinen E. Trace element partitioning during coal gasification. Fuel 1996;75:931–9.
[8] Argent BB, Thompson D. Thermodynamic equilibrium study of trace element mobilisation under air blown gasification conditions. Fuel 2002;81:75–89.
[9] Shen F, Liu J, Zhang Z, Yang Y. Temporal measurements and kinetics of selenium release during coal combustion and gasification in a fluidized bed. J Hazard Mater 2016;310: 40–7.
[10] Liu S, Wang Y, Yu L, Oakey J. Volatilization of mercury, arsenic and selenium during underground coal gasification. Fuel 2006;85:1550–8.
[11] Liu S-Q, Liu J-H, Yu L. Environmental benefits of underground coal gasification. J Environ Sci 2002;14:284–8.
[12] Lu H, Chen H, Li W, Li B. Transformation of arsenic in Yima coal during fluidized-bed pyrolysis. Fuel 2004;83:645–50.
[13] Wheeldon J. A review of PFBC power plant designs. In: Pittsburgh coal conference, Pittsburgh, PA (United States); 1996.
[14] Reed GP, Dugwell DR, Kandiyoti R. Control of trace elements in gasification: distribution to the output streams of a pilot scale gasifier. Energy Fuels 2001;15:794–800.
[15] Richaud R, Lachas H, Collot A-G, Mannerings AG, Herod AA, Dugwell DR, Kandiyoti R. Trace mercury concentrations in coals and coal-derived material determined by atomic absorption spectrophotometry. Fuel 1998;77:359–68.
[16] Vidic RD, Chang M-T, Thurnau RC. Kinetics of vapor-phase mercury uptake by virgin and sulfur-impregnated activated carbons. J Air Waste Manag Assoc 1998;48:247–55.
[17] Bunt JR, Waanders FB. Trace element behaviour in the Sasol–Lurgi MK IV FBDB gasifier. Part 2 – the semi-volatile elements: Cu, Mo, Ni and Zn. Fuel 2009;88:961–9.
[18] Bunt JR, Waanders FB. Trace element behaviour in the Sasol-Lurgi fixed-bed dry-bottom gasifier. Part 3 – the non-volatile elements: Ba, Co, Cr, Mn, and V. Fuel 2010;89:537–48.
[19] Dyk JCV, Waanders FB. Manipulation of gasification coal feed in order to increase the ash fusion temperature of the coal enabling the gasifiers to operate at higher temperatures. Fuel 2007;86:2728–35.
[20] Liu S, Wang Y, Yu L, Oakey J. Thermodynamic equilibrium study of trace element transformation during underground coal gasification. Fuel Process Technol 2006;87: 209–15.
[21] Wilde KA, Halbrook ME. Thermodynamic distribution of trace elements by minimization of free energy. Ind Eng Chem Fundam 1977;16:489–92.
[22] Frandsen F, Dam-Johansen K, Rasmussen P. Trace elements from combustion and gasification of coal—an equilibrium approach. Prog Energy Combust Sci 1994;20:115–38.
[23] Li W, Lu H, Chen H, Li B. Volatilization behavior of fluorine in coal during fluidized-bed pyrolysis and CO-gasification. Fuel 2005;84:353–7.
[24] Guo S, Yang J, Liu Z. The fate of fluorine and chlorine during thermal treatment of coals. Environ Sci Technol 2006;40:7886–9.

[25] Pinto F, Lopes H, André RN, Gulyurtlu I, Cabrita I. Effect of catalysts in the quality of syngas and by-products obtained by co-gasification of coal and wastes. 2: heavy metals, sulphur and halogen compounds abatement. Fuel 2008;87:1050–62.

[26] Weinell CE, Jensen PI, Dam-Johansen K, Livbjerg H. Hydrogen chloride reaction with lime and limestone: kinetics and sorption capacity. Ind Eng Chem Res 1992;31:164–71.

[27] Reed GP, Ergüdenler A, Grace JR, Watkinson AP, Herod AA, Dugwell D, Kandiyoti R. Control of gasifier mercury emissions in a hot gas filter: the effect of temperature. Fuel 2001;80:623–34.

Trace element partition in a coal-feed industry furnace

6

Zequn Yang [1], *Bengen Gong* [2], *Jianping Yang* [3], *Yongchun Zhao* [2], *Junying Zhang* [2]
[1]Department of Civil Engineering, The University of Hong Kong, Hong Kong, China;
[2]State Key Laboratory of Coal Combustion, Huazhong University of Science and Technology, Wuhan, China;
[3]School of Energy Science and Engineering, Central South University, Changsha, China

6.1 Introduction

The emission of trace elements (TEs) from coal combustion is important because of their association with environmental issues and the health of plants, animals, and humans. In recent years, more and more toxic incidents related to the adverse effects of TEs on public health and the ecological environment have been reported. More than 60% of total primary energy consumption is from coal in China, and it is predicted that this proportion will increase to 65% by 2050 [1]. Some TEs present in coal are known to be harmful to the environment and human health. TEs such as Cr, Ni, Pb, Zn, Cu, V, and Cd are carcinogenic, and As is one of the most common and hazardous carcinogenic elements [2–4]. Thus it is important to study the emission characteristics of TEs and develop effective control technologies for TE emissions. Understanding the behavior of these elements and controlling their emissions will provide basic information for the development of extraction methods.

6.2 Emission characteristics and removal of trace elements in coal-fired industrial boilers

China utilizes the majority of industrial boilers and kilns in the world. There were 460,000 coal-fired industrial boilers in China in 2011, which consumed about 700 million t of raw coal per year, accounting for 22.4% of Chinese total coal consumption. Different coal-feed industry furnaces can cause unique emissions of TEs. China's coal-fired industrial boilers are mainly the layer-burning boiler, which chiefly consists of the chain-grate boiler (CGB). Nearly 60% of industrial boilers produced by industrial boiler enterprises are CGB, and the circulating fluidized-bed (CFB) boiler has the advantages of a wide range of coal types and stable operation at low load. A large number of studies have been conducted on TE emissions from coal-fired power plant boilers, but there are few studies on TE emissions

Emission and Control of Trace Elements from Coal-Derived Gas Streams
https://doi.org/10.1016/B978-0-08-102591-8.00006-4

from coal-fired industrial boilers. Although the flue gas emissions from industrial boilers are much lower than coal-fired power plants, industrial boilers are mostly distributed in densely populated cities and suburban industrial areas. The control technology of industrial boilers is relatively simple, and its synergistic control effect is poor for heavy metal pollutants. Thus it is necessary to carry out research into the emissions of TEs from coal-fired industrial boilers [5].

6.2.1 Emission characteristics of trace elements in different coal-feed industry furnaces

Lopez-Anton et al. [6] studied Hg behavior in the industrial combustion equipment of a 50 MW CFB boiler. The first combustion sample is the blend of 36–41 wt% bituminous coal, 6 wt% limestone, and 51–56 wt% coal waste from the old disposal sites that contain coal; another combustion sample is the blend of 32.2 wt% coal, 5.4 wt% limestone, and 54.5 wt% coal waste. Coal blends, limestone, bottom ashes, fly ashes, airborne particulates, and flue gases were sampled and analyzed (Fig. 6.1).

The combustion equipment allows the burning of combustible blends with high ash content, which is about 63%–65%. The validity of the Hg content in the samples can be confirmed by the percent relative standard deviation value. It was found that Hg in

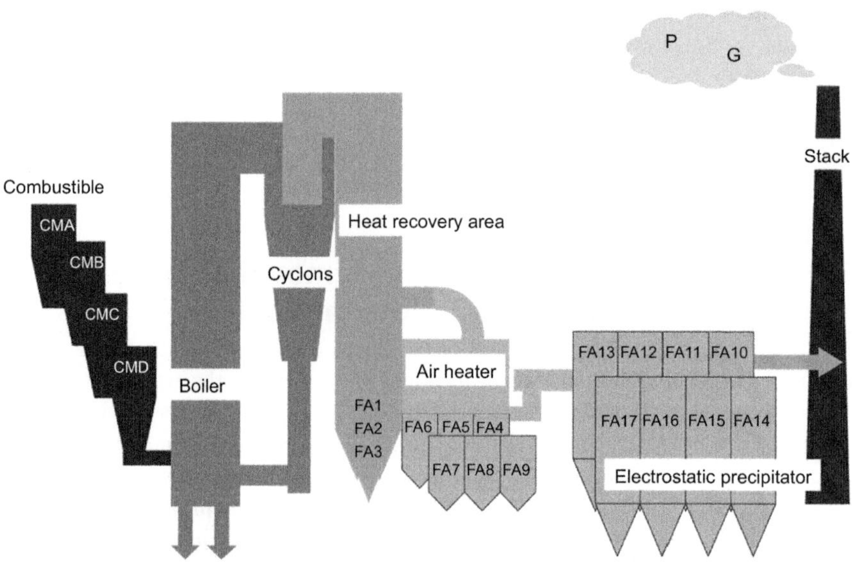

Figure 6.1 Schematic diagram of the sampling sites. *BA*, bottom ashes; *CM*, coal blends; *FA*, fly ashes; *G*, flue gases; *P*, airborne particulates.
From López-Antón MA, Díaz-Somoano M, Diaz L, Martínez-Tarazona MR. Avoiding mercury emissions by combustion in a Spanish circulating fluidized-bed combustion (CFBC) plant. Energy Fuels 2011;25:3002–8.

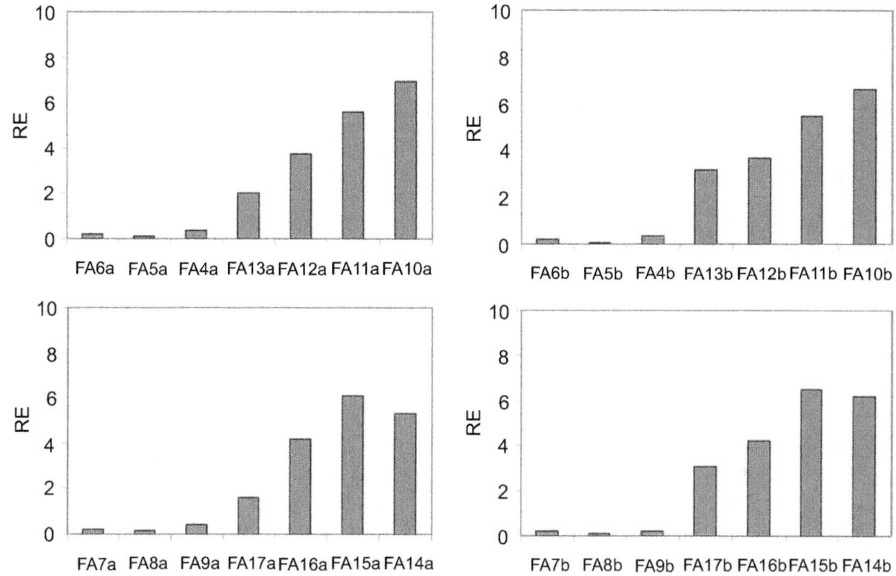

Figure 6.2 The relative enrichments (REs) of mercury in fly ash from the air heater hoppers and electrostatic precipitator.
From López-Antón MA, Díaz-Somoano M, Diaz L, Martínez-Tarazona MR. Avoiding mercury emissions by combustion in a Spanish circulating fluidized-bed combustion (CFBC) plant. Energy Fuels 2011;25:3002–8.

feed coal was distributed in combustion products. Hg concentration in the bed ash sampled from two strippers showed that the Hg content in bottom ash is not significant. In contrast, fly ash contained a high Hg concentration, which indicated that fly ashes played a very important role in capturing mercury. The fly ashes collected from the hoppers contained different Hg levels, among which the fly ash from the electrostatic precipitator (ESP) had the highest Hg concentration, especially for FA10, FA11, FA14, and FA15. Since the mercury was higher in the second campaign of combustible blend, the corresponding fly ash from this campaign contained higher mercury.

Fig. 6.2 shows the relative enrichment (RE) factor values of fly ash samples obtained from two sampling campaigns a and b. It is found that the fly ashes from the last two hoppers of ESP have the highest REs, indicating that these two fly ashes are highly enriched with Hg. Moreover, the value of REs of the fly ashes collected from the air heater hopper are very low, whereas the value of REs of the fly ashes collected from ESP are higher than 1; in particular, for FA10, FA11, FA14, and FA15, the values of REs are higher than 3. This is consistent with the strongest capture ability of mercury by the fly ashes from the last two hoppers of ESP, which have the characteristics of low particle size, high carbon content, and high surface area.

Table 6.1 Distribution of mercury in the input and output.

	Input	Output				
	CM Hg (g/h)	BA Hg (g/h)	FA Hg (g/h)[a]	P Hg (g/h)	G Hg (g/h)	Hg_{out}/Hg_{in}
Sampling 1	10.9	0.01	9.24	0.01	0.062	0.85
Sampling 2	14.5	0.01	14.1	0.02	0.041	0.99

BA, bottom ashes; *CM*, coal blends; *FA*, fly ashes; *G*, flue gases; *P*, airborne particulates.
[a]FA Hg (g/h) represents the weighted average of all FAs (FA1-FA17).
From López-Antón MA, Díaz-Somoano M, Diaz L, Martínez-Tarazona MR. Avoiding mercury emissions by combustion in a Spanish circulating fluidized-bed combustion (CFBC) plant. Energy Fuels 2011;25:3002–8.

Table 6.2 Mercury speciation in flue gases.

	Hg_{total} (μg/m^3)	Hg^{2+} (μg/m^3)	Hg^0 (μg/m^3)
Sampling 1	0.28	0.12	0.16
Sampling 2	0.21	0.04	0.17

From López-Antón MA, Díaz-Somoano M, Diaz L, Martínez-Tarazona MR. Avoiding mercury emissions by combustion in a Spanish circulating fluidized-bed combustion (CFBC) plant. Energy Fuels 2011;25:3002–8.

The mass balance of the input and output of the two sampling campaigns is shown in Table 6.1. It can be observed that the Hg content in particulate matter (PM) of flue gas is relatively high. However, the mass of such PM is very low. Thus the output rate of particulate Hg in flue gas is relatively low. Most of the mercury in this fluidized-bed power plant is retained in fly ash. Mercury emissions from the gas phase are significantly low with less than 1% total mercury (Hg^t) emissions from the gas phase.

Table 6.2 shows the concentration of gas-phase Hg speciation in flue gas. The results showed that there were oxidized and elemental mercury (Hg^0) in the flue gas, and Hg^0 had the highest proportion. The emission concentration of Hg^{2+} was lower than 0.02 μg/m^3, while Hg^0 is around the order of 0.2 μg/m^3.

Fluidized-bed furnace (FBF) and stoker-fired boiler (SFB) are widely used in small coal-fired power plants. Although the emission of Hg, As, and Se from industrial boilers is not growing rapidly, industrial boilers remain the largest Hg, As, and Se emission source related to coal utilization. The total amount of Hg, As, and Se from industrial boilers increased from 40.8, 382.67, and 364.81 t in 1980 to 155.46, 1348.70, and 1322.78 t in 2007, respectively [7]. The average release rates of Hg, As, and Se in FBF were 98.92%, 75.6%, and 98.05%, respectively, those in SFB were 85.00%, 77.18%, and 80.95%, respectively, and those in the coke furnace (CF) were 83.15%, 30.00%, and 40.00%, respectively.

The historical emission trend and composition of Cd, Cr, and Pb by different sectors are shown in Fig. 6.3 [8]. The emission of the industrial sector was the leading

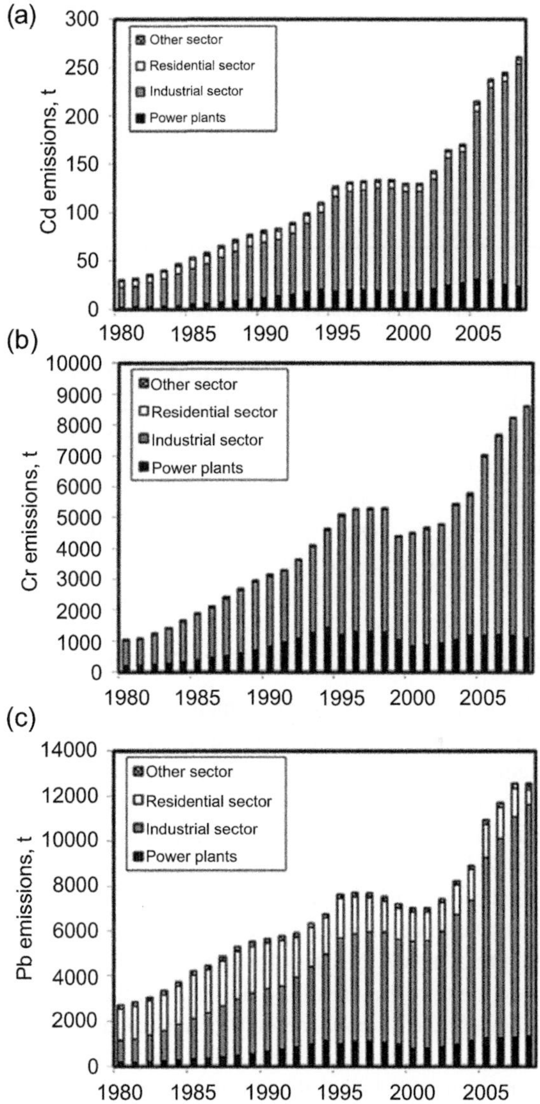

Figure 6.3 Trend of (a) Cd, (b) Cr, and (c) Pb emissions by different sectors in 1980–2008. From Tian HZ, Cheng K, Wang Y, Zhao D, Lu L, Jia WX, Hao JM. Temporal and spatial variation characteristics of atmospheric emissions of Cd, Cr, and Pb from coal in China. Atmos Environ 2012;50:157–63.

source of the three elements. The emission amounts of Cd, Cr, and Pb from the industrial sector increased from 19.89, 788.58, and 965.47 t in 1980 to 230.88, 7454.26, and 10271.45 t in 2008, which accounted for ~88.3%, ~86.7%, and ~81.8% of the total emission, respectively. The growth of the Cd, Cr, and Pb

Table 6.3 The release rates of Cd, Cr, and Pb during coal combustion in different kinds of industrial boilers (%).

Boiler type	Cd	Cr	Pb
Stoker-fired boiler	42.53	26.74	40.10
Fluidized-bed furnace	91.50	81.33	77.33
Coke furnace	20.00	24.00	31.50

From Tian HZ, Cheng K, Wang Y, Zhao D, Lu L, Jia WX, Hao JM. Temporal and spatial variation characteristics of atmospheric emissions of Cd, Cr, and Pb from coal in China. Atmos Environ 2012;50:157–63.

emissions by industrial sector increased most rapidly, with annual growth rates of 9.2%, 8.4%, and 8.8%, respectively. Table 6.3 summarizes the release rates of Cd, Cr, and Pb during coal combustion in FBF, SBF, and CF [8]. The release rates of the three elements in FBF were 91.50%, 81.33%, and 77.33%, respectively, those in SFB were 42.53%, 26.74%, and 40.10%, respectively, and those in CF were 20.00%, 24.00%, and 31.50%, respectively. The release rates of Cd, Cr, and Pb in FBF were much higher than SFB and CF.

6.2.2 Distribution of trace elements in combustion products

Mercury is a volatile element in coal, most of which will evaporate into the atmosphere in the process of coal combustion. However, it was found that the residual carbon in fly ash can keep a certain amount of mercury, and it was more easily distributed in the fly ash than in the bottom ash. Yang et al. [9] investigated the Hg emission characteristics in a 135 MW CFB in China. The growth factor of Hg is higher than 2 in fly ash, while the development of the Hg factor in lower ash is less than 1, indicating that mercury was purged by fly ash and washed ash in the coal combustion CFB unit. Mercury has the potential to concentrate on solid products due to its unique combustion properties. Hg enrichment that causes fly ash before ESP is usually less than fly ash in ESP fields. This is because the temperature of the flue gas before ESP is higher than that of the ESP, which does not fit the gaseous mercury of adsorbed particulate mercury (Hg^P) from the perspective of chemical thermodynamics. The delay in ESP living time can lead to increased contact time between fly ash and gaseous mercury, which increases the proportion of Hg. It shows that the conversion between the gaseous and solid phases of Hg in flue gas is controlled by certain factors, such as reaction rate and reaction time. Thus ESP can help improve Hg gas adsorption by fly ash and reduce Hg emissions in the atmosphere.

Li et al. [10] studied TE distribution during bituminous coal combustion in a 50 kW stoker-fired unit. The bottom ash contained high Sc, Cs, Ba, and Rb, while the fly ash is enriched with As, Sb, Se, Br, Hg, Mn, Zn, and W. After feed coal combustion, Ba, V, Zr, Rb, Sr, Th, F, La, Ce, Nb, Ga, Y, Co, Cr, and Ni are mainly distributed in the bottom ash (Fig. 6.4). These elements were probably related to the minerals in feed coal, and are most likely strongly bonded or encapsulated in a silicate and aluminosilicate glass matrix.

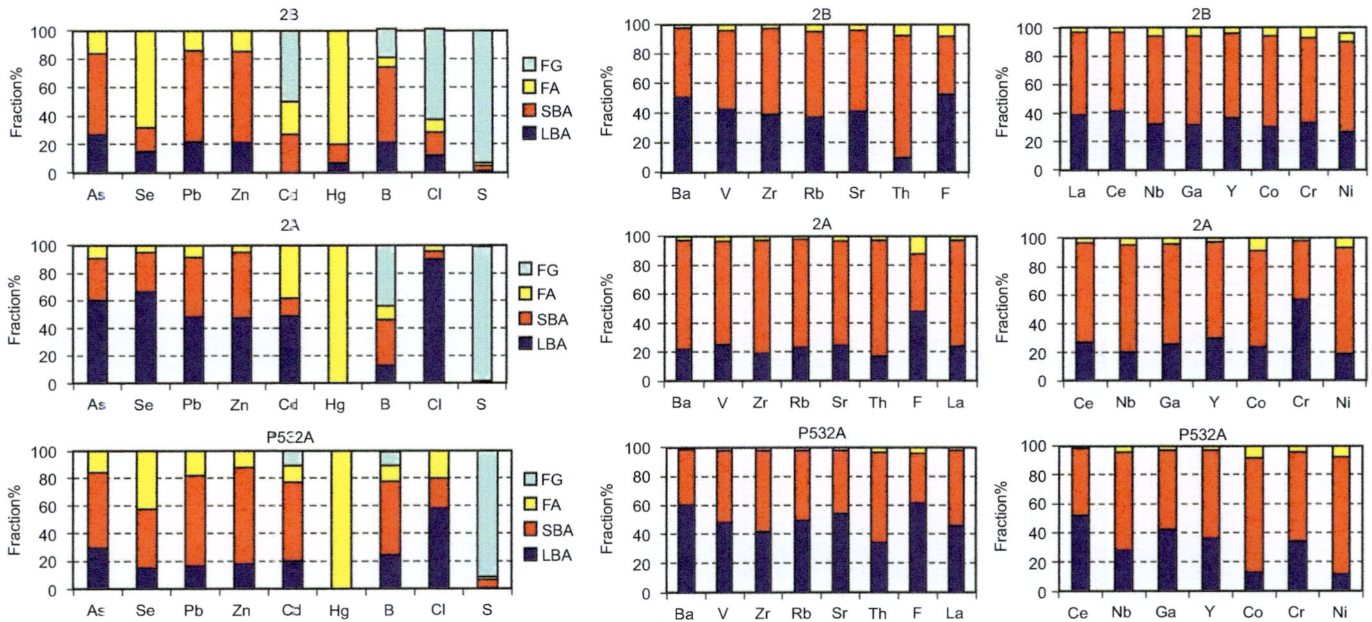

Figure 6.4 Partitioning of trace elements in large bottom ash (LBA), small bottom ash (SBA), fly ash (FA), and flue gas (FG).
From Li Z, Clemens AH, Moore TA, Gong D, Weaver SD, Eby N. Partitioning behaviour of trace elements in a stoker-fired combustion unit: an example using bituminous coals from the Greymouth coalfield (Cretaceous), New Zealand. Int J Coal Geol 2005;63:98–116.

As, Se, Pb, and Zn show a moderate growth rate (emission factor (EF) = 2−16) in the lower ash and large fly ash development (EF = 11−25). Although the contents of these elements are high in fly ash, the ratio of TEs usually does not reach 20%, and over 80% is left in ashes. In the lower ash, these elements can often be distributed within the aluminosilicate melt, oxide, or glass matrix. As for fly ashes, the elements pass over the particles due to large surface area and electrostatic effect. Although most common element can easily produce, 84%−90% of them failed to escape the firebox, and only 10%−16% of the As is fly ash. The distribution of Se is different in samples 2B (from the bottom of the seam) and 2A (from the surface). When sample 2B was burned, 68% of Se was released from the firebox and remained on the fly ash. When sample 2A was burned, only 5% of the Se was concentrated on the fly ash, and 95% was given ash. Sample 2A has higher ash, higher Si and K, and lower Fe, which produces additional Si-Al glass; it helps the Se acquire lower ash. In contrast, the ash and Si and K contents of sample 2B are lower, while the Fe content is higher, so the lower ash of sample 2B has a limited ability to extract Se, and most Se remains on the fly ash.

The elements Hg, B, Cd, and Cl showed a significant enrichment with fly ash (EF = 10−40) and moderate enrichment with bottom ash (EF < 10). These elements were generally highly volatile, and had a high partitioning proportion in the flue gas (up to 97%) and fly ash by adsorption or condensation. There was about 90% of Hg, 50% of Cd, 64% of Cl, and 44% of B distributed in the flue gas.

Clemens et al. [11] investigated the partitioning behavior of TEs during the combustion of two coal samples (CRL#61/208 and 61/209) in a 50 kW underfeed stoker combustion unit. As shown in Table 6.4, Ca can affect the partitioning of volatile

Table 6.4 Partitioning of trace elements.

	Coal sample 61/208			Coal sample 61/209		
Element	**Bottom ash (mg/g)**	**Cyclone ash (mg/g)**	**Flue gas (mg/dsm^3)**	**Bottom ash (mg/g)**	**Cyclone ash (mg/g)**	**Flue gas (mg/dsm^3)**
As	11.6 (58)	230 (29)	Not detected	24 (39)	510 (51)	Not detected
B	1,700 (60)	5,400 (4)	202 (1)	6,800 (99)	600 (2)	23 (0.2)
Se	0.75 (7)	34 (8)	0.32 (1)	0.95 (8)	35 (18)	0.71 (2)
V	425 (73)	180 (1)	Not detected	250 (90)	110 (3)	Not detected
F	Not detected	230 (10)	949 (41)	Not detected	260 (1)	750 (28)
Ti	52,000 (95)	14,300 (1)	Not detected	37,000 (100)	14,400 (3)	Not detected

dsm^3, dry standard cubic meter.
From Clemens AH, Deely JM, Gong D, Moore TA, Shearer JC. Partitioning behaviour of some toxic trace elements during coal combustion—the influence of events occurring during the deposition stage. Fuel 2000;79:1781−4.

Table 6.5 Underfeed stoker combustion.

Element	Bottom ash (μg/g)	Cyclone ash (μg/g)	Filter ash (μg/g)	Flue gas (μg/dsm^3)
B	3,470 (60)	5,270 (2)	nm	698 (2)
As	5.5 (35)	220 (28)	nm	0.82 (1)
Se	0.75 (11)	22.0 (6)	nm	0.77 (2)
Hg	0.03 (2)	7.3 (10)	nm	3.7 (47)
S	7,080 (21)	29,800 (2)	nm	121,000 (47)

dsm^3, dry standard cubic meter; *nm*, not measured.
From Clemens AH, Damiano LF, Gong D, Matheson TW. Partitioning behaviour of some toxic volatile elements during stoker and fluidised bed combustion of alkaline sub-bituminous coal. Fuel 1999;78:1379–85.

elements As and Se, and form calcium arsenate and calcium selenate above 500°C. The enrichment of As and Se in bottom ash can be found in both samples. Coal from the center of the seam contained lower Ca content, and As and Se contents were fewer in the bottom ash of this coal sample. Coal from the center of the seam contained higher F content, and the concentration of F in flue gas was higher. Similarly, coal with higher V content had higher V content in its bottom ash.

Clemens et al. [12] studied the distribution of B, As, Hg, and Se in the process of low ash content and high alkaline sub-bituminous combustion in stoker and fluidized-bed conditions. Both the stoker and fluidized-bed combustion units are about 50 kW, and the combustion temperature of coal in the fluid bed is 800, 875, and 950°C, which is higher than in the underfeed stoker (>1000°C). Table 6.5 shows the partitioning behavior of B, As, Hg, and Se in underfeed stoker combustion, and Table 6.6 shows fluidized-bed combustion at 800, 875, and 950°C. More than 95% of B, Se, and Hg come from coal in fluidized-bed combustion, while 66% of As comes from coal and 34% of As comes from fluidizing sand.

Arsenic shows a similar distribution behavior for both the fluidized-bed and stoker systems. A large amount of As can be recovered from bottom ash and cyclone ash, and a small amount of As is recovered from the flue gas. There is a significant amount of As recovered from the fluidizing sand but not from the coal for the fluidized-bed combustor. However, for the stoker combustor, there is no ambiguity and 35% of As comes from the bottom ash. A lot of chemical compositions can retain As in the bottom ash, and As has an affinity with the silicate matrix. Gaseous As can react with the Ca-bearing matrix in coal, and the following arsenate-forming reaction can occur above 500°C during the combustion of high-Ca coal.

$$3CaO + As_2O_3 + O_2 = Ca_3(AsO_4)_2$$

The selenium balance is poor for the stoker system (19%), while it is better for the fluidized-bed system (51%–58%). An amount of Se is distributed in the bottom ash in the underfeed experiment because it has a larger particle size and higher temperatures

Table 6.6 Fluid bed combustion.

Temperature	Element	Bottom ash (μg/g)	Cyclone ash (μg/g)	Filter ash (μg/g)	Flue gas (μg/dsm^3)
800°C	B	350 (10)	4,600 (51)	12,000 (5)	188 (1)
	As	5.5 (39)	11 (30)	38 (4)	0.49 (0.4)
	Se	0.0 (0)	3.7 (34)	49 (17)	0 (0)
	Hg	0.007 (1)	0.18 (8)	0.71 (1)	4.5 (55)
	S	8,320 (40)	11,040 (21)	nm	57,000 (30)
875°C	B	520 (14)	4,100 (62)	11,000 (4)	212 (1)
	As	5.9 (39)	9.5 (30)	33 (3)	nm
	Se	0 (0)	3.5 (43)	52 (15)	0 (0)
	Hg	0.02 (2)	0.16 (9)	0.75 (1)	4.7 (55)
	S	5,560 (25)	9,760 (25)	nm	96,300 (48)
950°C	B	310 (9)	2,700 (70)	7,400 (3)	265 (1)
	As	4.8 (31)	8.9 (49)	34 (3)	11 (7)
	Se	0 (0)	2.1 (45)	27 (8)	0 (0)
	Hg	0.02 (2)	0.06 (6)	0.70 (1)	5.2 (59)
	S	160 (1)	4,760 (21)	nm	159,000 (79)

dsm^3, dry standard cubic meter; *nm*, not measured.
From Clemens AH, Damiano LF, Gong D, Matheson TW. Partitioning behaviour of some toxic volatile elements during stoker and fluidised bed combustion of alkaline sub-bituminous coal. Fuel 1999;78:1379–85.

in this mode of combustion. It is suggested that selenate formation is from the following reaction:

$$CaO + SeO_2 + 0.5O_2 = CaSeO_4$$

Selcuk et al. [13] studied the partitioning behavior of TEs during high-ash lignite combustion in a pilot-scale 0.3 MW_t atmospheric bubbling fluidized-bed combustor. The EFs of As, Cr, Ni, Pb and Zn were larger in fly ash than in bottom ash. The EF of Hg was lower than 1, and the EFs of As, Ba, V, and P in bottom and cyclone ashes were lower than 0.7. Fig. 6.5 shows the partitioning characteristics of TEs in samples for Runs 1 and 2. For the two runs, As, Cr, Ni, V, and Zn were mainly distributed in cyclone ashes, while Ba, Hg, Li, Sn, and Mo were mainly enriched with fly ash. The capture of As and Hg by fly ash was due to condensation at low temperature (350°C), which resulted from the presence of a cooler at the exit of the combustor. The elements of Cr and Ni were mainly enriched with fly ash. The results with or without the addition of limestone showed that limestone addition could shift the Ba, Cr, Mo, Ni, Sn, V, and Zn from bottom ash to fly ash.

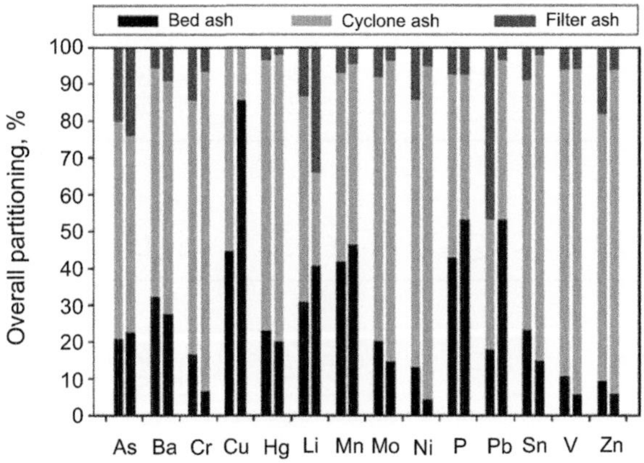

Figure 6.5 Partitioning of trace elements (Run 1: *left bar*, Run 2: *right bar*).
From Selcuk N, Gogebakan Y, Gogebakan Z. Partitioning behavior of trace elements during pilot-scale fluidized bed combustion of high ash content lignite. J Hazard Mater 2006;137: 1698–703.

Pöykiö et al. [14] investigated TE distribution in bottom ash and fly ash from a 77 MW multifuel boiler in Finland. As shown in Table 6.7, in general, the concentration of Mo is 8.2 μg/g and sulfate is 15.9 μg/g, which exceeds the limit of waste disposal costs for waste landfills. In the fly ash, the extractable Cr concentration of 1.8 μg/g and sulfate of 7200 μg/g also exceeds the limit of costs for receiving waste in the inert waste basement. In addition, the extractable concentration of Mo of 37.0 μg/g exceeds the limit of the cost of receiving waste from nonhazardous waste landfills. It has been found that the total concentration of the fly ash element is 1.1 (Sb) and 3.5 (Pb) times higher than those in the lower ash except Cu, S, and Se because of the fact that the temperature between 1000 and 1100°C in multiboilers is sufficient to vaporize certain elements. Additionally, most of these classes form compounds that surround the surface of the flue gas particles that lead to the development of certain elements of the fly ash fraction.

6.3 Emission characteristics and removal of trace elements in cement production processes

Cement production processes are another TE emission source. Compared to power plants, cement plants are quite different with regard to flue gas composition, temperature, residence time, and material circulation [15]. Cement kiln systems have an inherent ability to retain TEs in the solid materials, while kiln dust recirculation to the kiln will cause the release of captured TEs [16]. Thus TE migration in the cement kiln is complicated. A comprehensive understanding of mercury behavior in the cement production process will help to accurately estimate mercury emissions and

Table 6.7 Total concentrations of elements (μg/g; dw), extractable concentrations of elements (μg/g; dw) in acid soluble (BCR1), reducible (BCR2), and oxidizable (BCR3) fractions, and recovery values (*R*%) of the extraction procedure.

	Total concentration		Bottom ash				Fly ash			
Element	**Bottom ash**	**Fly ash**	**BCR 1**	**BCR 2**	**BCR 3**	***R* (%)**	**BCR 1**	**BCR 2**	**BCR 3**	***R* (%)**
As	17	23	3.2	7	1.3	67.6	<0.6	10.8	2.3	(>59.6)
Ba	650	1200	3.5	141	227	57.2	9.9	248	394	54.3
Be	2	4	0.3	0.2	0.5	50	0.4	1	0.7	52.5
Cd	0.6	1.2	0.3	<0.1	<0.1	(>83.3)	0.3	<0.1	0.1	(>41.7)
Co	17	29	2.5	0.6	0.8	22.9	2.8	2.4	2.1	25.2
Cr	49	69	2.5	1.2	2	11.6	6.5	5.4	6.2	26.2
Cu	120	110	10	14.1	40.2	53.6	10.2	15.6	16.6	38.5
Mn	570	750	69.7	66	47.4	32.1	100	121	138	47.9
Mo	27	63	8.2	6.5	3	65.5	7.3	9.4	18.6	56
Ni	170	240	26.6	5	6.9	22.6	30.6	12.3	12.1	22.9
Pb	26	91	1.3	9.3	4.1	56.5	0.6	10.3	17.2	30.9
Sb	6	7	<0.6	<0.6	<0.8	(>33.3)	0.7	<0.6	<0.8	(>30.0)
Se	<4.0	<4.0	<0.6	<0.6	<0.8	n.d.	<0.8	<0.8	1.8	n.d.
V	280	390	53.1	71.8	33.2	56.5	12.5	144	36.8	49.6
Zn	100	190	24	4	8.4	36.4	36.7	15.9	14.8	35.5
Hg	<0.04	0.09	n.d.	n.d.	n.d.	n.d.	n.d.	n.d.	n.d.	n.d.

n.d., no data.
From Pöykiö R, Manskinen K, Nurmesniemi H, Dahl O. Comparison of trace elements in bottom ash and fly ash from a large-sized (77 MW) multi-fuel boiler at the power plant of a fluting board mill, Finland. Energy Explor Exploit 2011;29:217–34.

develop control strategies. In this section, the emission characteristics of TEs in cement plants are discussed.

6.3.1 Cement production process

Cement production processes can be categorized as dry, semidry, semiwet, and wet processes depending on the handling of raw material before being fed to the rotary kiln. Nowadays, almost all new plants are based on the dry process and many old wet plants are also remodeled to dry or semidry processes. Dry cement manufacture has three fundamental stages: preparation of feedstocks, production of clinker, and preparation of cement [15,16].

(1) Preparation of feedstock. This stage includes the process of siege, crushing, and prehomogenization. Typical raw materials used for cement production have 85% cayenne, 13% clay or blackboard, and under 1% each of materials such as silica, alumina, and iron ore. These feedstocks are crushed into particles with a diameter of less than 20 mm and mixed with a prehomogenization pile [17].

(2) Clinker production. As shown in Fig. 6.6, the feedstock first enters the raw material to make a fine powder (raw feed), where 85% of the material is slightly smaller than 88 μm. Then, the feed is transferred to the homogenizing silos to impair the material difference. After that, the meal is thrown into the precalciner tower to start the chemical change to cement. Precyclone towers intermix with raw feed and almost 1000°C of exhaust gas to recover energy, preheat feed, and initiate chemical reactions that result in cement. Gases straight from the kiln, but in precalciner facilities, gasoline, and air are provided by a combustion vessel inside the tower and kiln. Typically, 60% of burned calciner and more than 90% calcination have been reached before the material enters the rotary kiln. Inside the kiln, temperatures reach approximately 1400°C to complete the process of chemical reactions and produce calcium silicates, called clinkers, with a diameter of 10–25 mm. The gas from the preheater tower is usually blended in a rawmill, which will help stabilize the future feedstock. After flowing through the rawmill, the gases are eventually released by a dust collector, which also obtains good particles when feedstuffs are milled. The dust will then be recycled into homogenizing silos and served as part of the kiln feed.

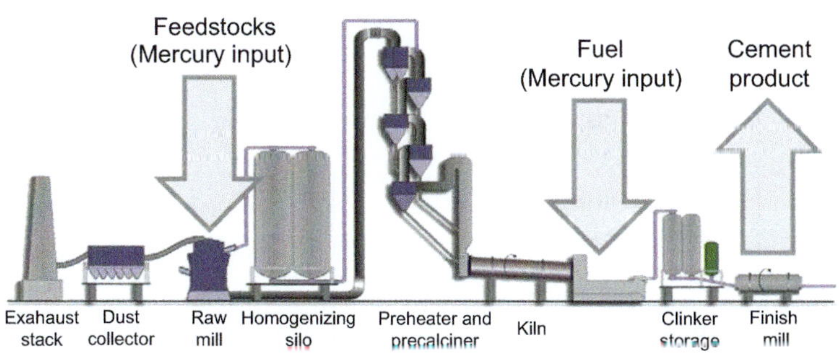

Figure 6.6 Overview of the dry cement manufacturing process.
From Sikkema JK, Alleman JE, Ong SK, Wheelock TD. Mercury regulation, fate, transport, transformation, and abatement within cement manufacturing facilities: review. Sci Total Environ 2011;409:4167–78.

(3) Preparation of cement. This stage completes the manufacturing process where clinker nodules are milled into cement. Following clinker milling, the cement is ready for use as a binder in various concrete mixes.

6.3.2 Mercury emission characteristics in cement plants

6.3.2.1 Mercury concentrations in solid samples

Mercury concentrations in different solid samples are shown in Table 6.8 [18]. The mercury concentrations in limestone were 16–42 ppb. Other materials, including sandstone, shale, clay, and coal gangue, had mercury concentrations of 6–47 ppb. Iron-containing materials had a high mercury concentration of 402 ppb. Coal was the only fuel used in these cement plants, the concentration of which was in the range 20–40 ppb. The mercury concentrations of the raw meal were 296 and 37 ppb in Plant

Table 6.8 Hg concentration of solid samples in the cement manufacturing process.

Materials from the process		Plant 1 (Sichuan Province)	Plant 2 (Sichuan Province)	Plant 3 (Shandong Province)
Raw materials (ppb)	Limestone	16 ± 3	18 ± 1	42 ± 39
	Sandstone	19 ± 0	47 ± 6	6 ± 1
	Shale	35 ± 8	–	–
	Clay	–	14 ± 4	–
	Coal gangue	–	–	49 ± 40
	Iron-rich materials	402 ± 9	27 ± 4	187 ± 80
	Nickel-rich materials	–	10 ± 1	–
Fuel (ppb)	Coal	25 ± 3	36 ± 18	26 ± 4
Samples in process (ppb)	Coal powder	544 ± 10	113 ± 4	–
	Raw meal	296 ± 3	37 ± 2	–
	Particulate matter at the kiln tail	1992 ± 951	428 ± 36	–
	Particulate matter at the kiln head	572 ± 243	14 ± 3	–
Product (ppb)	Clinker	1 ± 1	3 ± 2	3 ± 6
	Gypsum	499 ± 83	473 ± 95	–
	Cement	61 ± 0	35 ± 1	–

"–" means no sample or not used in the manufacturing process.
From Wang F, Wang S, Zhang L, Yang H, Wu Q, Hao J. Mercury enrichment and its effects on atmospheric emissions in cement plants of China. Atmos Environ 2014;92:421–8.

1 and Plant 2, respectively. The mercury concentrations of the PM from the fabric filter (FF) and ESP were 1992 and 572 ppb in Plant 1, respectively, and 428 and 14 ppb in Plant 2, respectively. PM from the FF and ESP is the main mercury source in raw meal. The concentrated mercury in the recycled PM results in a higher mercury content in the cement production process. The variation of mercury concentrations in the raw meal in Plants 1 and 2 is mainly from the different processes of mercury enrichment.

The mercury concentrations in coal powder from the FF of the coal mill were 544 ppb in Plant 1 and 113 ppb in Plant 2. The mercury concentrations of clinker were less than 4 ppb for all three cement plants. However, the mercury concentrations in cement are relatively high: 61 ppb in Plant 1 and 35 ppb in Plant 2, which was mainly due to adding the flue-gas desulfurization (FGD) gypsum from power plants. The mercury concentrations in gypsum were 499 ppb in Plant 1 and 473 ppb in Plant 2. The temperature of the process of adding gypsum to clinker is below 100°C, resulting in negligible mercury emissions from this process.

Won and Lee [19] determined the mercury content in feedstocks and fuels in the largest cement manufacturing facility (CMF) in Korea. The Hg contents of limestone, clay, fly ash, silica stone, bituminous coal, and solid waste-derived fuel were analyzed, the results of which are shown in Fig. 6.7. The inhomogeneity of feedstocks and fuels indicates the inhomogeneity of the Hg introduced into the cement kiln. The highest average Hg content (221 ng/g) was obtained from solid waste-derived fuel with a range 100–570 ng/g. The Hg contents of fly ash, limestone, silica stone, clay, and bituminous coal were in the range 30–390, 4.9–79.1, 3.4–21.6, 5.6–12.8, and 4.0–7.7 ng/g, respectively.

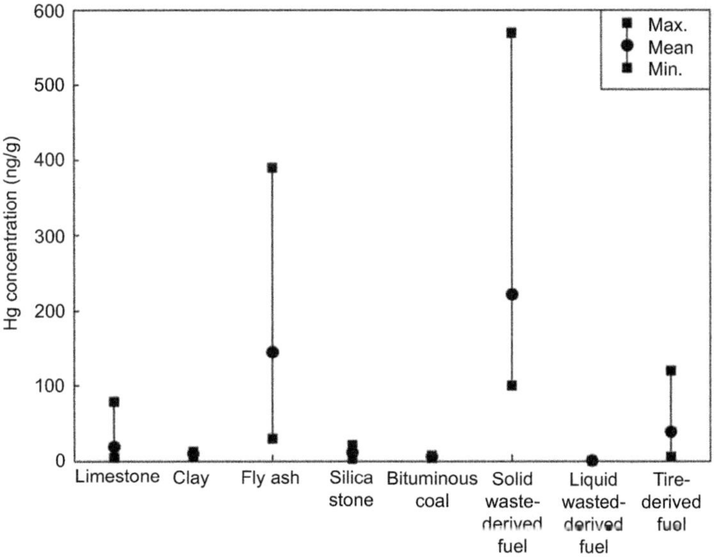

Figure 6.7 Hg concentrations of feedstocks and fuels.
From Won JH, Lee TG. Estimation of total annual mercury emissions from cement manufacturing facilities in Korea. Atmos Environ 2012;62:265–71.

The Hg contents of liquid waste-derived fuel and tire-derived fuel that are used in other CMFs were then analyzed to examine their relative contributions to the total annual Hg input to all CMFs in Korea. The corresponding kiln uses liquid waste-derived fuel only when the kiln is shut down for maintenance and the burner needs to be reignited. Liquid waste-derived fuel was not used in the corresponding kiln. As shown in Fig. 6.7, the liquid waste-derived fuel showed an average Hg content of 0.6 ng/g with a range 0.2−1.0 ng/g. The Hg content of tire-derived fuel was 6.1−120.1 ng/g with an average value of 39.2 ng/g.

6.3.2.2 Fate and transformation of mercury in the cement production process

Mercury species distribution in a cement kiln preheater was investigated using thermodynamic calculations by Larsen et al. [20]. To simulate a preheater environment, chloride, sulfide, and sulfate compounds were included, and the alkaline dust was represented by CaO. The dominant mercury species below 180°C are HgO and $HgCl_2$, while all mercury compounds thermodynamically preferred above 200°C are gaseous Hg^0. On the other hand, SO_2 is assumed available for reaction with Hg species. As shown, the dominant species below 200°C is $HgSO_4$, while a certain amount of $HgCl_2(g)$ is formed above 200°C. The $HgSO_4(g)$ decomposes at around 450°C, thus the dominant species above 450°C are $Hg^0(g)$ and $HgCl_2(g)$. These results are completely different from the condition containing HCl, SO_2, and CaO.

In the real cement manufacturing process, mercury transformation includes three stages: stripping, sorbing, and recycling (shown in Fig. 6.8).

(1) Cutting stage. In the kiln, mercury is burned from feedstock, while gas falls into the preheater tower and collides with the flue gas. HgS is the dominant type of mercury in feedstock, while other mercury compounds are molded over the feedstock when gas emissions are internally looped to preheat raw feed. Because different points of different mercury compounds vary, mercury emissions occur in many preheater stages. About 58%−82% of mercury is stripped off in the first preheater in the material stream and 97% of mercury is stripped off in the second preheater [21].

(2) Sorbing stage. When the exhaust gas comes out of the preheater tower, mercury gas flows into raw meal and can be traced to raw feed. This stage includes rawmills and dust collectors. The mill moves at 90−120°C and causes increased mercury gas to cool PM. Most particle-bound mercury of raw feed will flow into the trunk and transition to stage #3. The lesser particle-bound mercury of the flue gas will flow to the dust collector. Mercury contained in the dust collector is associated with lost bread in the bagh cloth.

(3) Internal mercury loop stage. The continuing operation of the facility causes the mercury to undergo a continuous process of stripping, sorbing, and recycling, hence enriching mercury in the raw meal many fold. These compounds may flow and fluctuate from the oven before reaching the temperature of decay. Completion of this mercury loop adds complication to an already imperfect explanation of speciation. The mercury compounds that sorb onto raw meal and collected dust are not necessarily broken down into elemental form upon entry into the preheater tower. Instead, these compounds maysimply vaporize and flow in reverse away from the kiln prior to reaching a decomposition temperature. At present, it is unknownwhich species may be transported in this manner, or if this type of transport occurs at all. Addressing these speciation-specific transport questions represents a major research goal for the industry.

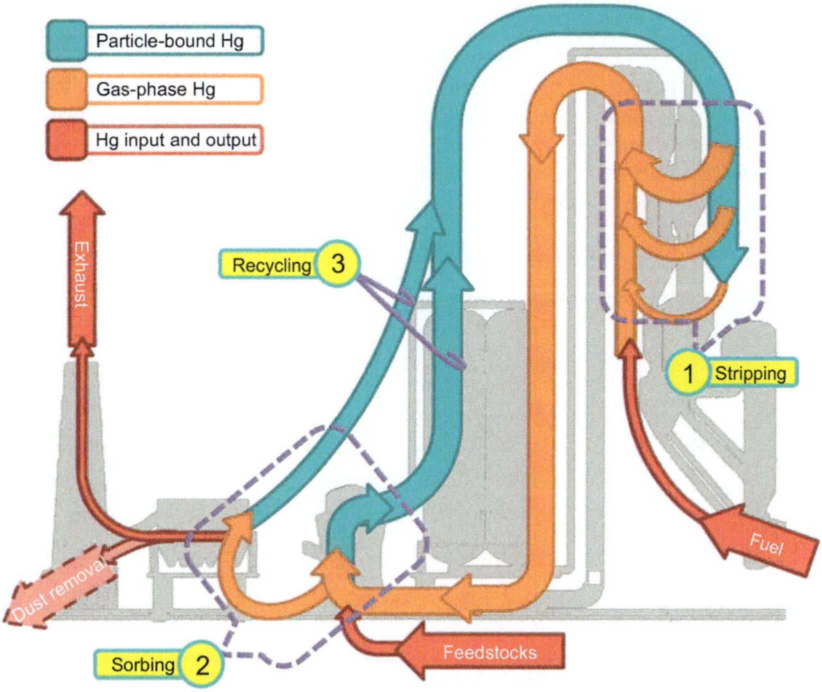

Figure 6.8 Mercury transport and fate within a cement manufacturing facility.
From Sikkema JK, Alleman JE, Ong SK, Wheelock TD. Mercury regulation, fate, transport, transformation, and abatement within cement manufacturing facilities: review. Sci Total Environ 2011;409:4167–78.

6.3.2.3 Mercury concentration and speciation in flue gas

Wang et al. [18] studied the mercury emission characteristics in three typical Chinese cement plants. As shown in Table 6.9, the mercury concentrations in the flue gas from the kiln tail were the highest in Plant 1 and Plant 3, with average values of 28.1 and 19.8 mg/m^3, respectively, whereas the concentrations at the kiln head were only 4.0 and 2.7 mg/m^3, respectively. However, the mercury concentration at the kiln head reached 31.0 mg/m^3, higher than that at the kiln tail of Plant 2. This might be mainly caused by the high Hg concentration in the flue gas of the kiln head, as well as poor Hg cobenefit performance of the ESP in Plant 2. Hg^{2+} was the main species in exhausted flue gas. More than 60% of the mercury was in the oxidized form except at three sampling sites in the tested cement plants: after the FF of the rawmill at the kiln tail in Plant 2, after the FF of the coal mill, and after the ESP at the kiln head in Plant 3.

PM recycling also impacts mercury speciation in flue gas. As shown in Fig. 6.9, about 64% and 76% of mercury in the flue gas from Plants 1 and 2 was Hg^{2+}, while Hg^p accounted for about 27% and 20% in Plants 1 and 2. Hg^0 only contributed a very small part of the Hg^t in flue gas. Mercury oxidation is mainly affected by the

Table 6.9 Hg concentration in the flue gas [18].

Hg concentration (μg/m³)		Before rawmill	After FF of rawmill at kiln tail	After FF of coal mill at kiln tail	After ESP at kiln head
Plant 1	Hg^0	18.9 ± 18.1	2.4 ± 0.3	0.8 ± 0.8	0.6 ± 0.4
	Hg^{2+}	124.5 ± 18.5	25.7 ± 19.6	9.2 ± 3.9	3.4 ± 2.2
	Hg^p	52.3 ± 24.7	0.0 ± 0.0	0.0 ± 0.0	0.0 ± 0.0
	Hg^t	195.6 ± 30.8	28.1 ± 19.9	9.9 ± 4.4	4.0 ± 2.0
Plant 2	Hg^0	1.5 ± 0.7	3.8 ± 1.3	1.0 ± 1.5	9.2 ± 3.2
	Hg^{2+}	33.0 ± 10.1	1.0 ± 0.4	1.7 ± 1.1	21.9 ± 5.9
	Hg^p	8.9 ± 4.4	0.0 ± 0.0	0.0 ± 0.0	0.0 ± 0.0
	Hg^t	43.4 ± 11.5	4.8 ± 1.4	2.6 ± 1.8	31.0 ± 8.9
Plant 3	Hg^0	–	3.0 ± 2.9	7.7 ± 7.7	2.3 ± 3.1
	Hg^{2+}	–	16.6 ± 5.7	5.9 ± 5.2	0.2 ± 0.1
	Hg^p	–	0.2 ± 0.2	0.6 ± 0.7	0.2 ± 0.2
	Hg^t	–	19.8 ± 3.1	14.2 ± 11.2	2.7 ± 3.2

"–" means no sampling at this site or no such facility. *ESP*, electrostatic precipitator; *FF*, fabric filter; Hg^0, elemental Hg; Hg^P, particulate Hg; Hg^t, total Hg.
From Wang F, Wang S, Zhang L, Yang H, Wu Q, Hao J. Mercury enrichment and its effects on atmospheric emissions in cement plants of China. Atmos Environ 2014;92:421–8.

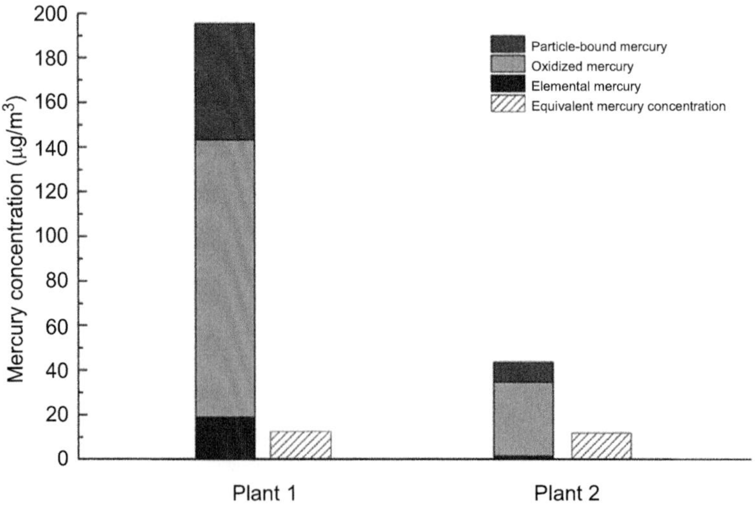

Figure 6.9 Mercury speciation before the rawmill and comparison with equivalent mercury concentration according to raw materials and coal.
From Wang F, Wang S, Zhang L, Yang H, Wu Q, Hao J. Mercury enrichment and its effects on atmospheric emissions in cement plants of China. Atmos Environ 2014;92:421–8.

compositions of flue gas and fly ashes as well as the cooling rate of flue gas. High PM concentration can promote mercury oxidation through a heterogeneous reaction occurring on the fly ashes' surface. The lower temperature at the rawmill inlet (about 200°C) and longer residence time were also beneficial for mercury oxidization.

Yan et al. [22] studied the distribution of Hg in flue gas from preheater–precalciner cement kilns (PPKs) and vertical shaft cement kilns (VSKs) in China. As shown in Table 6.10, the Hg^t concentrations were in the range 3.1–8.2 and 6.6–37 μg/m^3 for VSKs and PPKs, respectively. The Hg species mainly present as Hg^{2+} in PPKs and Hg^0 in VSKs, respectively. This might be attributed to the oxidizing conditions present in the PPKs with an O_2 concentration of 5.6% and negligible CO concentration at the first preheater cyclone outlet. More reducing conditions were present in the VSK process with O_2 and CO levels at 2.8% and 1.2%, respectively. Moreover, Hg^t concentrations increased when the PPKs were running in direct modes, that is, the flue gas was directly routed to the main FF or ESP without adsorbing onto the raw meal in the rawmill. The flue gas temperature was generally higher in direct mode operations due to less heat exchange. Earlier studies regarding Hg^t emission concentrations from PPKs gave ranges 0.89–18 μg/m^3 [19], 4.8–28 μg/m^3 [18], 1–12 μg/m^3 [21], and 4.0–25 μg/m^3 [23].

Won et al. [19] measured the Hg concentrations in stack gases at the largest cement plant in Korea. During the 4-week measurement period, a total of 14 measurements were performed using the Ontario Hydro Method. Hg concentrations in stack gases were in the range 0.89–17.89 μg/m^3 (Fig. 6.10). The average Hg^t concentration was 10.14 μg/m^3, and the average Hg^{2+} and Hg^0 concentrations were 1.52 and 8.62 μg/m^3, respectively. A negligible amount of Hg^p was detected from the stack gases. The results indicated that most of the Hg was in the form of Hg^0. Figs. 6.11 and 6.12 show the Hg concentrations measured by online Hg analyzers. The average Hg concentrations measured using the $SnCl_2$ and thermal conversion systems were 13.3 and 12.9 μg/m^3, respectively.

6.3.2.4 Mass flows of mercury in the cement manufacturing process

The input of mercury in the cement production process mainly originated from feedstocks and fuels, while the mercury output streams mainly consisted of exhaust, dust, and clinker (Fig. 6.13). Many facilities collected and recycled the dust, which decreased the output of mercury. The mercury in clinker is generally not included in the mass balances model because of the kiln's high temperature caused by mercury in the clinker in the flue gas [24]. However, one study reported an average mercury concentration of 5.2 ppb with a 122% standard deviation for a clinker sample [21]. The mercury in clinker may be present as stable mercury silicates like $HgSiO_3$ or $Hg_6Si_2O_7$, while the decomposition temperatures of these mercury compounds are still unknown [24]. A mass balance approach is valid for all process types over long periods of operation, but it is not accurate in the case of preheater and precalciner kilns with inline rawmills.

Table 6.10 Particles, Hg, As, and Se concentrations in flue gas.

Sample[a]	Flue gas temperature (°C)	Particles[b] (mg/Nm^3)	Concentration (μg/Nm^3)			
			GOM	Hg^0	Hg^P	Hg^t
VSK1	70	1007	0.8 ± 1.4	4.1 ± 0.90	0.14 ± 0.13	5.1 ± 1.1
VSK2a	97	1.7	0.8 ± 0.7	3.9 ± 0.10	0.14 ± 0.01	4.8 ± 0.8
VSK2b	90	3.8	0.2 ± 0.18	2.9 ± 0.70	0.01 ± 0.002	3.1 ± 0.9
VSK2c	65	0.2	0.6 ± 0.18	7.5 ± 0.20	0.014 ± 0.005	8.2 ± 0.4
PPK1	85	2.2	4.4 ± 3.2	2.2 ± 1.3	0.05 ± 0.06	7.0 ± 4
PPK1-MS	90	N.A.	10.5	3.2	0.16	13.8
PPK2a	115	3.9	4.1 ± 0.08	1.9 ± 0.5	0.05 ± 0.01	6.0 ± 0.4
PPK2a-MS	120	N.A.	4.9	3.5	0.23	8.7
PPK2b	90	1.1	13 ± 2.8	12.0 ± 1.2	0.04 ± 0.001	25 ± 4
PPK2b-MS	90	N.A.	22	15	0.21	37

GOM, the oxidized form; Hg^0, elemental Hg; Hg^P, particulate Hg; Hg^t, total Hg.

[a]One sample was taken at the plants running in direct mode (mill shut off); PPK1-MS, PPK2a-MS, and PPK2b-MS.

[b]*N.A.*, not analyzed for samples collected in direct mode.

From Yan D, Peng Z, Ding Q, Karstensen KH, Engelsen CJ, Li L, Ren Y, Jiang C. Distribution of Hg, As and Se in material and flue gas streams from preheater-precalciner cement kilns and vertical shaft cement kilns in China. J Air Waste Manag Assoc 2015;65:1002–10.

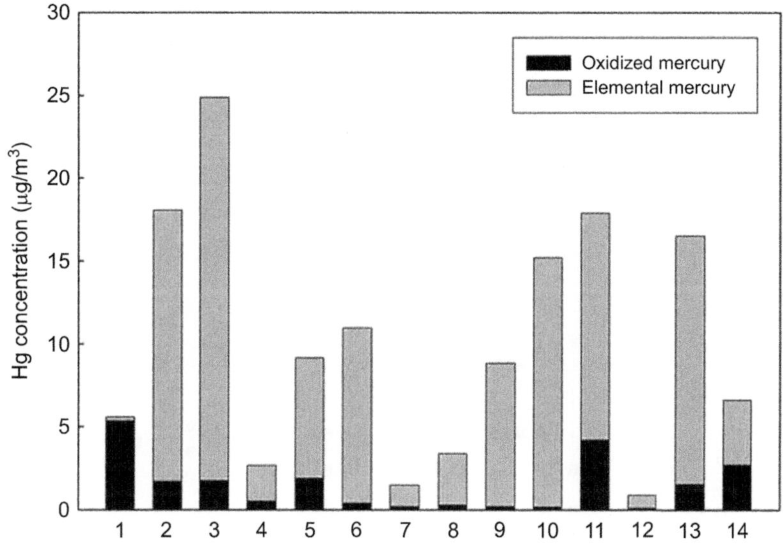

Figure 6.10 Hg concentrations in stack gases at a cement kiln measured by the Ontario Hydro Method.
From Won JH, Lee TG. Estimation of total annual mercury emissions from cement manufacturing facilities in Korea. Atmos Environ 2012;62:265–71.

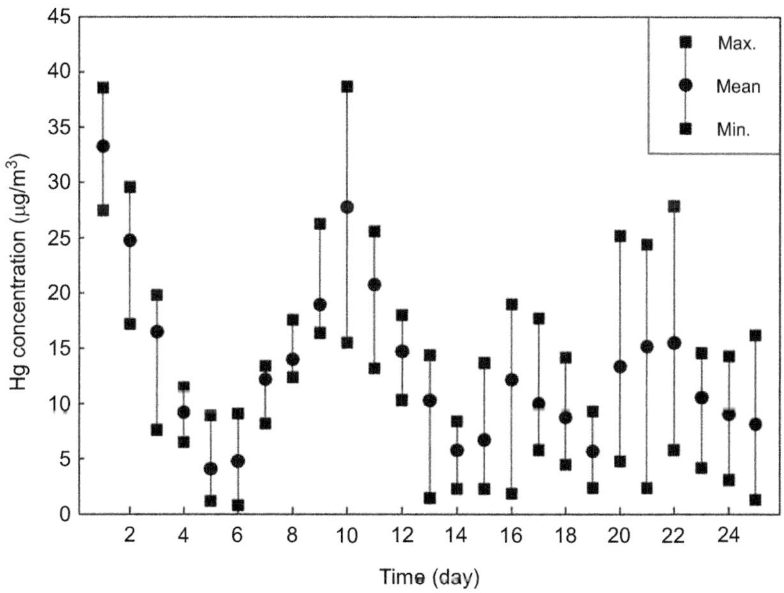

Figure 6.11 Hg concentrations in stack gases at a cement kiln measured by an online Hg analyzer with an $SnCl_2$ conversion system attached to the front.
From Won JH, Lee TG. Estimation of total annual mercury emissions from cement manufacturing facilities in Korea. Atmos Environ 2012;62:265–71.

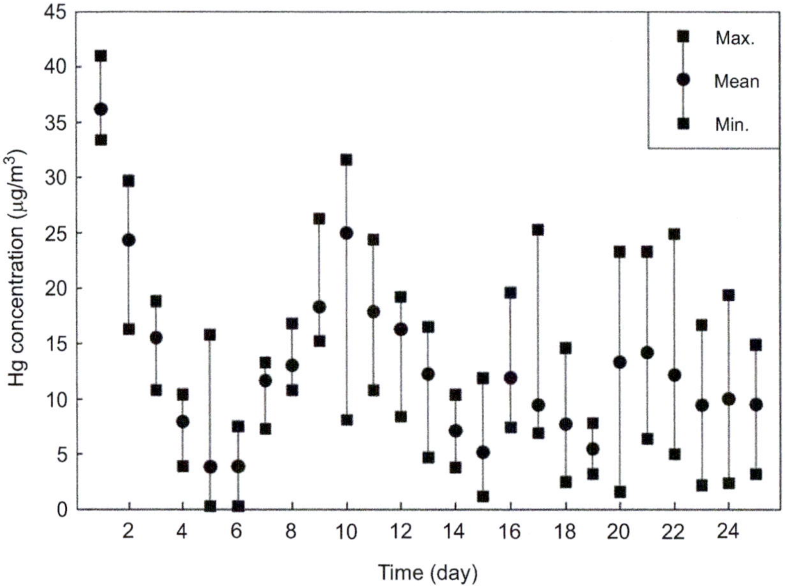

Figure 6.12 Hg concentrations in stack gases at a cement kiln measured by an online Hg analyzer with a thermal conversion system attached to the front.
From Won JH, Lee TG. Estimation of total annual mercury emissions from cement manufacturing facilities in Korea. Atmos Environ 2012;62:265–71.

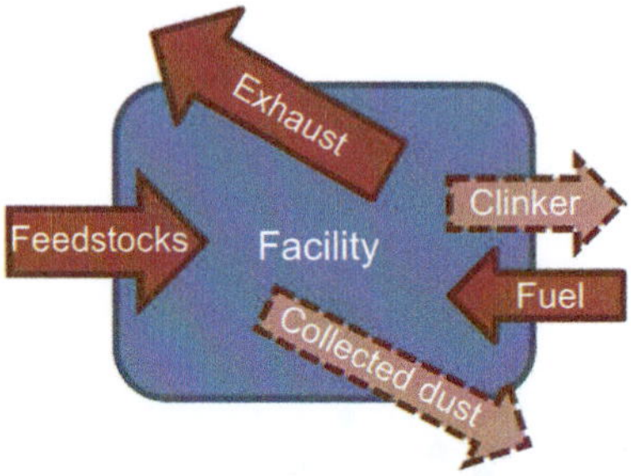

Figure 6.13 Cement manufacturing facility mercury mass balance.
From Sikkema JK, Alleman JE, Ong SK, Wheelock TD. Mercury regulation, fate, transport, transformation, and abatement within cement manufacturing facilities: review. Sci Total Environ 2011;409:4167–78.

Wang et al. [25] studied the mercury mass flows in the whole cement production process for two Chinese plants, the results of which are shown in Figs. 6.14 and 6.15. There were three mercury cycles between the kiln system and the rawmill system, coal mill system, and kiln head system. The rawmill system contributed to the largest proportion of mercury in the whole system, accounting for 73% and 57% of Hg^t mass flows in Plants 1 and 2, respectively. Thus mercury cycling in the rawmill system should be given more attention for mercury emission control of cement plants.

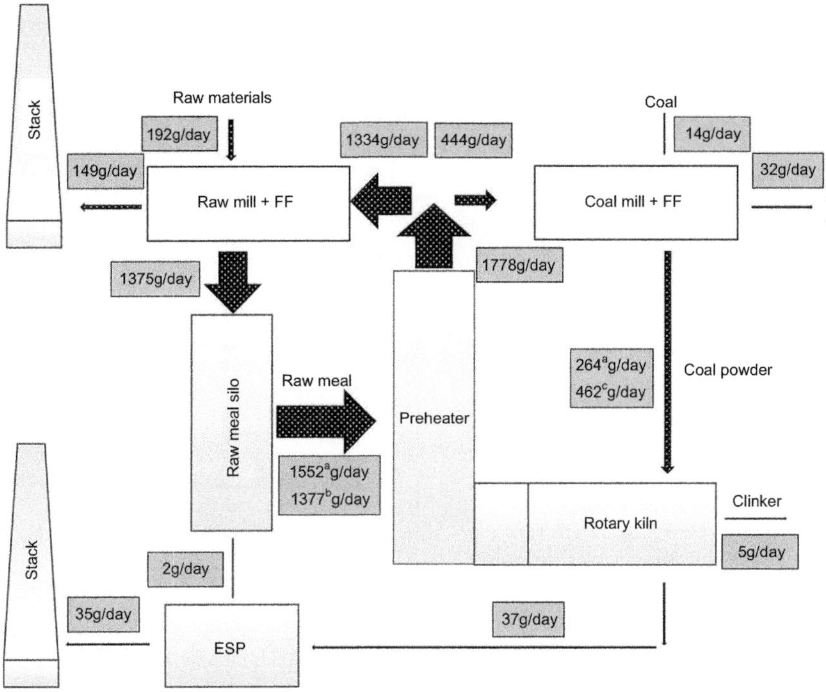

Figure 6.14 Mercury mass flows in Plant 1. (Note: the width of the *arrows* indicates the relative amount of mercury mass flows; a—calculated results for kiln system; b—calculated results for rawmill system; c—calculated results for coal mill system.) *ESP*, Electrostatic precipitator; *FF*, fabric filter.
From Wang F, Wang S, Zhang L, Yang H, Wu Q, Hao J. Characteristics of mercury cycling in the cement production process. J Hazard Mater 2016;302:27−35.

Compared to the amount of mercury input from raw materials and coal (206 g/day in Plant 1 and 156 g/day in Plant 2), the mercury was enriched in the cement production process, especially in the kiln system and the rawmill system (1334 g/day in Plant 1 and 302 g/day in Plant 2). Mercury enrichment primarily depended on the high proportions of mercury captured and cycled in the rawmill system. Only 11% and 22% of the mercury that was brought into the rawmill system was finally emitted into the atmosphere in Plant 1 and Plant 2. The concentration of mercury emitted from the stacks obviously increased when the rawmill shut off. Over 90% of the mercury contributed by raw materials and coal was emitted into the atmosphere [22].

The mercury mass flows in a cement production plant of the Salonit Anhovo Joint-Stock Co., a middle-sized production unit, situated in the western part of Slovenia, with a current cement production of about 900,000 t per year, was also studied [21]. As shown in Fig. 6.16, the emitted mercury mass flow was between 0.1 and 2.4 g/h, while the internal mercury mass flows were enriched up to 20 g/h due to cycling processes. The most important contribution to mercury comes from the raw materials (about 16 kg), but their contribution is less than twice the contribution of fuels (petrol

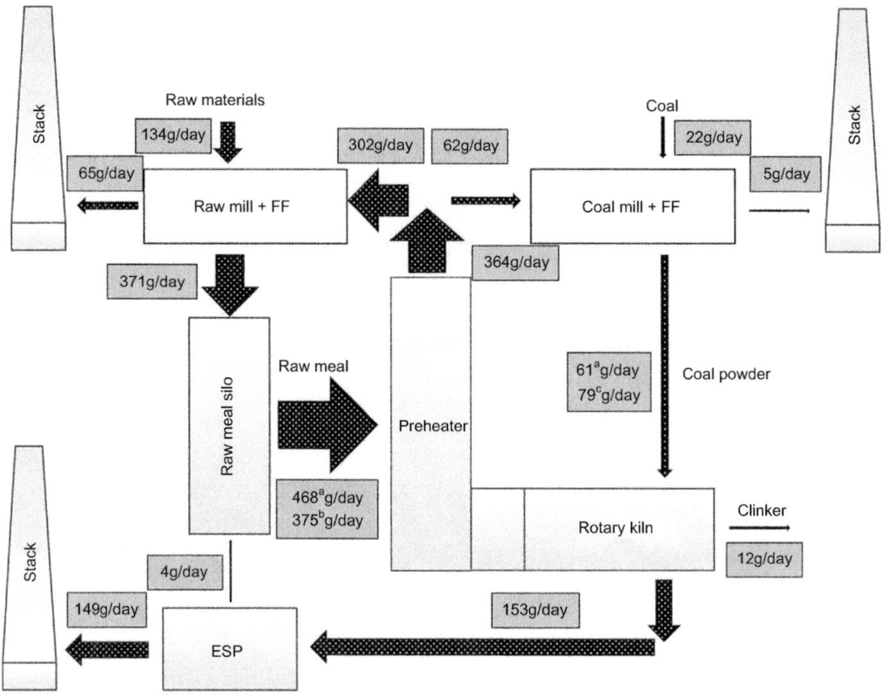

Figure 6.15 Mercury mass flows in Plant 2. (Note: the width of the *arrows* indicates the relative amount of mercury mass flows; a—calculated results for kiln system; b—calculated results for raw mill system; c—calculated results for coal mill system.) *ESP*, electrostatic precipitator; *FF*, fabric filter.
From Wang F, Wang S, Zhang L, Yang H, Wu Q, Hao J. Characteristics of mercury cycling in the cement production process. J Hazard Mater 2016;302:27–35.

coke 8.3 kg, waste tires 2.6 kg, waste oil 0.1 kg). This could be attributed to the enrichment of Hg in petrol coke. A similar combination of fuels was used during all the measurements, i.e., 1.9 t/h of tires and 3.8 t/h of petrol coke in the first sampling period and 5.8–7.2 t/h in other sampling periods.

6.3.3 Mercury control technologies for cement plants

Mercury emission reduction technologies for cement plants include coal and cement raw materials cleaning, feedstocks and fuels switching, wet scrubbing, cleaning of mercury-enriched dust, dry sorbent injection, and dry and semidry scrubbing. The existing air pollution control devices (APCDs) like ESP and FF could also somewhat reduce the emission of mercury. During the cement production process, the operating conditions, including kiln loads and flue gas temperature, can affect the cobeneficial removal capacity of the existing APCDs. Hence it could reduce mercury emission by optimizing the operating conditions.

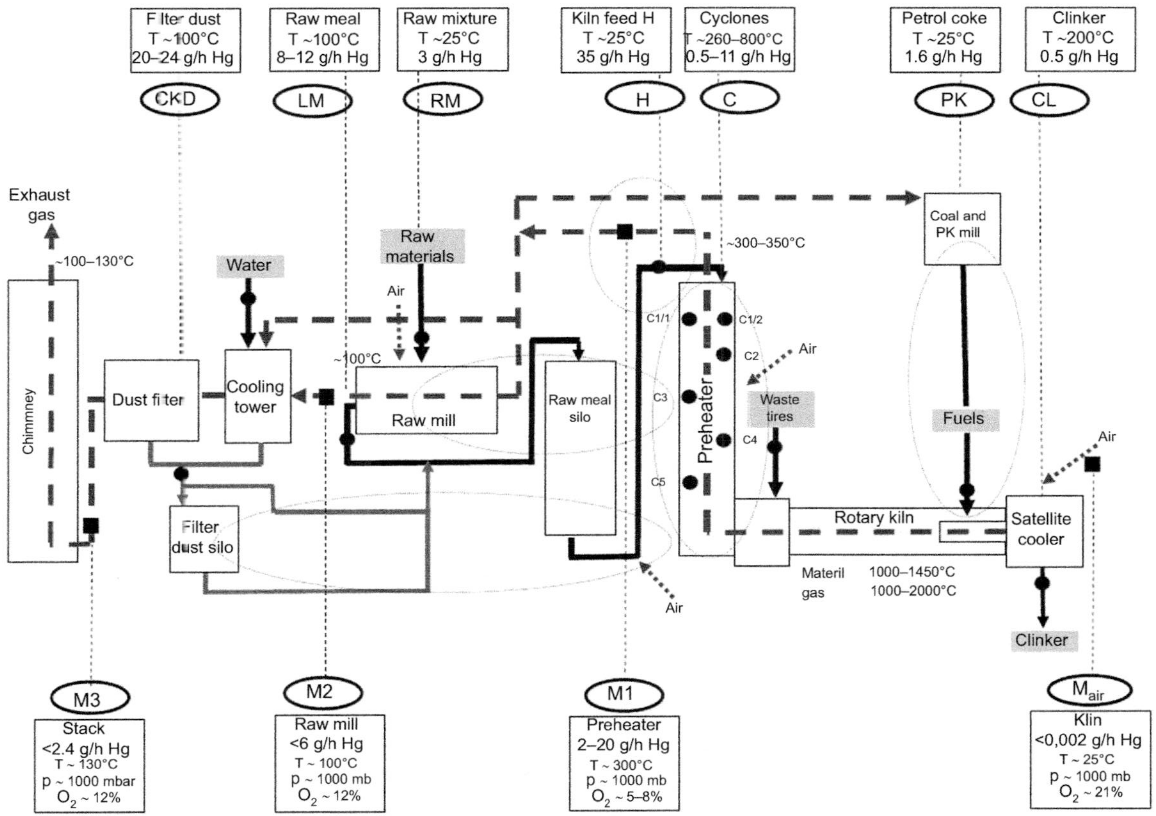

Figure 6.16 Schematic representation of the process of cement clinker production, sampling sites (d—materials, j—flue gases), typical Hg mass flows (*black line*—materials, *gray line*—filter dust, *dotted line*—flue gases) and labels (in *ellipses*). Hg cycling processes are indicated by *dotted circular lines*. *ESP*, electrostatic precipitator; *FF*, fabric filter.

From Yan D, Peng Z, Ding Q, Karstensen KH, Engelsen CJ, Li L, Ren Y, Jiang C. Distribution of Hg, As and Se in material and flue gas streams from preheater-precalciner cement kilns and vertical shaft cement kilns in China. J Air Waste Manag Assoc 2015;65:1002–10.

6.3.3.1 Coal and cement raw materials cleaning

Coal cleaning is used to remove ash and pyritic sulfur in coal [26,27]. Mercury that coexisted as pyrite can be removed by coal cleaning before combustion. Average mercury reduction by coal cleaning is estimated to be 20%–37% [28]. However, waste water for coal cleaning contains high-concentration mercury, which is classified as a hazardous waste. The cost of waste water treatment is very high. Because most of the mercury in cement production is from raw material, the extent of mercury removal through coal cleaning is very limited.

Washing and gasification can be used for removing mercury from cement raw materials before feeding to the kiln [29,30]. The water-soluble mercury in the raw materials can be removed by water washing. In the gasification process, raw materials are introduced into a heating furnace and the mercury in the raw materials is volatilized at high temperature. The gas containing mercury is passed thorough an activated carbon adsorption tower. Hence the mercury can be adsorbed and separated before feeding into the kiln. However, since large amounts of raw materials are applied in cement production, the cost of raw material cleaning is very high, which limits its wide application in cement plants.

The internal mercury loop enriches mercury in both raw meal and collected dust. Thus removing dust from the system will reduce stack emissions. A dust-cleaning technology is being tested on a pilot scale by FLSmidth (Bethlehem, Pennsylvania). This approach begins with removal of collected dust to disrupt the internal mercury loop. The dust is shuttled to a roaster where mercury is released to the gas phase at high temperature, and the dust is separated from gaseous mercury and returned to the kiln. Gaseous mercury in the roaster is quenched and condenses on a sorbent stream [31]. However, stack measurements indicate that the internal loop controls approximately 90% of emissions only when the rawmill is on [24]. The roaster does not control exhaust emissions downstream of the dust collector, raising the possibility of high mercury emissions when the mill is shut down.

6.3.3.2 Fuel and feedstocks switching

Switching clean fuels and feedstocks can minimize mercury input and thus reduce mercury emissions. Coal is the main fuel source for cement kilns. Many facilities used tire-derived fuel or other alternative fuels such as natural gas to replace coal. An experiment was carried out with a pulverized coal combustor of a pilot scale of 55 kW cofired fuel derived from tires with sub-bituminous coal. As a result, it was shown that fuel derived from co-combustion tires has a great influence on the type of mercury in the exhaust gas [32]. There was only 16.8% Hg^{2+} in the flue gas with 100% coal firing, while 47.7% and 84.8% Hg^{2+} was formed when 5 wt% and 10% tire-derived fuels were cofired. The increasing Hg^{2+} amount might be attributed to the additional chlorine from tire-derived fuel combustion, where the chlorine content in tire-derived fuel is about 600 ppm. Enhanced mercury oxidation is beneficial for mercury emission control, since Hg^{2+} is easily captured in the scrubber.

Clean fuel switching for coal might result in direct and significant reductions in mercury [33]. However, fuel is usually not the dominant source of mercury in the cement production process. Raw materials, particularly limestone, contribute the largest proportion of mercury. Thus the switching of fuels has limited impact on the reduction of mercury emission. Substitution of additives in feedstocks like iron ore and fly ash may be possible because they typically make up less than 1% of feedstock input. If mercury input is attributed to limestone or shale, selective quarrying in areas of low mercury materials could be instituted.

6.3.3.3 Sorbent injection and scrubbing

Activated carbon injection upstream of an ESP/FF is considered to be an effective technology for mercury removal. The dust collector can capture the sorbent and remove it from the gas stream. The mercury in flue gas can be adsorbed by activated carbon and removed from flue gas. This sorbent is most effective for Hg^{2+}, while bromine can enhance removal by oxidizing Hg^0 [34]. However, in activated carbon injection processes, a large C/Hg ratio (10,000–100,000 wt/wt) is required to achieve high (>90%) Hg^0 removal efficiency. The high operating cost limits its wide application in industry.

Wet scrubbers (WSs) are already utilized by the power industry and a few US cement facilities to control SO_2. Because of its water solubility, Hg^{2+} can be easily captured by the WSs. Scrubbers can remove about 90% of Hg^{2+}, while Hg^0 control was ineffective [34]. Although mercury removals of 60%–90% were calculated when the rawmill was shut down (approximately 10% of operating time), less than 50% removal occurred while the mill was operating, and at two facilities negative removal was measured [35]. Dry and semidry scrubbing systems, where sorbents are injected with little or no water, were also widely used in the power industry and also have cobeneficial mercury removal capacity. Mercury capture in dry and semidry scrubbers is less dependent on mercury speciation than in WSs, which only control Hg^{2+}. In addition, sorbent injection is hindered by high temperatures. Both of these issues are not present in a dry or semidry scrubber [36].

6.3.3.4 Cobeneficial removal of mercury by cleaning devices

Existing cleaning devices have an inherent ability to control mercury emissions. However, there was a significant difference in Hg concentrations and speciation between different cleaning systems (e.g., measurement periods 1 and 2—ESP, lower temperature; measurement period 3—FF, higher temperature). As shown in Table 6.11,

Table 6.11 Mercury removal capacity of a filter device fabric filter calculated as the portion of mercury mass flow removed per hour.

Mode	Hg^t	Hg^g	Hg^p	Hg^{2+}	Hg^0
Combined	98%	90%	100%	79%	94%
Direct	95%	94%	100%	86%	98%

From Mlakar TL, Horvat M, Vuk T, Stergaršek A, Kotnik J, Tratnik J, Fajon V. Mercury species, mass flows and processes in a cement plant. Fuel 2010;89:1936–45.

compared to the older ESP, the new FF efficiently lowered the Hg^t, Hg^p, and Hg^g mass flows. Removal capacity was very good for both Hg^{2+} and Hg^0. There was a slight difference between removal efficiency in direct and combined mode. Hg removal efficiency in cement plants is generally good, but to influence the overall Hg mass balance in the system and to lower the overall emission, it is necessary to remove some of the Hg from the hot part of the cement production process and thus interrupt certain cycling processes.

Wang et al. [18] investigated the removal efficiencies of rawmill + FF (RF combination) and coal-mill + FF (CF combination). Mercury in the flue gas is removed as the flue gas flows through the RF or CF complex. The removed mercury enters a raw meal for PM recycling. The removed mercury returns to the kiln system and causes mercury condensation in the cement manufacturing process. Consequently, mercury enrichment is related to the removal efficiency of the facility. Higher mercury removal efficiency of ESP or FF means higher mercury enrichment. For Plant 1 and Plant 2, the removal efficiency of the RF combination was 86% and 89%, but the CF combination removal efficiency was 95% and 94%, respectively. In Plant 1, all mercury species were efficiently removed from the flue gases in each combination. However, in Plant 2, only a fraction of Hg^0 was removed and Hg^0 increased with RF combination. In both Plant 1 and Plant 2, the CF combination removal efficiency was higher than the RF combination removal efficiency. Coal characteristics and operating parameters of each facility, including residence time and cooling rate of the flue gases, may be the main reasons. Despite similar mercury input from raw materials and coal, there is a big difference between mercury enrichment at Plant 1 and Plant 2.

Operating parameters and material chemical components can significantly affect mercury emissions [21]. The Hg concentration in raw meal passing the cyclones decreases with temperature, which is the opposite to the trend of chloride and sulfate. As presented in Fig. 6.17, the temperature in the system could influence different processes and reactions, and consequently different ratios of individual Hg species in the system could be obtained. A lower temperature of the flue gases would produce more Hg bound to particulates, which could be efficiently removed by the filter. Thus regulation of temperature in the system could control Hg emission.

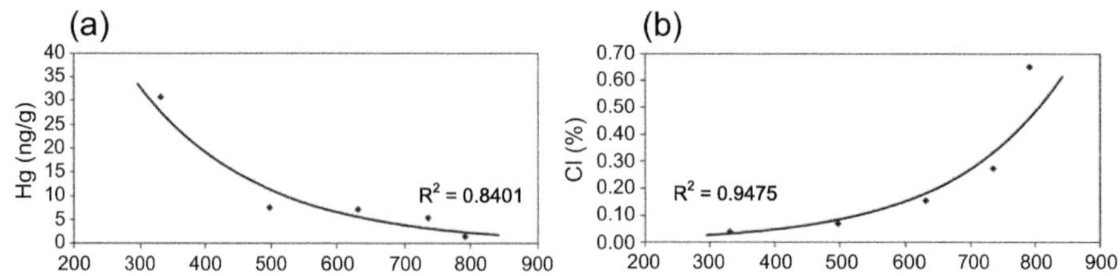

Figure 6.17 Examples of correlations between (a) Hg and (b) Cl concentrations in raw materials and the temperature of material in the cyclones of the preheater (measurements from the first sampling period).
From Mlakar TL, Horvat M, Vuk T, Stergaršek A, Kotnik J, Tratnik J, Fajon V. Mercury species, mass flows and processes in a cement plant. Fuel 2010;89:1936–45.

6.3.4 Emission and removal of other trace elements in cement plants

6.3.4.1 Trace element concentration in raw materials and products

Yan et al. [22] studied the distribution of As and Se in material and flue gas streams from PPKs and VSKs in China. The characteristics and location of cement plants and test details are summarized in Table 6.12. The total element contents are shown in Table 6.13. In general, the concentrations depended on the available raw materials and fuels in different regions, as expected. It can also be seen that the element contents of the feedstock and fuel significantly differ from kiln to kiln. Limestone is the main material component in cement clinker production and constitutes around 72%–86% of the raw material mix.

Concentrations of As were higher in the final raw meal than in the raw material mix. This tendency was not observed for Se. Regarding the total contents in the cement clinker, low concentrations of As and Se were found in most of the samples. Typical average concentrations of these constituents in final cements are around 0.05, 11, and <2 mg/kg, respectively. It can be seen that As concentrations of 65 and 90 mg/kg were found in PPK2a and VSK2a/b, respectively. In the latter case, high concentrations of As were also found in the raw materials and in the final raw meal. The high value obtained for PPK2a could not be explained by the raw material input and showed somehow the challenge of obtaining consistent field data from a high-temperature incineration process. However, the results indicated that significant quantities of As and Se are transferred into clinker.

Gupta et al. [37] studied the elemental emissions from a cement kiln, which is located in India and is one of the largest indigenous producers and exporters of cement. As shown in Table 6.14, Fe and Ca were generally higher in concentration than other metals. As, Cd, Cr, Ni, and Pb showed higher concentrations in the stack effluent. Only small amounts of elements associated with the particles were released through the stack (Table 6.15). Most of the elements were trapped by the bag filter and showed removal efficiency up to 99%. Of all the elements, Ca was the highest concentration among cyclone and baghouse, followed by Fe. The raw meal was composed mainly of limestone and laterite/red ochre cut at a ratio of 5:1, so the superiority of Ca and Fe was justified by the rawmill at the entrance to the cyclone and baghouse. The superiority of Ca and Fe was also found in the raw material, but the release of Ca through the PM through the stack is much lower than that of Fe, presumably because the majority of Ca is possibly retained by the cyclone. Sulphates and As–Ca compounds are produced due to mineral transformation that occur during coal combustion and increased from cyclone inlet to stack fly ash as a result of reaction between CaO and gaseous products such as SO_x [35].

6.3.4.2 Trace element concentration in flue gas

Yan et al. [22] studied the distribution of Hg, As, and Se in flue gas streams from PPKs and VSKs in China. Regarding the concentrations of As and Se in the flue gas, small

Table 6.12 Characteristics and location of cement plants and test details.

Sample name	Line number	City, province	Kiln type[a]	Pollution control[a,e]	Feeding rate (t/h)[b]	Yield of clinker (t/h)	Flow flue gas (Nm^3/h)	Operational mode[c]
VSK1	1	Xingtai, Hebei	VSK	WS	32	19	28,500	Normal
VSK2a	1	Zibo, Shandong	VSK	FF	28	17	24,711	Normal
VSK2b	2	Zibo, Shandong	VSK	FF	35	21	32,246	Normal
VSK2c[d]	2	Zibo, Shandong	VSK	FF	42	25	32,568	Normal
PPK1	1	Zhengzhou, Henan	PPK	FF	365 + 31	229	450,979	Compound
PPK1-MS	1	Zhengzhou, Henan	PPK	FF	365 + 31	229	450,979	Direct
PPK2a	1	Zibo, Shandong	PPK	ESP	350 + 28	217	394,993	Compound
PPK2a-MS	1	Zibo, Shandong	PPK	ESP	350 + 28	217	394,993	Direct
PPK2b	2	Zibo, Shandong	PPK	ESP	361 + 31	227	347,720	Compound
PPK2b-MS	2	Zibo, Shandong	PPK	ESP	361 + 31	227	347,720	Direct

[a]Vertical shaft kiln (VSK), preheater and precalciner kiln (PPK), wet scrubber (WS), fabric filter (FF), electrostatic precipitator (ESP).
[b]VSK: Total feeding rate of raw meal and fuel. PPK: Feeding rate of raw meal + fuel.
[c]Compound mode means that the raw meal mill is running. Direct mode means that the raw meal mill is shut off (MS).
[d]Samples taken 2 years afterward for VSK2b.
[e]The pollution control devices refer to location after the rawmill.
From Yan D, Peng Z, Ding Q, Karstensen KH, Engelsen CJ, Li L, Ren Y, Jiang C. Distribution of Hg, As and Se in material and flue gas streams from preheater-precalciner cement kilns and vertical shaft cement kilns in China. J Air Waste Manag Assoc 2015;65:1002–10.

Table 6.13 Material component compositions (%) in the raw material mix and the total concentrations of Hg, As, and Se in the solid fractions (mg/kg).[a]

Sample	Parameter	Limestone	Clay	Coal gangue	Fluorite	Silica stone	Coal fly ash	Iron-rich[b] material	Lignite	Coal	Raw material mix[c]	Recycled PM[d]	Final raw meal[e]	Clinker[f]
VSK1	Composition (%)	76	10	–	0.5	–	–	1.5	–	11	–	–	–	–
	Hg	0.007	N.D.	–	0.002	–	–	0.001	–	0.003	0.01	14	0.1	0.001
	As	5.4	12	–	13	–	–	3.2	–	16	7.2	–	5.4	6.0
	Se	1.1	0.2	–	0.6	–	–	0.9	–	24	3.5	–	0.4	0.3
VSK2a/b	Composition (%)	75	10	–	0.8	–	3.5	3.2	–	8.0	–	–	–	–
	Hg	0.02	0.02	–	N.D.	–	0.63	0.45	–	0.13	0.07	1.8	0.08	N.D.
	As	8.6	21	–	N.D.	–	32	835	–	8.2	37	3829	63	90
	Se	<0.4	<0.4	–	N.D.	–	1.3	4.4	–	4.8	0.9	241	2.6	2.8
VSK2c	Composition (%)	74	8.0	–	0.0	–	3.0	7.0	–	8.0	–	–	–	–
	Hg	0.001	0.02	–	0.06	–	0.27	0.001	–	0.14	0.02	4.5	0.05	N.D.
	As	3.2	13	–	16	–	22	3.4	–	4.7	4.7	47	5.0	7.4
	Se	0.4	0.5	–	1.4	–	36	3.3	–	3.4	2.0	137	2.9	1.1
PPK1	Composition (%)	86	–	6.3	–	6.2	–	1.8	–	–	–	–	–	–
	Hg	0.002	–	N.D.	–	N.D.	–	N.D.	N.D.	–	0.002	0.2	0.02	0.001
	As	1.6	–	6	–	9.2	–	9.6	1.6	–	2.5	14	3.2	6.0
	Se	N.D.	–	0.4	–	0.28	–	0.8	1.9	–	0.2	0.1	0.2	0.5

Continued

Table 6.13 Continued

Sample	Parameter	Limestone	Clay	Coal gangue	Fluorite	Silica stone	Coal fly ash	Iron-rich[b] material	Lignite	Coal	Raw material mix[c]	Recycled PM[d]	Final raw meal[e]	Clinker[f]
PPK2a	Composition (%)	79	–	6.0	–	13	–	2.0	–	–	–	–	–	–
	Hg	N.D.	–	N.D.	–	N.D.	–	0.4	–	0.01	0.01	4.6	0.3	N.D.
	As	4.5	–	17	–	N.D.	–	N.D.	–	N.D.	4.6	–	N.D.	65
	Se	0.6	–	5.0	–	N.D.	–	N.D.	–	N.D.	0.8	–	N.D.	3.1
PPK2b	Composition (%)	82	–	8.0	–	6.6	–	3.0	–	–	–	–	–	–
	Hg	0.04	–	0.08	–	0.03	–	0.08	–	0.273	0.05	4.3	0.3	N.D.
	As	4.6	–	12	–	6.2	–	13	–	4.7	5.5	87	10	38
	Se	0.6	–	1.7	–	1.8	–	15	–	3.4	1.2	–	0.3	1.0

[a]*N.D.*, not detected.
[b]Iron dust from the cobalt melting industry was used for VSK2a/b and VSK2c. Pyrite cinder was used in the other plants.
[c]Estimated concentrations as discussed in text.
[d]Estimated concentrations in recycled particulate matter (PM).
[e]Raw meal mixed with recycled PM constituted the final raw meal (for PPK and VSK).
[f]For VSK2a/b, target elements in VSK2a were determined.
From Yan D, Peng Z, Ding Q, Karstensen KH, Engelsen CJ, Li L, Ren Y, Jiang C. Distribution of Hg, As and Se in material and flue gas streams from preheater-precalciner cement kilns and vertical shaft cement kilns in China. J Air Waste Manag Assoc 2015;65:1002–10.

Table 6.14 Metal/ion contents in raw materials and finished products.

Material	Metal/ion concentration (mg/g)												
	Ca^{2+}	NO_3^-	SO_4^{2-}	As	Cd	Co	Cr	Cu	Fe	Mn	Ni	Pb	Zn
Coal + coke	NA	NA	NA	0.06	0.17	0.02	0.02	0.01	1.86	0.39	0.16	0.04	0.02
Gypsum	NA	NA	NA	0.12	0.08	0.03	0.01	0.01	2.50	0.07	0.03	0.13	0.03
Fly ash	NA	NA	NA	0.59	0.09	0.03	0.04	0.03	4.12	0.11	0.03	0.12	0.06
Laterite	0.42	1.39	0.01	0.01	0.03	0.06	0.03	0.01	10.6	0.44	0.02	0.03	0.01
Red ochre	1.42	1.64	0.07	0.04	0.01	0.03	NA	0.03	6.59	NA	0.01	0.04	0.02
Limestone	5.18	1.77	1.54	0.28	0.09	0.04	0.02	0.01	6.65	0.75	0.04	0.16	0.04
Clinker	NA	NA	NA	0.86	0.21	0.17	0.06	0.05	35.38	0.55	0.08	0.31	0.10
Cement	NA	NA	NA	0.84	0.19	0.15	0.06	0.04	28.38	0.62	0.07	0.29	0.10

NA, not analyzed.
From Gupta RK, Majumdar D, Trivedi JV, Bhanarkar AD. Particulate matter and elemental emissions from a cement kiln. Fuel Process Technol 2012;104:343–51.

Table 6.15 Concentration of cations/anions and metals from various discharge points.

Ion/metal	Concentration (μg/Nm^3)			Bag filter efficiency[a] (%)
	Cyclone inlet air	Bag filter inlet air	Kiln stack	
Ca^{2+}	1,066,796	523,461	349	99.93
NO_3^-	223,441	109,934	649	99.41
SO_4^{2-}	78,875	36,767	573	98.44
As	38,653	13,266	37	99.72
Cd	16,661	5,130	14.6	99.71
Cr	7,164	1,120	4.9	99.56
Co	7,164	2,064	15.4	99.25
Cu	2,666	707	2.1	99.71
Fe	847,374	241,507	3,062	98.73
Mn	78,473	24,941	76	99.70
Ni	7,497	2,417	5.1	99.79
Pb	41,486	13,502	34	99.75
Zn	15,495	2,771	13.5	99.51

[a]Efficiency in removing respective ions/metals.
From Gupta RK, Majumdar D, Trivedi JV, Bhanarkar AD. Particulate matter and elemental emissions from a cement kiln. Fuel Process Technol 2012;104:343–51.

concentrations were generally observed for Se, as can be seen in Table 6.16. In the case of As, most of the measurements were below 5 μg/m^3. Varying results of As were found in earlier studies of PPKs. Gupta et al. [37] reported 37 μg/m^3 and Conesa et al. [23] reported emissions of 0.9–3.7 μg/m^3 when waste tires or sewage sludge was used as cofuel.

Conesa at al [23]. studied the emissions of different pollutants when partially replacing the fuel type used in a cement kiln. Characterization of the sludges and tires was carried out. Table 6.17 shows details of the analysis performed. Two different sewage sludges were used in the study period due to the inability to obtain the same sludge continuously. Table 6.18 presents the amount of fuels fed to the furnace in each series, and the ratio corresponding to each waste fuel.

As shown in Table 6.19 and Fig. 6.18, the arrival of a number of metals has increased as the replication of the broken wheels. Various concentrations of metals obtained in each sampling series, with the corresponding percentage of wheels and waste debris, undergo analysis on the main part (PC). As shown in Fig. 6.19, the first PC (41.5% of the difference) is positively related to the concentration of V, Co, Ni, Pb, and % of the casting kiln. The second PC (29.9% of the difference) relates to the reaches of Cr, Cu, and Cd. The wheel is located near V, Co, Ni, and Pb, which shows that these changes have a similar nature. Multivariate analyses show that % wheels are

Table 6.16 Particles, Hg, As, and Se concentrations in flue gas.

Sample[a]	Flue gas temperature (°C)	Particles[b] (mg/Nm^3)	As	Se
VSK1	70	1007	73 ± 27	21 ± 5
VSK2a	97	1.7	2.7 ± 0.7	N.D.
VSK2b	90	3.8	2.3 ± 0.2	N.D.
VSK2c	65	0.2	N.D.	N.D.
PPK1	85	2.2	0.02 ± 0.01	N.D.
PPK1-MS	90	N.A.	N.D.	0.05
PPK2a	115	3.9	0.9 ± 0.7	N.D.
PPK2a-MS	120	N.A.	4.4	N.D.
PPK2b	90	1.1	4.3 ± 0.3	N.D.
PPK2b-MS	90	N.A.	0.07	N.D.

N.D., not detected.
[a]One sample was taken at the plants running in direct mode (mill shut off); PPK1-MS, PPK2a-MS, and PPK2b-MS.
[b]*N.A.*, not analyzed for samples collected in direct mode.
From Yan D, Peng Z, Ding Q, Karstensen KH, Engelsen CJ, Li L, Ren Y, Jiang C. Distribution of Hg, As and Se in material and flue gas streams from preheater-precalciner cement kilns and vertical shaft cement kilns in China. J Air Waste Manag Assoc 2015;65:1002–10.

bundled with a number of metals at high PC1 values, indicating that the metal in this component increases when the wheels are fed up. All the metals in classes II and III in HM are present in their toxicity. The metal of the class did not allow an increase in the number of circular ratios. On the other hand, the percentage of tar in the tar does not correspond to the metal casting. Ni and Pb are in the wheels and add to the stack when feeding the wheels. However, the metals in the cages (Cu, Cr, Mn, Ni, and V) do not add to their concentration when increasing the amount of pots. There are also metals, such as Hg and Sb, that have a very different nature with respect to other heavy metals. The global impact of each metal depends on its growth and the specific value of the wheels and sewage debris. This effect has been linked to feeding point differences in two materials.

Giannopoulos et al. [38] studied the heavy metal stack emissions from the coincineration of cutting oil emulsions in cement plants. Severe metal emissions are summarized excluding the quantities of cadmium. As shown in Figs. 6.20 and 6.21, the estimated heavy metal emissions caused by cutting oil emulsion injection are lower than the EU limit ($Cd + Tl < 0.05$ mg/m^3, $Hg < 0.05$ mg/m^3, $Sb + As + Pb + Cr + Co + Cu + Mn + Ni + V < 0.5$ mg/m^3). In a worst-case scenario involving a 3000 L/h volumetric rate and a 100% increase in emission rates, estimated Pb, Ni, Cu, Cr, and V emissions are 50 times lower and Cd emissions are 400 times lower than the EU directive limit. It also outlines the metal inputs and factors affecting heavy metal stack emissions linearly. However, the results of estimated

Table 6.17 Characterization of sludges and tires used in the cement kiln.

	Sewage sludge 1	Sewage sludge 2	Waste tires
Weight loss (105°C, %)	10.0	7.03	0.9
Weight loss (550°C, %)	27.6	31.2	65.5
NCV (MJ/kg)	16.99	11.77	33.08
Total chloride (mg/kg)	1,804	3,890	5,023
Total sulfur (mg/kg)	12,408	17,820	17,106
Metal analysis (mg/kg)			
Cd	13.4	2.9	–
Tl	<0.5	0.1	–
Hg	<0.5	1.2	–
Sb	4.2	0.15	–
As	<0.5	2.4	–
Pb	69	72	–
Co	1.6	4.5	–
Cu	217.7	562.3	–
Cr	178.1	572.7	–
Mn	128.1	182.2	–
Ni	35.8	223.2	–
V	11.3	17.4	–
Sn	<0.5	55.5	–

NCV, net calorific value.
From Conesa JA, Galvez A, Mateos F, Martin-Gullon I, Font R. Organic and inorganic pollutants from cement kiln stack feeding alternative fuels. J Hazard Mater 2008;158:585–92.

Table 6.18 Mean mass flows of the fuels fed to the furnace in each series.

	Sampling series					
	A	B	C	D	E	F
Petroleum coke (kg/h)	12,000	9,800	9,000	11,800	11,450	11,300
Waste tires (kg/h)	4,500	1,500	1,000	500	1,100	–
Sewage sludge 1 (kg/h)	–	1,500	3,000	4,000	–	–
Sewage sludge 2 (kg/h)	–	–	–	–	5,200	5,600
Weight tires (%)	27	12	8	3	6	0
Weight sludge (%)	0	12	23	25	30	33

From Conesa JA, Galvez A, Mateos F, Martin-Gullon I, Font R. Organic and inorganic pollutants from cement kiln stack feeding alternative fuels. J Hazard Mater 2008;158:585–92.

Table 6.19 Results for heavy metal emissions.

	Sampling series					
	A	B	C	D	E	F
	Metals (mg/Nm^3)					
V	0.0036	0.0062	0.0062	0.00130	0.00196	
Cr	0.0031	nd	0.0016	0.00240	0.00051	
Mn	0.011	0.0177	0.0197	0.00430	0.08546	
Co	0.0033	0.0034	0.0006	0.00030	0.00123	
Ni	0.0045	0.004	0.0041	0.00180	0.00261	
Cu	0.0124	0.0114	0.0167	0.01960	0.00165	
Zn	0.0153	0.0209	0.0071	<0.01	0.32963	
As	0.0012	0.0014	0.0037	0.00090	0.00102	
Cd	0.0002	0.0003	0.0002	0.00030	0.00002	
Sn	0.0068	nd	0.0007	0.00090	0.00047	
Sb	0.0004	0.0003	0.0004	0.00170	na	
Tl	0.0003	0.0011	0.0012	0.00010	0.00009	
Pb	0.0504	0.0206	0.0391	0.00800	0.00065	
Hg	<0.005	<0.004	0.0091	0.02500	0.00837	
Cd + Tl	0.00045	0.00147	0.0014	0.00040	0.00011	
Sb + As + Pb + Cr + Cu + Co + Mn + Ni + V	0.08990	0.06500	0.09210	0.04030	0.09509	

nd, not detected; *na*, not analyzed. *HM*, heavy metal.
From Conesa JA Galvez A, Mateos F, Martin-Gullon I, Font R. Organic and inorganic pollutants from cement kiln stack feeding alternative fuels. J Hazard Mater 2008;158:585–92.

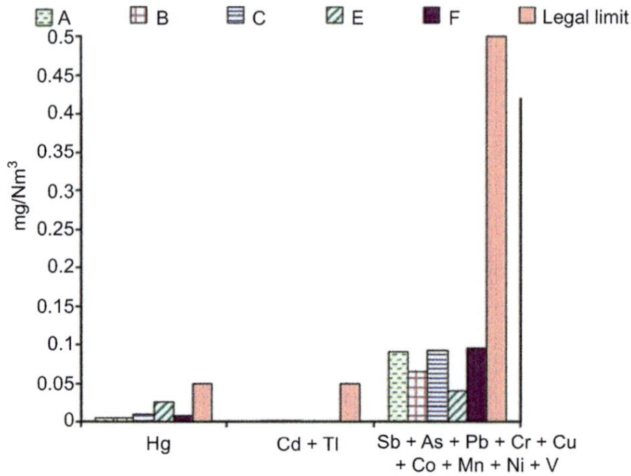

Figure 6.18 Metal emissions compared to the legal limit.
From Conesa JA, Galvez A, Mateos F, Martin-Gullon I, Font R. Organic and inorganic pollutants from cement kiln stack feeding alternative fuels. J Hazard Mater 2008;158:585–92.

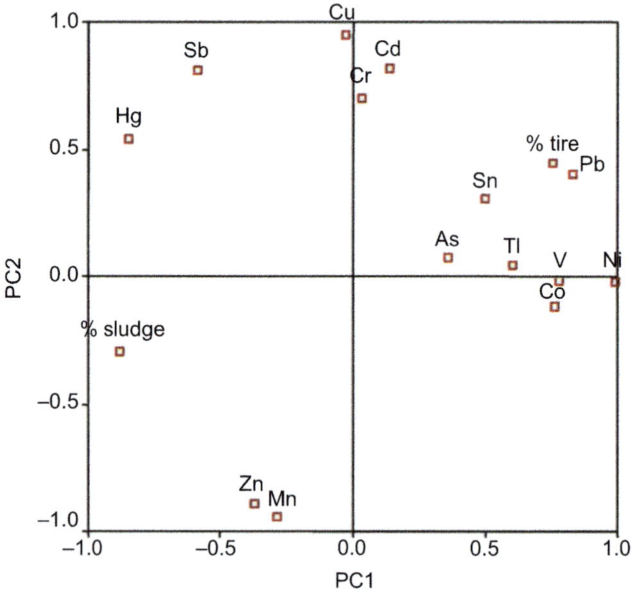

Figure 6.19 Principal component (PC) analysis for heavy metal emissions.
From Conesa JA, Galvez A, Mateos F, Martin-Gullon I, Font R. Organic and inorganic pollutants from cement kiln stack feeding alternative fuels. J Hazard Mater 2008;158:585–92.

emissions are two or three orders of magnitude less than the specified limits. This great difference provides enough support to claim that the cutting of heavy metal compositions and passing through the feeder seem to have nothing to do with the nature of the production plant in cement.

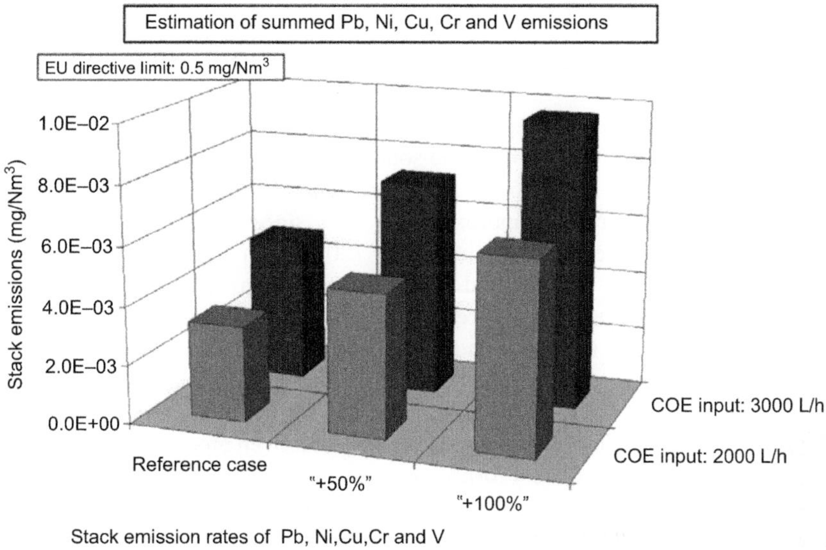

Figure 6.20 Estimation of summed Pb, Ni, Cu, Cr, and V stack emissions for cutting oil injection in the cement kiln. *COE*, cutting oil emulsion.
From Giannopoulos D, Kolaitis DI, Togkalidou A, Skevis G, Founti MA. Quantification of emissions from the co-incineration of cutting oil emulsions in cement plants – part II: trace species. Fuel 2007;86:2491–501.

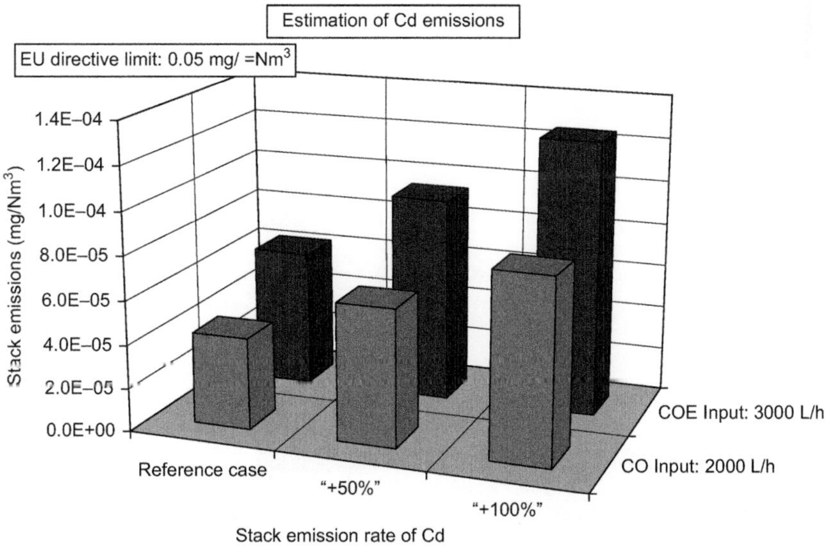

Figure 6.21 Estimation of Cd stack emissions for cutting oil injection in the cement kiln.
From Giannopoulos D, Kolaitis DI, Togkalidou A, Skevis G, Founti MA. Quantification of emissions from the co-incineration of cutting oil emulsions in cement plants – part II: trace species. Fuel 2007;86:2491–501.

6.4 Emission characteristics and removal of trace elements in iron and steel production processes

Besides cement production processes, iron and steel production (ISP) is considered to be one of the predominant anthropogenic sources of many hazardous TEs, like atmospheric mercury (shown in Fig. 6.22) [17,35]. China's ISP has been increasing at an average annual growth rate of approximately 15% since 2000. As the largest iron and steel producer in the world, China made 780 million t of crude steel in 2013. Therefore it is important to study and control TE emissions from ISP in China [39]. The iron and steel manufacturing process is notably complicated, in which more emission points are contained than other industrial processes. ISP is classified into two major types: converter furnace processes and electric furnace processes. The crude steel production of converter furnace processes accounts for more than 90% of production in China. Converter furnace processes include the coke-oven process, sintering furnace process, pelletizing process, blast furnace process, and converter procedure [40]. The process uses coal and coke as the main energy sources, and iron ore is smelted into molten iron or pig iron in a blast furnace, and then the molten iron is smelted into molten steel into a billet in a converter. Due to the large scale, high productivity, and wide variety of products, China's steel-making process is dominated by blast furnace-converter steel making. The scrap-electric furnace process uses scrap steel as raw material and electric energy as the main energy source, and smelts it into a molten steel billet in an electric furnace. The direct reduction-electric furnace process uses noncoking coal and iron ore resources for solid reduction in a direct reduction reactor to obtain direct-reduced iron. Because direct-reduced iron is not liquid molten iron, generally, alternative scrap is used for electric furnace smelting. Noncoking coal

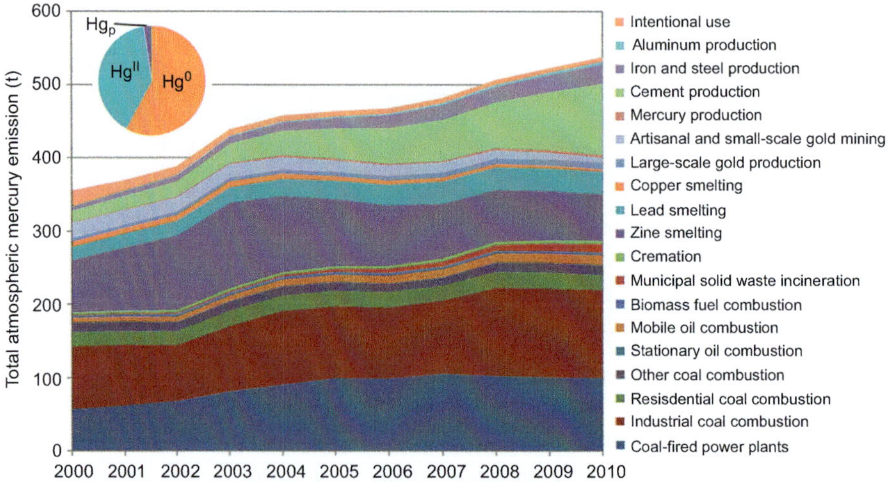

Figure 6.22 Temporal trend and species distribution of mercury emissions from China. From Zhang L, Wang S, Wang L, Wu Y, Duan L, Wu Q, Wang F, Yang M, Yang H, Hao J, Liu X. Updated emission inventories for speciated atmospheric mercury from anthropogenic sources in China. Environ Sci Technol 2015;49:3185–94.

and iron ore resources are obtained in a smelting reduction reactor to obtain liquid molten iron, which is subjected to converter steel making. Unless in a particulate state, TEs may be another important issue for pollution abatement in the iron and steel industry. Thus investigating TE emissions is of great importance to better comprehend pollution abatement from the iron and steel-making industrial.

6.4.1 Trace element partition and emission from the steel-making process

6.4.1.1 Emission of mercury from the steel-making process

Hg is the most persistent hazard emitted from the steel production process due to its high volatility and insolubility in water, and has attracted wide exploration of its emission properties. Hg input to ISP increased from 21.6 t in 2000 to 94.5 t in 2015 (shown in Fig. 6.23). During 2000–14, the average annual growth rate (AAGR) of Hg input was 11%, while Hg input reduced by 3% from 2014 to 2015. Of the various types of raw materials, coking coal and iron concentrates contributed the largest amount of Hg input, accounting for 35%–46% and 25%–32% of the total, respectively. Hg input due to the use of coking coal increased from 9.9 t in 2000 to 33.5 t in 2015. Hg input with iron concentrates increased at an AAGR of 12% from 2000 and reached 29.2 t in 2015. Pulverized coal injection coal brought approximately 6%–9% of Hg to ISP. Hg in the additives (including limestone and dolomite) contributed 12%–18% of total Hg input. Hg in iron blocks was in the range 0.6–3.6 t. Hg input due to the use of steel scrap and alloy was 4.1 t in 2015, accounting for 4% of the national total. However, steel scrap and alloy contributed 7% to total crude steel production in 2015 [29].

Atmospheric Hg emissions from ISP increased from 11.5 t in 2000 to 32.7 t in 2015 (Fig. 6.24). The peak in emissions was in 2013 when they reached 35.6 t. In 2015,

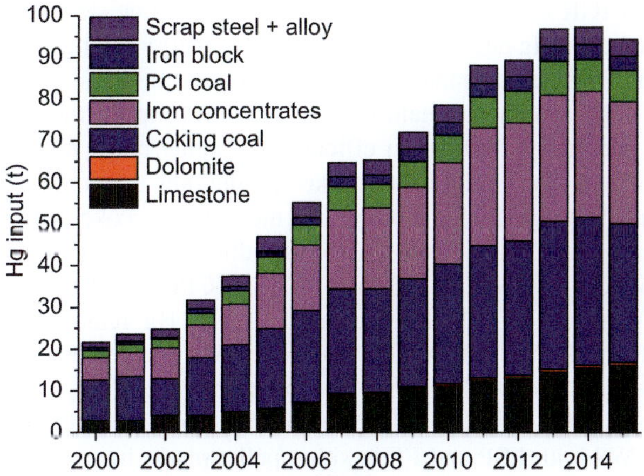

Figure 6.23 Hg input trends by material. *PCI*, pulverized coal injection.
From Wu Q, Gao W, Wang S, Hao J. Updated atmospheric speciated mercury emissions from iron and steel production in China during 2000–2015. Atmos Chem Phys 2017;17:10423–33.

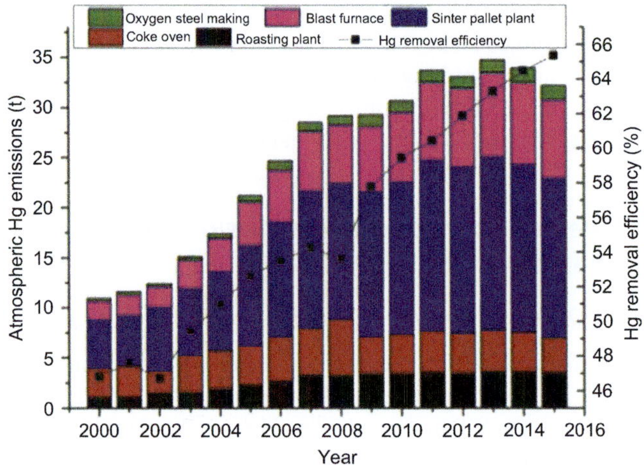

Figure 6.24 Hg emission trends by material.
From Wu Q, Gao W, Wang S, Hao J. Updated atmospheric speciated mercury emissions from iron and steel production in China during 2000–2015. Atmos Chem Phys 2017;17:10423–33.

emissions from long-process steel-making and short-process steel-making methods were 32.2 and 0.5 t, accounting for 98.3% and 1.7% of the national total, respectively. Thus emissions from long-process steel making were still the dominant emission process of China's ISP. Among the processes, emissions from sinter/pellet plants accounted for 42%–49% of the annual total. Their emissions increased from 4.8 t in 2000 at an AAGR of 10.1% to 15.9 t in 2015. Blast furnaces were also significant Hg emitters. Their emissions increased from 1.9 t in 2000 to 7.9 t in 2015 at an AAGR of 10.0%. The AAGR for roasting plants and coke ovens was 8.3% and 1.2%. In 2015, emissions from both roasting plants and coke ovens were 7.0 t.

The slower AAGR of Hg emissions (7.2%) than of crude steel production (13%) reflected the impact on Hg emission reduction due to energy saving and environmental protection in ISP. On the one hand, Hg input to produce unitary crude steel decreased from 0.17 to 0.12 g/t, which was mainly the result of the improvement of coke production efficiency and energy utilization efficiency of sinter/pellet plants and blast furnaces. Since 2004, the indigenous coke production method with high coal consumption has been gradually replaced by the machine coke production method. The coke ratio in sinter/pellet plants has been reduced from approximately 388 kg/t of pig iron produced in 2000 to 363 kg/t in 2015 (CISIA, 2001–16). On the other hand, the improvement of APCDs increased overall Hg removal efficiency from 47% in 2000 to 65% in 2015. APCDs for coke ovens have shown the largest Hg removal efficiencies (64%–87%), while pollution control in sinter/pellet plants contributed most to the rapid Hg reduction speed during 2000–15. The replacement of cyclone and WS with ESP and FF in sinter/pellet plants improved Hg removal efficiency from 21% in 2000 to 44% in 2010. The application of FGD in addition to dust collectors was the main driver of Hg reduction in the sinter/pellet process during 2011–15. Hg removal efficiency in sinter/pellet plants was 53% in 2015.

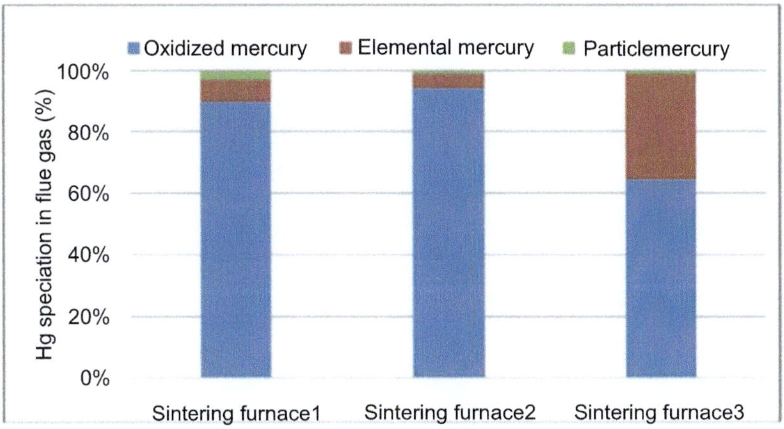

Figure 6.25 Hg speciation in the flue gas.
From Xu W, Shao M, Yang Y, Liu R, Wu Y, Zhu T. Mercury emission from sintering process in the iron and steel industry of China. Fuel Process Technol 2017;159:340–4.

Mercury emission is classified as gaseous Hg^0, gaseous Hg^{2+}, and Hg^P. Pacyna et al. reported that the percentages of Hg^0, Hg^{2+}, and Hg^P were 80%, 15%, and 5%, respectively [41]. This result was similar to that of a previous report, which suggested that the percentages of Hg^0, Hg^{2+}, and Hg^P from steel–iron production were 81.2%, 14.6%, and 4.2%, respectively [24]. However, oxidized mercury was found to be the predominant species in a recent study. The proportion of Hg^{2+} in flue gas reached as high as 59%–73% and the proportion of Hg^P was under the detection limit because of the installation of ESPs for the examined iron and steel plants [42]. Wu et al. also reported that Hg^{2+} dominated the mercury species in flue gas (shown in Fig. 6.25) [40]. In addition, mercury concentrations varied widely in gas samples from ISP. Yue et al. measured mercury concentration in the flue gas of a sintering machine upstream, and the value was 18–100 μg/m^3 before the FGD device [43]. The reported mercury concentration in the gas from a sintering furnace in Korea was 13.27–114.05 μg/m^3 at the inlet of the ESP and 10.18–19.12 μg/m^3 at the exit [27]. Furthermore, another study reported mercury emission concentrations of 4.95 and 9.95 μg/m^3 from electric and sintering furnaces, respectively [28].

Due to the complicated ISP processes and limitations of data availability, the process combinations considered in different inventories were divided into four types (shown in Fig. 6.26) [29,31,32,42,44]. The first type (green symbols) included sinter plants and blast furnaces, which were the basic assumption in the emission inventories of Zhang et al. and AMAP/UNEP [17,35]. In these studies, a unique emission factor of 0.0400 g/t was applied, and their emissions were similar in the same inventory year. Wu et al.'s emissions for this process combination were almost the same as the foregoing estimations around 2005. However, the gap grew with time when FGD was gradually applied in sinter/pellet plants. Therefore the emission factor for this type of combination was reduced from 0.0527 g/t in 2000 to 0.0296 g/t in 2015. The second type (orange symbols) also consisted of steel making in addition to the first type.

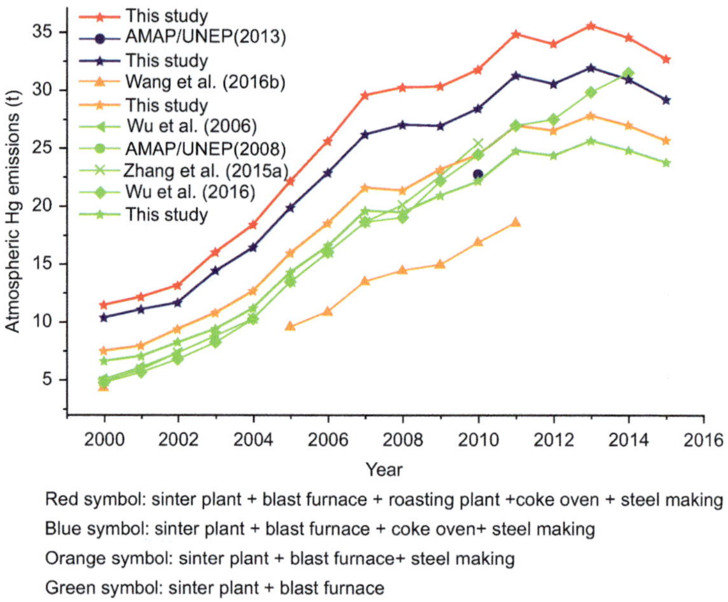

Figure 6.26 Atmospheric Hg emissions from iron and steel production in different studies. From Wu Q, Gao W, Wang S, Hao J. Updated atmospheric speciated mercury emissions from iron and steel production in China during 2000–2015. Atmos Chem Phys 2017;17:10423–33.

However, the technology applied in Europe may be better than in China. For example, the emission factor of 0.00019 g/t applied to blast furnaces with FF was used as the best emission.

6.4.1.2 Partitions of other trace elements during the steel production process

Because other TEs like As, Se, F, and Cr are less hazardous and volatile to Hg, Tsai et al. investigated their partition patterns in particulates as by-products of the steel-making flue gas process [34]. As shown in Table 6.20, As showed accumulation in

Table 6.20 Elemental concentrations in particulate emissions from various processes in a steel plant.

Concentration (mg/g)	Coke making	Sintering	Cold forming	Hot forming
As	0.1252 ± 0.13	0.0750 ± 0.013	0.3702 ± 0.098	0.2032 ± 0.094
Se	0.0677 ± 0.01	0.3984 ± 0.020	0.1565 ± 0.011	Not detected
F	1.524 ± 0.076	3.178 ± 0.029	0.520 ± 0.010	4.862 ± 0.105
Cr	Not detected	0.2106 ± 0.125	0.5226 ± 0.172	0.1819 ± 0.181

From Tsai J, Lin K, Chen C, Ding J, Choa C, Chiang H. Chemical constituents in particulate emissions from an integrated iron and steel facility. J Hazard Mater 2007;147:111–19.

particulates in all four steps, including coke-making, sintering, cold-forming, and hot-forming processes during steel making, while Se exhibited accumulation at the former parts. F and Cr both established accumulation in the latter parts of the steel-making process. Besides this study, the emission properties of As, Se, F, and Cr were rarely investigated from iron and steel factories.

6.4.2 Mercury pollution control from the steel-making industry

6.4.2.1 Cobenefit from APCDs

Fig. 6.27 shows a typical treatment process for steel-making flue gas. The desulfurization methods of the three sintering furnaces are: circulating fluidized-bed flue gas desulfurization (CFB-FGD), dense flow absorber flue gas desulfurization (DFA-FGD), and ammonia flue gas desulfurization (AFGD). As shown, the flue gas is emitted into the atmosphere after the ESP/FF and FGD. Mercury was simultaneously tested at three sampling points: up- and downstream of the ESP and outlet of the FGD, which is at the stack.

Mercury in the flue gas is removed when the flue gas flows through the APCDs. The removal efficiencies of APCDs are calculated according to the mercury concentration at the inlet and outlet of the APCDs. Table 6.21 shows the mercury removal efficiencies of the APCDs. The Hg^t removal efficiencies of the APCDs were 97.5%−98.3%, including 30.0%−58.2% for ESP and 40.1%−67.5% for FGD. The results are much higher than those in previous studies for coal-fired power plants. Wang et al. reported that the mercury removal efficiencies for ESP and ESP + FGD were 6%−39% and 13%−81%, respectively [39]. Lee et al. showed that the tested results were 21.9%−57.9% for ESP and 16.3%−64.4% for ESP + WFGD (wet flue gas desulfurization) [45]. Zhang et al. showed that the average mercury removal efficiency of the ESP system of six coal-fired power plants was 11.5% and that of the ESP + FGD system was 13.7%, which were lower than the average removal efficiencies of pollution control device systems in US plants [46].

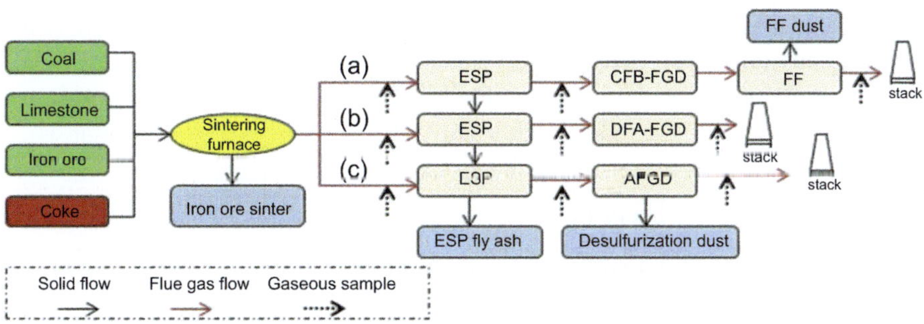

Figure 6.27 Schematic diagram of the selected sintering furnace. *AFGD*, ammonia flue gas desulfurization; *CFB-FGD*, circulating fluidized-bed flue gas desulfurization; *DFA-FGD*, dense flow absorber flue gas desulfurization; *ESP*, electrostatic precipitator; *FF*, fabric filter. From Xu W, Shao M, Yang Y, Liu R, Wu Y, Zhu T. Mercury emission from sintering process in the iron and steel industry of China. Fuel Process Technol 2017;159:340−4.

Table 6.21 Hg removal efficiencies of APCDs.

Sintering furnace	APCDs	Mercury removal efficiency (%)		
		ESP	FGD	ESP + FGD
1	ESP + CFB-FGD	45.9	52.0	97.9
2	ESP + DFA-FGD	58.2	40.1	98.3
3	ESP + AFGD	30.0	67.5	97.5

AFGD, ammonia flue gas desulfurization; *APCDs*, air pollution control devices; *CFB-FGD*, circulating fluidized-bed flue gas desulfurization; *DFA-FGD*, dense flow absorber flue gas desulfurization; *ESP*, electrostatic precipitator.
From Xu W, Shao M, Yang Y, Liu R, Wu Y, Zhu T. Mercury emission from sintering process in the iron and steel industry of China. Fuel Process Technol 2017;159:340–4.

With the increase in Hg^{2+} proportion, the removal efficiency of the ESP also increases, indicating that Hg^{2+} concentration in the flue gas affects the mercury removal efficiency of the ESP. According to previous reports, Hg^{2+} tended to be adsorbed on ash particulates, thus a portion of Hg^{2+} could be removed by the ESP. In addition, because the large proportion of Hg^{2+} in the flue gas is more chemically reactive and water soluble, it can be fully removed by the FGD, particularly using wet desulfurization. Calcium-based sorbents (CaO, $Ca(OH)_2$, etc.) provide strong Hg^{2+} adsorption capacity [47], whereas SO_2 could weaken Hg^{2+} capture because of the loss of alkaline sites through Ca-SO_2 [48]. In all, the Hg^t removal efficiency of ESP + FGD in the sintering flue gas was more than 97%, which is higher than that for flue gas treatment in coal-fired power plants.

The production rates of iron ore sinter in the three sintering furnace processes were 187.5, 356.4, and 580.0 t/h, respectively. As known, the mercury emission factors were calculated by dividing gaseous mercury emissions (466.9, 966.0, and 744.0 mg/h) by the production rates of sinter, and it was calculated to be 2.49, 2.71, and 1.28 mg/t sinter (Table 6.22), which are much lower than those reported by previous studies. The mercury emission factors of Pirrone et al. and Fukuda et al. were 0.04 g/t crude steel and 0.0488 g/t crude steel, respectively. The mercury concentration of the raw material, pollution control devices, and production process can affect the mercury emission factor. There is substantial uncertainty in the estimations of the mercury emission factor. Furthermore, the ISP process is notably complicated, and

Table 6.22 Mercury emission factors from the sintering furnace process.

Sintering furnace	APCDs	Hg emission factors (mg/t sinter)
1	ESP + CFB-FGD	2.49
2	ESP + DFA-FGD	2.71
3	ESP + AFGD	1.28

AFGD, Ammonia flue gas desulfurization; *APCDs*, air pollution control devices; *CFB-FGD*, circulating fluidized-bed flue gas desulfurization; *DFA-FGD*, dense flow absorber flue gas desulfurization; *ESP*, electrostatic precipitator.
From Xu W, Shao M, Yang Y, Liu R, Wu Y, Zhu T. Mercury emission from sintering process in the iron and steel industry of China. Fuel Process Technol 2017;159:340–4.

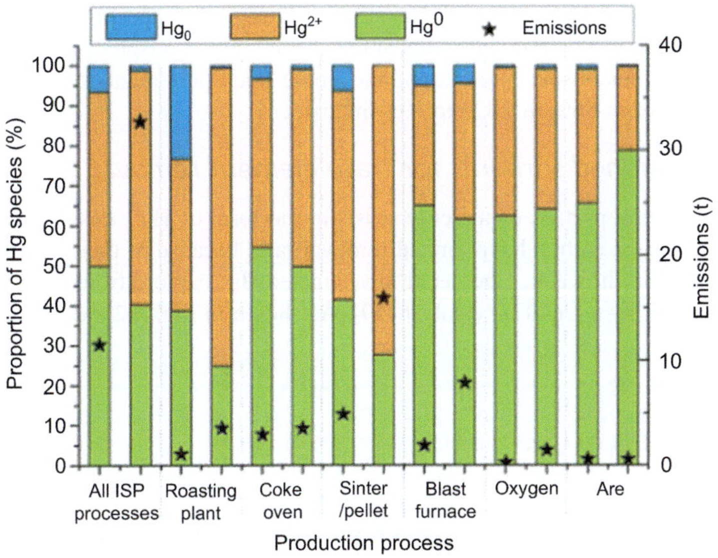

Figure 6.28 Proportion of different Hg species. *ISP*, iron and steel production.
From Wu Q, Gao W, Wang S, Hao J. Updated atmospheric speciated mercury emissions from iron and steel production in China during 2000–2015. Atmos Chem Phys 2017;17:10423–33.

the sintering furnace process is only one important process. The comprehensive mercury emissions from the iron and steel industries in China must be further monitored and studied for the coke-oven process, blast furnace process, and converter process.

In addition, Wu et al. also investigated Hg emissions in different sections of the steel-making flue gas treatment processes (shown in Fig. 6.28) [29]. Overall, the Hg speciation profile of ISP experienced great change during the study period from 50/44/6 (gaseous Hg^0/gaseous Hg^{2+}/Hg^p) in 2000 to 40/59/1 in 2015. The proportion of Hg^{2+} increased by 15%, whereas both the Hg^0 and Hg^p proportion showed a decreasing trend. Such a shift indicated a higher deposition proportion of Hg around the emission points since Hg^{2+} has a larger deposition velocity and higher water solubility. For long-process steel making, the Hg speciation profile shifted from 49/44/7 in 2000 to 39/60/1. The speciation shift in roasting plants was mainly impacted by the replacement of WS and cyclone with FF, which increased the emitted Hg^{2+} proportion from 38% to 75%. Initially, the replacement of the indigenous coke production method by the machine coke production method mainly contributed to the Hg^{2+} proportion increase from 42% to 52%. However, the gradual installation of a WS in addition to a cooler for air pollution control of the machine coke production method further washed Hg^{2+} and reduced the Hg^{2+} proportion to 49% in 2015. The Hg^{2+} proportion in the exhaust gas of sinter/pellet plants has increased by 20%. The increase in the Hg^{2+} proportion in sinter/pellet plants was mainly impacted by the substitution of WS by ESP, FF, ESP + WFGD, or the ESP + DFGD (dry flue gas desulfurization) + FF system, which generally emitted gas with a higher Hg^{2+} proportion. An increase in the Hg^{2+} emission proportion in blast furnaces was due to a higher

Hg^{2+} emission proportion after FF than venturi. In the oxygen steel-making process, the Hg speciation profile remained almost unchanged. For short-process steel making, Hg^0 was the dominant speciation during the whole study period and the proportion of Hg^0 increased from 66% in 2000 to 79% in 2015.

6.4.2.2 Designed sorbents for trace element removal

Since a substantial portion of mercury exists in the form of Hg^0, desulfurization devices and activated carbon have limited removal rates because of their high volatility and low solubility in water. Chen et al. investigated Hg^0 capture from iron and steel-making flue gas by a Cu-BTC catalyst [49]. As shown in Fig. 6.29(a), the impact of

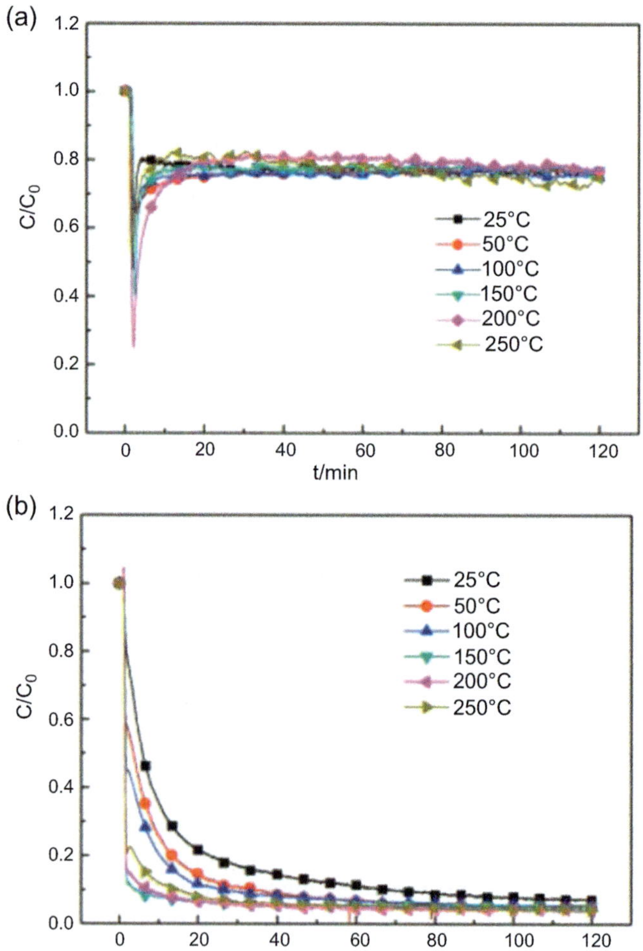

Figure 6.29 The Hg^0 removal performance at different temperatures: (a) without HCl; (b) with the existence of 15 ppm HCl at different temperatures.
From Chen D, Zhao S, Qu Z, Yan N. Cu-BTC as a novel material for elemental mercury removal from sintering gas. Fuel 2018;217:297–305.

temperature on Hg^0 removal performance is not significant, and the Hg^0 removal efficiency was around 24% at various temperatures. In this gas condition, the mercury removal ability of Cu-BTC was not very good. However, when 15 ppm HCl was added, the Hg^0 removal efficiency was significantly increased, as shown in Fig. 6.29(b). When temperature increased from 25 to 150°C, the Hg^0 removal efficiency slightly increased from 93% to 96%, and then maintained approximately 96% in the temperature range 150–250°C. In all, from 25 to 250°C, Cu-BTC showed excellent Hg^0 removal capacity in the presence of HCl. Since HCl exists in sintering gas, Cu-BTC is well suited for Hg^0 removal.

Moreover, the influence of SO_2 (100 ppm), NO (100 ppm), and H_2O was studied in typical sintering gas conditions (150°C, 15 ppm HCl, 10% O_2), as shown in Fig. 6.30. When SO_2 was introduced, the Hg^0 removal efficiency slightly decreased (Fig. 6.30(a)), and when NO was in the inlet gas, no significant change in Hg^0 removal efficiency was found (Fig. 6.30(b)). Also, the Hg^0 removal efficiency did not change significantly when H_2O was in the inlet gas (Fig. 6.30(c)). This may be because the Hg^0 removal efficiency is already high due to the high concentration of HCl and O_2 and the influence of SO_2, NO, and H_2O was small. Therefore another experiment was conducted when HCl and O_2 were at low concentration (5 ppm HCl, 2% O_2). Furthermore, the influence of the coexistence of HCl, SO_2, NO, and H_2O on Hg^0 removal was investigated. In typical sintering gas conditions (150°C, 15 ppm HCl, 10% O_2), the Hg^0 removal rate was slightly decreased (Fig. 30(d)). As shown in Fig. 6.30(e), when SO_2 was passed, the Hg^0 removal efficiency decreased more significantly, and an increase in the Hg^0 removal efficiency was observed when NO was in the inlet gas (Fig. 6.30(f)). When there was H_2O in the gas, no significant change in Hg^0 removal efficiency was found (Fig. 6.30(g)). A similar result was also found when HCl and O_2 were at low concentration (5 ppm HCl, 2% O_2) (Fig. 6.30(h)). In conclusion, SO_2 has an inhibition effect on the removal of Hg^0, while NO can promote Hg^0 removal. As for H_2O, it has almost no effect on mercury removal. When HCl and O_2 are at high concentration, the influence of SO_2, NO, and H_2O are not significant. The influence on Hg^0 removal under the coexistence of HCl, SO_2, NO, and H_2O is also not significant. This means that in typical sintering gas conditions, Cu-BTC can still have high Hg^0 removal efficiency with the existence of other gases.

Moreover, Zhao et al. used metal–organic frameworks, i.e., Uio-66 and MIL-101, as efficient catalysts to remove Hg^0 from sintering flue gas [50,51]. Fig. 6.31 shows the Hg^0 and Hg^t concentration change curves passing through various adsorbents at 250°C. The difference value for Hg^t and Hg^0 is the amount of Hg^{2+}. The concentrations of Hg^t and Hg^0 both decreased after passing through adsorbents. However, there was very little Hg^{2+} in flue gas, indicating that Hg^0 was captured by adsorbents or the generated Hg^{2+} combined with adsorbent. Furthermore, Mil-101(Cr) had the highest Hg^0 removal efficiency compared with UiO-66, Cu-BTC, and activated carbon, indicating that the open metal site Cr^{3+} was more active than that of UiO-66 and Cu-BTC.

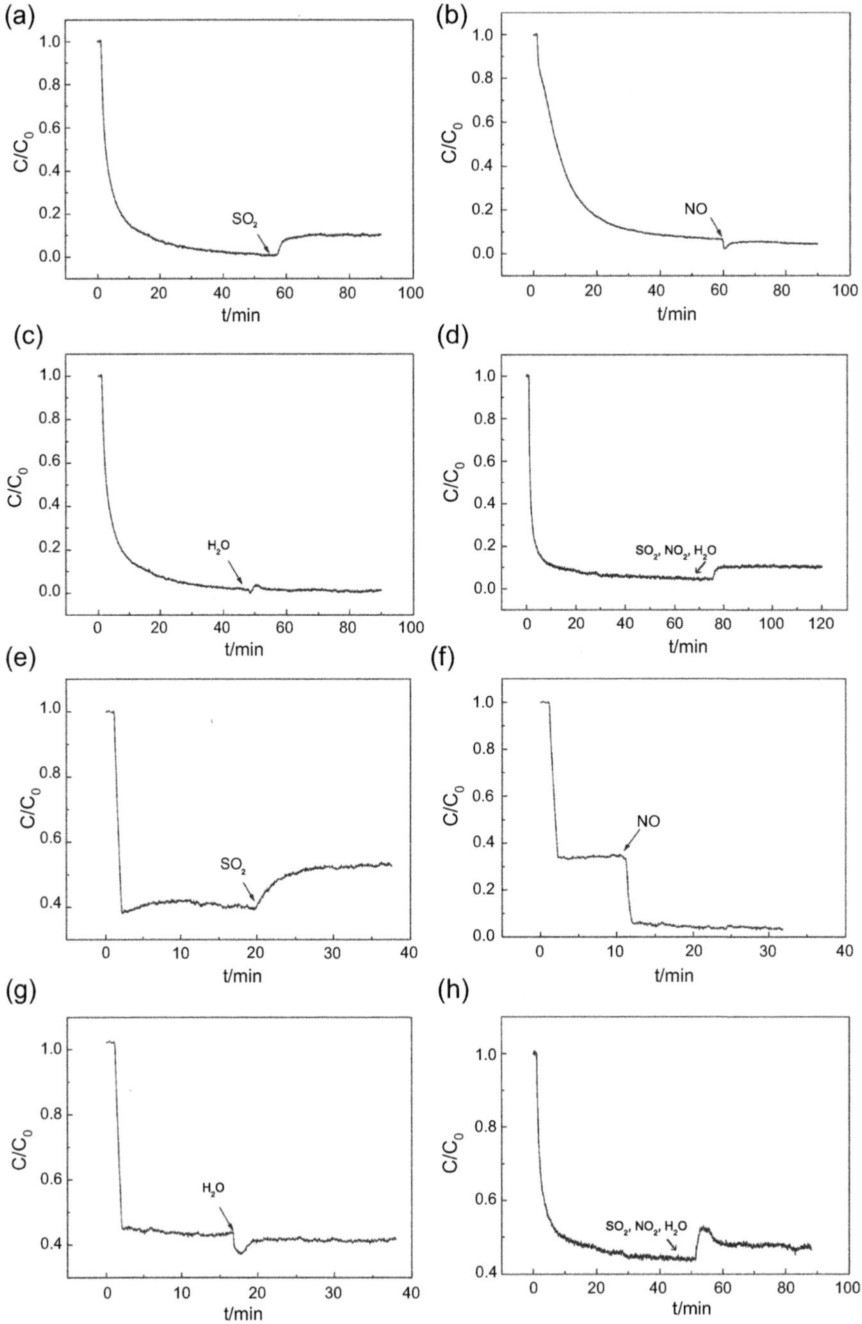

Figure 6.30 The influence of gas composition: (a) SO_2, (b) NO, (c) H_2O, and (d) coexistence of HCl, SO_2, NO, and H_2O in the condition of 15 ppm HCl and 10% O_2; (e) SO_2, (f) NO, (g) H_2O, and (h) coexistence of HCl, SO_2, NO, and H_2O in the condition of 5 ppm HCl and 2% O_2.

From Chen D, Zhao S, Qu Z, Yan N. Cu-BTC as a novel material for elemental mercury removal from sintering gas. Fuel 2018;217:297−305.

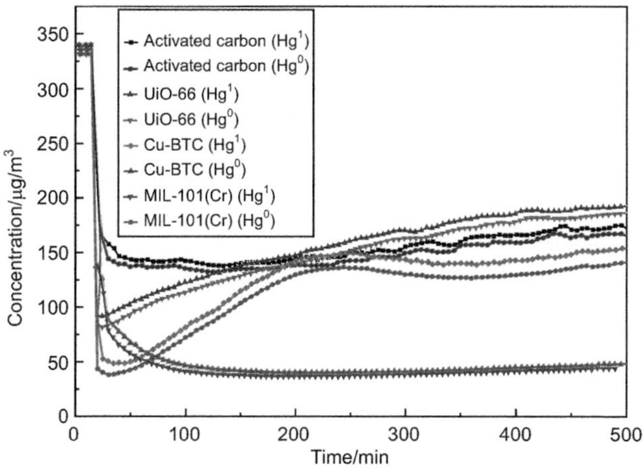

Figure 6.31 Elemental mercury and total mercury concentration change curves passing through various adsorbents at 250°C.
From Zhao S, Mei J, Xu H, Liu W, Qu Z, Cui Y, Yan N. Research of mercury removal from sintering flue gas of iron and steel by the open metal site of Mil-101(Cr). J Hazard Mater 2018; 351:301–7.

6.5 Summary

The emission characteristics and effects of APCDs for TEs from the coal-feed industry furnace, i.e., industrial boiler, cement production, and steel-making process, were discussed. Hg is the most volatile and hazardous element during coal burning in the industrial processes, whose behavior has been widely investigated in previous studies. The partition and emission patterns of Hg were highly dependent on many factors like other raw materials, burning conditions, and flue gas treatment strategies. The equipment of APCDs partially transformed Hg^0 into Hg^{2+}/Hg^p and removed the Hg^{2+}/Hg^p from the flue gas. However, due to high volatility, Hg^0 accounts for a very high proportion of total Hg in the flue gas and emitted to the atmosphere. Thus Cu-BTC and MIL-101 were developed as catalysts to degrade Hg^0 from coal-fed industrial furnace flue gas, which were demonstrated to exhibit decent Hg^0 oxidation performance. Other representative TEs like As, Se, and Cr were proven to be semivolatile and largely accumulate in particulates as coal burning by-products. These TEs are easily captured by particulate control facilities like FF/ESP systems. Thus limited studies have focused on their emission patterns during coal-fed industrial processes, while their partition characterizations have been adequately reviewed in this chapter.

Acknowledgments

This project was supported by the National Natural Science Foundation of China (51776227), Natural Science Foundation of Hunan Province, China (2018JJ1039, 2018JJ3675).

References

[1] Zhao YC, Yang JP, Ma SM, Zhang SB, Liu H, Gong BG, Zhang JY, Zheng CG. Emission controls of mercury and other trace elements during coal combustion in China: a review. Int Geol Rev 2017;59:1−33.

[2] Clarke LB, Sloss LL. Trace elements- missions from coal combustion and gasification. London: IEA Coal Research; 1992. p. 111.

[3] Xu MH, Yan R, Zheng CG, Qiao Y. Status of trace element emission in a coal combustion process: a review. Fuel Process Technol 2003;85:215−37.

[4] Zhu WY, Xu MH, Sui JC, Han J. Simulation of the vaporization of trace element during coal combustion. Coal Technol 2005;24:115−7.

[5] Guo L, Zhang C, Wang X, Liu Y, Wang F, Zhang F, Miao J, Qian F. Study on emission of heavy metals from the coal-fired industrial boilers. Environ Eng 2015;8:47−51.

[6] López-Antón MA, Díaz-Somoano M, Diaz L, Martínez-Tarazona MR. Avoiding mercury emissions by combustion in a Spanish circulating fluidized-bed combustion (CFBC) plant. Energy Fuels 2011;25:3002−8.

[7] Tian HZ, Wang Y, Xue ZG, Cheng K, Qu YP, Chai FH, Hao JM. Trend and characteristics of atmospheric emissions of Hg, As, and Se from coal combustion in China, 1980−2007. Atmos Chem Phys 2010;10:11905−19.

[8] Tian HZ, Cheng K, Wang Y, Zhao D, Lu L, Jia WX, Hao JM. Temporal and spatial variation characteristics of atmospheric emissions of Cd, Cr, and Pb from coal in China. Atmos Environ 2012;50:157−63.

[9] Yang LG, Duan YF, Wang YJ, Jiang YM, Yang XH, Zhao CS. Influence of boiler capacities on enrichment law of mercury. J Power Eng 2008;28:302−7.

[10] Li Z, Clemens AH, Moore TA, Gong D, Weaver SD, Eby N. Partitioning behaviour of trace elements in a stoker-fired combustion unit: an example using bituminous coals from the Greymouth coalfield (Cretaceous), New Zealand. Int J Coal Geol 2005;63: 98−116.

[11] Clemens AH, Deely JM, Gong D, Moore TA, Shearer JC. Partitioning behaviour of some toxic trace elements during coal combustion - the influence of events occurring during the deposition stage. Fuel 2000;79:1781−4.

[12] Clemens AH, Damiano LF, Gong D, Matheson TW. Partitioning behaviour of some toxic volatile elements during stoker and fluidised bed combustion of alkaline sub-bituminous coal. Fuel 1999;78:1379−85.

[13] Selcuk N, Gogebakan Y, Gogebakan Z. Partitioning behavior of trace elements during pilot-scale fluidized bed combustion of high ash content lignite. J Hazard Mater 2006;137: 1698−703.

[14] Pöykiö R, Manskinen K, Nurmesniemi H, Dahl O. Comparison of trace elements in bottom ash and fly ash from a large-sized (77 MW) multi-fuel boiler at the power plant of a fluting board mill, Finland. Energy Explor Exploit 2011;29:217−34.

[15] Zheng Y, Jensen AD, Windelin C, Jensen F. Review of technologies for mercury removal from flue gas from cement production processes. Prog Energy Combust Sci 2012;38: 599−629.

[16] Sikkema JK, Alleman JE, Ong SK, Wheelock TD. Mercury regulation, fate, transport, transformation, and abatement within cement manufacturing facilities: review. Sci Total Environ 2011;409:4167−78.

[17] Global Mercury Assessment 2013: sources, emissions, releases and environmental transport. In: U.N.E. Programme; 2013.

[18] Wang F, Wang S, Zhang L, Yang H, Wu Q, Hao J. Mercury enrichment and its effects on atmospheric emissions in cement plants of China. Atmos Environ 2014;92:421–8.
[19] Won JH, Lee TG. Estimation of total annual mercury emissions from cement manufacturing facilities in Korea. Atmos Environ 2012;62:265–71.
[20] Larsen MB, Schmidt I, Paone P, Salmento J, Petersen A, Jørgensen AW. Mercury in cement production-a literature review. FLSmidth internal report. 2007.
[21] Mlakar TL, Horvat M, Vuk T, Stergaršek A, Kotnik J, Tratnik J, Fajon V. Mercury species, mass flows and processes in a cement plant. Fuel 2010;89:1936–45.
[22] Yan D, Peng Z, Ding Q, Karstensen KH, Engelsen CJ, Li L, Ren Y, Jiang C. Distribution of Hg, as and Se in material and flue gas streams from preheater-precalciner cement kilns and vertical shaft cement kilns in China. J Air Waste Manag Assoc 2015;65: 1002–10.
[23] Conesa JA, Galvez A, Mateos F, Martin-Gullon I, Font R. Organic and inorganic pollutants from cement kiln stack feeding alternative fuels. J Hazard Mater 2008;158: 585–92.
[24] Pirrone N, Costa P, Pacyna J, Ferrara R. Mercury emissions to the atmosphere from natural and anthropogenic sources in the Mediterranean region. Atmos Environ 2001;35: 2997–3006.
[25] Wang F, Wang S, Zhang L, Yang H, Wu Q, Hao J. Characteristics of mercury cycling in the cement production process. J Hazard Mater 2016;302:27–35.
[26] Yue T, Wang F, Han BJ, Zuo PL, Zhang F. Analysis on mercury emission and control technology of typical industries in China. Appl Mech Mater 2013:859–71. Trans. Tech. Publ.
[27] Park K, Seo YC, Lee S, Lee J. Emission and speciation of mercury from various combustion sources. Powder Technol 2008;180:151–6.
[28] Kim JH, Park JM, Lee SB, Pudasainee D, Seo YC. Anthropogenic mercury emission inventory with emission factors and total emission in Korea. Atmos Environ 2010;44: 2714–21.
[29] Wu Q, Gao W, Wang S, Hao J. Updated atmospheric speciated mercury emissions from iron and steel production in China during 2000–2015. Atmos Chem Phys 2017;17: 10423–33.
[30] Wu Q, Wang S, Li G, Liang S, Lin CJ, Wang Y, Cai S, Liu K, Hao J. Temporal trend and spatial distribution of speciated atmospheric mercury emissions in China during 1978–2014. Environ Sci Technol 2016;50:13428–35.
[31] Wu Y, Wang S, Streets DG, Hao J, Chan M, Jiang J. Trends in anthropogenic mercury emissions in China from 1995 to 2003. Environ Sci Technol 2006;40:5312–8.
[32] Zhang Y, Zhang C, Wang D, Luo C, Yang X, Xu F. Characteristic of mercury emissions and mass balance of the typical iron and steel industry. Huanjing Kexue 2015;36:4366–73.
[33] Pacyna J, Munthen J, Wilson S, Maxon P, Sundseth K, Pacyna E, Harper E, Kindbom K, Wangberg I, Panasiuk D. Technical background report to the global atmospheric mercury assessment. In: A.M.a.A.P.U.C. Branch; 2008.
[34] Tsai JH, Lin KH, Chen CY, Ding JY, Choa CG, Chiang HL. Chemical constituents in particulate emissions from an integrated iron and steel facility. J Hazard Mater 2007;147: 111–9.
[35] Zhang L, Wang S, Wang L, Wu Y, Duan L, Wu Q, Wang F, Yang M, Yang H, Hao J. Updated emission inventories for speciated atmospheric mercury from anthropogenic sources in China. Environ Sci Technol 2015;49:3185–94.

[36] Ghorishi SB, Sedman CB. Low concentration mercury sorption mechanisms and control by calcium-based sorbents: application in coal-fired processes. J Air Waste Manag Assoc 1998;48:1191−8.
[37] Gupta RK, Majumdar D, Trivedi JV, Bhanarkar AD. Particulate matter and elemental emissions from a cement kiln. Fuel Process Technol 2012;104:343−51.
[38] Giannopoulos D, Kolaitis DI, Togkalidou A, Skevis G, Founti MA. Quantification of emissions from the co-incineration of cutting oil emulsions in cement plants − part II: trace species. Fuel 2007;86:2491−501.
[39] Wang S, Zhang L, Wang L, Wu Q, Wang F, Hao J. A review of atmospheric mercury emissions, pollution and control in China. Front Environ Sci Eng 2014;8:631−49.
[40] Xu W, Shao M, Yang Y, Liu R, Wu Y, Zhu T. Mercury emission from sintering process in the iron and steel industry of China. Fuel Process Technol 2017;159:340−4.
[41] Pacyna EG, Pacyna JM. Global emission of mercury from anthropogenic sources in 1995. Water Air Soil Pollut 2002;137:149−65.
[42] Wang F, Wang S, Zhang L, Yang H, Gao W, Wu Q, Hao J. Mercury mass flow in iron and steel production process and its implications for mercury emission control. J Environ Sci (China) 2016;43:293−301.
[43] Tang X, Zhong W, Zhuang D, Li C, Liu Y. Analysis on mercury emission and control technology of typical industries in China. Appl Mech Mater 2013. Trans. Tech. Publ.
[44] Arctic Monitoring and Assessment Programme (AMAP) and United Nations Environment Programme (UNEP). Technical background report to the global atmospheric mercury assessment. 2008.
[45] Lee S, Seo Y, Jang H, Park K, Baek J, An H, Song K. Speciation and mass distribution of mercury in a bituminous coal-fired power plant. Atmos Environ 2006;40:2215−24.
[46] Zhuang Y, Thompson JS, Zygarlicke CJ, Pavlish JH. Impact of calcium chloride addition on mercury transformations and control in coal flue gas. Fuel 2007;86:2351−9.
[47] Ghorishi B, Gullett BK. Sorption of mercury species by activated carbons and calcium-based sorbents: effect of temperature, mercury concentration and acid gases. Waste Manag Res 1998;16:582−93.
[48] Yang H, Xu Z, Fan M, Bland AE, Judkins RR. Adsorbents for capturing mercury in coal-fired boiler flue gas. J Hazard Mater 2007;146:1−11.
[49] Chen D, Zhao S, Qu Z, Yan N. Cu-BTC as a novel material for elemental mercury removal from sintering gas. Fuel 2018;217:297−305.
[50] Zhao S, Mei J, Xu H, Liu W, Qu Z, Cui Y, Yan N. Research of mercury removal from sintering flue gas of iron and steel by the open metal site of Mil-101(Cr). J Hazard Mater 2018;351:301−7.
[51] Zhao S, Xu H, Qu Z, Liu P, Cui Y, Yan N. Combined effects of Ag and UiO-66 for removal of elemental mercury from flue gas. Chemosphere 2018;197:65−72.

Trace element emissions from coal-fired power plants

7

Jianping Yang[1], *Qin Li*[1], *Yongchun Zhao*[2], *Junying Zhang*[2]
[1]School of Energy Science and Engineering, Central South University, Changsha, China;
[2]State Key Laboratory of Coal Combustion, Huazhong University of Science and Technology, Wuhan, China

7.1 Introduction

Coal-fired power plants are the largest consumers of coal, accounting for over 50% of the consumption by the whole country, and therefore, large amounts of trace elements (TEs) are released into the atmosphere. In 2016, the US Environmental Protection Agency (EPA) proposed a new policy for controlling TE emissions, stating that the emission of 10 kinds of TEs, including As, Se, and Cd, from coal-fired power plants should be limited. New or reconstructed units must comply with stricter limits, by which As, Se, and Cd emissions should be below 1.36, 22.67, and 0.18 g/GWh, respectively. The State Council of China also published the 12th Five-Year Plan for the comprehensive treatment of TE pollution in 2011.

To control air pollutants, such as NO_x, particulate matter (PM), and SO_2, emitted from power plants, air pollution control devices (APCDs), such as selective catalytic reduction (SCR) units, electrostatic precipitators (ESPs), wet flue gas desulfurization (WFGD) systems, wet electrostatic precipitators (WESPs), etc., have been installed. These APCDs can synergistically remove TEs to some extent. To effectively reduce TE emission from coal-fired power plants, understanding the distribution and migration of TEs across the whole system comprehensively is a very important step.

7.2 Emission and migration characteristics of trace elements across APCDs

7.2.1 Mercury emission and migration across APCDs

Most of the mercury present in coal is released in elemental form (Hg^0) during the combustion process. As the flue gas cools, the gaseous Hg^0 and Hg^{2+} can condense or adsorb onto the fly ashes. The distribution and speciation of mercury in the flue gas across the APCD is important for removal efficiency. Therefore, the behavior of Hg across each APCD is discussed separately as follows.

Emission and Control of Trace Elements from Coal-Derived Gas Streams
https://doi.org/10.1016/B978-0-08-102591-8.00007-6

7.2.1.1 Behavior of Hg at the selective catalytic reduction system

SCR denitration is applied to remove NO_x from the flue gas of coal-fired power plants [1]. Extensive studies show that the SCR catalyst can have a significant impact on the mercury speciation in flue gas [2–6]. Xu et al. [7] studied the behavior of Hg across the SCR systems of 20 typical coal-fired power plants in China. The results show that SCR had no obvious effect on total Hg (Hg^T) release. After passing through the SCR unit, the Hg^p concentration in the flue gas did not change significantly. However, the Hg^0 concentration decreased and the Hg^{2+} concentration increased accordingly. Thus, the SCR catalyst can oxidize Hg^0 into Hg^{2+}. However, the Hg^0 oxidation efficiencies in SCR systems at different power plants varied significantly. The exhaust composition, operating conditions, catalyst age, and boiler loading all have significant effects on Hg^0 oxidation over the SCR catalyst.

In 2014, the Chinese government issued the *Coal Energy Saving and Emissions Reduction Reform and Improvement Plan*, which limited the emissions of PM, NO_x, and SO_2 to 5, 35, and 50 mg/m^3, respectively, in some provinces, such as Jiangsu and Zhejiang. The SCR, ESP, and WFGD systems have been reformed in several power plants to meet these ultralow emission (ULE) standards. Fig. 7.1 shows the mercury concentrations and species across SCR systems of different boilers at a

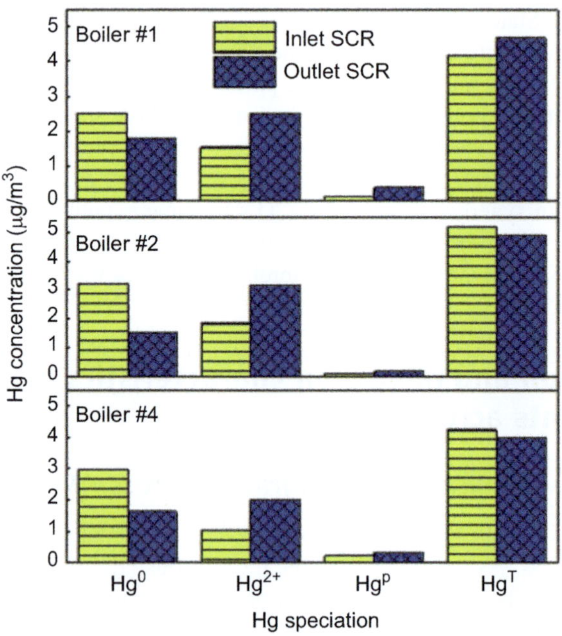

Figure 7.1 Hg concentration at the inlet and outlet of selective catalytic reduction (*SCR*) systems.
From Zhang Y, Yang J, Yu X, Sun P, Zhao Y, Zhang J, Chen G, Yao H, Zheng C. Migration and emission characteristics of Hg in coal-fired power plant of China with ultra low emission air pollution control devices. Fuel Process. Technol. 2017;158:272–80.

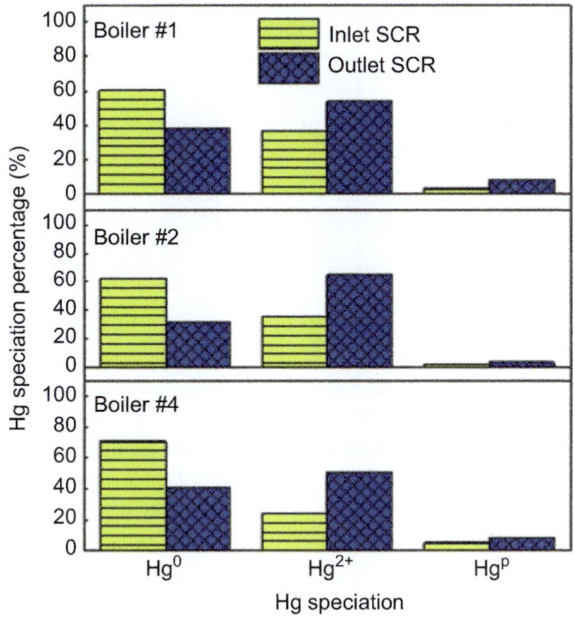

Figure 7.2 Hg speciation percentage at the inlet and outlet of selective catalytic reduction (*SCR*) systems.
From Zhang Y, Yang J, Yu X, Sun P, Zhao Y, Zhang J, Chen G, Yao H, Zheng C. Migration and emission characteristics of Hg in coal-fired power plant of China with ultra low emission air pollution control devices. Fuel Process. Technol. 2017;158:272–80.

ULE coal-fired power plant in China [8]. The Hg^T concentrations of boilers 1, 2, and 4 were 4.16, 5.17, and 4.23 μg/m^3 at the SCR inlet, while they were 4.67, 4.89, and 3.97 μg/m^3 at the SCR outlet, respectively. The Hg^T concentration was balanced between the entrance and the exit of SCR (86.87%–104.71%). This suggests that SCR had little effect in removing total Hg.

Fig. 7.2 shows that the concentration and ratio of Hg^{2+} clearly increase with SCR. The proportion of Hg^0 decreased from 71.97%–77.95% to 31.69%–46.45% via SCR, suggesting that 41.43%, 51.61%, and 47.68% of the Hg^0 was oxidized to Hg^{2+} across the SCR system. Hg^{2+} was water soluble, tended to be trapped by fly ash, and could be removed by WFGD and ESP. Therefore, although SCR does not play a direct role in Hg elimination, the Hg^0 oxidation role of SCR has a potential impact on the co-benefit removal efficiency of other APCDs.

The behavior of Hg in the SCR system was influenced by a number of factors, such as flue gas components, catalyst uptime, and boiler loading. The influence of each factor is systematically discussed as follows.

Effect of flue gas components on mercury migration

The flue gas from coal burning is a complex mixture containing fly ash particles, oxygen, moisture, carbon monoxide, and many acid gases. Many studies have

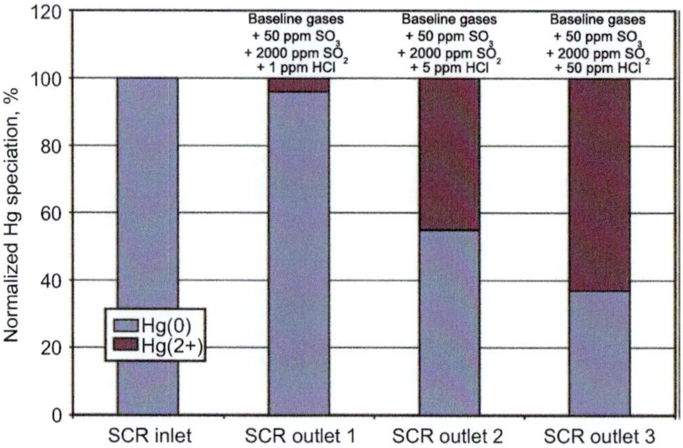

Figure 7.3 Effect of HCl on Hg oxidation across the selective catalytic reduction (*SCR*) system. From Zhuang Y, Laumb J, Liggett R, Holmes M, Pavlish J. Impacts of acid gases on mercury oxidation across SCR catalyst. Fuel Process. Technol. 2007;88(10):929–34.

demonstrated that flue gas components such as SO_2, SO_3, NO, NO_2, and HCl significantly affect Hg^0 oxidation over SCR catalysts [3]. HCl is considered to be the most important component for promoting Hg^0 oxidation [2–4]. Zhuang et al. [9] reported that an addition of 5 ppm HCl in the flue gas resulted in 45% of the Hg^0 oxidation efficiency over the SCR catalyst, which was 4% when 1 ppm HCl was in the flue gas. As the HCl concentration increased to 50 ppm, 63% of Hg^0 oxidation was obtained (Fig. 7.3).

The promotion effect of HCl on Hg^0 oxidation can be attributed to the formation of active Cl species originating from the adsorbing HCl at the active site. Hg^0 can then be oxidized by active Cl to produce $HgCl_2$. The HCl concentration in flue gas has a positive correlation with the chlorine content in feed coal [10]. When the chlorine content in the feed coal increased from 109 to 876 mg/kg, the Hg^0 oxidation efficiency of the SCR catalyst increased from 2% to 89% [7]. The chlorine content of bituminous coal is higher than that of sub-bituminous coal. Thus, the Hg^0 oxidation efficiency of SCR catalysts in a power station burning bituminous coal is generally higher than in those burning sub-bituminous coal [2].

The effect of SO_2/SO_3 on Hg^0 oxidation over the SCR catalyst is unclear. Zhuang et al. [9] studied the influence of SO_2/SO_3 on Hg^0 oxidation when passed through the SCR catalyst (Fig. 7.4). The results show that adding 2000 ppm SO_2 could induce 7% Hg^{2+} at the SCR outlet and 50 ppm SO_3 improved Hg^0 oxidation; 20% Hg^0 oxidation was obtained at the SCR exit in the absence of acid gas, while the Hg^0 oxidation efficiency was 3% in baseline flue gas without acid gas. The combination of SO_2/SO_3 (2000/50 ppm) resulted in an additional 4% Hg^0 oxidation.

In 2013, Li et al. [11] conducted a systematic survey on the effect of SO_2 on Hg^0 oxidation over a CeO_2-TiO_2 catalyst. SO_2 had different or opposite effects on Hg^0 oxidation under different conditions. In a pure N_2 atmosphere, SO_2 suppressed Hg^0 oxidation, while in the N_2 plus O_2 atmosphere, SO_2 at a low concentration

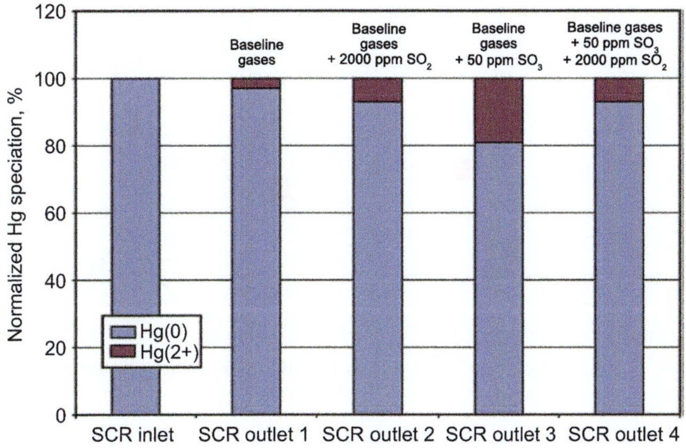

Figure 7.4 Effect of SO_2/SO_3 on mercury oxidation across the selective catalytic reduction (*SCR*) system.
From Zhuang Y, Laumb J, Liggett R, Holmes M, Pavlish J. Impacts of acid gases on mercury oxidation across SCR catalyst. Fuel Process. Technol. 2007;88(10):929–34.

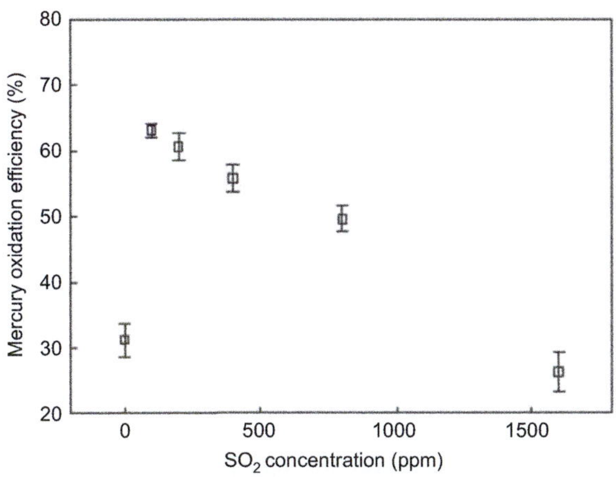

Figure 7.5 Effect of SO_2 concentration on Hg^0 oxidation in the presence of 4% O_2.
From Li H, Wu C, Li Y, Li L, Zhao Y, Zhang J. Impact of SO_2 on elemental mercury oxidation over CeO_2–TiO_2 catalyst. Chem. Eng. J. 2013;219:319–26.

promoted Hg^0 oxidation and SO_2 at a high concentration lowered Hg^0 oxidation (Fig. 7.5). The promoting effect of SO_2 was probably due to the generation of SO_3 and the inhibitory effect of SO_2 was due to competitive adsorption between SO_2 and Hg^0 (Fig. 7.6).

The effect of NO on Hg^0 oxidation over the SCR catalyst is also not entirely definite. Li et al. [12] studied the oxidation and capture of Hg^0 on SiO_2–TiO_2–V_2O_5 catalysts. The results show that 100 ppm NO led to 15.38% Hg^0 oxidation (E_{oxi}) and

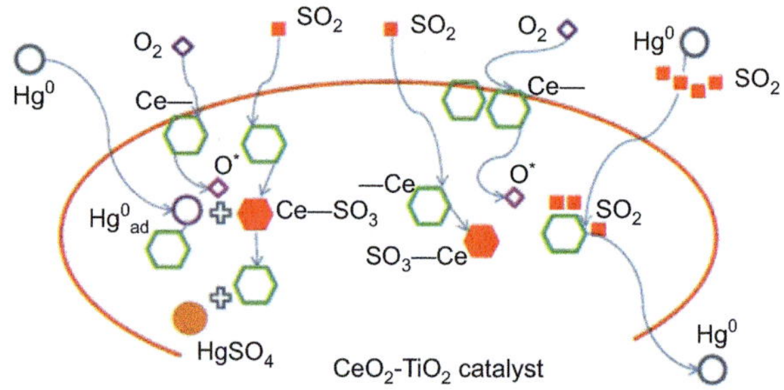

Figure 7.6 Illustration of the effect of SO_2 on Hg^0 oxidation.
From Li H, Wu C, Li Y, Li L, Zhao Y, Zhang J. Impact of SO_2 on elemental mercury oxidation over $CeO_2{-}TiO_2$ catalyst. Chem. Eng. J. 2013;219:319–26.

12.31% Hg^0 capture (E_{cap}) at 350 °C. The E_{oxi} and E_{cap} increased to 26.15% and 16.92% when the NO concentration was further increased to 300 ppm. NO can be adsorbed to a vanadium-based catalyst in molecular form, yielding active species such as NO^+, NO^{2+}, and NO_2 for Hg^0 oxidation. In the presence of 4% O_2, 300 ppm of NO improved E_{oxi} and E_{cap} to 52.92% and 21.08%, respectively, due to the reoxidation of the reducing surface by the gas-phase O_2. However, when NO and O_2 coexist, nitrites, which are inert for Hg^0 oxidation, are likely to be produced. The active sites are blocked by these substances and can inhibit the activity of the SCR catalyst for Hg^0 oxidation.

Since NH_3 is used as a reducing agent for SCR de-NO_x, the influence of NH_3 on Hg^0 oxidation should be considered. As the ratio of NH_3/NO increases, the Hg^0 oxidation rate decreases. NH_3 strongly inhibits Hg^0 oxidation over SCR, and the inhibitory effect increases as the NH_3/NO ratio increases. The inhibition by NH_3 on Hg^0 oxidation can be due to: (1) NH_3 competing with Hg^0 for the active sites; (2) NH_3 reducing the adsorbed HCl on the catalyst, which is an important intermediate for Hg^0 oxidation; (3) NH_3 consuming HCl involved in Hg^0 oxidation; and/or (4) NH_3 inducing Hg^{2+} reduction into Hg^0 [13].

Effect of catalyst service time on mercury migration

In addition to the flue gas components, the catalyst service time of the SCR catalyst is an important factor for determining Hg^0 oxidation. Eswaran et al. [10] reported that the activity of an SCR catalyst decreased as the catalyst service time increased. The results of Kamata et al. [14] show the Hg^0 oxidation efficiency of a fresh catalyst was about 40%, while it decreased to 15% when the catalyst had served 71,000 h. Zhao et al. [14] reported that commercial SCR catalyst equipment in a coal-fired power plant lost its Hg^0 oxidation capacity when the service time reached 35,000 h. Similar results were also obtained by Zhou et al. [15].

Effect of boiler load on mercury migration

The boiler load also significantly influences the Hg^0 oxidation of an SCR system [15]. As shown in Fig. 7.7 and Fig. 7.8, the Hg^0 oxidation efficiencies were 72.12%,

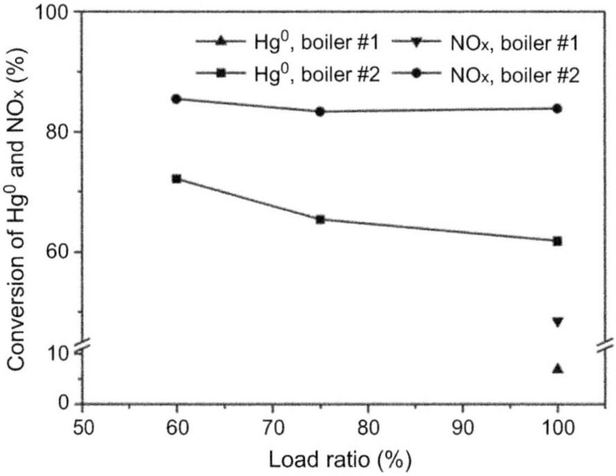

Figure 7.7 Conversion of NO_x and Hg^0 across selective catalytic reduction catalysts.
From Zhou Z, Liu X, Zhao B, Chen Z, Shao H, Wang L, Xu M. Effects of existing energy saving and air pollution control devices on mercury removal in coal-fired power plants. Fuel Process. Technol. 2015;131:99–108.

65.71%, and 61.78% when the boiler load was 60%, 75% and 100%, respectively. Therefore, the Hg^0 oxidation was inversely proportional to the load ratio. The Hg^0 oxidation was more sensitive than NO_x reduction to load fluctuations. When the load was 60%, the Hg^0 concentration obviously decreased as flue gas traversed the SCR catalyst. The Hg^0 conversion rate was as high as 72.12%. The proportion of Hg^0 before crossing the SCR catalyst was about 76.35%, but decreased to 22.69% when passing through the SCR system. At loading rates of 75% and 100%, the Hg^0 conversion was 65.71% and 61.78%, respectively.

Zhao et al. [16] also tested the mercury species at the SCR inlet and outlet at different boiler loads, as shown in Table 7.1. Hg^0 is the major species at the furnace exit. After passing through the SCR system, the concentrations of Hg^0 and Hg^{2+} decreased and the concentration of Hg^P increased. Due to the reduction in boiler load, the flow rate decreased, the gaseous hourly space velocity (GHSV) was lowered, and the residence time was lengthened. The low GHSV promoted the diffusion of gaseous HCl and Hg^0 to the catalyst surface and subsequent chemical reaction. As the boiler load decreased, the flue gas temperature decreased. Lower temperature promoted the adsorption of HCl and Hg^0 onto the catalyst surface and thus facilitated the Hg^0 conversion.

7.2.1.2 Transformation of Hg speciation at a low-temperature economizer

A low-temperature economizer (LTE) was installed at an ESP entrance with the intention to recover the heat generated from the flue gas. The temperature of the flue gas entering the ESP drops somewhat. The viscosity and speed of the flue gas, the

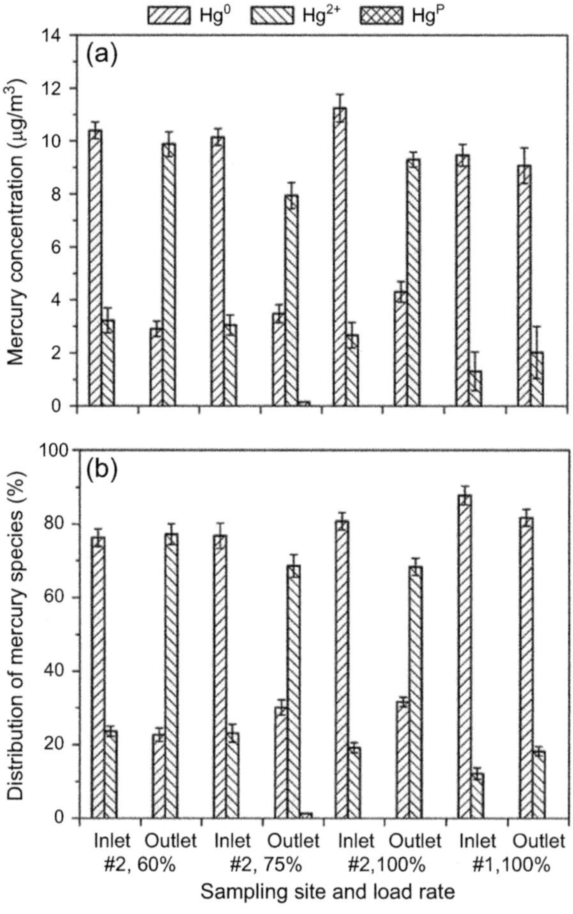

Figure 7.8 Mercury concentration and mass distribution at the selective catalytic reduction systems of two boilers, 1 and 2.
From Zhou Z, Liu X, Zhao B, Chen Z, Shao H, Wang L, Xu M. Effects of existing energy saving and air pollution control devices on mercury removal in coal-fired power plants. Fuel Process. Technol. 2015;131:99–108.

resistance of the fly ash, the composition of the flue gases, and the particles all can vary greatly with a decrease in flue gas temperature. Therefore, an LTE may affect the movement of mercury in the flue gas when passed through the LTE.

In recent years, the role of LTEs on the movement and emission of mercury in exhaust gas has been studied [8]. As shown in Figs. 7.9 and 7.10, when the flue gas passes through an LTE, the concentrations of Hg^0 and Hg^{2+} decrease and the Hg^P concentration increases. The flue gas temperature after passing the LTE decreased from 130°C (boiler 2) and 127°C (boiler 4) to 108°C (boiler 2) and 95°C (boiler 4). Thus, some flue gas components, like water vapor, SO_3, and HCl, may condense on

Table 7.1 Flue gas mercury speciation concentration at the inlet/outlet of an air pollution control device (6% O_2, μg/m^3).

		SCR	
Sampling load	**Hg speciation**	**Inlet**	**Outlet**
100%	Hg^0	3.77	1.88
	Hg^{2+}	1.77	0.89
	Hg^p	0.67	1.87
85%	Hg^0	4.27	1.38
	Hg^{2+}	2.00	0.65
	Hg^p	0.44	2.86
68%	Hg^0	4.23	1.62
	Hg^{2+}	1.98	0.80
	Hg^p	0.26	2.60

SCR, selective catalytic reduction.
From Zhao S, Duan Y, Chen L, Li Y, Yao T, Liu S, Liu M, Lu J. Study on emission of hazardous trace elements in a 350 MW coal-fired power plant. Part 1. Mercury. Environ. Pollut. 2017;229:863–70.

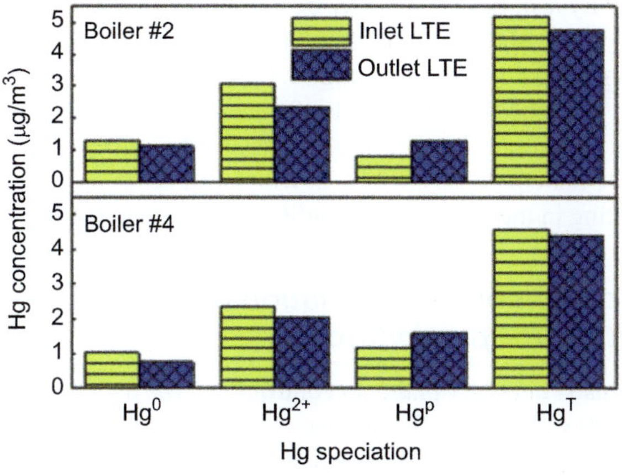

Figure 7.9 Hg concentration at the inlet and outlet of a low-temperature economizer (*LTE*). From Zhang Y, Yang J, Yu X, Sun P, Zhao Y, Zhang J, Chen G, Yao H, Zheng C. Migration and emission characteristics of Hg in coal-fired power plant of China with ultra low emission air pollution control devices. Fuel Process. Technol. 2017;158:272–80.

the fly ash surface to form a liquid film, which can form acid active adsorption sites to enhance Hg^0 adsorption.

According to the decomposition temperature of pure Hg compounds, it can be concluded that $HgCl_2$ (peaking at about 145°C) and three-way red HgS (peaking at about 300°C.) are major Hg species in the fly ash collected at LTE inlet

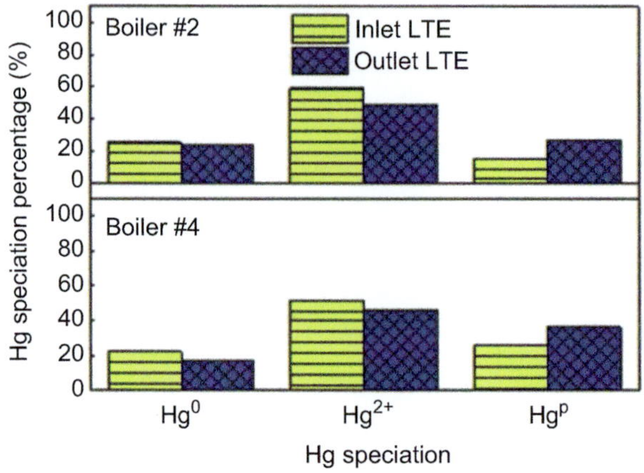

Figure 7.10 Hg speciation percentage at the inlet and outlet of a low-temperature economizer (*LTE*).
From Zhang Y, Yang J, Yu X, Sun P, Zhao Y, Zhang J, Chen G, Yao H, Zheng C. Migration and emission characteristics of Hg in coal-fired power plant of China with ultra low emission air pollution control devices. Fuel Process. Technol. 2017;158:272–80.

(Fig. 7.11A). HgS was due to the interaction between Hg and sulfur remaining in unburned carbon in fly ash, but $HgCl_2$ may be due to the adsorption of Hg on the chlorinated site. Fig. 7.11B shows that $HgSO_4$ is present in fly ash recovered at the LTE outlet in addition to $HgCl_2$ and HgS. Fig. 7.12 shows that the SO_3 concentration in flue gases varies from 64.8 mg/m^3 (boiler 2) and 77.9 mg/m^3 (boiler 4) to 51.3 mg/m^3 (boiler 2) and 48.5 mg/m^3 (boiler 4) after passing through the LTE. This fully demonstrates that SO_3 can condense on fly ash, which can act as an acid active site for Hg adsorption, resulting in the formation of $HgSO_4$ and the increase in Hg^P concentration.

7.2.1.3 Concentration and speciation of mercury across an electrostatic precipitator

ESPs are widely used in power plants to control PM. The average particle removal efficiency is above 99%. Thus, ESPs can effectively remove the mercury adsorbed on the fly ash. The US EPA tested the Hg removal behavior of 14 pulverized coal boilers equipped with cold-state ESPs (cs-ESPs) [17]. The Hg content and type distribution in coal, boiler type, and ESP operating temperature all significantly affect the Hg^T reduction of cs-ESPs at different power plants.

Effect of coal characteristics on mercury removal

The coal type is an important factor for reducing Hg^T in a cs-ESP. The reduction in Hg^T in the cs-ESP of sub-bituminous coal- and lignite-burning plants is −4% to 12%, and Hg^0 concentrations emitted into the atmosphere are higher than from those that burn bituminous coals. The Cl content in coal is an important factor affecting the distribution of mercury species. The higher the Cl content in the coal, the higher the

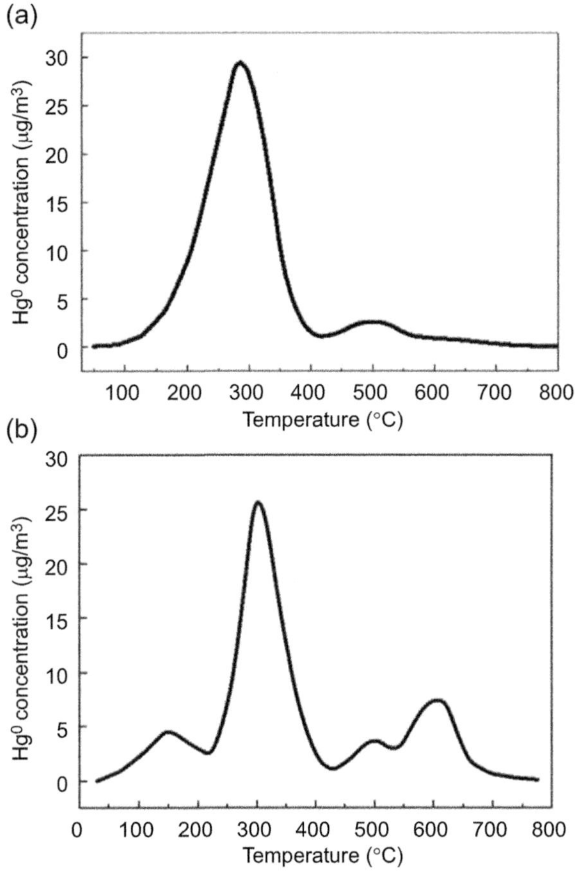

Figure 7.11 Hg species desorbed from fly ash (a) with and (b) without a low-temperature economizer (*LTE*).
From Zhang Y, Yang J, Yu X, Sun P, Zhao Y, Zhang J, Chen G, Yao H, Zheng C. Migration and emission characteristics of Hg in coal-fired power plant of China with ultra low emission air pollution control devices. Fuel Process. Technol. 2017;158:272–80.

proportion of Hg^{2+} in the flue gas (Fig. 7.13) [18]. Compared with Hg^0, Hg^{2+} tends to be adsorbed onto fly ash and easily captured by an ESP. Chlorine content in coal is also considered to be an important factor affecting Hg^0 oxidation (Fig. 7.14) [18]. Fly ash may react with chlorine and compete with Hg. Fig. 7.15 is a schematic diagram showing factors influencing mercury types in a coal-fired power plant. The important order influencing Hg oxidation is fuel properties and then operating conditions and particulate control equipment.

Effect of dust control device modes

The Hg removal efficiencies of different types of particle control devices vary significantly [17]. Compared with cs-ESP, the high-temperature-state ESP (hs-ESP) has

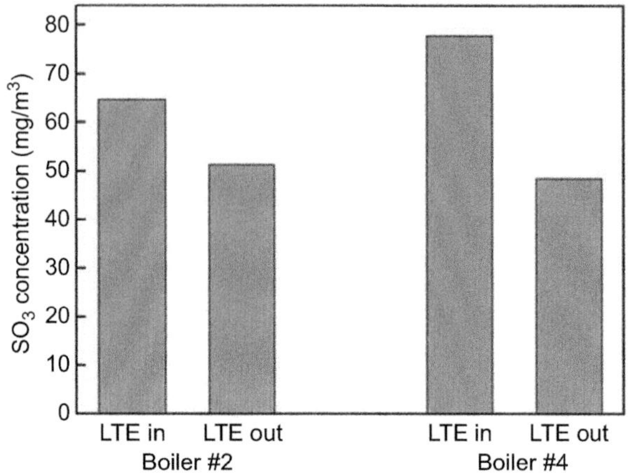

Figure 7.12 SO_3 concentration before and after passing through a low-temperature economizer (*LTE*).
From Zhang Y, Yang J, Yu X, Sun P, Zhao Y, Zhang J, Chen G, Yao H, Zheng C. Migration and emission characteristics of Hg in coal-fired power plant of China with ultra low emission air pollution control devices. Fuel Process. Technol. 2017;158:272–80.

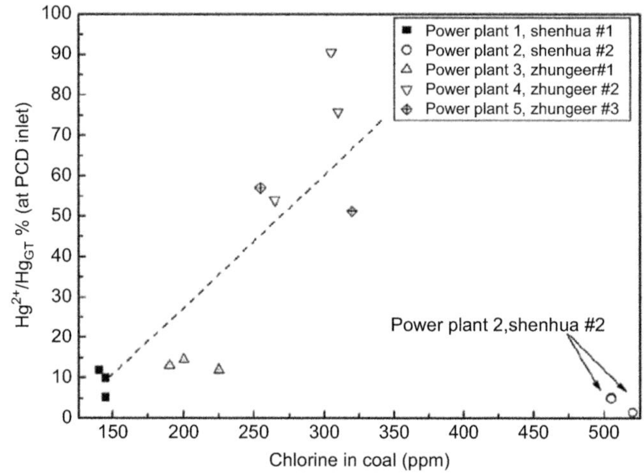

Figure 7.13 Mercury oxidation by chlorine in flue gas (at the electrostatic precipitator and fabric filter inlet) *PCD*, particulate control device.
From Chen L, Duan Y, Zhuo Y, Yang L, Zhang L, Yang X, Yao Q, Jiang Y, Xu X. Mercury transformation across particulate control devices in six power plants of China: The co-effect of chlorine and ash composition. Fuel 2007;86(4):603–10.

poor Hg removal capacity. The Hg reduction of an hs-ESP in a power plant that burns bituminous coal is 17%–27%, and that in a sub-bituminous coal–burning power plant decreases to 1%–6%. This is due to the high operating temperature of the hs-ESP, which results in little adsorption of Hg on fly ash. Fabric filters (FFs) show the highest

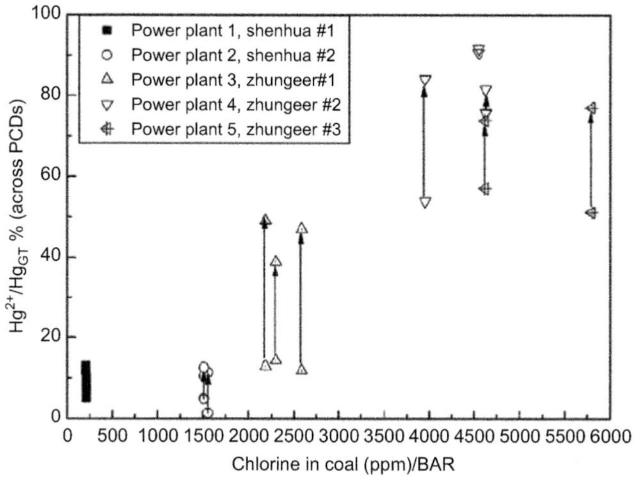

Figure 7.14 Mercury oxidation in the function of C, considering both chlorine in coal and basic ash compositions (at the electrostatic precipitator and fabric filter inlet). *C* is the quotient of chlorine in coal and BAR (base/acid ratio) in coal ash; *PCD,* particulate control device.
From Chen L, Duan Y, Zhuo Y, Yang L, Zhang L, Yang X, Yao Q, Jiang Y, Xu X. Mercury transformation across particulate control devices in six power plants of China: The co-effect of chlorine and ash composition. Fuel 2007;86(4):603−10.

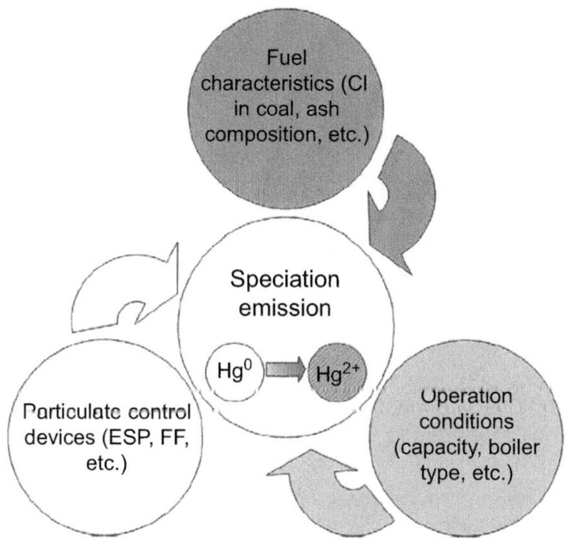

Figure 7.15 Schematic of factors affecting mercury speciation in this study. *ESP*, electrostatic precipitator; *FF*, fabric filter; *PCD*, particulate control device.
From Chen L, Duan Y, Zhuo Y, Yang L, Zhang L, Yang X, Yao Q, Jiang Y, Xu X. Mercury transformation across particulate control devices in six power plants of China: The co-effect of chlorine and ash composition. Fuel 2007;86(4):603−10.

Hg removal capacity, with an average efficiency of 72%–90%. The Hg removal capacity of a cs-ESP and FF is related to the feed coal type. In a bitumen-burning plant, the Hg^T reduction of a cs-ESP and FF is higher than that of the sub-bituminous coal- and lignite-burning plants. This is mainly because the contact time between Hg and fly ash in an FF becomes long.

Effect of fly ash characteristics

Enrichment of Hg in fly ash significantly determines the Hg removal efficiency, which is affected by combustion mode, boiler capacity, boiler load, particulate controller type, etc. Yang et al. [19] reported that the Hg^0 removal capacity of an ESP decreases with increasing boiler capacity. This is because the larger boiler capacity results in a longer retention time of the unburned carbon (UBC) particles in the boiler and the combustion of the coal particles is promoted. Thus, the UBC content in fly ash decreases with increasing boiler capacity and decreases the adsorption of Hg in fly ash.

Effect of APCD equipment

Xu et al. [7] tested the type and concentration of Hg before and after ESP/FF in 20 typical coal-fired power plants in China. Reduction of Hg^T mainly depended on the removal of Hg^P. The average Hg^P fraction in flue gas decreased from 28.4% to 5.3% after passing through the ESPs, and the Hg^P removal efficiency of the ESPs exceeded 90%. Other PM control devices (PMCDs) had a significant effect on the Hg removal by an ESP. After installation of an SCR system, the Hg removal capability of the ESP increased significantly. The Hg^P removal efficiency of an ESP can reach 80.6% with the assistance of SCR. This is because the SCR catalyst oxidizes Hg^0 to Hg^{2+}, while Hg^{2+} is easily adsorbed on the fly ash surface and subsequently removed by the ESP.

As energy-saving equipment, an LTE itself cannot directly reduce pollutants. However, during the operation of an LTE, the flue gas temperature decreases to some extent. Therefore, an LTE may affect mercury release and transition characteristics in an ESP. As shown in Fig. 7.16 and Fig. 7.17, part of the Hg^P decreased from 27.00%–34.11% to 1.57%–1.58% passing through an ESP equipped with an LTE [8]. Thus, along with the particulate removal, most of the Hg^P in the flue gas was removed. As stated before, ESPs generally have little effect on gaseous Hg conversion and cannot effectively remove gaseous Hg [15,20]. However, the concentration of gaseous Hg^{2+} is obviously reduced when equipped with an LTE. The Hg content in fly ash was 143.1 ng/g when an LTE was on and 81.22 ng/g when the LTE was off. This suggests that the reduced gaseous Hg was due to adsorption in the fly ash when the flue gas temperature decreased at the LTE. Therefore, LTEs play an important role in elimination by ESPs.

Zhou et al. [15] also compared the changes in Hg species and removal efficiency when the flue gas passed through the ESP under different LTE operating conditions, as shown in Fig. 7.18. When the LTE was off, only about 10% of the Hg^T was removed by the ESP. Hg^0 concentration decreased by 46%, Hg^{2+} concentration increased by 51%, and the ESP removed Hg^P only. When the LTE was on, the removal efficiency

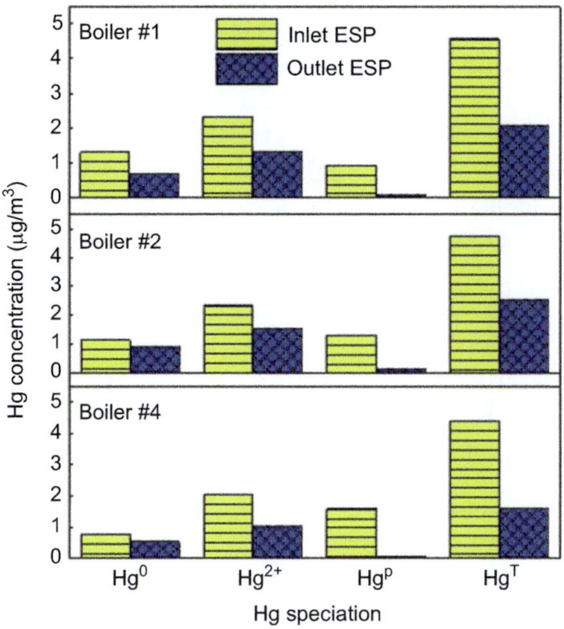

Figure 7.16 Hg concentration at the inlet and outlet of an electrostatic precipitator (*ESP*). From Zhang Y, Yang J, Yu X, Sun P, Zhao Y, Zhang J, Chen G, Yao H, Zheng C. Migration and emission characteristics of Hg in coal-fired power plant of China with ultra low emission air pollution control devices. Fuel Process. Technol. 2017;158:272–80.

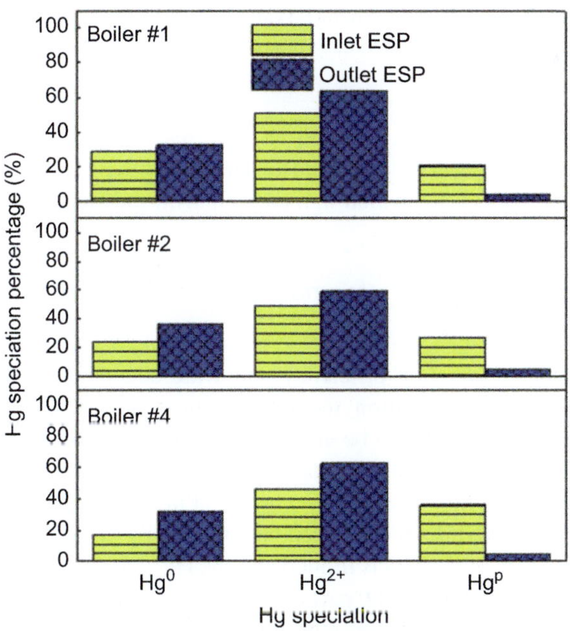

Figure 7.17 Hg speciation percentage at the inlet and outlet of an electrostatic precipitator (*ESP*).
From Zhang Y, Yang J, Yu X, Sun P, Zhao Y, Zhang J, Chen G, Yao H, Zheng C. Migration and emission characteristics of Hg in coal-fired power plant of China with ultra low emission air pollution control devices. Fuel Process. Technol. 2017;158:272–80.

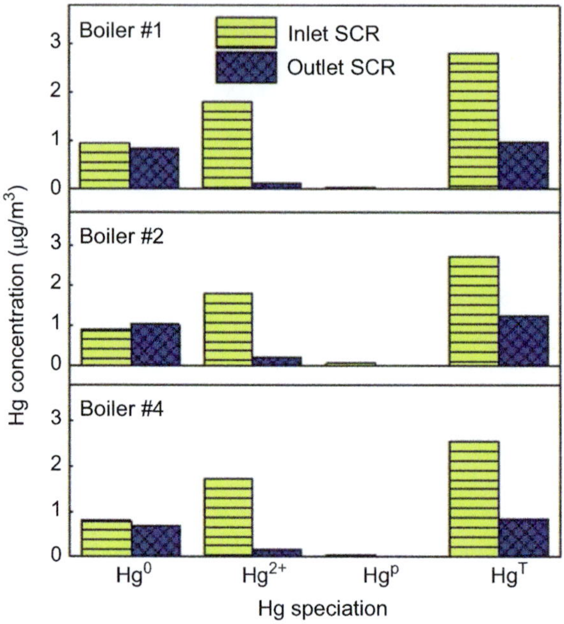

Figure 7.18 Mercury (a) concentration and (b) mass distribution of Hg measured at the electrostatic precipitator in boiler 1. *LTE*, low-temperature economizer.
From Zhou Z, Liu X, Zhao B, Chen Z, Shao H, Wang L, Xu M. Effects of existing energy saving and air pollution control devices on mercury removal in coal-fired power plants. Fuel Process. Technol. 2015;131:99–108.

of Hg^T and Hg^0 increased by 42.87% and 18.85%, respectively. The ESP itself could not effectively control gaseous Hg release. However, the LTE promoted gaseous mercury removal by the ESP, which agreed well with the study of Zhang et al. [8] as presented earlier.

7.2.1.4 Removal and transformation of Hg in wet flue gas desulfurization

Due to the high desulfurization efficiency and wide adaptability, WFGD has been selected as a major desulfurization method for most power plants. Water-soluble Hg^{2+} compounds in flue gases can be effectively removed by WFGD. In recent years, the influence of WFGD on Hg release control has received wide attention. A field test of Hg emissions at a 300-MW coal-fired power plant installed with an ESP and WFGD showed that 79.93%–90.35% of Hg^{2+} can be removed by WFGD [21]. However, the Hg^0 removal efficiency of WFGD is limited and the Hg^0 content instead shows a small increase. This is because in WFGD, part of the Hg^{2+} is reduced to Hg^0, which is promoted by increasing the SO_2 concentration in the flue gas. The greater proportion of Hg^{2+} in the flue gas is beneficial for a higher Hg removal efficiency of WFGD. Therefore, to improve the Hg removal capacity of WFGD, the proportion of Hg^{2+} in the exhaust gas should be increased.

To improve the Hg removal ability of WFGD, the influence of the liquid/gas ratio (L/G) and the pH value of the desulfurization slurry on the Hg removal efficiency was studied. The Hg removal efficiency of WFGD increased with L/G and radians of pH value [7]. In addition, the coal rank and dust remover mode also influence the Hg removal capability of WFGD. The US EPA examined the Hg release characteristics of several power plants with a WFGD system [17]. Most Hg^{2+} can be trapped by WFGD after passing through it. Different PMCDs have a major influence on the Hg capture by WFGD. For power plants equipped with an FF, cs-ESP, or hs-ESP, the Hg removal efficiency of WFGD is 72%, 66%, or 42%, respectively. Since an FF extends the contact time between the fly ash and the flue gas, more Hg^0 is oxidized into Hg^{2+} by an FF than by other PMCDs. As a result, the mercury removal efficiency of the FF + WFGD configuration is higher than those of cs-ESP + WFGD and hs-ESP + WFGD. Furthermore, suppression of Hg emissions in WFGD also strongly depends on the coal type. The WFGD in a power plant burning bituminous shows better Hg control capability than those in sub-bituminous coal- and lignite-burning plants. This is because the higher Cl content in the coal is beneficial for the conversion of gaseous Hg^0 to Hg^{2+}, and hence promotes the removal of Hg in WFGD.

To meet the ultralow SO_2 emission standard, the WFGD system has been converted at several power stations, especially in China's power plants. Zhang et al. [8] studied the Hg emissions and removal characteristics of WFGD in a ULE coal-fired power plant. As shown in Fig. 7.19 and Fig. 7.20, Hg^{2+} concentration decreased obviously when the flue gas passed through the WFGD system, and most Hg^{2+} (88.20%–92.80%) was removed by WFGD. However, the Hg^0 concentration decreased across WFGD in boilers 1 and 4 and slightly increased in boiler 2. The increase in Hg^0 concentration in WFGD was due to the reduction in Hg^{2+}, which is due to the following reasons [22,23]:

(1) The evaporation of lime or limestone seriflux may form an aqueous layer on the desulfurizer's outer surface. Gaseous Hg^{2+} and Hg^0 react with Hg_2^{2+} on the aqueous layer and Hg_2^{2+} could react with OH^- in the seriflux to form HgO and Hg^0. HgO can be reduced to Hg^0 by SO_2. This can be interpreted by Reactions (7.1) and (7.3).

(2) SO_2 could be dissolved in the scrubber solution to form sulfites and sulfates, which could interact with Hg^{2+} to form $HgSO_3$ and $HgSO_4$. Part of the $HgSO_3$ and $HgSO_4$ could be further reduced and released to Hg^0, interpreted by Reactions (7.4) and (7.5).

$$Hg^{2+} + Hg^0 \leftrightarrow Hg_2^{2+} \tag{7.1}$$

$$Hg_2^{2+} + 2OH^- \leftrightarrow H_2O + HgO + Hg^0 \tag{7.2}$$

$$HgO + SO_2 \leftrightarrow Hg^0 + SO_3 \tag{7.3}$$

$$Hg^{2+} + SO_3^{2-} \leftrightarrow HgSO_3 \rightarrow Hg^0 \tag{7.4}$$

$$Hg^{2+} + SO_4^{2-} \leftrightarrow HgSO_4 \rightarrow Hg^0 \tag{7.5}$$

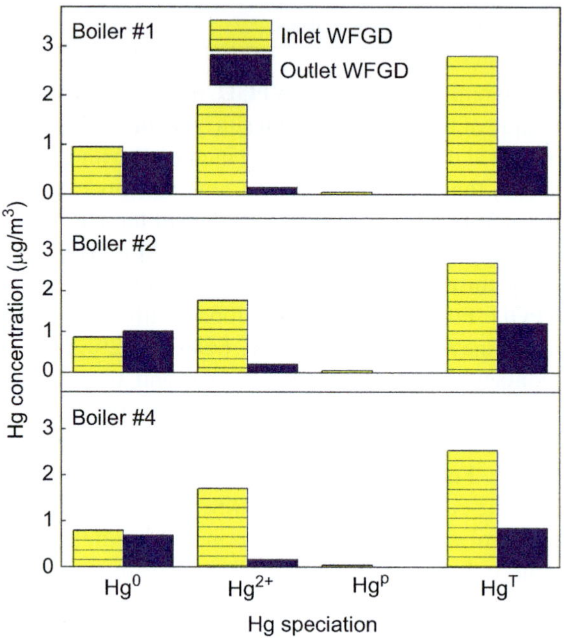

Figure 7.19 Hg concentration at the inlet and outlet of wet flue gas desulfurization. From Zhang Y, Yang J, Yu X, Sun P, Zhao Y, Zhang J, Chen G, Yao H, Zheng C. Migration and emission characteristics of Hg in coal-fired power plant of China with ultra low emission air pollution control devices. Fuel Process. Technol. 2017;158:272–80.

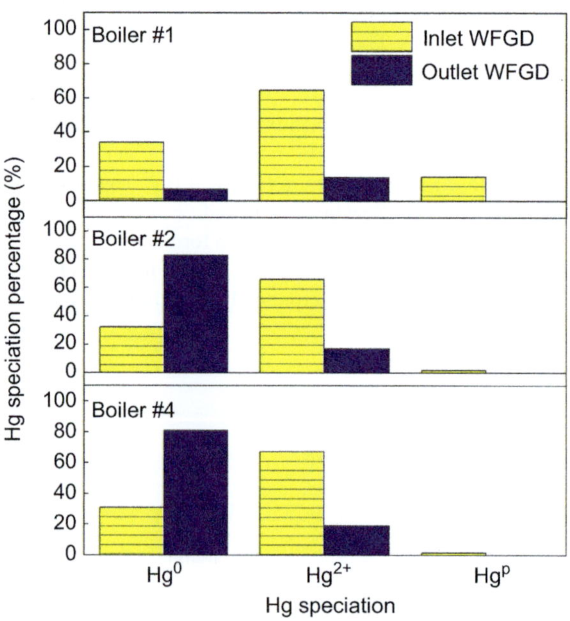

Figure 7.20 Hg speciation percentage at the inlet and outlet of wet flue gas desulfurization. From Zhang Y, Yang J, Yu X, Sun P, Zhao Y, Zhang J, Chen G, Yao H, Zheng C. Migration and emission characteristics of Hg in coal-fired power plant of China with ultra low emission air pollution control devices. Fuel Process. Technol. 2017;158:272–80.

The conversion of Hg is complicated and is determined by operating conditions. The gas temperature at the WFGD entrance of boilers 2 and 4 was 124.6°C and 109.84°C, respectively. An increase in flue gas temperature within the scrubber can result in an increase in Hg^0 reemission. Furthermore, the pH of the slurry is also an important factor affecting rerelease of Hg^0. The pH of the slurry was 5.02 and 5.39 in boilers 2 and 4, respectively. The slurry pH decreases as the O_2 concentration in the flue gas increases, resulting in an increase in Hg^0 reemission.

7.2.1.5 Migration and emission of mercury across wet electrostatic precipitators

Several coal-fired power plants in China have been equipped with WESPs to achieve the near zero emissions goal. WESPs play an important role in the removal of ultrafine particles and acrosol, which is widely used in power plant after ULE reformation. As shown in Fig. 7.21 and Fig. 7.22, after passing through a WESP, the concentration of Hg^0 decreased from 0.68−0.84 μg/m^3 to 0.45−0.48 μg/m^3 and the concentration of Hg^{2+} decreased from 0.13−0.16 μg/m^3 to 0.06−0.07 μg/m^3. Hg^{2+} in the exhaust gas is water soluble and may dissolve in the water film on the dust-collecting plate of a WESP. Since Hg^0 is water insoluble, the reduced Hg^0 concentration has to undergo the conversion process of Hg^0 to Hg^{2+}/Hg^p. The WESP electrode can generate corona electrons and several free radicals, such as O, OH, O_3, Cl, and Cl_2, from the HCl generated. These free radicals have high reactivity to other substances. Therefore, Hg^0 was oxidized to Hg^{2+} by free radicals and the formed Hg^{2+} could be removed by the WESP. This could be interpreted by Reactions (7.6)-(7.17) [24−28] and Fig. 7.23.

A previous study of the US Department of Energy [29] demonstrated that the removal efficiency of Hg^0, Hg^{2+}, and Hg^p in a WESP was 32%−40%, 72%−82%, and 33%−100%, respectively. The WESP electrode material affects the removal

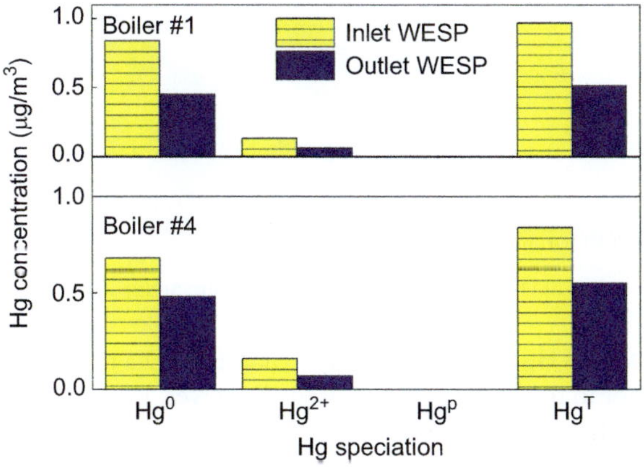

Figure 7.21 Hg concentration at the inlet and outlet of a wet electrostatic precipitator. From Zhang Y, Yang J, Yu X, Sun P, Zhao Y, Zhang J, Chen G, Yao H, Zheng C. Migration and emission characteristics of Hg in coal-fired power plant of China with ultra low emission air pollution control devices. Fuel Process. Technol. 2017;158:272−80.

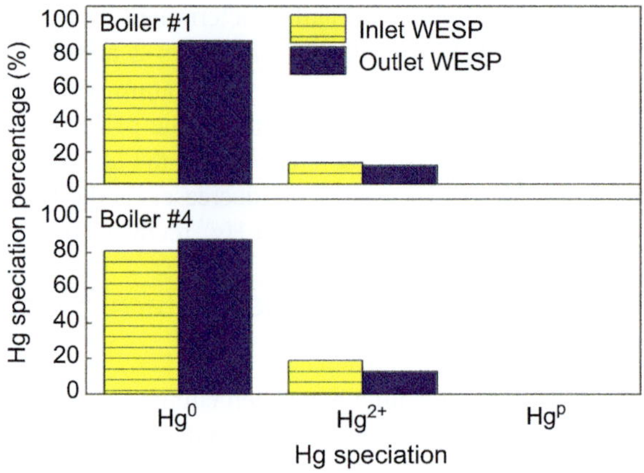

Figure 7.22 Hg speciation percentage at the inlet and outlet of a wet electrostatic precipitator. From Zhang Y, Yang J, Yu X, Sun P, Zhao Y, Zhang J, Chen G, Yao H, Zheng C. Migration and emission characteristics of Hg in coal-fired power plant of China with ultra low emission air pollution control devices. Fuel Process. Technol. 2017;158:272−80.

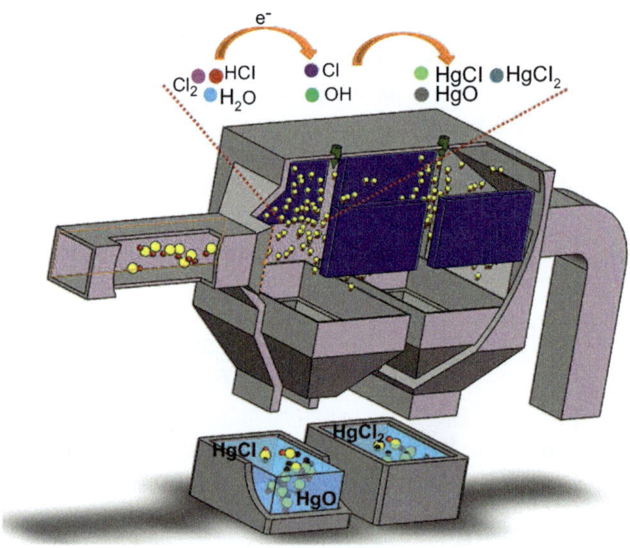

Figure 7.23 Mechanism of Hg oxidation in a wet electrostatic precipitator. *PM*, particulate matter.

efficiency of Hg^{2+} and Hg^{p}. Removal of Hg^{2+} and Hg^{p} by a nylon fiber material using a WESP is superior to the metal material used. However, the effect of the WESP electrode material on Hg^{0} removal is not clear.

$$O_2 + 2e^- \rightarrow 2O + 2e^- \quad (7.6)$$

$$O_2 + O \rightarrow O_3 \quad (7.7)$$

$$O + H_2O \rightarrow OH + OH \tag{7.8}$$

$$HCl + e^- \rightarrow H + Cl + e^- \tag{7.9}$$

$$Hg^0 + O \rightarrow HgO \tag{7.10}$$

$$Hg^0 + O_3 \rightarrow HgO + O_2 \tag{7.11}$$

$$Hg^0 + OH \rightarrow HgOH \tag{7.12}$$

$$Hg^0 + Cl \rightarrow HgCl \tag{7.13}$$

$$HgCl + Cl \rightarrow HgCl_2 \tag{7.14}$$

$$Hg^0 + Cl_2 \rightarrow HgCl + Cl \tag{7.15}$$

$$HgCl + Cl_2 \rightarrow HgCl_2 + Cl \tag{7.16}$$

$$Hg^0 + Cl_2 \rightarrow HgCl_2 \tag{7.17}$$

7.2.2 Other trace element emissions and migration across APCDs

7.2.2.1 Other trace element transformations across selective catalytic reduction

A survey of As emissions in nine coal-fired power plants equipped with various APCDs in China and the United States was conducted (Table 7.2) [30]. In units 1, 2, and 9, equipped SCR systems (see Table 7.3), the recovery of As was low; while As recovery was higher in unit 5 (without an SCR system) and unit 6 (hs-ESP). The recovery indices ($RI = M_{out}/M_{in}$, where M_{out} and M_{in} refer to the overall output and input of arsenic) of units 5 and 6 were 88.20% and 95.20%, respectively. The RI of units 9 and 2 were 71.63% and 81.92%, respectively. Partial gaseous As_2O_3 solidifies on the SCR catalyst and reduces the catalyst activity. Approximately 15% of As can be retained in the SCR system compared with other units. Unit 6 has the highest As recovery, which may be due to the application of an hs-ESP downstream of the air preheater. Most As compounds could be trapped by an hs-ESP before the flue gas entered the SCR system.

Zhao et al. [31] studied the emission characteristics of As, Cr, Ba, Mn, and Pb at three different loads (100%, 83%, and 71% output) and different coal types on a 350-MW coal-fired power plant equipped with SCR, ESP + FF, and WFGD. As shown in Table 7.4, particulate TEs occupy a proportion greater than 99.82% at the furnace outlet. As the load decreases, the concentration of particulate TEs decreases except for As. This is because the lower boiler load decrease the furnace temperature, which inhibits the volatilization of some TEs. In addition, the ash content decreases as

Table 7.2 Electric power plant (EPP) data[a].

Unit ID	Unit 1	Unit 2	Unit 3	Unit 4	Unit 5	Unit 6	Unit 7	Unit 8	Unit 9
Unit capacity (MW)	200	300	200	300	300	500	508	125	300
Country	China	China	China	China	USA	USA	USA	USA	China
Boiler	PC	PC	PC	CFB	PC	PC	PC	PC	PC
WESP	No	No	No	No	No	No	No	No	Yes
SO_2 control	FGD	FGD	FGD	Injection of limestone into the furnace	No	No	FGD	No	FGD
Particulate control	Cold ESP	Cold ESP	Cold ESP	Cold ESP	Cold ESP	Hot ESP	Cold ESP	Cold ESP	Cold ESP
Low-temperature economizer	No	No	No	No	No	No	No	No	Yes
NO_x control	SCR	SCR	SCR	No	No	No	SCR	No	SCR
Fuel type	Bituminous coal								

[a]*CFB*, circulating fluidized-bed furnace; *ESP*, electrostatic precipitator equipment; *FGD*, flue gas desulfurization equipment; *PC*, pulverized coal furnace; *SCR*, selective catalytic reduction equipment; *WESP*, wet electrostatic precipitator.

From Wang J, Zhang Y, Liu Z, Norris P, Romero C, Xu H, Pan W. Effect of coordinated air pollution control devices in coal-fired power plants on arsenic emissions. Energ. Fuel. 2017;31(7): 7309–316.

Table 7.3 Recovery values for arsenic at different units.

	Unit 1	Unit 2	Unit 5	Unit 6	Unit 9
Recovery index 1 (%)		81.92	88.20	95.20	71.63
Recovery index 2 (%)	72.24				73.11

From Wang J, Zhang Y, Liu Z, Norris P, Romero C, Xu H, Pan W. Effect of coordinated air pollution control devices in coal-fired power plants on arsenic emissions. Energ. Fuel. 2017;31(7):7309–316.

Table 7.4 Concentrations of As, Cr, Ba, Mn, and Pb in the flue gas at the inlet/outlet of each air pollution control device.

		SCR			
		Inlet		Outlet	
Sampling load	HTE	Con.	Prop.	Con.	Prop.
100%	As_g	0.10	0.01	0.24	0.03
	As_p	7.97×10^2	99.99	7.46×10^2	99.97
	Cr_g	1.22	0.07	0.71	0.05
	Cr_p	1.63×10^3	99.93	1.52×10^3	99.95
	Ba_g	7.39	0.03	8.63	0.04
	Ba_p	2.11×10^4	99.97	2.26×10^4	99.96
	Mn_g	2.49	0.02	1.94	0.02
	Mn_p	1.07×10^4	99.98	7.89×10^3	99.98
	Pb_g	0.11	0.01	0.12	0.01
	Pb_p	1.44×10^3	99.99	1.11×10^3	99.99
83%	As_g	0.05	0.01	0.10	0.01
	As_p	9.43×10^2	99.99	9.62×10^2	99.99
	Cr_g	0.95	0.10	0.46	0.05
	Cr_p	9.63×10^2	99.90	9.07×10^2	99.95
	Ba_g	3.16	0.02	1.70	0.01
	Ba_p	1.47×10^4	99.98	1.57×10^4	99.99
	Mn_g	0.97	0.01	1.46	0.02
	Mn_p	9.40×10^3	99.99	8.62×10^3	99.98
	Pb_g	0.11	0.02	0.43	0.06
	Pb_p	7.11×10^2	99.98	6.98×10^2	99.94

Continued

Table 7.4 Continued

		SCR			
		Inlet		Outlet	
Sampling load	**HTE**	**Con.**	**Prop.**	**Con.**	**Prop.**
71%	As_g	0.03	0.00	0.08	0.01
	As_p	7.63×10^2	100.00	7.86×10^2	99.99
	Cr_g	1.64	0.18	0.76	0.08
	Cr_p	9.15×10^2	99.82	9.06×10^2	99.92
	Ba_g	6.96	0.08	3.45	0.03
	Ba_p	9.16×10^3	99.92	1.14×10^4	99.97
	Mn_g	6.29	0.12	1.54	0.03
	Mn_p	5.43×10^3	99.88	4.87×10^3	99.97
	Pb_g	0.09	0.02	0.07	0.01
	Pb_p	5.15×10^2	99.98	5.16×10^2	99.99

Subscript g indicates gas phase; subscript p indicates particulate bound. *Con.*, concentration; *HTE*, hazardous trace element; *Prop.*, proportion; *SCR*, selective catalytic reduction.
From Zhao S, Duan Y, Chen L, Li Y, Yao T, Liu S, Liu M, Lu J. Study on emission of hazardous trace elements in a 350 MW coal-fired power plant. Part 2. arsenic, chromium, barium, manganese, lead. Environ. Pollut. 2017;226:404–11.

load is reduced. Since the concentration of particulate TEs is much higher than the concentration of gas TEs, there is no obvious trend for SCR fluctuations at three loads.

Pan et al. [32] studied the effects of SCR de-NO_x on the TE distribution in PM at two 300-MW pulverized coal-fired boiler units. In the process of characterizing and analyzing the TEs (Cr, As, Cd, Pb) in the particles, the results of the 0.2- to 0.5-μm samples ($PM_{0.2-0.5}$) collected by the dekati gravimetric impactor (DGI) represent sub-micrometer modes. For the PM, the ultrafine microparticles were represented by samples of 0.5–1.0 μm ($PM_{0.5-1.0}$). Comparing the TE contents in the SCR inlet and outlet smoke PM, it was found that the TE content in $PM_{0.2-0.5}$ changed significantly, while the TE content in $PM_{0.5-1.0}$ increased only slightly.

The TEs show different trends after the SCR process, in which the contents of As and Cd change most significantly, while the contents of Pb and Cr change little. After the SCR, the contents of As, Cd, and Pb in the $PM_{0.2-0.5}$ increased by 104.15%, 12.26%, and 12.18%, respectively, and the Cr content decreased by 23.38%. The content of TEs in PM is affected by both PM quality and TE quality. In the flue gas at the inlet of the SCR reactor, some TEs still exist in a gaseous form. Part of the gaseous TEs are converted into PM by condensation during the cooling process in the SCR reactor, so that the TE content in the PM rises. Moreover, the presence of SO_3 and H_2O

substances generated during the action of SCR will further promote the migration of gaseous TEs to the PM. The gasification characteristics of As, Cd, and Pb are stronger than those of Cr, so more As and Cd is converted into PM during the SCR action. However, according to the previous analysis, the NH_4^+, SO_4^{2+}, and other subcomponents produced by some SCR processes are converted into PM, which dilutes the TE content in the PM. During the experiment, the conversion of As and Cd to $PM_{0.2-0.5}$ was stronger than that of NH_4^+, SO_4^{2+}, and other subcomponents, so the content of As and Cd increased. The conversion of Cr to $PM_{0.2-0.5}$ is weak, so the content is reduced.

7.2.2.2 Other trace element emissions and migration across electrostatic precipitator/fabric filter

At the high-temperature combustion zone in the boiler, TEs may volatilize and release into flue gases. As the flue gas temperature decreases, some gaseous metals condense or adsorb onto fly ash particles and form particulate-bound TEs. Elements bound within solid particles can be efficiently trapped by an ESP and FF. As shown in Table 7.5, the concentrations of Cd, Pb, and Mn in the gas phase are 1−2 orders of magnitude lower than those of the particulate forms prior to the particulate precipitant [33]. This is attributed to the low volatility of these metals.

Table 7.5 Concentrations of Pb, Cd, and Mn in flue gas along the flue gas path and their removal.

Power plant	Element in flue gas	Particulate capture device		
		In (μg/Nm^3)	Out (μg/Nm^3)	Removal rate (%)
Plant 1	Cd_p	132.6 ± 15.4	0.2 ± 0.0	99.9
	Cd_g	13.8 ± 2.5	0.3 ± 0.0	97.6
	Pb_p	3060.5 ± 142.1	134.5 ± 23.7	95.6
	Pb_g	440.2 ± 32.6	181.9 ± 20.3	58.5
	Mn_p	12,787.4 ± 801.3	18.4 ± 2.6	99.9
	Mn_g	587.6 ± 92.5	58.0 ± 4.7	90.1
Plant 2	Cd_p	137.6 ± 27.13	0.5 ± 0.0	99.6
	Cd_g	1.7 ± 0.6	0.2 ± 0.0	87.2
	Pb_p	3630 ± 243.8	108.9 ± 46.3	97.0
	Pb_g	570 ± 62.4	45.7 ± 6.9	92.0
	Mn_p	6652.2 ± 431.5	5.8 ± 0.7	99.9
	Mn_g	56.6 ± 3.9	24.0 ± 3.5	57.6

Continued

Table 7.5 Continued

Power plant	Element in flue gas	Particulate capture device		
		In ($\mu g/Nm^3$)	Out ($\mu g/Nm^3$)	Removal rate (%)
Plant 3	Cd_p	39.6 ± 2.5	0.1 ± 0.0	99.9
	Cd_g	3.4 ± 0.2	0.8 ± 0.1	76.3
	Pb_p	1450 ± 89.3	40.8 ± 6.4	97.2
	Pb_g	230 ± 56.70	28.9 ± 7.1	87.4
	Mn_p	4237.5 ± 532.8	19.4 ± 2.9	99.5
	Mn_g	302.2 ± 19.7	49.6 ± 6.5	83.6
Plant 4	Cd_p	22.8 ± 3.4	0.2 ± 0.0	99.4
	Cd_g	5.5 ± 0.6	0.3 ± 0.0	94.4
	Pb_p	2530 ± 94.8	103.4 ± 7.8	95.9
	Pb_g	240 ± 35.6	52.2 ± 6.8	78.0
	Mn_p	2347.4 ± 107.5	22.4 ± 3.5	99.1
	Mn_g	153.4 ± 26.7	26.1 ± 3.4	83.0
Plant 5	Cd_p	21.5 ± 3.2	0.1 ± 0.0	99.8
	Cd_g	6.2 ± 0.4	0.4 ± 0.0	93.5
	Pb_p	2370 ± 64.3	118.1 ± 23.5	95.0
	Pb_g	360. ± 4.7	153.0 ± 21.6	57.9
	Mn_p	2464.2 ± 175.6	6.3 ± 0.1	99.7
	Mn_g	165.6 ± 23.1	38.9 ± 1.8	76.5
Plant 6	Cd_p	204.6 ± 37.5	0.1 ± 0.0	99.9
	Cd_g	14.3 ± 2.6	2.2 ± 0.3	84.4
	Pb_p	5760 ± 253.2	336.4 ± 17.3	94.2
	Pb_g	570 ± 121.7	295.6 ± 16.9	48.5
	Mn_p	10,156.6 ± 382.9	2.4 ± 0.5	99.9
	Mn_g	461.8 ± 57.4	158.5 ± 23.6	65.7

Subscript g indicates gas phase; subscript p indicates particulate bound.
From Deng S, Shi Y, Liu Y, Zhang C, Wang X, Cao Q, Li S, Zhang F. Emission characteristics of Cd, Pb and Mn from coal combustion: Field study at coal-fired power plants in China. Fuel Process. Technol. 2014;126:469–75.

Effect of Cl and unburned carbon content on trace element removal

The chlorine content in coal plays an important role in the distribution of TEs. Fig. 7.24 shows the change in percentage of Cd_p, Pb_p, and Mn_p in chlorine-containing flue gas [33]. The correlation coefficient of Pb_p in the flue gas with chlorine

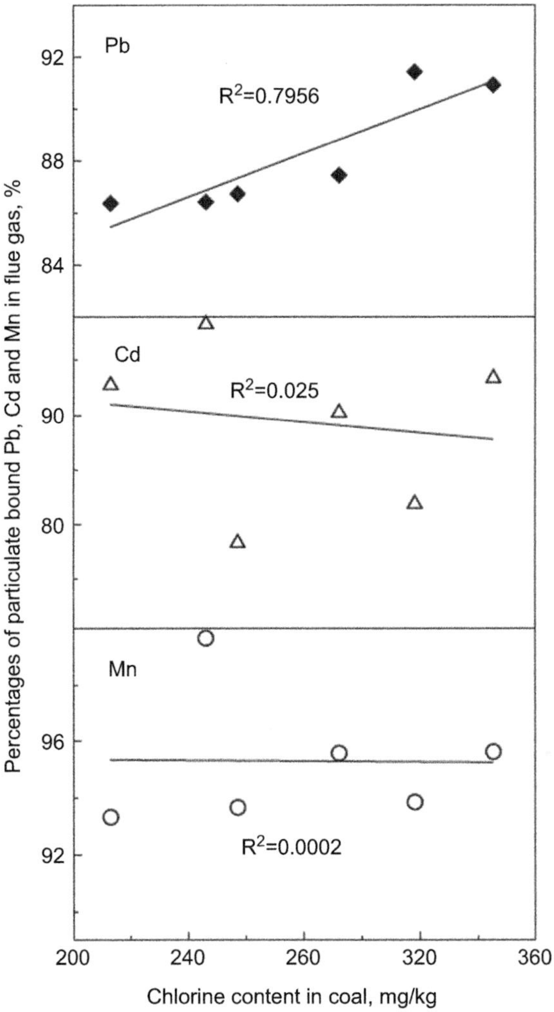

Figure 7.24 Correlations between percentages of Cd_p, Pb_p, and Mn_p (subscript p indicates particulate bound) in flue gas and Cl content in coal.
From Deng S, Shi Y, Liu Y, Zhang C, Wang X, Cao Q, Li S, Zhang F. Emission characteristics of Cd, Pb and Mn from coal combustion: Field study at coal-fired power plants in China. Fuel Process. Technol. 2014;126:469–75.

content in coal reached 0.7952, indicating that chlorine can have a significant effect on Pb_p formation. However, the effect of chlorine content in coal was negligible for the distribution of Cd and Mn.

Studies on the retention of TEs on unburned carbon particles are rare except for Hg. As shown in Fig. 7.25, only Pb shows a significant relationship with loss on ignition (LOI) of the ash particle (0.5049). The LOI may not be significant for the retention of

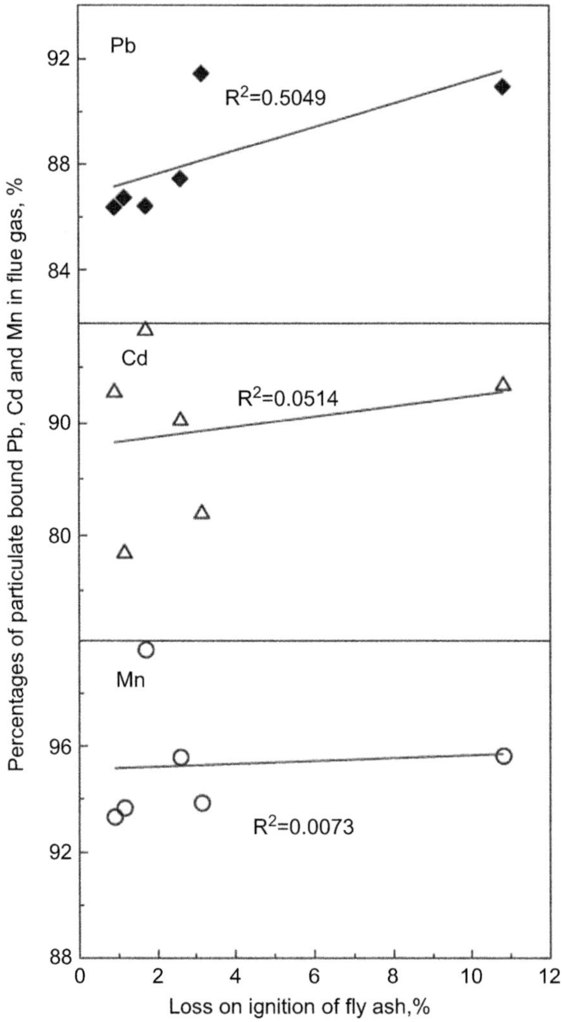

Figure 7.25 Correlations between percentages of Cd_p, Pb_p, and Mn_p (subscript p indicates particulate bound) in flue gas and the loss on ignition of fly ash.
From Deng S, Shi Y, Liu Y, Zhang C, Wang X, Cao Q, Li S, Zhang F. Emission characteristics of Cd, Pb and Mn from coal combustion: Field study at coal-fired power plants in China. Fuel Process. Technol. 2014;126:469–75.

Cd and Mn. Pb, Cd, and Mn are not highly volatile elements and cannot evaporate completely in the combustion chamber. Elements associated with certain minerals in the coal tend to be retained in the ash during combustion, and the elements associated with the volatile matrix are prone to volatilize. Therefore, the relationship between Mn_p and Cd_p with LOI of fly ash is insignificant. Pb is more volatile than Cd and Mn. Thus, it shows a relatively strong correlation with the aforementioned parameters.

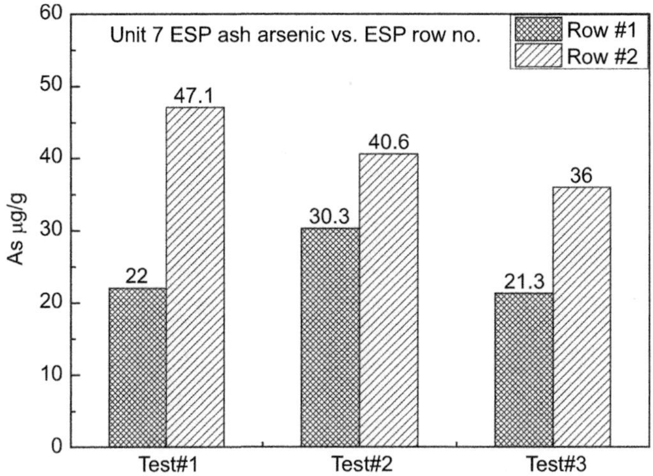

Figure 7.26 As concentration in electrostatic precipitator (*ESP*) fly ash at the hoppers for different rows.
From Wang J, Zhang Y, Liu Z, Norris P, Romero C, Xu H, Pan W. Effect of coordinated air pollution control devices in coal-fired power plants on arsenic emissions. Energ. Fuel. 2017; 31(7):7309–316.

Effect of fly ash particle size on trace element removal

Fly ash particle size is another factor affecting the capture efficiency. Wang et al. [30] compared the As concentration in the ashes from the first two rows of hoppers of an ESP. Fig. 7.26 shows that the As concentration was higher in the ash samples collected from the second row than from the first row. This is because, as the fly ash surface area increases with the decrease in particle size, the absorption capacity increases.

Effect of dust removal device mode

The dust removal device mode also significantly affects the removal efficiency of TEs. Wang et al. [30] investigated the As capture efficiency from nine units in China and the United States. As shown in Fig. 7.27, the hs-ESP of unit 5 and unit 6 had a higher As removal efficiency, where the particulate precipitant collected about 90% of the arsenic. Unit 9 had undergone ULE reformation, and the As removal efficiency of the ESP in unit 9 was 98.22%. Increasing the particle capture efficiency can improve the As removal efficiency.

Zheng et al. [34] studied the partitioning of hazardous TEs among APCDs in five ULE coal-fired power plants. Results showed that most of the TEs were enriched in fly ash at 58.0%–93.3% (Hg), 75.2%–95.3% (As), 78.2%–94.9% (Cd), 79.4%–96.6% (Se), 73.8%–89.2% (Cr), and 86.5%–99.5% (Pb). A low-low temperature ESP (LLT-ESP) and electric FF (EFF) greatly increased the relative enrichment factors (REFs) of Hg, As, and Se in fly ash, up to 0.78–1.23, 0.85–1.04, and 0.83–0.99, respectively.

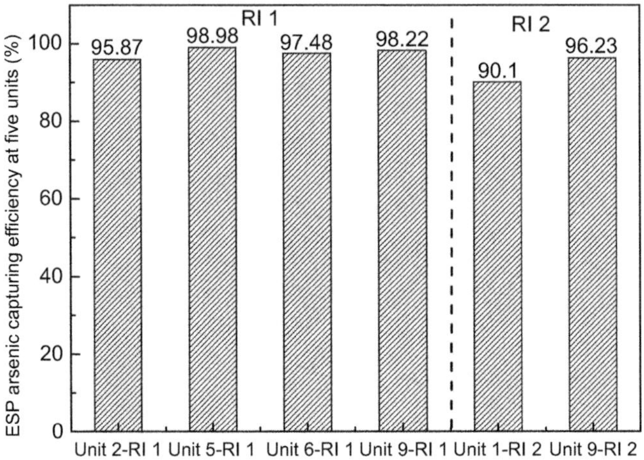

Figure 7.27 Electrostatic precipitator (*ESP*) arsenic capturing efficiency at five units. *RI*, recovery index.
From Wang J, Zhang Y, Liu Z, Norris P, Romero C, Xu H, Pan W. Effect of coordinated air pollution control devices in coal-fired power plants on arsenic emissions. Energ. Fuel. 2017; 31(7):7309–316.

7.2.2.3 Other trace element emissions and migration across wet flue gas desulfurization

In the WFGD system, circulation of the desulfurization slurry accelerates adsorption of SO_2 and washes away TEs such as Hg, As, Pb, and Cr. Córdoba et al. [35] found that WFGD has a significant effect on TE distribution (Fig. 7.28). Residence in gypsum and dissolution in gypsum slurry water can be regarded as the main output of the WFGD system. Sulfur, F, Se, and As are mainly retained in gypsum sludge, while Cl and B are dissolved in gypsum slurry water. A high gas retention capacity of 92%–100% for S, Cl, F, Se, As, and B is achieved with WFGD.

Elements escaping from the ESP are released mainly in the flue gas phase as gas species. Moreover, TEs in those fly ash particles not removed by the particle control system will enter the WFGD system and interact with the limestone slurry present in the FGD scrubber [36]. Table 7.6 shows the TE concentrations in limestone, FGD gypsum, process water, and filtration wastewater.

In Fig. 7.29, the abatement efficiencies of As, Cl, F, S, and Se are in the range of 96%–100%. The removal behavior of Se from the flue gas is similar to the process for S. Selenates and sulfates have similar structural and chemical properties and can play similar geochemical roles. Gypsum and calcium selenate dihydrate ($CaSeO_4 \cdot 2H_2O$) are isomorphic with similar lattice constants. Selenate can be replaced by gypsum sulfate. Regarding As, its retention by limestone slurry occurs by the formation of $Ca_3(AsO_4)_2$. The high removal of F is known to be caused by the precipitation of CaF_2.

Removal of Se across the WFGD in the study of Senior et al. [37] had an average value of 61% in a 3-day test at a 900-MW coal-fired power plant

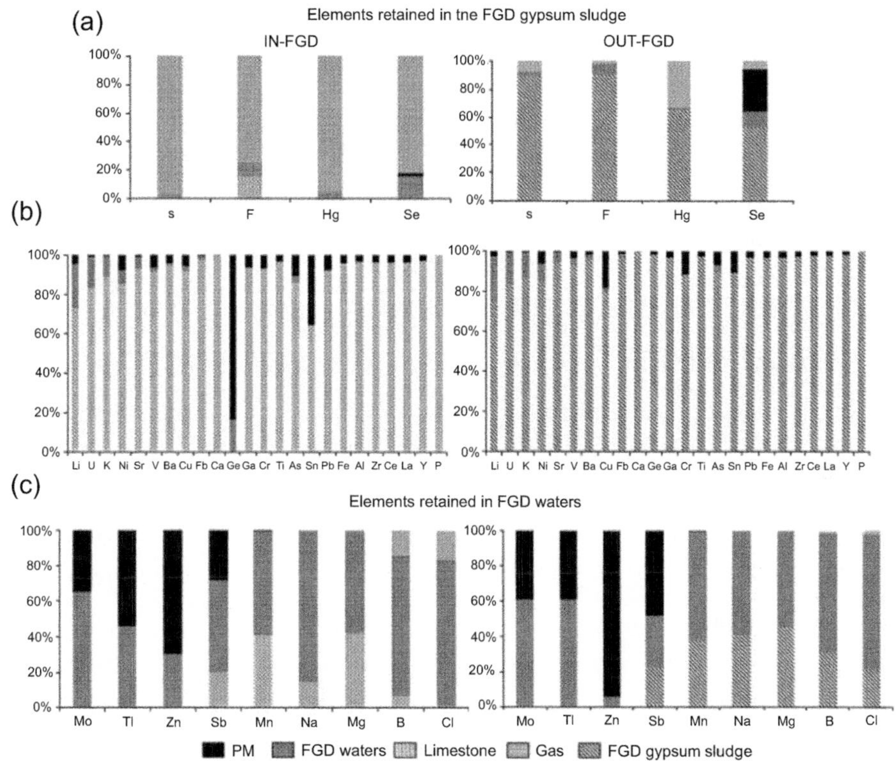

Figure 7.28 Distribution and partitioning of the element IN and OUT flue gas desulfurization (*FGD*) streams. (a) IN-FGD (left) and OUT-FGD (right) partitioning of the highly and moderately volatile elements retained in FGD gypsum sludge. (b) IN-FGD (left) and OUT-FGD (right) partitioning of elements associated with the Al–Si of limestone and those as fly ash (*PM* IN-FGD) retained in the FGD gypsum sludge. (c) IN-FGD (left) and OUT-FGD (right) partitioning of elements retained in FGD waters.
From Córdoba P, Ochoa-Gonzalez R, Font O, Izquierdo M, Querol X, Leiva C, López-Antón MA, Díaz-Somoano M, Rosa Martinez-Tarazona M, Fernandez C, Tomás A. Partitioning of trace inorganic elements in a coal-fired power plant equipped with a wet Flue Gas Desulphurisation system. Fuel 2012;92(1):145–57.

(Table 7.7). Approximately 80% of the granular bond Se was removed with WFGD when injected with hydrated lime. However, without hydrated lime injection only 40%–70% was removed.

The particle-bound Se concentration entering the scrubber was highest for particles larger than 2 μm (Fig. 7.30). When hydrated lime was injected upstream of the inlet measurement position, the Se concentration was higher compared with the test without adding hydrated lime. At the scrubber outlet, the particle-bound Se concentration increased compared with that at the scrubber inlet. For the particles of about 0.3–1 μm, the concentration increased more than 10 times from the entrance to the exit of the scrubber. Thus, even though a significant portion of the particles were

Table 7.6 Element contents in limestone, flue gas desulfurization gypsum, process water, and filtered wastewater involved in the desulfurization step following the coal combustion process.

	Limestone (mg/kg)	FGD gypsum (mg/kg)	Process water (mg/L)	Filtered wastewater (mg/L)
As	1.0	2.6	0.003	0.024
Cl	148	307	98.6	1137
F	152	778	0.536	26.4
Hg	0.001	0.212	<0.001	0.007
S	509	175,379	332	7120
Se	2.3	4.0	0.003	0.030
Ba	15.4	9.21	0.089	0.164
Cd	0.05	0.04	<0.001	0.001
Cr	5.8	4.0	<0.001	0.076
Cu	2.6	2.8	0.032	0.043
Mo	0.14	0.11	0.002	0.004
Ni	8.7	5.7	0.008	0.047
Pb	0.88	0.64	<0.001	0.002
Sb	0.09	0.06	<0.001	0.001
Zn	9.9	7.9	0.029	0.134

FGD, flue gas desulfurization.
From Alvarez-Ayuso E, Querol X, Tomas A. Environmental impact of a coal combustion-desulphurisation plant: abatement capacity of desulphurisation process and environmental characterisation of combustion by-products. Chemosphere 2006; 65(11):2009–17.

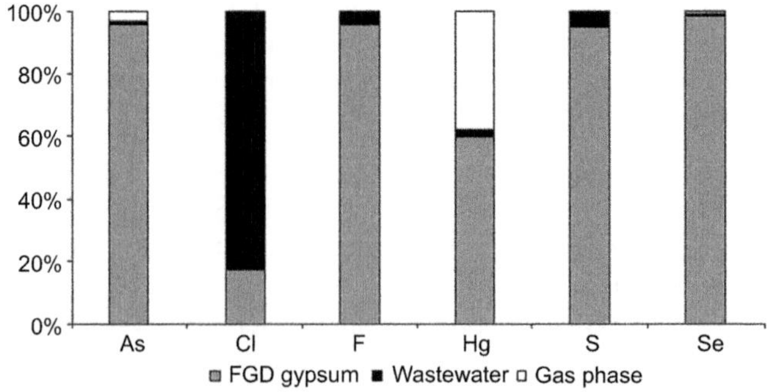

Figure 7.29 Abatement capacity of the flue gas desulfurization (*FGD*) plant for hazardous emissions arising from the coal-combustion power plant.
From Alvarez-Ayuso E, Querol X, Tomas A. Environmental impact of a coal combustion-desulphurisation plant: abatement capacity of desulphurisation process and environmental characterisation of combustion by-products. Chemosphere 2006;65(11):2009–17.

Table 7.7 Gas and particulate Se removal across a scrubber, synthesized from a cascade impactor Method 29 test results.

	Process conditions			Se concentrations in mg/dscm						Removal across scrubber		
				Solid (impactor)	Gas (M29)	Total	Solid (impactor)	Gas (M29)	Total			
Sample	Lime, lb/h	Scrubber pH	Scrubber LS, cm	Inlet	Inlet	Inlet	Outlet	Outlet	Outlet	Solid	Gas	Total
Day 1 a.m.	185	5.2	21.3	0.075	0.0217	0.0967	0.015	0.0142	0.0292	80%	35%	70%
Day 1 p.m.	388	5.2	21.1	0.106	0.055	0.161	0.015	0.0283	0.0433	86%	49%	73%
Day 2 a.m.	0	5.7	20.9	0.023	0.0762	0.0992	0.014	0.0598	0.0738	39%	22%	26%
Day 2 p.m.	381	5.9	20.8	0.09	0.0454	0.1354	0.014	0.0249	0.0389	84%	45%	71%
Day 3 a.m.	0	5.2	21.2	0.042	0.1484	0.1904	0.014	0.0514	0.0654	67%	65%	66%
Day 3 p.m.	0	5.2	21.0	0.031	0.1176	0.1486	0.014	0.0502	0.0642	55%	57%	57%

From Senior C, Tyree C, Meeks N, Acharya C, McCain J, Cushing K. Selenium Partitioning and Removal Across a Wet FGD Scrubber at a Coal-Fired Power Plant. Environ. Sci. Technol. 2015; 49(24):14376−382.

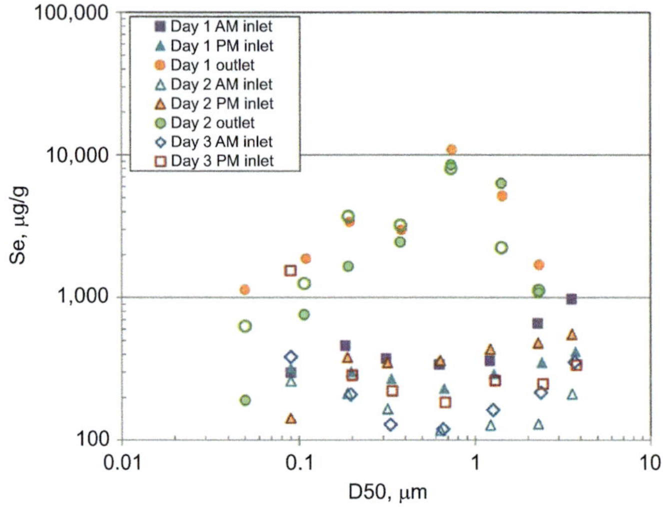

Figure 7.30 Concentration of Se (µg/g) in fly ash particles as a function of particle size. *Open symbols* correspond to periods with no hydrated lime injection. *Closed symbols* correspond to periods with hydrated lime injection.
From Senior C, Tyree C, Meeks N, Acharya C, McCain J, Cushing K. Selenium Partitioning and Removal Across a Wet FGD Scrubber at a Coal-Fired Power Plant. Environ. Sci. Technol. 2015; 49(24):14376–382.

removed by the scrubber, the remaining submicrometer particles had much higher concentrations of Se. Across the scrubber, the Se concentration increased at the exit from the scrubber for entrance of particles less than about 0.5 µm (Fig. 7.31). This increase was most pronounced in measurements made without hydrated lime injection.

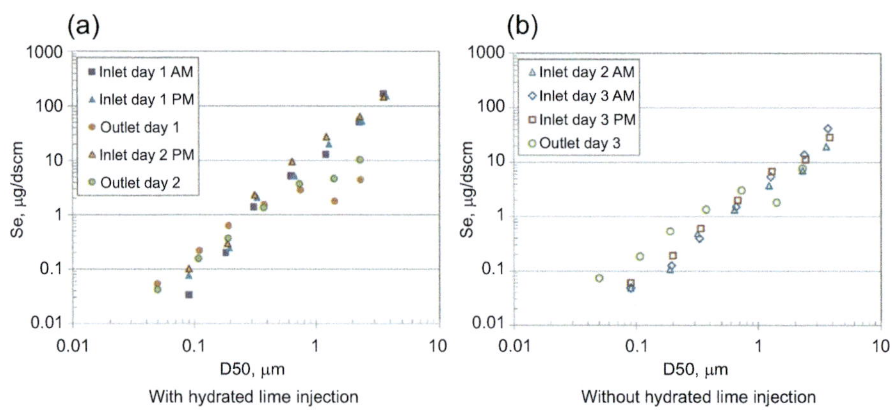

Figure 7.31 (a) Concentration of particulate-bound Se in flue gas as a function of particle size. *Closed symbols* correspond to periods with hydrated lime injection. (b) *Open symbols* correspond to periods with no hydrated lime injection.
From Senior C, Tyree C, Meeks N, Acharya C, McCain J, Cushing K. Selenium Partitioning and Removal Across a Wet FGD Scrubber at a Coal-Fired Power Plant. Environ. Sci. Technol. 2015; 49(24):14376–382.

Furthermore, Córdoba et al. [38] systematically studied the Se behavior in WFGD. In WFGD, fly ash components can be dissolved in the aqueous phase of the slurry or retained in the gypsum sludge. Selenium chemisorbed in fly ashes (e.g., SeO_3^{2-}) may then be dissolved in the aqueous phase of the slurry to form an array of selenium aqueous complexes and/or solid species (Fig. 7.32). If SeO_2 remains in the gas phase, a portion of the gaseous SeO_2 reacts with moisture in the flue gas to produce H_2SeO_3, after which the flue gas is quenched to a temperature of 50°C or 60°C.

Most of Se in the WFGD system is removed with water-based wastewater (filtered water) and gypsum, and only a little Se remains in the exhaust gas. Enrichment of Se by continuous water recirculation within the scrubber was promoted by operating conditions affecting WFGD [38]. Release of Se particles and droplets from gypsum slurry with OUT-FGD gas was the most important result of high enrichment of inorganic trace contaminants in recycled water.

7.2.2.4 Other trace element emissions and migration across a wet electrostatic precipitator

WESPs play an important role in the removal of ultrafine particles and acrosol. Since arsenic is more easily enriched in particulates, the particulate removal efficiency has a significant impact on the removal of arsenic. Wang et al. [30] reported that the As concentrations in flue gas were 0.130 and 0.098 μg/m^3 at the WESP inlet and outlet of unit 9, respectively. The corresponding As removal efficiency of WESP was 24.6%.

Zheng et al. [34] studied the TE behavior when passed through the WESP. The concentrations of Hg, As, Cd, Se, Cr, and Pb in WESP wastewater slag were in the ranges 8.67−17.5, 45.2−133, 4.51−9.26, 61.3−121, 274−1303, and 74.9−255 mg/kg. Pb was not enriched in WESP wastewater slag compared with fly ash. The concentrations of TEs in the WESP wastewater were extremely low. The WESP removed TEs mainly by trapping the fine particles, while a small amount of TEs (0.2%−26%) were retained in the washing water.

7.3 Mercury concentration and enrichment in combustion products

7.3.1 Trace element distribution in fly ash and bottom ash

Fig. 7.33 shows three distinct element groups rich in ash depending on volatility.

Group I contains the high-volatility elements of As, Hg, and Zn, which exist in sulfide minerals or organics in coal and exhibit volatile behavior.

Group II contains semivolatile elements, including Mg, Li, Cd, and Mn, most of which are expected to be divided mainly by fly ash rather than bottom ash.

Group III contains the low-volatility elements, including Co, Ni, Cr, Fe, Pb, and Cu, which show higher affinity with the ash phase than with the vapor phase.

Tang et al. [39] studied the distribution and fate of As, Hg, Sb, and Se at coal-fired power plants in Huainan, Anhui Province, China. Table 7.8 shows that the Se

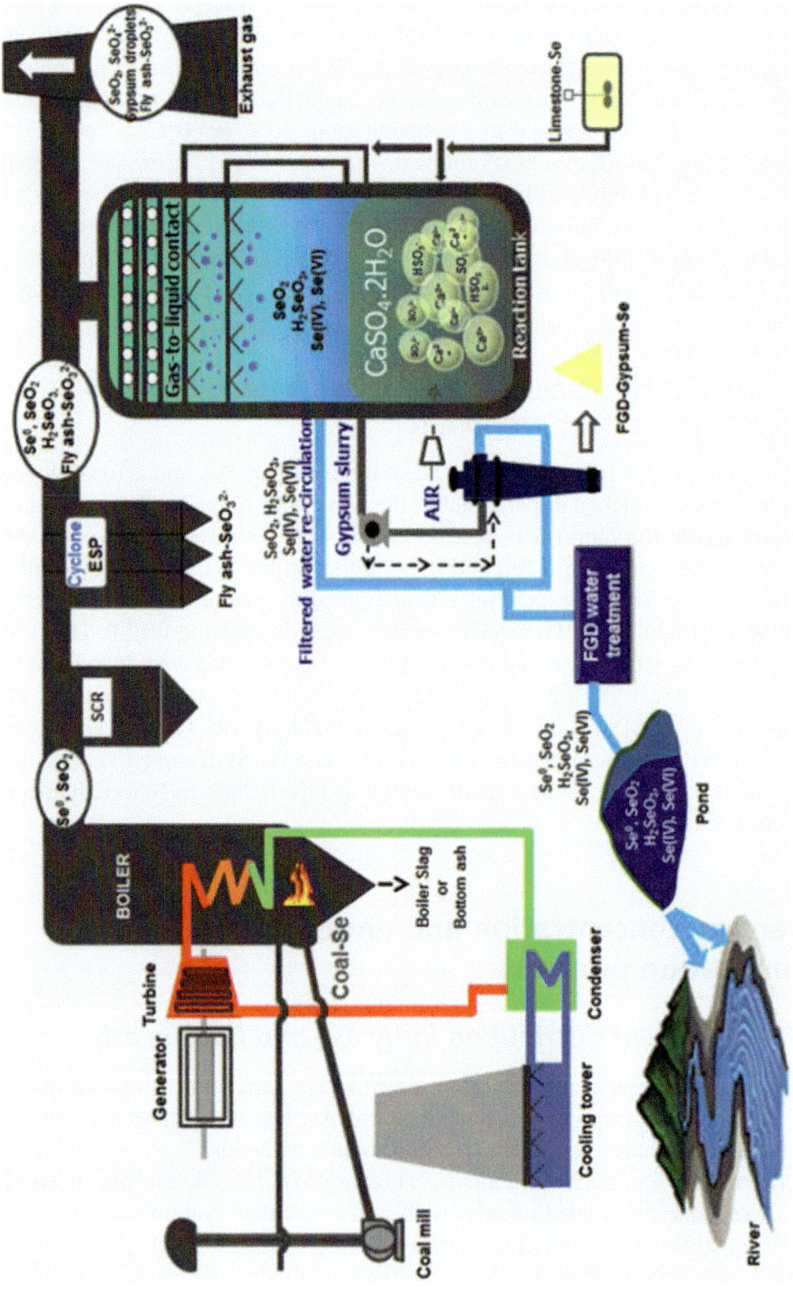

Figure 7.32 Selenium chemistry in wet flue gas desulfurization systems. *ESP*, electrostatic precipitator; *FGD*, flue gas desulfurization; *SCR*, selective catalytic reduction.

From Córdoba P. Partitioning and speciation of selenium in wet limestone flue gas desulphurisation systems: A review. Fuel 2017;202:184–95.

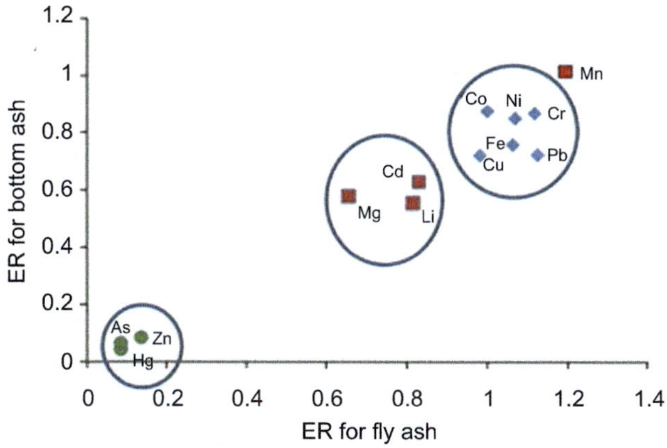

Figure 7.33 Enrichment behavior of elements in ash samples. *ER*, enrichment ratio.
From Bhangare R, Ajmal P, Sahu S, Pandit G, Puranik V. Distribution of trace elements in coal and combustion residues from five thermal power plants in India. Int. J. Coal. Geol. 2011;86(4):349–356.

Table 7.8 Concentrations of As, Hg, Sb, and Se in feed coal, bottom ash, and fly ash (in μg/g) and relative enrichment factors in bottom ash and fly ash.

	Coal	Bottom ash	Fly ash	RE (bottom ash)	RE (fly ash)
Luohe #2					
As	1.07	2	3.14	0.11	0.72
Hg	0.085	0.001	0.039	0.001	0.11
Sb	0.495	2.26	4.1	0.28	2.02
Se	3.2	0.69	9.41	0.01	0.72
Luohe #3					
As	1.575	2.02	4.7	0.08	0.75
Hg	0.051	0	0.032	0	0.16
Sb	2.15	2.51	7.97	0.07	0.94
Se	3.6	0.28	10.1	0.005	0.71
Pingwei #1					
As	1.69	1.7	7	0.05	0.87
Hg	0.05	0	0.043	0	0.18
Sb	0.88	1.8	4.04	0.11	0.97
Se	3.712	0.5	12.6	0.01	0.71
Pingwei #2					
As	1.69	2.7	5.6	0.08	0.70
Hg	0.05	0.001	0.041	0.001	0.17
Sb	0.88	1.86	4	0.11	0.96
Se	3.712	0.29	12.31	0.004	0.70

RE, relative enrichment.
From Tang Q, Liu G, Yan Z, Sun R. Distribution and fate of environmentally sensitive elements (arsenic, mercury, stibium and selenium) in coal-fired power plants at Huainan, Anhui, China. Fuel 2012;95:334–39.

Table 7.9 Concentration and relative enrichment factor of arsenic in coal and coal ashes (for $n = 3$).

	Concentration (μg/g)	RE
Coal	3.60	
EF-1	6.03	0.50
EF-2	25.6	2.12
EF-3	37.7	3.11
EF-4	45.6	3.79
EF-5	66.9	5.53
Slag	0.54	0.045

EF, ESP fly ash; RE, relative enrichment.
From Guo X, Zheng C, Xu M. Characterization of arsenic emissions from a coal-fired power plant. Energy Fuels 2004; 18(6):1822–26.

concentration in bottom ash is lower than that in raw coal, while the Se concentration in fly ash is relatively high. This suggests that Se tends to volatilize during combustion and can be enriched in fly ash particles. The As and Sb concentrations in bottom ash and fly ash were higher than those in coal. However, the Hg concentration in bottom ash and fly ash was lower than that in raw coal. Hg is mostly volatile but Sb has the lowest volatility. The relatively high partition in fly ash of As and Se is due to the affinity for calcium oxide to form $Ca_3(AsO_4)_2$ and $CaAsO_4$ and $CaSeO_4$.

Guo et al. [40] reported that the relationship between As concentration and ash particle size is an important factor in determining the behavior of components. As shown in Table 7.9, As is significantly enriched in smaller particles. For the finest particles, RE = 5.53, and for the coarsest particles, RE = 0.5. This phenomenon could be attributed to the following aspects: (1) the sum of the available surface area following volatilization of As is larger with small particles and (2) the calcium present in the fly ash is a solid–gas surface reaction.

The mechanism of TE enrichment of fine particles has been studied by many researchers, since PM10 microparticles are highly concentrated in TEs and are likely to enter the respiratory tract, which is very harmful to human health. Wang et al. [41] studied the distribution characteristics of TEs in fly ash with different particle size distributions in a 660-MW coal-fired power plant. The concentrations of As, Cr, Cd, and Pb on each particle size segment increased with the decrease in particle size. This is because these four elements are all volatile, and a considerable part of the gas is burned during the combustion process. In the process of cooling of the flue gas, homogeneous nucleation forms fine particles, or heterogeneous condensation on other fine particles causes enrichment in those fine particles. In addition, the concentrations of As, Cr, Cd, and Pb are much larger in submicrometer particles than in ultramicrometer particles, indicating that these elements are controlled by substituting mechanisms in submicrometer and ultramicrometer particles. The concentration of

Table 7.10 Hg concentration in combustion products.

Boiler	Solid samples (μg/kg)					Liquid samples (μg/L)	
	Fly ash	Bottom ash	Gypsum (fine)	Gypsum (coarse)	WESP ash	FGD effluent	WESP effluent
1	123.9	1.3	125.4	37.2	1945.2	78.2	39.2
2	109.5	0.8	171.2	40.6		89.4	
4	143.1	1.1	137.7	55.1	1541.7	96.9	35.8

FGD, flue-gas desulfurization; *WESP*, wet electrostatic precipitator.
From Zhang Y, Yang J, Yu X, Sun P, Zhao Y, Zhang J, Chen G, Yao H, Zheng C. Migration and emission characteristics of Hg in coal-fired power plant of China with ultra low emission air pollution control devices. Fuel Process. Technol. 2017; 158:272–80.

Mn on the PM10 level is greater than that on the coarse fly ash at the cyclone, indicating that Mn also has a certain volatility during the combustion process. For PM10, the concentration of Mn is almost constant as the particle size decreases.

In recent years, LTE-ESPs have been applied to decrease the ESP inlet gas temperature below the acid dew point using LTE or heat transfer gas, gas heat exchangers, and gas. The temperature is usually below 95°C after installing an LTE. Several studies have shown that LTEs could affect TE release. Zhang et al. [8] studied the migration and emission characteristics of Hg in Chinese coal-fired power plants using an LTE-ESP. As shown in Table 7.10, the Hg concentrations in the bottom ash and fly ash are in the range of 0.8–1.3 and 109.5–143.1 μg/kg, respectively. As shown in Fig. 7.34, the REF of the fly ash is much higher than the REF of the bottom ash.

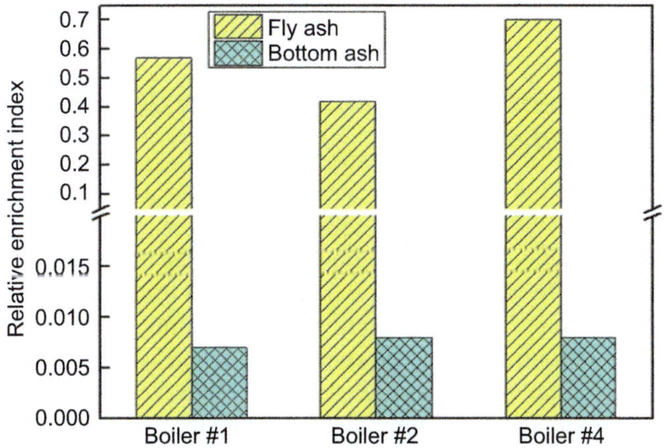

Figure 7.34 Relative enrichment factor in fly ash and bottom ash.
From Zhang Y, Yang J, Yu X, Sun P, Zhao Y, Zhang J, Chen G, Yao H, Zheng C. Migration and emission characteristics of Hg in coal-fired power plant of China with ultra low emission air pollution control devices. Fuel Process. Technol. 2017;158:272–80.

Compared with the REF of fly ash reported in the literature [39,42,43], the REF value in this study is relatively high. Since the inlet temperature of an LLT-ESP was below 95°C, SO_3 and HCl may condense on the fly ash surface to form active adsorption sites and promote adsorption of Hg on the fly ash.

Wang et al. [41] studied the mass distribution of TEs in bottom ash and fly ash in systems equipped with an LTE. Table 7.11 shows that the amounts of TEs in the ESP ash account for over 90% of the total amount in the solid residue, which could be attributed to the high particulate capture efficiency of the ESP. These five elements are volatile during coal combustion, resulting in their consumption in bottom ash. The TE amount at the ESP exit when the LTE is turned on is less than that when the LTE is turned off, suggesting that LTE reduces the TE emissions. This rule is very obvious for particles with sizes >10, 1.0–2.5, and 0.5–1.0 μm, which correspond to the particle mass in each size segment that is turning the LTE on and off.

Mn shows similar REFs at PM0.5–1.0 and PM0.2–0.5, but the REFs for As, Cd, Cr, and Pb at PM0.2–0.5 are significantly higher than those for PM0.5–1.0 (Fig. 7.35). Mn is a nonvolatile element, and the enrichment properties of Mn do not have significant differences in the segments of each particle size. As, Cd, Cr, and Pb are volatile during coal combustion, which is controlled by the evaporative condensation mechanism, and therefore tend to be enriched in the microparticles. The REFs of Mn are almost the same whether the LTE is on or not. The same phenomenon occurs for the REFs of As, Cd, Cr, and Pb at PM0.5–1.0. However, for PM0.2–0.5, enrichment of As, Cd, Cr, and Pb is lower with the LTE turned on compared with when the LTE is turned off. Since As, Cd, Cr, and Pb are mostly retained in the solid residue around 150°C, it is impossible to effectively increase the concentrations of As, Cd, Cr, and Pb in PM at an ESP. The fraction of fine particles in the 0.2–0.5 μm size segment at the ESP outlet with the LTE on is significantly smaller than that with the LTE off, which could lead to the lower enrichment of As, Cd, Cr, and Pb at PM0.2–0.5.

7.3.2 Trace element distribution in flue gas desulfurization effluent and gypsum

Several soluble TEs in the exhaust gas could be removed from the scrubber solution because of the reduction in exhaust gas temperature [39]. As shown in Table 7.12, compared with Hg and Sb, the concentrations of As and Se in gypsum and wastewater are relatively high. In the WFGD system, Se is removed from the gaseous effluent by absorption into limestone slurry as selenite, which has similar structural and chemical properties compared with sulfate. Gypsum and calcium selenate dihydrate ($CaSeO_4 \cdot 2H_2O$) are isomorphic and have very similar lattice constants. Thus, selenate can replace gypsum sulfate. Regarding As, the capture mechanism by limestone slurry is attributed to the formation of $Ca_3(AsO_4)_2$.

To meet the ultralow SO_2 emission standard, the WFGD systems have been converted at several power stations, especially in China's power plants. Zhang et al. [8]. studied Hg concentrations in FGD effluent and gypsum. Hg is apparently enriched

Table 7.11 The mass distribution of trace elements in solid combustion residues (mg/t coal).

Element	LTE	Bottom ash	Fly ash collected by ESP	Fly ash emitted from ESP				
				>10 μm	2.5–10 μm	1.0–2.5 μm	0.5–1.0 μm	0.2–0.5 μm
Solid residues	Off	3.57×10^7	2.03×10^8	12,313	1417	5385	3366	1240
	On			5689	1655	1182	590	1113
As	Off	304.3	9092.6	0.334	0.083	0.766	1.953	3.536
	On			0.155	0.086	0.196	0.370	2.389
Cd	Off	9.2	93.0	0.004	0.002	0.013	0.015	0.021
	On			0.002	0.001	0.002	0.003	0.014
Cr	Off	673.8	8961.8	0.429	0.195	1.224	1.321	0.873
	On			0.199	0.203	0.302	0.257	0.682
Pb	Off	970.6	9336.4	0.490	0.200	1.071	0.729	0.621
	On			0.227	0.210	0.179	0.146	0.442
Mn	Off	5975.7	67,345.6	3.922	0.993	4.008	2.773	0.859
	On			1.818	1.375	0.791	0.440	0.904

ESP, electrostatic precipitator; *LTE*, low-temperature economizer.

From Wang C, Liu X, Li D, Si J, Zhao B, Xu M. Measurement of particulate matter and trace elements from a coal-fired power plant with electrostatic precipitators equipped the low temperature economizer. P. Combust. Inst. 2015;35(3):2793–2800.

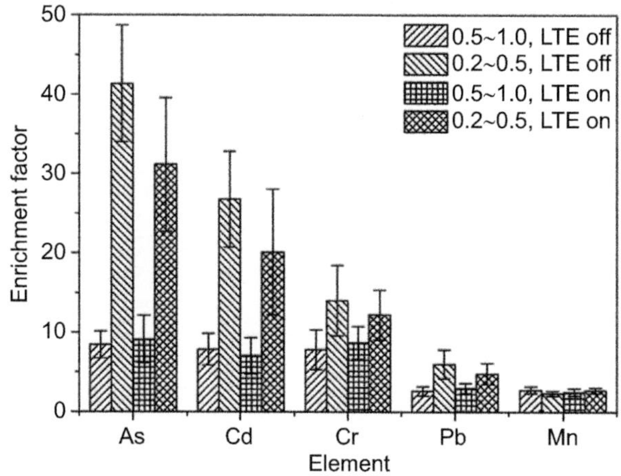

Figure 7.35 The relative enrichment factors of trace elements in PM0.5−1.0 and PM0.2−0.5. *LTE*, low-temperature economizer.
From Wang C, Liu X, Li D, Si J, Zhao B, Xu M. Measurement of particulate matter and trace elements from a coal-fired power plant with electrostatic precipitators equipped the low temperature economizer. P. Combust. Inst. 2015;35(3):2793−2800.

in gypsum, where the concentration in fine gypsum is about 2.51−4.28 times that in coarse gypsum. Despite the fact that the mass fraction of fine gypsum parts is only about 5%, the Hg content accounts for about 12.21%−13.05% of the crude FGD product, which is close to the coarse part (14.16%−19.35%). This is precisely because the microparticles have a higher BET surface area and Hg^{2+} is easily adsorbed. Furthermore, it is possible to capture a part of the fine PM with an FGD system and mix it with gypsum (fine). Generally, fine PM contains a lot of Hg, and the Hg concentration of plaster becomes high. However, as shown in Fig. 7.36, most of the Hg (67.6%−72.9%) remains in FGD effluent. This suggests that the Hg adsorption capacity of gypsum is limited.

Zheng et al. [34] also reported that the concentration of TEs in the fine fraction was higher than the TE concentration in coarse fraction. For plants 2 and 5, the slurry water retained 67% and 72% of the total Hg in the WFGD, respectively. In other plants, the Hg in the liquid phase was in the range of only 0.6%−7%. A part of the Hg^{2+} absorbed in the slurry formed a low-solubility sulfate and was cured in the solid phase. The other part was preserved in the liquid phase, presumably in the $Hg(SO_3)_2^{2-}$, $HgCl_2$, and $Hg(OH)^+$ forms.

Most of As, Cr, and Pb in the slurry migrated to the solid phase (80%−99%), while Cr (62%−76%) and Pb (46%-8%) were distributed in the fine fractions. The particulate-bound Cr and Pb was concentrated in the fine fly ash particles, and the fine ashes escaping from the ESP were then mixed into the solid stream in the WFGD. Since a certain amount of As in the WFGD system was derived from limestone and As_2O_5 in the flue gas tended to react with calcium oxide to form $Ca_2(As_2O_3)$,

Table 7.12 Concentrations of As, Hg, Sb, and Se in samples from the flue gas desulfurization process.

	Limestone (μg/g)	Gypsum (μg/g)	Sludge (μg/g)	Process water (μg/g)	Wastewater (μg/g)
Luohe 2					
As	0.006	2.5	0.36	1.9	9.4
Hg	0.002	0.68	0.6	0.1	51
Sb	0.11	0.115	0.075	0.9	14
Se	2.14	7.4	7.3	1.4	280
Luohe 3					
As	0.005	3.04	3	1.1	14
Hg	0.003	0.65	0.67	0.1	39
Sb	0.1	0.11	0.06	1.3	45
Se	1.98	12.41	13.03	3.6	270
Pingwei 1					
As	1.2	3.21	2	2.4	2
Hg	0.001	0.64	0.65	1.6	4.8
Sb	0.12	0.12	0.07	0.2	8.2
Se	2	11	13	1.2	21
Pingwei 2					
As	1.2	4.81	6	2.4	3.8
Hg	0.001	0.7	0.64	1.6	1.2
Sb	0.12	0.13	0.17	0.2	27
Se	2	12	14	1.2	62

From Tang Q, Liu G, Yan Z, Sun R. Distribution and fate of environmentally sensitive elements (arsenic, mercury, stibium and selenium) in coal-fired power plants at Huainan, Anhui, China. Fuel 2012;95:334–39.

most of the As was concentrated in the coarse proportion. Part of the Se was distributed in the solid phase in significant amounts in slurry water (7%–38%). Cd, which was between 44% and 91%, was highest in the liquid phase compared with other TEs. Water circulation in FGD results in the concentration of Cd. Over 90% of the As, Cr, and Pb was distributed in the gypsum. The amounts of Hg, Se, and Cd accumulated in the waste sludge ranged from 2.6% to 35%, 1%–45%, and 0%–17.8%, respectively. The waste-water retained 0.01%–43% of mercury, 0.8%–7% of Se, and 6%–81% of Cd.

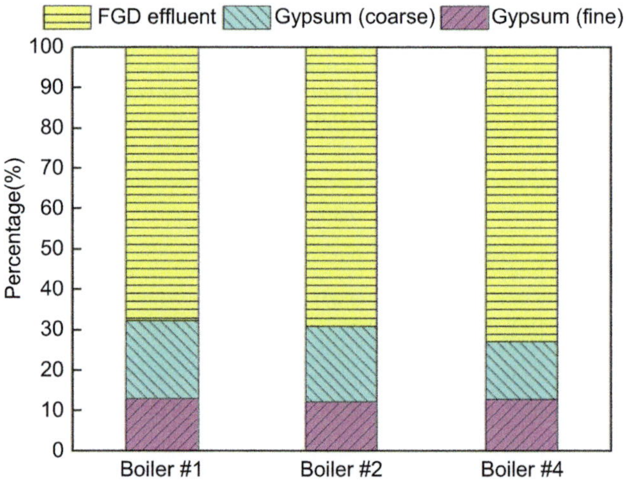

Figure 7.36 Hg partitioning in wet flue gas desulfurization (*FGD*) slurry.
From Zhang Y, Yang J, Yu X, Sun P, Zhao Y, Zhang J, Chen G, Yao H, Zheng C. Migration and emission characteristics of Hg in coal-fired power plant of China with ultra low emission air pollution control devices. Fuel Process. Technol. 2017;158:272–80.

7.3.3 Trace element distribution in wet electrostatic precipitator effluent and ash

WESPs are applied to control the emission of microparticles. WESPs can generate corona electrons that promote the generation of free radicals such as O, OH, O_3, Cl, and Cl_2 from O_2, H_2O, and HCl. These free radicals can oxidize Hg^0 to Hg^{2+}. Therefore, WESPs have a good removal effect on Hg^0, Hg^{2+}, and Hg^p. The Hg concentration in WESP ash ranges from 1541.7 to 1945.2 μg/kg, much higher than in fly ash and gypsum. The WESP ash contains mainly submicrometer PM and unburned carbon particles and Hg is easily enriched in these particle parts. This results in significant enrichment of Hg in WESP ash. Moreover, most of the Hg is included in the WESP effluent due to the solubility of gaseous Hg^{2+}.

Zheng et al. [34] also measured the concentrations of Hg, As, Cd, Se, Cr, and Pb in WESP wastewater and corresponding slag. The concentrations of Hg, As, Cd, Se, Cr, and Pb in the WESP slag were in the ranges 8.67−17.5, 45.2−133, 4.51−9.26, 61.3−121, 274−1303, and 74.9−255 mg/kg, respectively. The Hg concentration in the WESP slag was 16.8−60.1 times the Hg concentration in the fly ash. Clear enrichment of Pb was not observed in WESP slag compared with fly ash. However, the concentration of TEs in WESP wastewater was extremely low, consistent with the work of Wang et al. [44].

7.3.4 Partitioning of Hg and other trace elements in the whole system

Zhao et al. [16] investigated the mercury emissions at three different loads (100%, 85%, 68%) using different types of coal at 350-MW pulverized coal−burning plants equipped with SCR, ESP + FF, and WFGD. As shown in Fig. 7.37, the ratios of

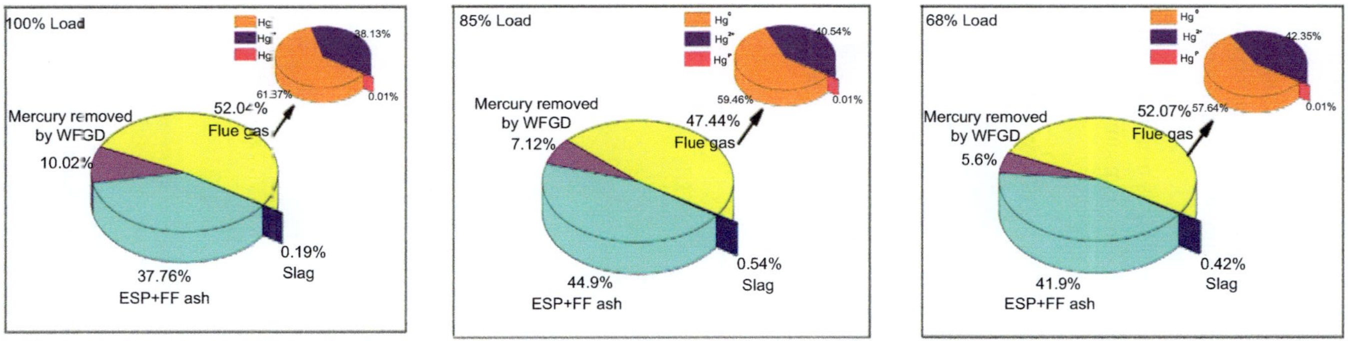

Figure 7.37 Mercury mass distribution across the power plant. *ESP + FF*, electrostatic precipitator plus fabric filter; *WFGD*, wet flue gas desulfurization.
From Zhao S, Duan Y, Chen L, Li Y, Yao T, Liu S, Liu M, Lu J. Study on emission of hazardous trace elements in a 350 MW coal-fired power plant. Part 1. Mercury. Environ. Pollut. 2017;229:863–70.

mercury in flue gas, slag, ESP + FF ash, and WFGD product had the same trend at the three different loads. Compared with the 100% load, more mercury existed in the slag at the loads of 85% and 68%. The lower combustion temperature at the lower loads made the ratio of slag to ash in the coal higher. In addition, it is possible to suppress the release of mercury from the slag and promote the adsorption of mercury to the fly ash at a lower furnace temperature. The ratio of Hg^{2+} in the exhaust gas at 100%, 85%, and 68% loads increased from 38.13% to 42.35%.

Fig. 7.38 shows that the total output of Hg is mainly fly ash (45.3%–57.8%), FGD drainage (11.2%), WESP ash (7.3%–8.4%), gypsum (rough) (2.6%–3.2%), gypsum

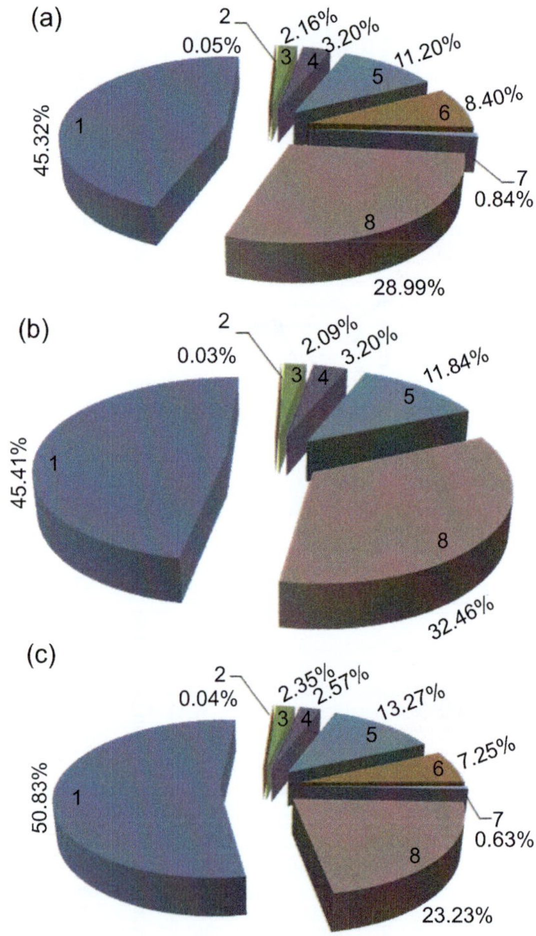

Figure 7.38 Partitioning of Hg in combustion products of a whole system. (a) Boiler 1, (b) boiler 2, (c) boiler 4. *1*, Fly ash; *2*, bottom ash; *3*, gypsum (fine); *4*, gypsum (coarse); *5*, wet electrostatic precipitator (WESP) ash; *6*, flue gas desulfurization effluent; *7*, WESP effluent; *8*, stack gas. From Zhang Y, Yang J, Yu X, Sun P, Zhao Y, Zhang J, Chen G, Yao H, Zheng C. Migration and emission characteristics of Hg in coal-fired power plant of China with ultra low emission air pollution control devices. Fuel Process. Technol. 2017;158:272–80.

(fine) (2.1%–2.4%), WESP ash (0.6%–0.8%), and bottom ash (0.03%–0.05%) [8]. The Hg trapped in ESP fly ash is much higher than that in conventional ESP ash [15,45,46]. This is due to the decrease in flue gas temperature with the LTE equipment, which can improve adsorption of Hg in the fly ash. When the flue gas passes through the WFGD system, about 5% Hg is captured in the gypsum, but still a large portion remains in the FGD effluent. This suggests that most of the Hg^{2+} dissolved in the desulfurized slurry cannot be adsorbed in gypsum. Hg^{2+} trapped in the desulfurized slurry forms an insoluble sulfate. A part of the sulfate can be mixed into the gypsum; the rest remains in the gypsum. Capture of Hg by WESP remains mainly in the effluent, while the relative mass fraction of Hg in WESP ash is low. This is because the concentration of particulates in exhaust gas after passing through ESP and WFGD is very low. Therefore, Hg^{P} captured by WESP occupies only a small percentage. However, the capture of this part of the microparticles is very important, since Hg is considerably enriched in WESP ash and will pose a major risk to human health.

The application of LLT-ESP/EFF, WFGD, and WESP systems was conducive to controlling the emission of TEs. Zheng et al. [34] reported that most of the Hg was concentrated in the range of 58.0%–93.3%. In addition, the dust removal device mode also significantly affects the distribution of mercury in coal combustion products. The third plant's cs-ESP showed the lowest abatement capacity against Hg (58%). This is thought to be due to the maximum inlet temperature of cs-ESP of plant 3 (125°C) among all the plants tested. The ability to remove Hg by cs-ESP was limited, but WESP in the third plant promoted the solidification of mercury in gypsum (40.9%) and a limited amount was retained in the wastewater and sludge. Other TEs were also enriched in fly ash in the ranges 75.2%–95.3% (As), 78.2%–94.9% (Cd), 79.4%–96.6% (Se), 73.8%–89.2% (Cr), and 86.5%–99.5% (Pb).

Fig. 7.39 shows the mass distribution of TEs (Zn, Sb, Pb, Cd, As, Cr, Mn, Ba) in a conventional 320-MW coal-fired power plant equipped with SCR, ESP, and WFGD [47]. There were many TEs in the ESP ash and bottom ash, and the sum of them was 98.75%–99.86%. Most TEs were released into the atmosphere in the exhaust gas and were less than 0.05%. The mass of TEs in the bottom ash was 1.31%–16.39%. In this study, the amounts of Mn and Ba in the bottom ash were greater than the amounts of other studied TEs. The amounts of Zn, Sb, Pb, Cd, and As in the ESP ash were larger than the amounts of Mn and Ba.

In a ULE coal-fired power plant, TEs were mainly distributed in the bottom ash and the ESP ash, in the range 2.11%–12.15% and 87.83%–97.83%, respectively [48]. These results are close to the studies reported by Zheng et al. [34]. Small amounts of TEs are distributed in WFGD and WESP systems, as well as escaping from the stack, the proportions being −0.08%–0.05%, 0.00%–0.06%, and 0.00%–0.10%, respectively. As, Mo, Cd, Pb, etc., are usually bonded with sulfur minerals in coal. During the combustion of coal, sulfur minerals are first decomposed and discharged, and when the flue gas cools, it condenses in fly ash, resulting in a small amount of ash present in the bottom ash but a high proportion of ESP ash. By contrast, Ba, Mn, etc., can be present in separate minerals highly enriched in the ash matrix that can be used to account for small amounts of Cr, Mn, and Ba in ESP ash, distributed among themselves in a large amount of bottom ash.

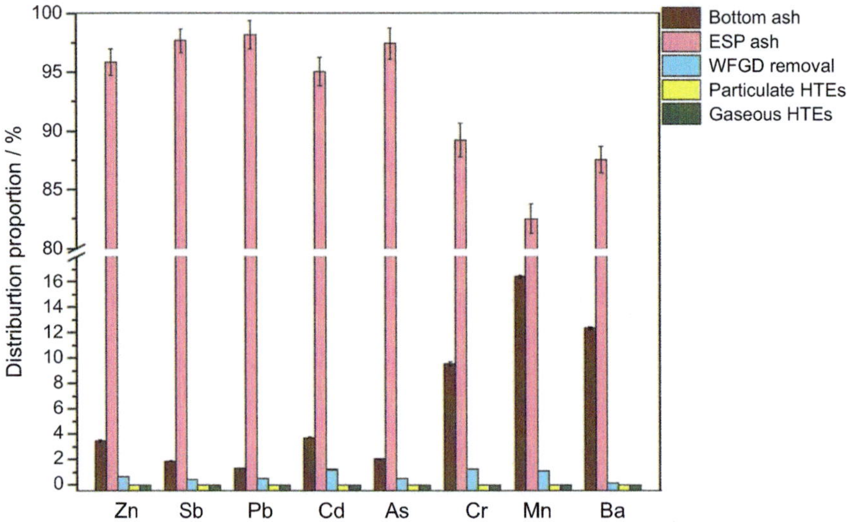

Figure 7.39 Mass distribution of hazardous trace elements (*HTEs*) in a coal-fired power plant. *ESP*, electrostatic precipitator plus fabric filter; *WFGD*, wet flue gas desulfurization.
From Zhao S, Duan Y, Li Y, Liu M, Lu J, Ding Y, Du M. Emission characteristic and transformation mechanism of hazardous trace elements in a coal-fired power plant. Fuel 2018; 214:597–606.

7.4 Co-benefit removal efficiency of trace elements over APCDs

The total elimination efficiencies of the ESP and WFGD systems for Hg, As, and Se are estimated to be 71.4%, 97.3%, and 93.4%, respectively. The higher volatility of Hg compared with As and Se results in a lower removal efficiency. However, little research on the removal efficiency of APCDs for Hg, As, and Se in China has been conducted. Simultaneous mercury removal efficiency of APCDs in a 350-MW pulverized coal–burning power plant equipped with SCR, ESP + FF, and WFGD is shown in Fig. 7.40 [16]. The Hg^0 oxidation efficiency of SCR in power plants at three loads was in the range of 50.13%–67.68%. ESP + FF had a high removal effect on Hg^P at a value of 99.95%–99.97%. WFGD was able to capture 12.64%–26.23% Hg^{2+}. Total removal efficiency with the help of SCR + (ESP + FF) + WFGD at three loads ranged from 58.78% to 73.32%.

The mercury removal in a coal-fired power plant depends on the mercury content in the coal, available oxygen or halogen concentration, residence time of the flue gas, and mercury removal rate of the APCDs. Zhao et al. [49] reported that, in a 660-MW ULE power plant equipped with SCR + ESP + WFGD + WESP, the Hg^0 oxidation efficiency across the SCR is 45.47%. Fig. 7.41 shows that ESPs and WESPs are beneficial for Hg^0 removal. WFGD exhibited an adverse effect on Hg^0 removal due to the rerelease of Hg^0, and its value is −8.27%. For Hg^{2+} removal, WFGD has the best effect

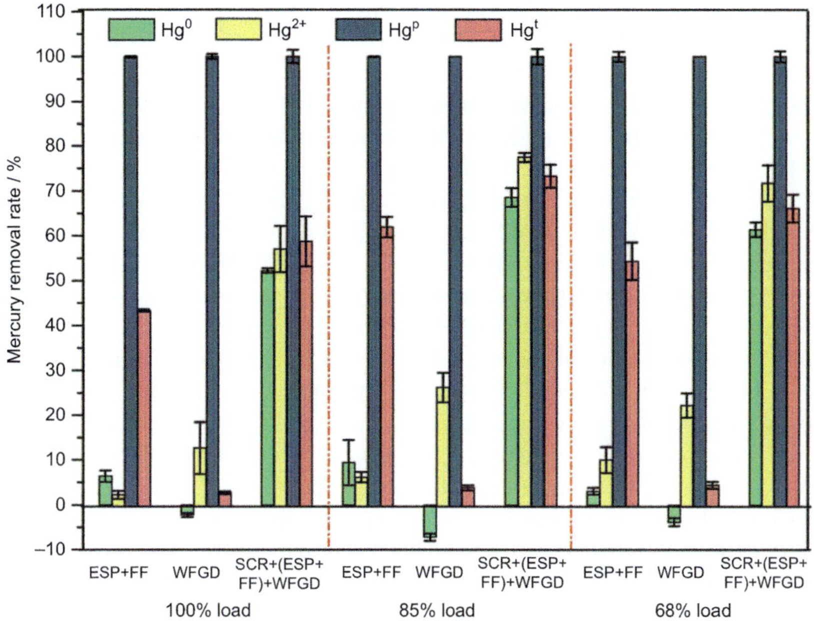

Figure 7.40 Mercury co-removal efficiency across the air pollution control devices in a power plant. *ESP*, electrostatic precipitator; *FF*, fabric filter; *SCR*, selective catalytic reduction; *WFGD*, wet flue gas desulfurization.
From Zhao S, Duan Y, Chen L, Li Y, Yao T, Liu S, Liu M, Lu J. Study on emission of hazardous trace elements in a 350 MW coal-fired power plant. Part 1. Mercury. Environ. Pollut. 2017;229: 863–70.

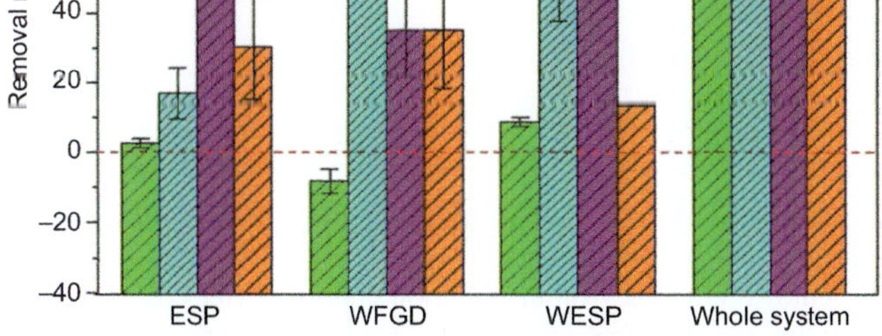

Figure 7.41 The removal rate for gaseous mercury across air pollution control devices. *ESP*, electrostatic precipitator; *WESP*, wet electrostatic precipitator; *WFGD*, wet flue gas desulfurization.
From Zhao S, Duan Y, Yao T, Liu M, Lu J, Tan H, Wang X, Wu L. Study on the mercury emission and transformation in an ultra-low emission coal-fired power plant. Fuel 2017;199:653–61.

with a value of 83.45%. The total system of Hg^{2+} removal is 74.06%. Most Hg^P will be deleted by ESP and WESP. The Hg^T removal rate across the APCDs follows the order of WFGD > ESP > WESP.

Compared with the aforementioned mercury removal efficiency reported by Zhao et al. [49], higher Hg removal efficiencies (88.5%−89.6%) were obtained in a ULE power plant equipped with SCR + LTE + ESP + WFGD + WESP [8]. The total Hg removal efficiencies of SCR + ESP, SCR + ESP + WFGD, and SCR + ESP + WFGD + WESP were 37.2%−49.5%, 74.5%−83.3%, and 88.5%−89.6%, respectively, as shown in Fig. 7.42. The Hg removal efficiency of SCR + ESP + WFGD in boilers 2 and 4 before ULE reformation was 72.8% and 82.2%, which is close to that after ULE reformation. However, after ULE reformation, the total Hg removal efficiency increased by about 15%. The Hg removal efficiency of SCR + ESP clearly increased after ULE reformation. This is thought to be due to the relatively high removal efficiency of ESP, as the SCR did not affect the total Hg removal. Improvement of the Hg removal capability of an ESP should be due to installation of an LTE. As the temperature of the flue gas decreases, the fly ash contains SO_3 and HCl, which is advantageous for adsorption of Hg in the fly ash. This results in an increase in Hg^P concentration and subsequent removal by ESP. The removal efficiency of SCR + ESP + WFGD in power plants in China, the United States, Japan, and

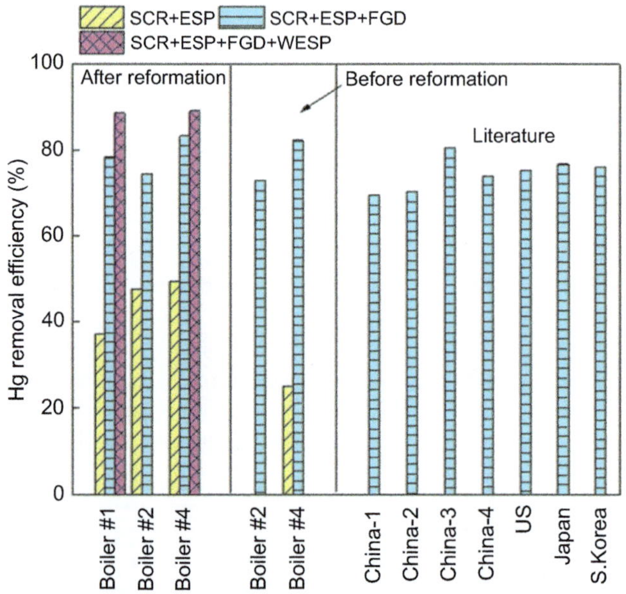

Figure 7.42 Hg removal efficiency of air pollution control devices. *ESP*, electrostatic precipitator; *FGD*, flue gas desulfurization; *SCR*, selective catalytic reduction; *WESP*, wet electrostatic precipitator.
From Zhang Y, Yang J, Yu X, Sun P, Zhao Y, Zhang J, Chen G, Yao H, Zheng C. Migration and emission characteristics of Hg in coal-fired power plant of China with ultra low emission air pollution control devices. Fuel Process. Technol. 2017;158:272−80.

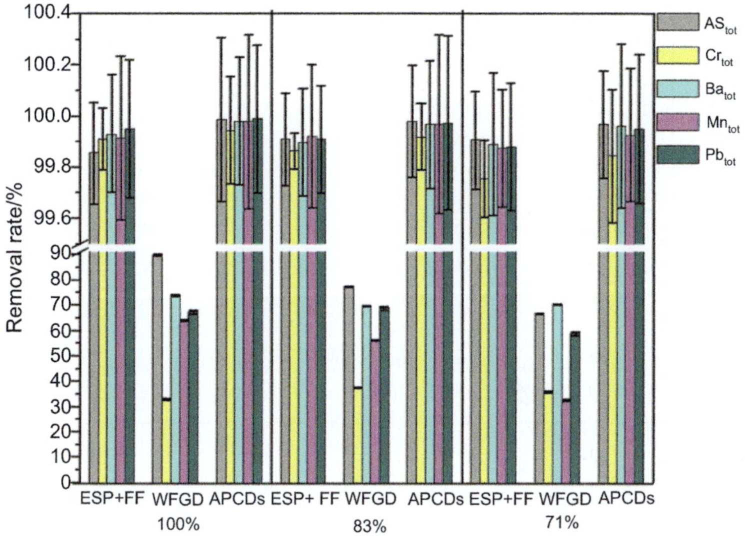

Figure 7.43 Co-removal rates of As, Cr, Ba, Mn, and Pb across the air pollution control devices (*APCDs*) in a power plant. *ESP*, electrostatic precipitator; *FF*, fabric filter; *WFGD*, wet flue gas desulfurization.
From Zhao S, Duan Y, Chen L, Li Y, Yao T, Liu S, Liu M, Lu J. Study on emission of hazardous trace elements in a 350 MW coal-fired power plant. Part 2. arsenic, chromium, barium, manganese, lead. Environ. Pollut. 2017;226:404–11.

Korea was 69.5%–80.5% [20,45,50–53]. This is close to the power plant before the reformation of the ULE. Therefore, in addition to NO_x, PM, and SO_2, co-benefit removal of Hg by APCDs via ULE reformation can be greatly improved.

Co-removal rates of As, Cr, Ba, Mn, and Pb across APCDs in a 350-MW pulverized coal combustion power plant equipped with SCR, ESP + FF, and WFGD are shown in Fig. 7.43 [31]. With three loads, ESP + FF showed a high removal effect in the range of 99.75%–99.95%. WFGD could capture 66.67%–89.91% As, 32.85%–37.30% Cr, 69.49%–73.79% Ba, 32.57%–63.85% Mn, and 58.73%–68.75% Pb. The removal rate of SCR + (ESP + FF) + WFGD was in the range of 99.84%–99.99% and showed a good performance in the simultaneous capture of tested TEs.

Deng et al. [33] investigated the emission of Cd, Pb, and Mn by coal burning at six coal-fired power stations in China. Likewise, WFGD and a particle control device were combined to remove 94.9%–98.2% of Pb, 98.9%–99.9% of Cd, and 99.9% of Mn in the flue gases. As shown in Table 7.13, the particle-bound trace metals are predominant in the exhaust gas before the particulate settling agent (ESP or FF). After passing through the particulate sediment, the particle-bound trace metals in the flue gases are significantly reduced. The removal efficiency of Cd_p, Pb_p, and Mn_p of the precipitator is 94% or more. The removal rate of gaseous metal is relatively low, ranging from 48.5% to 97.6%. In the case of WFGD, in particular with Pb and Cd, the retention of gaseous metals is relatively high compared with the granular form.

Table 7.13 Concentrations of Pb, Cd, and Mn in flue gas along the flue gas path and their removal.

Power plant	Element in flue gas	Particulate capture device			Wet FGD		Overall removal rate (%)
		In (μg/Nm^3)	Out (μg/Nm^3)	Removal rate (%)	Out (μg/Nm^3)	Removal rate (%)	
Plant 1	Cd_p	132.6 ± 15.4	0.2 ± 0.0	99.9	0.1 ± 0.0	50.0	99.9
	Cd_g	13.8 ± 2.5	0.3 ± 0.0	97.6	0.1 ± 0.0	66.7	
	Pb_p	3060.5 ± 142.1	134.5 ± 23.7	95.6	49.3 ± 5.6	63.4	98.0
	Pb_g	440.2 ± 32.6	181.9 ± 20.3	58.5	20.9 ± 3.8	88.5	
	Mn_p	12,787.4 ± 801.3	18.4 ± 2.6	99.9	4.8 ± 0.5	73.9	99.9
	Mn_g	587.6 ± 92.5	58.0 ± 4.7	90.1	10.1 ± 2.3	82.6	
Plant 2	Cd_p	137.6 ± 27.13	0.5 ± 0.0	99.6	–	–	99.5
	Cd_g	1.7 ± 0.6	0.2 ± 0.0	87.2	–	–	
	Pb_p	3630 ± 243.8	108.9 ± 46.3	97.0	–	–	96.3
	Pb_g	570 ± 62.4	45.7 ± 6.9	92.0	–	–	
	Mn_p	6652.2 ± 431.5	5.8 ± 0.7	99.9	–	–	99.9
	Mn_g	56.6 ± 3.9	24.0 ± 3.5	57.6	–	–	
Plant 3	Cd_p	39.6 ± 2.5	0.1 ± 0.0	99.9	–	–	99.9
	Cd_g	3.4 ± 0.2	0.8 ± 0.1	76.3	0.3 ± 0.0	56.6	
	Pb_p	1450 ± 89.3	40.8 ± 6.4	97.2	24.8 ± 1.3	39.2	98.2
	Pb_g	230 ± 56.70	28.9 ± 7.1	87.4	5.6 ± 1.5	80.6	
	Mn_p	4237.5 ± 532.8	19.4 ± 2.9	99.5	10.1 ± 0.2	47.9	99.9
	Mn_g	302.2 ± 19.7	49.6 ± 6.5	83.6	17.8 ± 0.9	64.1	

Plant 4	Cd_p	22.8 ± 3.4	0.2 ± 0.0	99.4	0.1 ± 0.0	50.0	98.9
	Cd_g	5.5 ± 0.6	0.3 ± 0.0	94.4	0.2 ± 0.0	33.3	
	Pb_p	2530. ± 94.8	103.4 ± 7.8	95.9	79.7 ± 5.3	22.9	96.4
	Pb_g	240. ± 35.6	52.2 ± 6.8	78.0	20.4 ± 1.4	60.9	
	Mn_p	2347.4 ± 107.5	22.4 ± 3.5	99.1	3.4 ± 0.7	84.8	99.9
	Mn_g	153.4 ± 26.7	26.1 ± 3.4	83.0	30.3 ± 1.8	–	
Plant 5	Cd_p	21.5 ± 3.2	0.1 ± 0.0	99.8	0.1 ± 0.0	93.0	99.3
	Cd_g	6.2 ± 0.4	0.4 ± 0.0	93.5	0.1 ± 0.0	75.0	
	Pb_p	2370 ± 64.3	118.1 ± 23.5	95.0	91.4 ± 10.3	22.6	94.9
	Pb_g	360. ± 4.7	153.0 ± 21.6	57.9	47.2 ± 11.5	69.2	
	Mn_p	2464.2 ± 175.6	6.3 ± 0.1	99.7	4.2 ± 0.7	33.3	99.9
	Mn_g	165.6 ± 23.1	38.9 ± 1.8	76.5	52.1 ± 35.4	–	
Plant 6	Cd_p	204.6 ± 37.5	0.1 ± 0.0	99.9	–	–	99.4
	Cd_g	14.3 ± 2.6	2.2 ± 0.3	84.4	1.2 ± 0.2	45.4	
	Pb_p	5760. ± 253.2	336.4 ± 17.3	94.2	254.7 ± 36.8	24.3	95.0
	Pb_g	570. ± 121.7	295.6 ± 16.9	48.5	65.6 ± 7.4	77.8	
	Mn_p	10,156.6 ± 382.9	2.4 ± 0.5	99.9	1.6 ± 0.5	33.3	99.9
	Mn_g	461.8 ± 57.4	158.5 ± 23.6	65.7	6.3 ± 0.3	96.0	

Subscript g indicates gas phase; subscript p indicates particulate bound. *FGD*, flue gas desulfurization.
From Deng S, Shi Y, Liu Y, Zhang C, Wang X, Cao Q, Li S, Zhang F. Emission characteristics of Cd, Pb and Mn from coal combustion: Field study at coal-fired power plants in China. Fuel Process. Technol. 2014;126:469–75.

Zhao et al. [48] studied the TE removal efficiency of SCR, ESP, WFGD, and WESP in a 660-MW ULE power plant. The total TE removal rate was higher than 99.90% except for Cr, which was 99.87%. In the flue gas before ESP, particulate TEs were the main form. Therefore, the high dust efficiency removal performance of ESP brings about a large TE removal effect. A TE removal rate exceeding 60% in WFGD was for Co, Ni, Zn, Sb, and Ba. Some TEs such as Co, Ni, Zn, and Sb are present in the chloride form, which is readily soluble in water. The oxidation state, e.g., CrO_3 and Cr_2O_3, is the main form of Cr in the flue gases. Therefore, the removal capacity of Cr by WFGD is limited. On the other hand, PM is trapped by WFGD, and TE removal efficiency improves. More than 60% of Mn, Co, As, Ag, Sb, Ba, and Pb was removed by WESP. This was under a lower temperature of about 50°C under conditions of more steam. All TEs in the flue gas after WESP were gaseous. The total removal rate of TEs over ESP + WFGD + WESP was greater than 99.90%, indicating that this ULE coal-fired power plant had an excellent trapping effect on TEs.

7.5 Atmospheric emission factor of trace elements

Emission factors (EFs), which are defined as the ratio of the final mercury released into the atmosphere to the calorific value from the coal supplied, were used to compare the emissions under different conditions. The EF for mercury at the three loads of the 350-MW coal-fired power plant with SCR, ESP + FF, and WFGD ranged from 0.92 to $1.17 g/10^{12}$ J [16]. The EF of this study was smaller than the average of China and the United States. This is because the mercury content of Chinese coal is low and the halogen content may be low, so related departments should make efforts to suppress mercury emitted from power plants.

The ULE reformation with the equipment of SCR + ESP + WFGD + WESP could significantly reduce the emission of Hg into the atmosphere [8]. As shown in Fig. 7.44, the EF of SCR + ESP + WFGD is 0.81 g/TJ (boiler 2) and of SCR + ESP + WFGD + WESP is 0.47 g/TJ (boiler 1) or 0.39 g/TJ (boiler 4). For comparison, the EFs of boilers 2 and 4 before ULE reformation were also examined. The EF of boiler 2 was 2.34 g/TJ and the EF of boiler 4 was 2.18 g/TJ. The EF of this plant before ULE reform in this study is close to the scope of the US power plant. After the ULE reform, the EF of the power plant dramatically decreased. The Hg emissions from power plants are complex and depend on the type of coal, the type of boiler, the operating conditions, and the APCD changes. Therefore, a more comprehensive search is needed to understand the sequestration and emissions of Hg from ULE power plants in China.

Fig. 7.45 gives the EFs of As, Cr, Ba, Mn, and Pb at three loads in a 350-MW power plant [31]. As and Pb have small EFs of 0.04–0.09 and $0.07–0.09 g/10^{12}$ J, respectively. Ba and Mn have a relatively large amount of light emission, and their EFs are in the range of 1.37–1.54 and $0.92–1.49 g/10^{12}$ J, respectively. The EF of Cr is in the range of $0.30–0.52 g/10^{12}$ J. Further attention should be paid to the estimation of TE emissions in China based on the fact that there is little work on the EFs of TEs emitted from power plants.

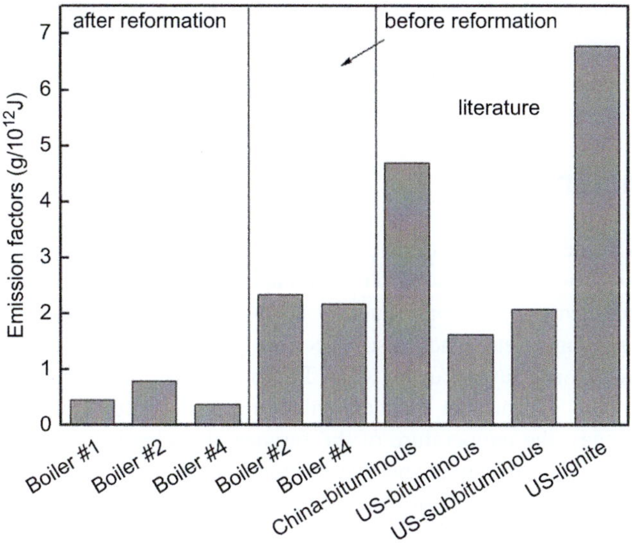

Figure 7.44 Hg emission factors from power plants before and after ultralow emission reformation and from US power plants.
From Zhang Y, Yang J, Yu X, Sun P, Zhao Y, Zhang J, Chen G, Yao H, Zheng C. Migration and emission characteristics of Hg in coal-fired power plant of China with ultra low emission air pollution control devices. Fuel Process. Technol. 2017;158:272–80.

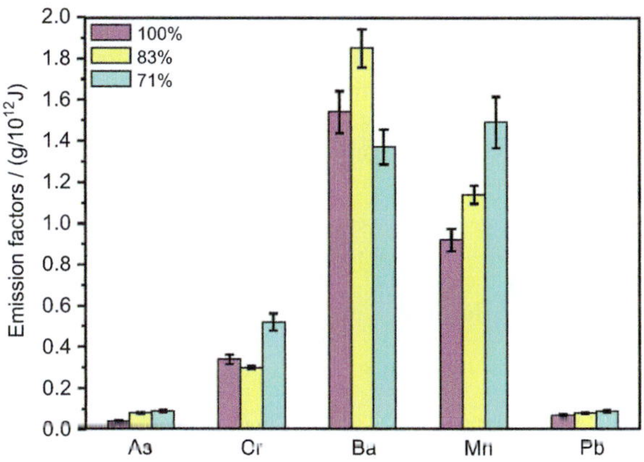

Figure 7.45 Emission factors at three loads in a power plant.
From Zhao S, Duan Y, Chen L, Li Y, Yao T, Liu S, Liu M, Lu J. Study on emission of hazardous trace elements in a 350 MW coal-fired power plant. Part 2. arsenic, chromium, barium, manganese, lead. Environ. Pollut. 2017;226:404–11.

ULE reformation for conventional power plants, like installing WESPs, could effectively reduce the EFs of TEs. Co, Ni, Cu, As, Mo, Ag, Cd, Sb, Ba, and Pb discharged via the stack are lower than 3 mg/t of coal; Zn is 6.58 mg/t of coal. Compared with other TEs, Cr and Mn have relatively high values of approximately 11 mg/t of coal [48].

In contrast to Pb, Cd, and Mn released from the six Chinese coal-fired power plants, the TEs emitted from this power plant are extremely low. A coal-fired power plant equipped with SCR + ESP + WFGD + WESP can effectively control TE emissions.

7.6 Summary

Many factors influence the movement and conversion behavior of TEs in the coal combustion process, such as the generation of TEs in coal, their evaporation characteristics, flue gas components, oxidation/reduction atmosphere, condensation/adsorption between TE compounds, etc. They also affect the potential fly ash surface and so on. In addition, reactions may occur between TE compounds. This problem leads to the complexity of the TE moving mechanism during coal combustion. As the load on the boiler changes, the temperature of the furnace also changes. It ultimately affects factors that affect TE emissions and distribution in coal-fired power plants. To better understand and remove the TEs of the coal-fired power plant, it is necessary to conduct even more research on the following:

(1) Distribution and enrichment characteristics of TEs in coal, ash, and flue gases: TE emissions characteristics in typical coal-fired power plants should be systematically tested, including emissions in flue gases. Concentration should be focused on ESP ash, FGD gypsum, and the generation and content of TEs in FGD wastewater.
(2) Conversion mechanisms of TEs traversing the furnace, SCE, ESP, WFGD, etc.
(3) Management methods of TEs in coal-fired power plants: We should investigate the removal effect of TEs by fly ash itself, additives added to coal, and reaction mechanisms between solid adsorbents and TEs. An integrated control technology for TEs and other pollutants with very efficient and low-input quantities should be developed.

Acknowledgments

This project was supported by the National Natural Science Foundation of China (51776227), Natural Science Foundation of Hunan Province, China (2018JJ1039, 2018JJ3675).

References

[1] Skodras G, Diamantopoulou I, Pantoleontos G, Sakellaropoulos G. Kinetic studies of elemental mercury adsorption in activated carbon fixed bed reactor. J. Hazard. Mater. 2008;158(1):1−13.
[2] Zhao L, Li C, Wang Y, Wu H, Gao L, Zhang J, Zeng G. Simultaneous removal of elemental mercury and NO from simulated flue gas using a CeO_2 modified $V_2O_5-WO_3$/TiO_2 catalyst. Catal. Sci. Technol. 2016;6:6076−86.
[3] Ding Z, Li L, Wade D, Gloyna E. Supercritical water oxidation of NH_3 over a MnO_2/CeO_2 catalyst. Ind. Eng. Chem. Res. 1998;37(5):1707−16.

[4] Zhao L, Li C, Li S, Wang Y, Zhang J, Wang T, Zeng G. Simultaneous removal of elemental mercury and NO in simulated fluegas over V_2O_5/ZrO_2-CeO_2 catalyst. Appl. Catal. B: Environ. 2016;198:420−30.

[5] Li H, Zhao J, Zhang W, Yang J, Wang J, Zhang M, Yang Z, Li L, Shih K. NH_3 inhibits mercury oxidation over low-temperature MnO_x/TiO_2 SCR catalyst. Fuel Process. Technol. 2018;176:124−30.

[6] Qi G, Yang R, Chang R. MnO_x-CeO_2 mixed oxides prepared by co-precipitation for selective catalytic reduction of NO with NH_3 at low temperatures. Appl. Catal. B: Environ. 2004;51(2):93−106.

[7] Xu Y, Xue J, Wang H, Li B, Guan Y, Liu J. Research on mercury collaborative control by conventional pollutants purification facilities of coal-fired power plants. Proc. Chin. Soc. Electrical. Eng. 2014;34:3924−31.

[8] Zhang Y, Yang J, Yu X, Sun P, Zhao Y, Zhang J, Chen G, Yao H, Zheng C. Migration and emission characteristics of Hg in coal-fired power plant of China with ultra low emission air pollution control devices. Fuel Process. Technol. 2017;158:272−80.

[9] Zhuang Y, Laumb J, Liggett R, Holmes M, Pavlish J. Impacts of acid gases on mercury oxidation across SCR catalyst. Fuel. Process. Technol. 2007;88(10):929−34.

[10] Eswaran S, Stenger H. Effect of halogens on mercury conversion in SCR catalysts. Fuel. Process. Technol. 2008;89(11):1153−9.

[11] Li H, Wu C, Li Y, Li L, Zhao Y, Zhang J. Impact of SO_2 on elemental mercury oxidation over CeO_2−TiO_2 catalyst. Chem. Eng. J. 2013;219:319−26.

[12] Li H, Li Y, Wu C-Y, Zhang J. Oxidation and capture of elemental mercury over SiO_2−TiO_2−V_2O_5 catalysts in simulated low-rank coal combustion flue gas. Chem. Eng. J. 2011; 169(1−3):186−93.

[13] Kamata H, Ueno S, Naito T, Yukimura A. Mercury oxidation over the $V_2O_5(WO_3)/TiO_2$ commercial SCR catalyst. Ind. Eng. Chem. Res. 2008;47(21):8136−41.

[14] Zhao B, Liu XW, Zhou ZJ, Shao HZ, Wang C, Xu MH. Mercury oxidized by V_2O_5−MoO_3/TiO_2 under multiple components flue gas: an actual coal-fired power plant test and a laboratory experiment. Fuel Process.Technol. 2015;134:198−204.

[15] Zhou Z, Liu X, Zhao B, Chen Z, Shao H, Wang L, Xu M. Effects of existing energy saving and air pollution control devices on mercury removal in coal-fired power plants. Fuel. Process. Technol. 2015;131:99−108.

[16] Zhao S, Duan Y, Chen L, Li Y, Yao T, Liu S, Liu M, Lu J. Study on emission of hazardous trace elements in a 350 MW coal-fired power plant. Part 1. Mercury. Environ. Pollut. 2017; 229:863−70.

[17] James D, Kilgroe C, Sedman R. Control of mercury emissions from coal-fired electric Utility boilers: interim report. 2002. EPA-600.

[18] Chen L, Duan Y, Zhuo Y, Yang L, Zhang L, Yang X, Yao Q, Jiang Y, Xu X. Mercury transformation across particulate control devices in six power plants of China: the co-effect of chlorine and ash composition. Fuel 2007;86(4):603−10.

[19] Zhang S, Zhao Y, Wang Z, Zhang J, Wang L, Zheng C. Integrated removal of NO and mercury from coal combustion flue gas using manganese oxides supported on TiO_2. J. Environ. Sci. 2017;53:141−50.

[20] Pudasainee D, Kim J, Yoon Y, Seo Y. Oxidation, reemission and mass distribution of mercury in bituminous coal-fired power plants with SCR, CS-ESP and wet FGD. Fuel 2012;93:312−8.

[21] Li H, Wu S, Li L, Wang J, Ma W, Shih K. CuO−CeO_2/TiO_2 catalyst for simultaneous NO reduction and Hg^0 oxidation at low temperatures. Catal. Sci. Technol. 2015;5(12): 5129−38.

[22] Ochoa González R, Díaz-Somoano M, López Antón MA, Martínez-Tarazona MR. Effect of adding aluminum salts to wet FGD systems upon the stabilization of mercury. Fuel 2012;96:568–71.
[23] Córdoba P, Font O, Izquierdo M, Querol X, Tobías A, López-Antón MA, Ochoa-Gonzalez R, Díaz-Somoano M, Martínez-Tarazona MR, Ayora C, Leiva C, Fernández C, Giménez A. Enrichment of inorganic trace pollutants in re-circulated water streams from a wet limestone flue gas desulphurisation system in two coal power plants. Fuel Process. Technol. 2011;92(9):1764–75.
[24] Ma S, Zhao Y, Yang J, Zhang S, Zhang J, Zheng C. Research progress of pollutants removal from coal-fired flue gas using non-thermal plasma. Renew. Sust. Energ. Rev. 2016;67:791–810.
[25] Wang Z, Jiang S, Zhu Y, Zhou J, Zhou J, Li Z, Cen K. Investigation on elemental mercury oxidation mechanism by non-thermal plasma treatment. Fuel. Process. Technol. 2010; 91(11):1395–400.
[26] Ko K, Byun Y, Cho M, Namkung W, Shin D, Koh D, Kim K. Influence of HCl on oxidation of gaseous elemental mercury by dielectric barrier discharge process. Chemosphere 2008;71(9):1674–82.
[27] Chen Z, Mannava D, Mathur V. Mercury oxidization in dielectric barrier discharge plasma system. Ind. Eng. Chem. Res. 2006;45(17):6050–5.
[28] An J, Shang K, Lu N, Jiang Y, Wang T, Li J, Wu Y. Oxidation of elemental mercury by active species generated from a surface dielectric barrier discharge plasma reactor. Plasma. Chem. Plasma. 2014;34:217–28.
[29] Reynold J, Bayless D, Caine J. Multi-pollutant control using membrane-based up-flow wet electrostatic precipitation. Akron. Ohio: Ohio University; 2004. DOE. Final. Report.
[30] Wang J, Zhang Y, Liu Z, Norris P, Romero C, Xu H, Pan W. Effect of coordinated air pollution control devices in coal-fired power plants on arsenic emissions. Energy Fuels 2017;31(7):7309–16.
[31] Zhao S, Duan Y, Chen L, Li Y, Yao T, Liu S, Liu M, Lu J. Study on emission of hazardous trace elements in a 350 MW coal-fired power plant. Part 2. arsenic, chromium, barium, manganese, lead. Environ. Pollut. 2017;226:404–11.
[32] Pan S, Xu Y, Wang X, Zhou D, Zhang K, Liu X. Effects of SCR DeNOx on the PM and trace elements. J. Chin. Coal. Soc. 2015;40:2702–7.
[33] Deng S, Shi Y, Liu Y, Zhang C, Wang X, Cao Q, Li S, Zhang F. Emission characteristics of Cd, Pb and Mn from coal combustion: field study at coal-fired power plants in China. Fuel Process. Technol. 2014;126:469–75.
[34] Zheng C, Wang L, Zhang Y, Zhang J, Zhao H, Zhou J, Gao X, Cen K. Partitioning of hazardous trace elements among air pollution control devices in ultra-low-emission coal-fired power plants. Energy Fuels 2017;31(6):6334–44.
[35] Córdoba P, Ochoa-Gonzalez R, Font O, Izquierdo M, Querol X, Leiva C, López-Antón MA, Díaz-Somoano M, Rosa Martinez-Tarazona M, Fernandez C, Tomás A. Partitioning of trace inorganic elements in a coal-fired power plant equipped with a wet Flue Gas Desulphurisation system. Fuel 2012;92(1):145–57.
[36] Alvarez-Ayuso E, Querol X, Tomas A. Environmental impact of a coal combustion-desulphurisation plant: abatement capacity of desulphurisation process and environmental characterisation of combustion by-products. Chemosphere 2006;65(11):2009–17.
[37] Senior C, Tyree C, Meeks N, Acharya C, McCain J, Cushing K. Selenium partitioning and removal across a wet FGD scrubber at a coal-fired power plant. Environ. Sci. Technol. 2015;49(24):14376–82.

[38] Córdoba P. Partitioning and speciation of selenium in wet limestone flue gas desulphurisation systems: a review. Fuel 2017;202:184–95.
[39] Tang Q, Liu G, Yan Z, Sun R. Distribution and fate of environmentally sensitive elements (arsenic, mercury, stibium and selenium) in coal-fired power plants at Huainan, Anhui, China. Fuel 2012;95:334–9.
[40] Guo X, Zheng C, Xu M. Characterization of arsenic emissions from a coal-fired power plant. Energy Fuels 2004;18(6):1822–6.
[41] Wang C, Liu X, Li D, Si J, Zhao B, Xu M. Measurement of particulate matter and trace elements from a coal-fired power plant with electrostatic precipitators equipped the low temperature economizer. P. Combust. Institute. 2015;35(3):2793–800.
[42] López-Antón M, Díaz-Somoano M, Ochoa-González R, Martínez-Tarazona M. Distribution of trace elements from a coal burned in two different Spanish power stations. Ind. Eng. Chem. Res. 2011;50(21):12208–16.
[43] Bhangare R, Ajmal P, Sahu S, Pandit G, Puranik V. Distribution of trace elements in coal and combustion residues from five thermal power plants in India. Inter. J. Coal. Geol. 2011; 86(4):349–56.
[44] Wang L, Zheng C, Zhang Y, Yue T, Weng W, Zhao H, Zhou J, Gao X. Speciation characteristics and mobility of trace elements across ultralow emission air pollution control devices. Energy Fuels 2017;31(12):13963–71.
[45] Ballestero D, Gomez-Gimenez C, Garcia-Diez E, Juan R, Rubio B, Izquierdo M. Influence of temperature and regeneration cycles on Hg capture and efficiency by structured Au/C regenerable sorbents. J. Hazard. Mater. 2013;260:247–54.
[46] Zhang L, Zhuo Y, Chen L, Xu X, Chen C. Mercury emissions from six coal-fired power plants in China. Fuel Process. Technol. 2008;89(11):1033–40.
[47] Zhao S, Duan Y, Li Y, Liu M, Lu J, Ding Y, Du M. Emission characteristic and transformation mechanism of hazardous trace elements in a coal-fired power plant. Fuel 2018; 214:597–606.
[48] Zhao S, Duan Y, Tan H, Liu M, Wang X, Wu L, Wang C, Lv J, Yao T, She M, Tang H. Migration and emission characteristics of trace elements in a 660 MW coal-fired power plant of China. Energy Fuels 2016;30(7):5937–44.
[49] Zhao S, Duan Y, Yao T, Liu M, Lu J, Tan H, Wang X, Wu L. Study on the mercury emission and transformation in an ultra-low emission coal-fired power plant. Fuel 2017; 199:653–61.
[50] Tang H, Duan Y, Zhu C, Li C, She M, Zhou Q, Cai L. Characteristics of a biomass-based sorbent trap and its application to coal-fired flue gas mercury emission monitoring. Inter. J. Coal Geol. 2017;170:19–27.
[51] Pudasainee D, Kim J, Seo Y. Mercury emission trend influenced by stringent air pollutants regulation for coal-fired power plants in Korea. Atmos. Environ. 2009;43(39):6254–9.
[52] Pudasainee D, Seo Y, Sung J, Jang H, Gupta R. Mercury co-beneficial capture in air pollution control devices of coal-fired power plants. Inter. J. Coal Geol. 2017;170:48–53.
[53] Tian H, Wang Y, Xue Z, Qu Y, Chai F, Hao J. Atmospheric emissions estimation of Hg, As, and Se from coal-fired power plants in China, 2007. Sci. Total. Environ. 2011;409(16): 3078–81.

Sorbents for trace elements in coal-derived flue gas

8

Jianping Yang[1], *Qin Li*[1], *Jiexia Zhao*[1], *Yongchun Zhao*[2], *Junying Zhang*[2]
[1]School of Energy Science and Engineering, Central South University, Changsha, China;
[2]State Key Laboratory of Coal Combustion, Huazhong University of Science and Technology, Wuhan, China

8.1 Introduction

China is the world's largest coal producer and consumer. The emission and pollution of trace elements (TEs) have raised wide attention worldwide. Mercury is a poisonous metallic material with bioaccumulation and persistency behaviors. Other TEs like Cr, Ni, Pb, Zn, Cu, V, and Cd are carcinogenic elements. Excess fluoride can cause fluorosis in bones and teeth. High concentrations of Se can damage human organisms, causing hair loss and nail loss, as well as nervous system disorders. Therefore it is important to develop effective TE control technologies. This chapter provides an overview of TE control strategies during the coal-burning process.

8.2 Sorbents for capturing Hg in coal-derived flue gas

8.2.1 Hg removal by activated carbon

Activated carbon (AC) is an excellent sorbent for mercury either in lab-scale or full-scale systems. The development and commercialization of activated carbon injection (ACI) technology has been achieved through close collaboration between the Department of Energy (DOE), the Electric Power Research Institute, and a group of utility companies through a series of pilot programs.

8.2.1.1 Scheme of activated carbon injection technology

Different strategies have been proposed for ACI technology:

> Scheme A: Powered AC is pneumatically injected from a storage silo into the flue gas ductwork. AC adsorbs mercury and is then collected with fly ash in the particulate collector. Fabric filters (FFs) are more effective than electrostatic precipitators (ESPs) for ACI technology, since FFs can increase the contact time between mercury and AC particles.
> Scheme B: The flue gas was first discharged by spray cooling downstream of the ESP. AC is injected into the duct and then collected in the FF. Since low temperature is favorable for Hg adsorption over virgin AC, cooling flue gas can lower the AC dosage for Hg removal.
> Scheme C: Flue gas is first sprayed downstream of the preheater. AC is injected into the duct for capturing mercury and then collected in the particulate collector (ESP/FF).

Emission and Control of Trace Elements from Coal-Derived Gas Streams
https://doi.org/10.1016/B978-0-08-102591-8.00008-8

Table 8.1 Brunauer–Emmett–Teller (BET) surface areas and bulk densities of various activated carbons (ACs).

AC	BET surface area (m^2/g)	Bulk density (g/cm^3)
TX-104T (coconut)	1192	0.4998
TX-104D (coconut)	1114	0.4529
TX-104G (coconut)	824	0.4512
Fruit shell	711	0.5200
Bamboo	863	0.2936
Coal	859	0.4848

From Wu S, Yang W, Zhou J, Wang H, Xie Z. Effects of properties of activated carbon on its activity for mercury removal and mercury desorption from used activated carbons. Energy Fuels 2015;29(3):1946–50.

The mercury adsorption capacity of AC significantly relies on physical and chemical properties. Flue gas components such as HCl, Cl_2, SO_x, NO_x, H_2O, and O_2 also affect mercury removal, with competitive or beneficial effects. To gain an overview of the impact factor for AC adsorption, each one will be discussed.

8.2.1.2 Effect of physical properties of AC

The physical–chemical properties of AC significantly affect mercury removal performance [1]. Table 8.1 shows the Brunauer–Emmett–Teller (BET) surface areas and bulk densities of various ACs. Although TX-104G, AC bamboo, and fruit shell AC had similar BET surface areas, TX-104G and coal AC showed higher mercury adsorption capacity than the bamboo and fruit shell ACs (Fig. 8.1). Fig. 8.2 shows that the BET surface area of AC had no significant impact on mercury adsorption capacity.

Besides the BET surface area, pore size distribution is another important factor affecting mercury adsorption on AC. As shown in Figs. 8.3 and 8.4, pore distribution is independent of the mercury removal capacity. In contrast, in the case of coconut ACs (Figs. 8.1 and 8.2), the order of mercury removal capacity is TX-104G > TX-104T > TX-104D, which is consistent with the pore volume order in the range 100–8000 nm. Combining these results with those of the foregoing BET surface area, it is apparent that both pore structure and BET surface have no significant relationship with mercury removal performance. Hsi et al. [2] also reported that the BET surface of ACs had no relationship with its mercury removal rate, and the adsorbent is mainly limited to chemical properties.

In the presence of acid flue gases, the physical properties of AC show different effects on mercury adsorption performance [3]. Mercury adsorption efficiency shows a linear increase to BET surface area under baseline gas conditions. The number of mesopores is less correlated with mercury adsorption than the BET surface area.

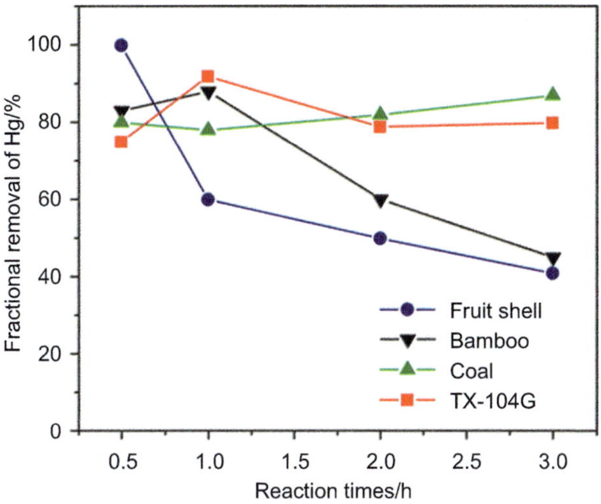

Figure 8.1 Mercury removal abilities of activated carbons derived from various sources.
From Wu S, Yang W, Zhou J, Wang H, Xie Z. Effects of properties of activated carbon on its activity for mercury removal and mercury desorption from used activated carbons. Energy Fuels 2015;29(3):1946–50.

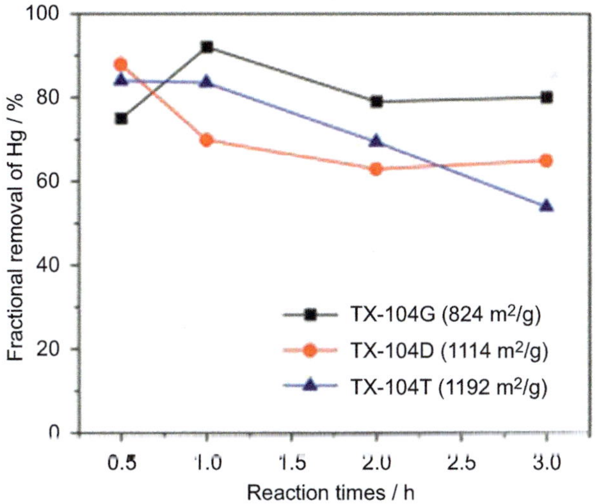

Figure 8.2 Effect of Brunauer–Emmett–Teller surface area on the mercury removal abilities of coconut activated carbons.
From Wu S, Yang W, Zhou J, Wang H, Xie Z. Effects of properties of activated carbon on its activity for mercury removal and mercury desorption from used activated carbons. Energy Fuels 2015;29(3):1946–50.

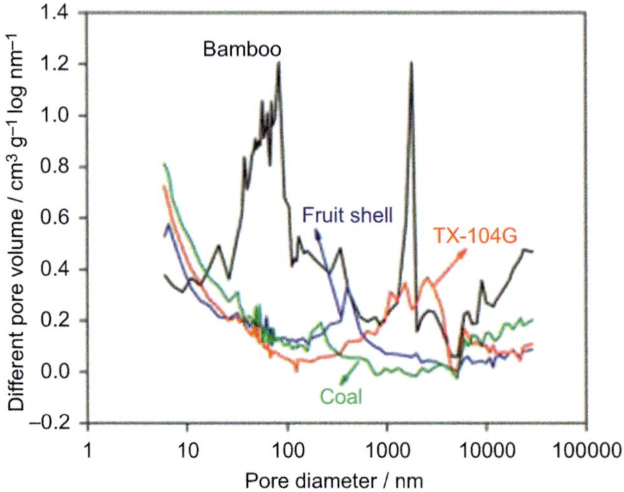

Figure 8.3 Pore size distribution of activated carbons derived from various sources.
From Wu S, Yang W, Zhou J, Wang H, Xie Z. Effects of properties of activated carbon on its activity for mercury removal and mercury desorption from used activated carbons. Energy Fuels 2015;29(3):1946–50.

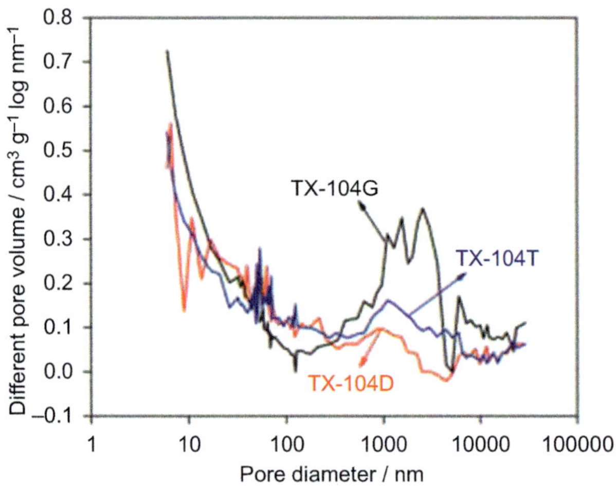

Figure 8.4 Pore size distribution of coconut activated carbons.
From Wu S, Yang W, Zhou J, Wang H, Xie Z. Effects of properties of activated carbon on its activity for mercury removal and mercury desorption from used activated carbons. Energy Fuels 2015;29(3):1946–50.

It shows that physical mercury adsorption is the primary mechanism of AC under baseline gas conditions. In the presence of SO_2 and NO, mercury adsorption efficiency increases with BET surface area, pore volume, and mesopore volume of AC. All physical properties of AC, such as BET surface area, micropore volume, and mesopore volume, determine the adsorption capacity of mercury in flue gas without HCl. However, HCl significantly decreases the effects of physical properties on elemental mercury (Hg^0) adsorption.

8.2.1.3 Effect of surface functional groups

Functional groups (e.g., carboxyl groups, carbonyl groups, hydroxyl groups, and lactones) on the AC surface affect Hg^0 adsorption capacity. The functional groups are mainly classified into acidic and basic functional groups. Representative surface acidic functional groups are carbonyl, carboxyl, phenolic hydroxyl, and lactone groups. So far, the structure of alkaline oxides is still unclear, and there is much controversy in this regard. Li et al. [4] studied the effect of chemical properties of ACs on mercury adsorption. The functional group amounts and mercury adsorption capacities of different samples, including heat treatment (1200 K) in N_2, air oxidation (693 K), and nitric acid (6 N HNO_3) in two ACs (BPL, WPL), are shown in Tables 8.2 and 8.3. WPL-AR was found to have a higher Hg^0 adsorption capacity than BPL-AR. There are lactones and carbonyl groups in WPL-AR compared to lactones and carbonyl groups in BPL-AR. The phenolic group in BPL-AR is more than the phenolic group in WPL-AR, indicating that the phenolic group can prevent Hg^0 adsorption. In addition, it has also been found that the high adsorption capacity of Hg^0 is related to the lower phenol/carbonyl ratio.

8.2.1.4 Effect of flue gas components

Currently, there are many unclear observations about the competitive or positive effects of gas components in mercury adsorption. O_2 causes a slight increase in mercury adsorption capacity on AC (Fig. 8.5) [5], which can be attributed to the formation of carbon−oxygen complexes as well as heterogeneous reactions between carbon, mercury, and oxygen.

Table 8.2 Surface acidic functional groups of different carbons.

Samples	Carboxyl (mmol/ 100g)	Lactone (mmol/ 100g)	Phenol (mmol/ 100g)	Carbonyl (mmol/ 100g)	Phenol/ carbonyl ratio
BPL AR	2	9	14	53	0.26
BPL-HNO_3	29	46	6	256	0.02
BPL-693	37	1	110	287	0.38
WPL-AR	7	39	5	153	0.03
WPL-HNO_3	13	64	3	232	0.01
WPL-693	42	18	116	359	0.32

From Li Y, Lee C, Gullett B. Importance of activated carbon's oxygen surface functional groups on elemental mercury adsorption. Fuel 2003;82(4):451−7.

Table 8.3 Mercury adsorption capacity of different carbons.

Samples	Hg^0 (μg/g)
BPL-AR	40
BPL-1200	0
BPL-HNO_3	1520
BPL-693	40
WPL-AR	925
WPL-1200	0
WPL-HNO_3	700
WPL-693	20

From Li Y, Lee C, Gullett B. Importance of activated carbon's oxygen surface functional groups on elemental mercury adsorption. Fuel 2003;82(4):451–7.

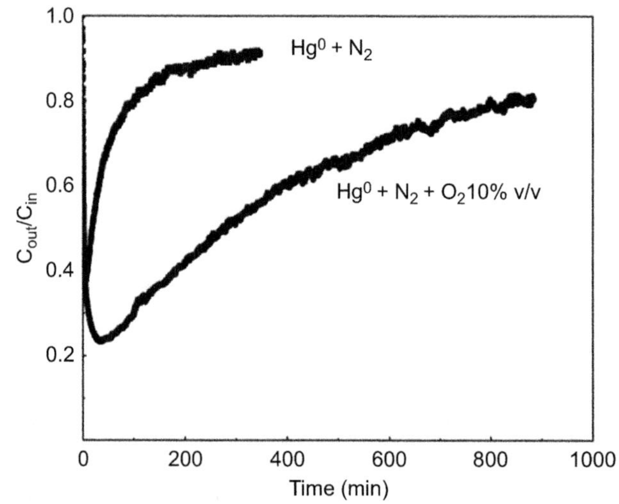

Figure 8.5 Effect of O_2 (10% v/v) on mercury adsorption on F400 at 150°C.
From Diamantopoulou I, Skodras G, Sakellaropoulos G. Sorption of mercury by activated carbon in the presence of flue gas components. Fuel Process Technol 2010;91(2):158–63.

Typically, CO_2 is inert to mercury adsorption on the adsorbent. However, the study by Diamantopoulou et al. [5] shows that in the presence of CO_2, the mercury breakthrough curve reaches equilibrium faster (Fig. 8.6). CO_2 inhibited the adsorption of mercury because CO_2 fills the micropores in AC. Since physisorption is limited at high temperatures, competitive adsorption on the same chemisorption sites between Hg^0 and CO_2 is a possible explanation.

Li et al. [6] studied the effect of moisture on AC surfaces. As shown in Figs. 8.7 and 8.8, the Hg^0 adsorption capacity of AC was significantly decreased after removing

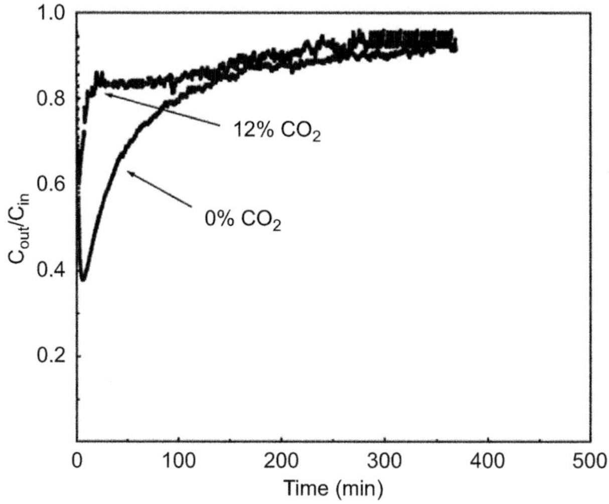

Figure 8.6 Effect of CO_2 (12% v/v) on mercury adsorption on F400 at 150°C.
From Diamantopoulou I, Skodras G, Sakellaropoulos G. Sorption of mercury by activated carbon in the presence of flue gas components. Fuel Process Technol 2010;91(2):158–63.

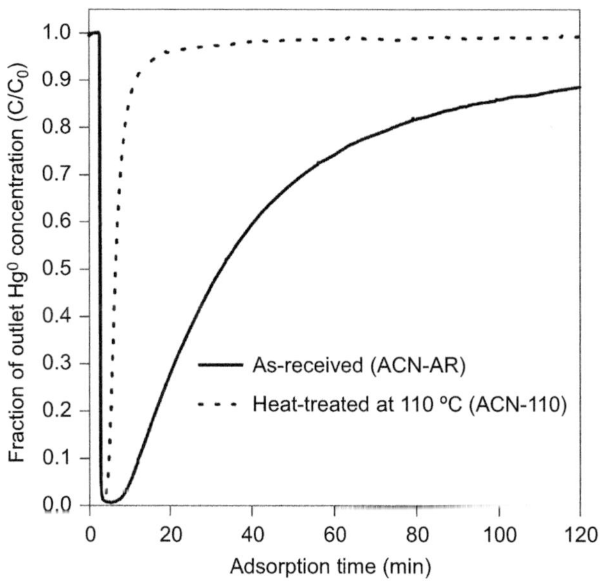

Figure 8.7 Mercury adsorption breakthrough curves of activated carbon nano-fiber (ACN) samples.
From Li Y. Lee C, Gullett B. The effect of activated carbon surface moisture on low temperature mercury adsorption. Carbon 2002;40(1):65–72.

moisture on the AC surface. The chemisorption of Hg^0 is the main process for Hg^0 adsorption of carbon monoxide samples. The interaction between H_2O and the carbon–oxygen complex may result in some active regions or surface conditions that favor Hg^0 adsorption.

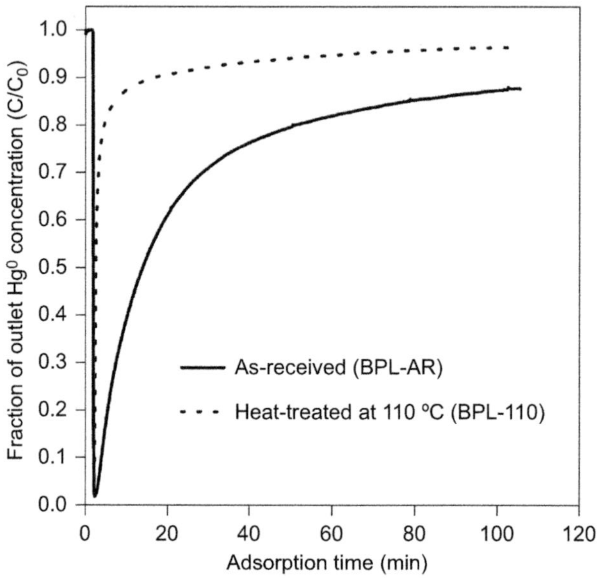

Figure 8.8 Mercury adsorption breakthrough curves of bituminous-coal-based activated carbon (BPL) samples.
From Li Y. Lee C, Gullett B. The effect of activated carbon surface moisture on low temperature mercury adsorption. Carbon 2002;40(1):65–72.

Acid flue gases usually affect the adsorption of mercury on AC. Among these gas components, HCl exhibited the most significant promotional effect on mercury removal. As shown in Fig. 8.9, F400 shows the mercury adsorption of F400 in pure N_2 and HCl/N_2 atmospheres. The mercury absorption capacity of F400 is greatly

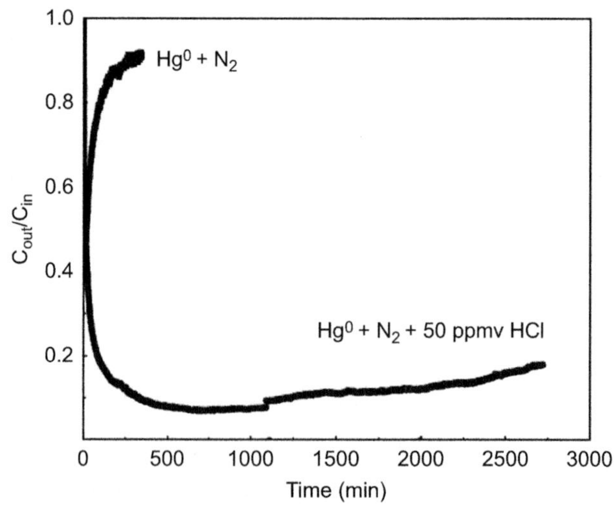

Figure 8.9 Effect of HCl (50 ppm) on mercury adsorption on F400 at 150°C.
From Li Y. Lee C, Gullett B. The effect of activated carbon surface moisture on low temperature mercury adsorption. Carbon 2002;40(1):65–72.

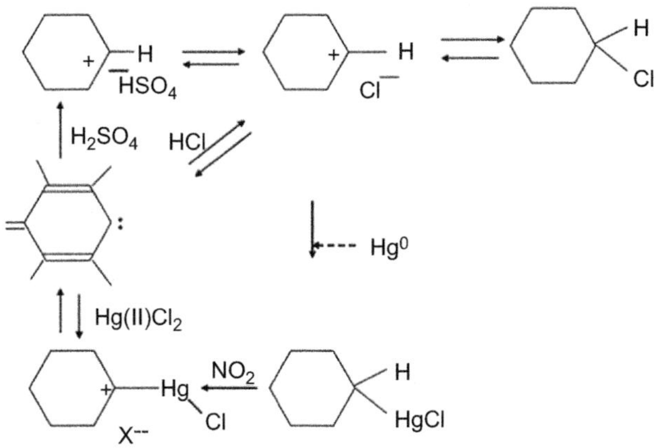

Figure 8.10 Oxidation mechanism of Hg^0 in the presence of HCl.
From Li Y. Lee C, Gullett B. The effect of activated carbon surface moisture on low temperature mercury adsorption. Carbon 2002;40(1):65–72.

increased by the presence of HCl, which could be attributed to the heterogeneous mercury oxidation mechanism. Fig. 8.10 shows a detailed mechanism for explaining the role of HCl in mercury oxidation. The carbonium ion reacts with the aromatic ring as a Lewis acid site, and the aromatic ring directly receives electrons from Hg^0, thereby forming intermediate organic mercury.

The effect of SO_2 on mercury adsorption over AC depends on the reaction conditions. Uddin et al. [7] reported that SO_2 was essential for Hg^0 removal by untreated AC. However, the AC proposed by SO_2 or H_2SO_4 also shows Hg^0 removal activity even without SO_2. For SO_2 containing flue gas, the presence of both O_2 and H_2O was necessary for Hg^0 removal. However, the presence of SO_2 suppressed the Hg removal of both H_2SO_4-preadsorbed AC and SO_2-pretreated AC. The reduction of Hg removal efficiency of AC by SO_2 was caused by the reduction of HgO to Hg^0 by SO_2 in the presence of H_2O. A reaction scheme that could explain the experimental results is shown in Fig. 8.11.

SO_3 exhibited a significantly inhibitive role in mercury adsorption over AC even at the lowest concentration tested (20 ppm) [8]. Fig. 8.12 shows the results of ACI demonstrations run by the US DOE and industry. FFs are more effective when used in conjunction with ACI than ESPs. Sites with low-halogen flue gas, including subbituminous coals from the Powder River Basin and those with spray dryer absorbers, can achieve high mercury removal levels using halogen-treated AC. ACI at sites firing Powder River Basin coals or lignite coals results in higher mercury removal than that firing bituminous coals. As the sulfur level of the coal increases, or when the SO_3 concentration increases, the mercury removal efficiency of ACI decreases.

Zhuang et al. [9] reported that as the SO_3 concentration increases, the mercury removal rate of ACI decreases (Fig. 8.13). This is because SO_3 competes for the same binding site as the mercury molecule on the AC surface. To achieve high mercury removal efficiency, the SO_3 concentration in flue gas should be controlled.

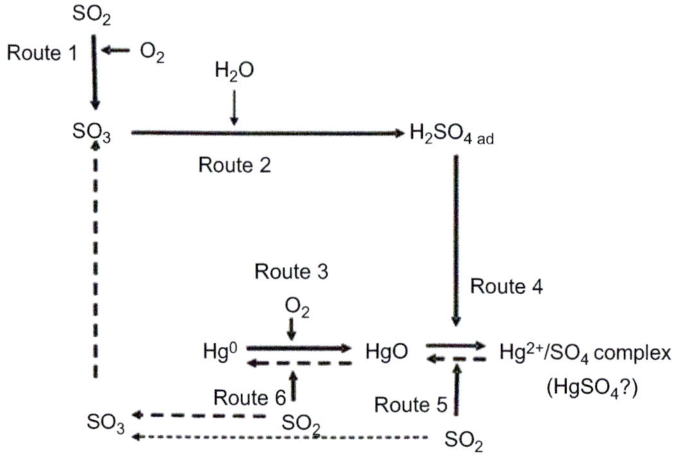

Figure 8.11 Mechanism of mercury removal by activated carbon in the presence of SO_2, O_2, and H_2O.
From Uddin M, Yamada T, Ochiai R, Sasaoka E. Role of SO_2 for elemental mercury removal from coal combustion flue gas by activated carbon. Energy Fuels 2008;22(4):2284–9.

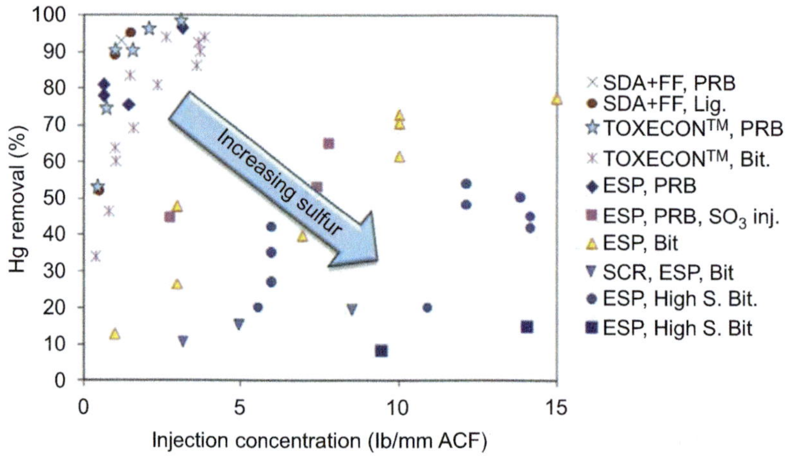

Figure 8.12 Compilation of results from Department of Energy mercury control programs. *Bit*, Bituminous; *ESP*, electrostatic precipitator; *FF*, fabric filter; *Lig*, lignite; *PRB*, the Powder River Basin; *SCR*, selective catalytic reduction; *SDA*, spray dryer absorbers.
From Sjostrom S, Durham M, Bustard C, Martin C. Activated carbon injection for mercury control: Overview. Fuel 2010;89(6):1320–2.

Current information indicates that the SO_3 concentration should be below 2 ppm when 80% mercury removal efficiency is achieved by injecting 0.08 g/m^3 AC. Utility plants that burn high-sulfur coal or use SO_3 flue gas conditioning should consider multiple pollutant reduction strategies to effectively eliminate SO_3 emissions from ACI applications.

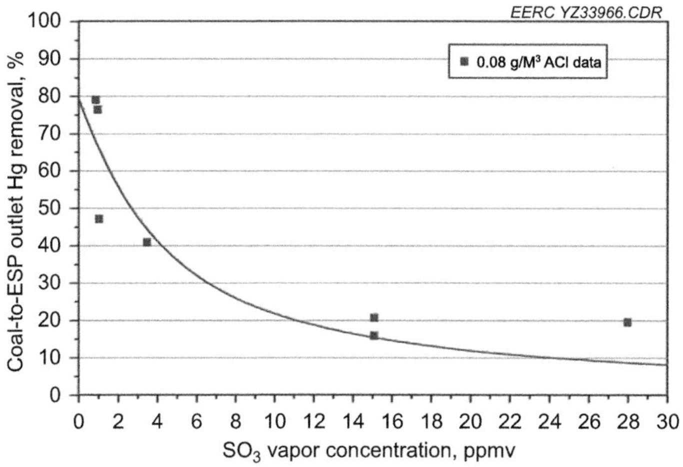

Figure 8.13 Correlation of activated carbon injection (ACI) performance and SO_3 vapor concentration in flue gas. *ESP*, Electrostatic precipitator.
From Zhuang Y, Martin C, Pavlish J, Botha F. Cobenefit of SO_3 reduction on mercury capture with activated carbon in coal flue gas. Fuel 2011;90(10):2998–3006.

8.2.1.5 Modification of activated carbon

Since virgin ACs generally show poor mercury removal performance, chemically treated ACs are often employed for mercury control. Chemically impregnated ACs by sulfur [10,11], chlorine (Cl) [12,13], bromine (Br) [14,15], or iodine (I) [16,17] are more expensive than virgin precursors. Therefore many researchers have studied ways to enhance the mercury adsorption ability of AC while reducing the cost of mercury removal.

Lee et al. [18] studied the mercury removal capacity of cupric chloride-impregnated carbon ($CuCl_2$-AC) and brominated AC (DARCO Hg-LH) in a fixed-bed system and a filter-added entrained-flow system. The $CuCl_2$-AC has higher Hg^0 oxidation ability than DARCO Hg-LH, and the adsorbent has good adsorption capacity. Fig. 8.14 shows the Hg^0 oxidation and adsorption mechanism proposed in $CuCl_2$-AC. $CuCl_2$-AC is capable of oxidizing Hg^0 and readsorbing the resultant oxidized mercury from the reaction between Hg^0 and $CuCl_2$.

Ghorishi et al. [19] developed a Cl-impregnated AC for entrained-flow capture of Hg^0. As seen in Fig. 8.15, Cl-flue gas desulfurization (FGD) shows a greater Hg^0 removal capacity than virgin FGD. Entrained-flow Hg^0 removal of more than 80% (carbon-to-mercury weight ratios of 1800–5000) could be obtained using the Cl-FGD produced in a large-scale batch. However, a much higher carbon-to-mercury weight ratio (>10,000) is required to achieve high Hg^0 removal efficiency using virgin FGD.

Zhou et al. [20] studied the mercury removal performance of raw activated carbon (R-AC) and bromine-modified activated carbon (AC-Br). As shown in Fig. 8.16, the Hg^0 breakthrough rate of R-AC increases with test time and flue gas temperature.

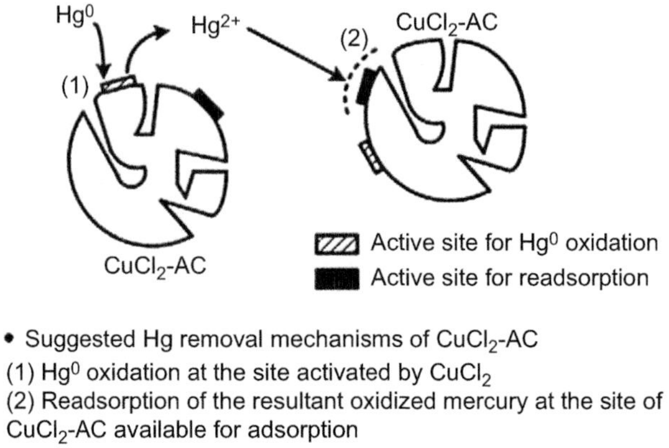

Figure 8.14 Suggested Hg^0 oxidation and adsorption mechanisms of $CuCl_2$-AC. *AC*, Activated carbon.
From Lee S, Lee J, Keener T. Mercury oxidation and adsorption characteristics of chemically promoted activated carbon sorbents. Fuel Process Technol 2009;90(10):1314–8.

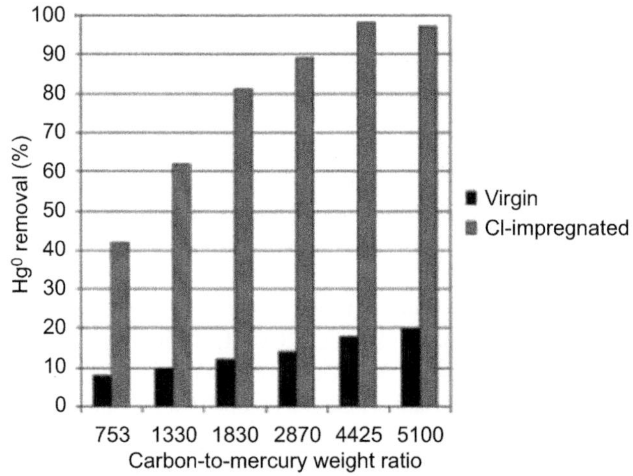

Figure 8.15 Hg^0 removal by virgin and Cl-impregnated flue gas desulfurization in a flow reactor. Inlet Hg^0 concentration was 86 ppb (in N_2); reactor temperature was 100°C; residence time was 3.8 s.
From Ghorishi S, Keeney R, Serre S, Gullett B, Jozewicz W. Development of a Cl-impregnated activated carbon for entrained-flow capture of elemental mercury. Environ Sci Technol 2002; 36(20):4454–9.

Higher temperature may reduce the mercury adsorption capacity of R-AC. However, the Hg^0 breakthrough rate of AC-Br continuously reduces with flue gas temperature (Fig. 8.17), indicating that the higher temperature helps to promote mercury adsorption over AC-Br. At 50°C, the mercury adsorption capacity of AC-Br is less than R-AC.

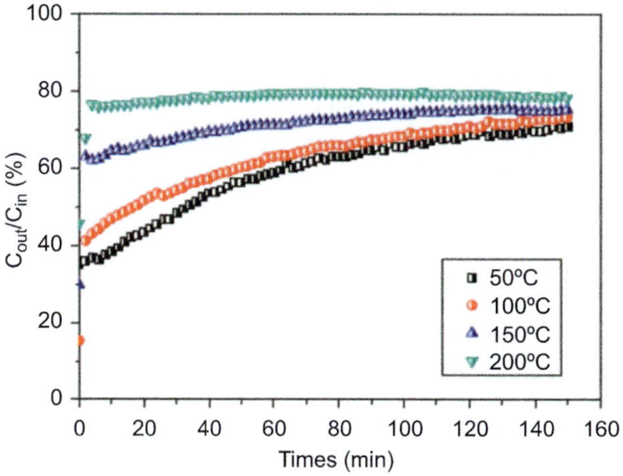

Figure 8.16 Mercury breakthrough curves of raw activated carbon.
From Zhou Q, Duan Y, Hong Y, Zhu C, She M, Zhang J, Wei H. Experimental and kinetic studies of gas-phase mercury adsorption by raw and bromine modified activated carbon. Fuel Process Technol 2015;134:325–32.

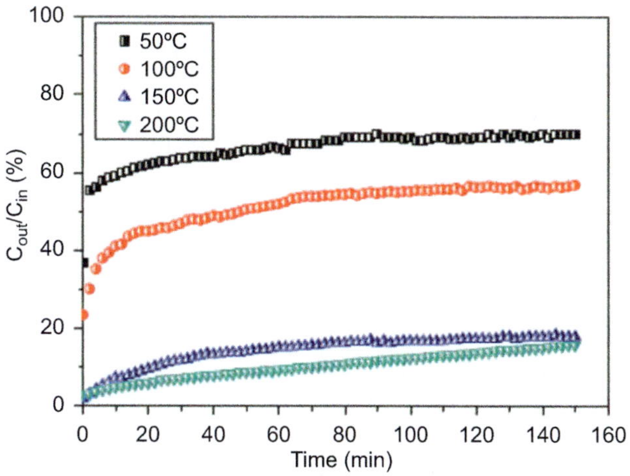

Figure 8.17 Mercury breakthrough curves of bromine-modified activated carbon.
From Zhou Q, Duan Y, Hong Y, Zhu C, She M, Zhang J, Wei H. Experimental and kinetic studies of gas-phase mercury adsorption by raw and bromine modified activated carbon. Fuel Process Technol 2015;134:325–32.

However, at 150 and 200°C, the mercury adsorption capacity of AC-Br is higher than that of R-AC. Thus modification by Br reduces physisorption at low temperature and enhances chemisorption at high temperature.

Hsi et al. [2] employed ACs and activated carbon fiber (ACF) adsorbents impregnated with sulfur for adsorbing and oxidizing Hg^0. Sulfur impregnation at 400°C is the optimum condition for preparing Hg adsorbents. The inherent oxygenated groups, which act as more effective sites of Hg^0 oxidation than the newly introduced organic sulfur groups at 650°C when H_2O exists in flue gases, accounted for the bulk of Hg^0 adsorption and oxidation occurring on raw ACs and ACF. The sulfur groups impregnated at 400°C might act as more effective oxidative and adsorptive sites in the presence of H_2O competition when compared to the inherent oxygenated groups or with sulfur impregnated at 650°C.

8.2.2 Hg removal by fly ash

Recent studies found that fly ash could adsorb and oxidize Hg^0 [21–23]. The content and type of unburned carbon are the key factors for determining the Hg adsorption capacity of fly ash [24–28]. However, recent studies show that mercury capture by fly ash also depends on the physical characteristics, the amount and type of carbon in the fly ash, and the inorganic components in fly ash.

8.2.2.1 Effect of physical characteristics on mercury removal

Dunham et al. [29] exposed 16 different fly ashes to a simulated flue gas containing either Hg^0 or $HgCl_2$. In Fig. 8.18, there is a certain correlation between the equilibrium capacity of ash and Hg^0 and the surface area, but there is considerable dispersion in the data. This may reflect the effect of flue gas composition and surface properties on Hg^0 adsorption. In contrast, there is a good correlation between $HgCl_2$ capacity and surface area of the ash sample (Fig. 8.19). The correlation between $HgCl_2$ capacity and loss on ignition (LOI) in ash is not strong, indicating that besides the carbon content, the nature of the carbon also affects the capture of $HgCl_2$.

Fig. 8.20 shows the effect of surface area on Hg^0 oxidation. The flue gas temperature had little effect on Hg^0 oxidation although there was considerable scatter in the data. The scatter in the data also suggested that surface area alone may not account for the oxidation of Hg^0.

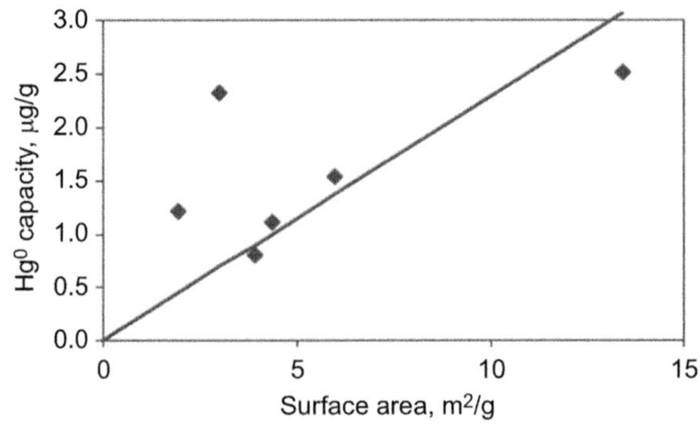

Figure 8.18 Equilibrium capacity (in Ag/g) for Hg^0 as a function of surface area at 121°C. From Dunham G, DeWall R, Senior C. Fixed-bed studies of the interactions between mercury and coal combustion fly ash. Fuel Process Technol 2003;82(2–3):197–213.

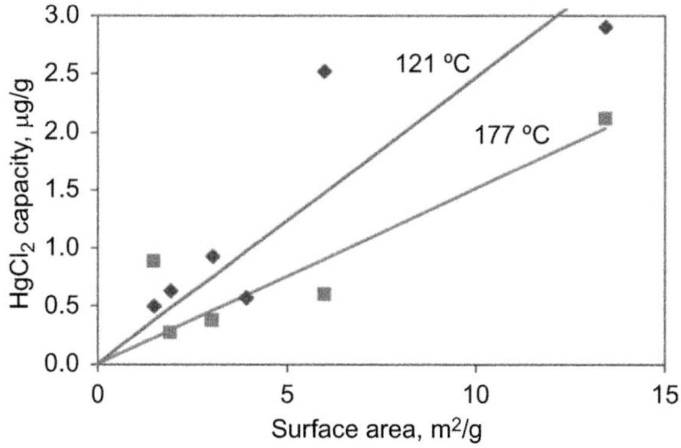

Figure 8.19 Equilibrium capacity (Ag/g) for $HgCl_2$ as a function of ash surface area.
From Dunham G, DeWall R, Senior C. Fixed-bed studies of the interactions between mercury and coal combustion fly ash. Fuel Process Technol 2003;82(2–3):197–213.

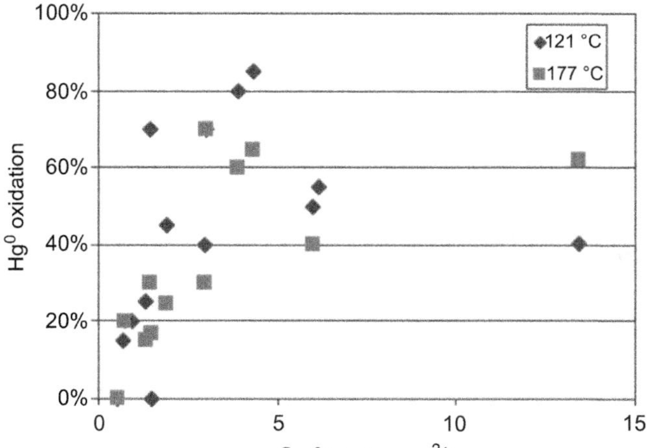

Figure 8.20 Effect of surface area on Hg^0 oxidation.
From Dunham G, DeWall R, Senior C. Fixed-bed studies of the interactions between mercury and coal combustion fly ash. Fuel Process Technol 2003;82(2–3):197–213.

Zhao et al. [23] studied the mercury capture performance of various fly ashes. The carbon content in the fly ash ranges from 2% to 7.2%, while the carbon content in the carbon-rich portion ranges from 17.8% to 54.2% (Table 8.4). There is no significant correlation between specific surface area and LOI, whether it is raw fly ash or carbon-rich fractions. Samples FA3 > 100 and FA2 > 80 have the largest BET surface area, which is up to 23.9 and 17.6 m^2/g. Comparing the relationship between mercury content of fly ash and LOI (Fig. 8.21) and specific surface area (Fig. 8.22), there is no clear correlation between them because fly ash comes from different coal grades and various carbons' structural and petrological components. Therefore there is a big

Table 8.4 Mercury content, loss on ignition (LOI), and surface area of fly ash.

	LOI (wt%)[a]	A (m^2/g)	Hg (μg/g)
FA1	5.7	1.6	0.39
FA1 > 150	22.4	4.2	0.4
FA2	7.2	9.4	1.8
FA3 > 80	54.2	17.6	4.95
FA3	5.6	4.1	0.42
FA3 > 100	35.4	23.9	0.71
FA4	2.0	1.9	0.04
FA4 > 200	17.8	13.4	0.10
FA5	3.8	6.7	
CAC	—	1183	

[a]LOI was determined by the combustion of the organic matter in air at 815°C.
From Zhao Y, Zhang J, Liu J, Diaz-Somoano M, Abad-Valle P, Martinez-Tarazona M, Zheng C. Experimental study on fly ash capture mercury in flue gas. Chinese Sci Bull 2010;53:976–83.

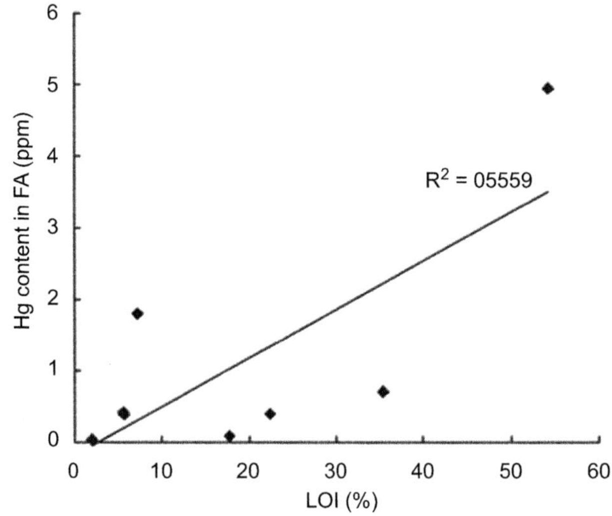

Figure 8.21 Relation between mercury content and loss on ignition (LOI) of fly ash (FA). From Zhao Y, Zhang J, Liu J, Diaz-Somoano M, Abad-Valle P, Martinez-Tarazona M, Zheng C. Experimental study on fly ash capture mercury in flue gas. Chinese Sci Bull 2010;53:976–83.

difference between the capture capacities. This also confirms that the carbon content of the fly ash is not a key factor in the ability to capture mercury.

It is generally believed that mercury capture by fly ash is enhanced by the decrease in the size of fly ash particles. However, some studies have reached the opposite

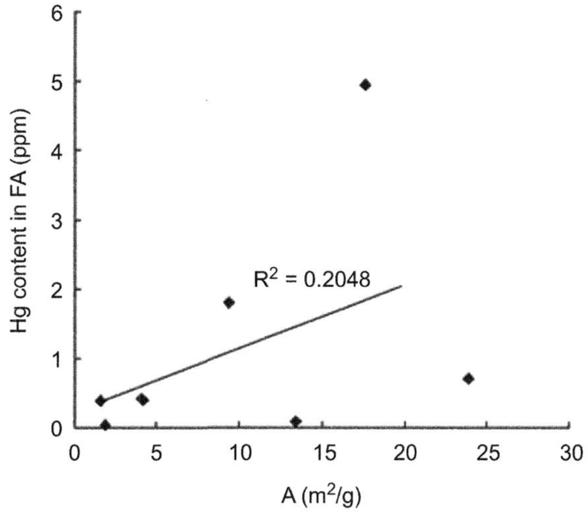

Figure 8.22 Relation between mercury content and surface area of fly ash (FA).
From Zhao Y, Zhang J, Liu J, Diaz-Somoano M, Abad-Valle P, Martinez-Tarazona M, Zheng C. Experimental study on fly ash capture mercury in flue gas. Chinese Sci Bull 2010;53:976–83.

conclusion when examining the effects of particle sizes on mercury removal performance of fly ash. The mercury removal performance of coarse-grained fly ash is significantly higher than that of fine particles, which may be caused by the presence of more unburned charcoal in the coarse-grained fly ash. However, compared with CTL > 100 μm and CTL 80–100 μm, the carbon content of the two is not much different, but the Hg removal capacity is quite different, which again proves that the carbon content is not the main factor affecting the Hg performance of fly ash.

8.2.2.2 Effect of unburned carbon content

Unburned carbons are involved in most of the interactions between Hg and fly ash. Unburned carbon not only plays an important role in mercury retention, but also acts as the main medium for Hg^0 oxidation. The carbon concentrate with the highest Hg retention capacity produces the highest Hg oxidation [24]. Hower et al. [26] analyzed the mercury content of wetted fly ash collected in the same ESP row of a Kentucky power plant for several months. They found that Hg is highly correlated with fly ash carbon (Fig. 8.23).

Hower et al. [26] also studied the mercury removal capacity of fly ash collected from mechanical separation prior to the FF array and FF hoppers at a western Kentucky power plant. The difference in Hg levels between the two parts of the system is notable (Fig. 8.24). The mechanically separated (cyclone separation) fly ash has higher carbon content than the FF fly ash because of the higher flue gas temperature. However, the Hg content of the mechanical fly ash is significantly lower than the FF fly ash.

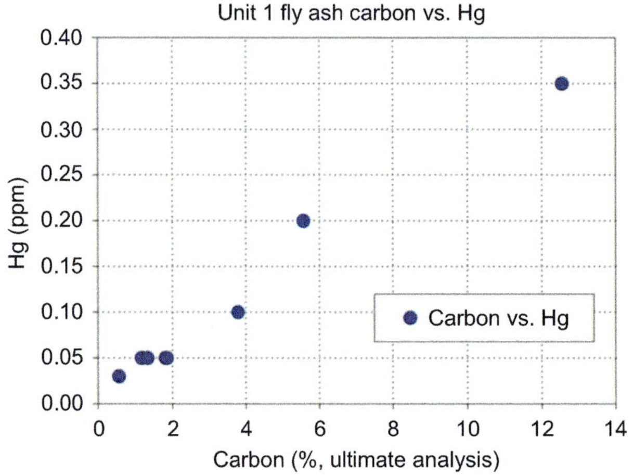

Figure 8.23 Fly ash carbon versus mercury content in wet-screened fly ashes from two collection times of the same row of a Kentucky power plant burning a highly volatile bituminous Illinois Basin coal blend.
From Hower J, Senior C, Suuberg E, Hurt R, Wilcox J, Olson E. Mercury capture by native fly ash carbons in coal-fired power plants. Prog Energy Combust 2010;36(4):510–29.

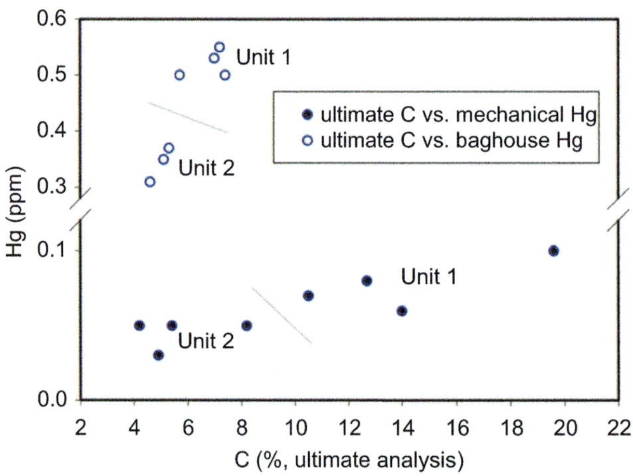

Figure 8.24 Fly ash carbon versus mercury for mechanical (cyclone) and baghouse (fabric filter) collection for two units burning highly volatile A bituminous Central Appalachian coal at the same power plant.
From Hower J, Senior C, Suuberg E, Hurt R, Wilcox J, Olson E. Mercury capture by native fly ash carbons in coal-fired power plants. Prog Energy Combust 2010;36(4):510–29.

Abad-Valle et al. [24] studied the role of unburned carbon concentrates from fly ashes in the oxidation and retention of mercury. Different behaviors were observed for different gas mixtures (Table 8.5). The amount of Hg retained in descending order is CTL-EC > CTEEC > CTL-O > CTE-O. Therefore a higher unburned carbon

Table 8.5 Mercury retention by fly ashes in different atmospheres[a].

Atmosphere	Hg retention (μg/g)			
	CTL-O	CTL-EC	CTE-O	CTE-EC
N_2	1.7	20	1.0	9.0
12.6% O_2	1.8	20	1.1	8.5
20% O_2	1.9	20	1.2	9.2
16% CO_2	2.2	20	1.0	8.7
0.2% SO_2	1.7	19	1.0	5.7
50 ppm HCl	21	250[a]	9.9	200[a]
3% H_2O	1.7	20	0.9	5.6
10% O_2 + 16% CO_2 + 0.2% SO_2	2.1	29	1.1	8.9
10% O_2 + 16% CO_2 + 0.2% SO_2 + 3% H_2O	1.7	310[a]	0.9	160[b]
10% O_2 + 16% CO_2 + 0.2% SO_2 + 3% H_2O + 50 ppm HCl	8.5	35	4.4	17

[a]CTL-O and CTE-O came from the burning of bituminous and sub-bituminous coals; CTL-EC and CTEEC were the fractions of these fly ashes enriched with unburned carbon particles.
[b]Retention at 67 h, no saturation.
From Abad-Valle P, Lopez-Anton M, Diaz-Somoano M, Martinez-Tarazona M. The role of unburned carbon concentrates from fly ashes in the oxidation and retention of mercury. Chem Eng J 2011;174(1):86–92.

content favors the retention of mercury. However, in the atmosphere of SO_2, HCl, water vapor, and O_2 + CO_2 + SO_2, this ratio was not precisely maintained in the carbon concentrates (CTL-EC and CTE-EC) (Table 8.5). It is inferred that certain characteristics of fly ash may cause different interactions between gas and carbon particles in the presence of certain gases.

For the original fly ash, CTL has the largest mercury capture capacity, which is up to 5.02 μg/g (Fig. 8.25). For carbon-rich fractions, CTL >100 and CTSR >80 have relatively high mercury capture capacities of 9.36 and 10.3 μg/g, respectively. For fly ash produced by bituminous coal (CTL, CTSR) combustion, the carbon-rich fraction has a higher mercury capture capacity than the original fly ash. For fly ash from high-grade anthracite and sub-bituminous coal combustion, the mercury-capturing capacity of the carbon-rich fraction is similar to the original sample. These results indicate that the carbon content of fly ash is not the only factor affecting the ability of mercury capture, and the type of carbon particles has a significant impact.

Recent studies have shown that the carbon type, the halogen content, the fly ash control device types, and the temperature of the fly ash control device all play major roles in mercury capture. Goodarzi et al. [25] examined the mercury capture capacity of fly ash from pulverized coal and fluidized-bed burners in Canadian power plants. As shown in Fig. 8.26, the mercury captured by the fly ash depends on the grade and mixture of the raw coal and the type of carbon in the fly ash. Isotropic glassy carbon

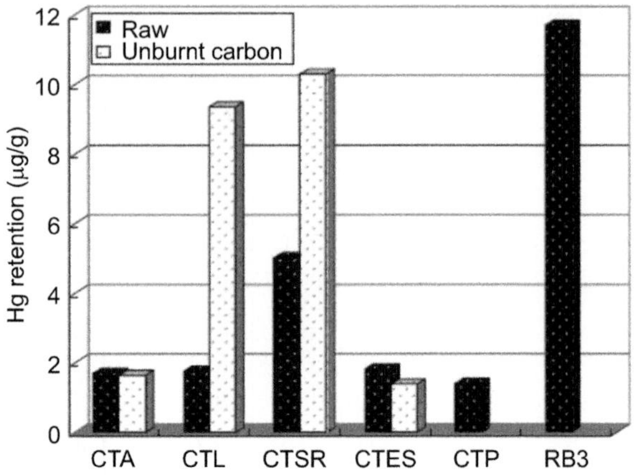

Figure 8.25 Mercury capture capacities of fly ashes.
From Abad-Valle P, Lopez-Anton M, Diaz-Somoano M, Martinez-Tarazona M. The role of unburned carbon concentrates from fly ashes in the oxidation and retention of mercury. Chem Eng J 2011;174(1):86–92.

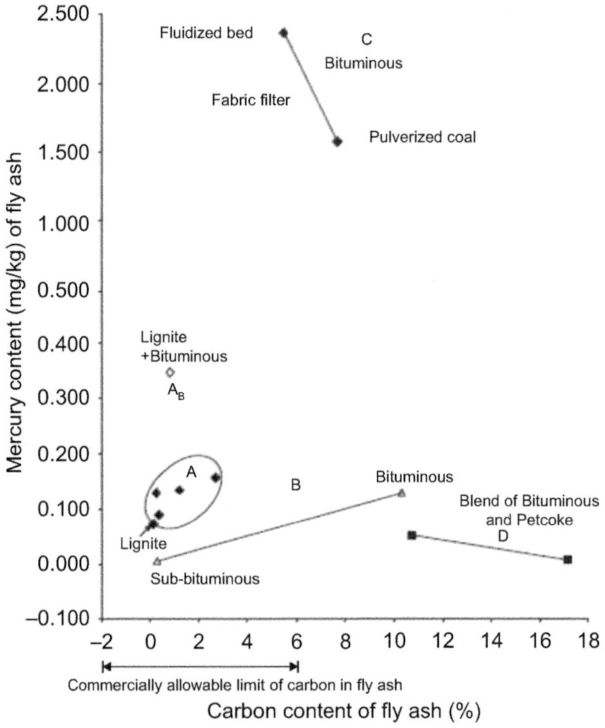

Figure 8.26 Variation in carbon and mercury content of fly ashes collected from various feed coal and collection systems.
From Goodarzi F Hower JC. Classification of carbon in Canadian fly ashes and their implications in the capture of mercury. Fuel 2008;87:1949–57.

is primarily responsible for capturing most of the mercury in fly ash. The unintentional increase in carbon content due to the mixing of coal and petroleum coke does not increase the amount of Hg captured by fly ash. The fly ash collected by the hot side electrostatic precipitator has a low Hg content, and no relationship between the Hg and carbon content of the ash is observed.

8.2.2.3 Relation between unburned carbon petrology components and mercury capture capacity of fly ash

To understand the effect of carbon particle type on mercury capture capacity, Zhao et al. [23] studied the relationship between unburned carbon petrology components and mercury capture capacity of fly ash (Figs. 8.27 and 8.28). The content of anisotropic particles, especially anisotropic porous carbon particles, has an important correlation with the mercury adsorption capacity of fly ash. Anisotropic carbon particles have a higher mercury capture capacity than isotropic carbon particles. Porous carbon particles have a large specific surface area so that it can adsorb more mercury.

López-Antón et al. [27] investigated mercury adsorption performance of fly ash in high-mercury concentration. They also suggested that the relation between anisotropic carbon and mercury capture capacity is very clear. Compared to anisotropic carbon particles, the relation between isotropic carbon particles and mercury capture capacity is not so clear, even in high-mercury concentrations (Fig. 8.29).

8.2.2.4 Effect of inorganic components

Besides the unburned carbon, inorganic components also play important roles in mercury removal of fly ash. Wang et al. [28] studied the roles of different fly ash compositions. Substitutes were selected based on compositional analysis to simulate the five metal oxides in the fly ash. As shown in Fig. 8.30, Al_2O_3, Fe_2O_3, and TiO_2 are capable of adsorbing mercury. Among these components, Al_2O_3 shows the largest adsorption capacity. No mercury adsorption occurred on the surface of CaO or MgO.

For the bituminous coals, which have a significant amount of iron oxide in the ash, oxidation may be catalyzed by iron [29]. Fig. 8.31 shows the relationship between magnetite content of the ash and mercury oxidation at 121 and 177°C. Ash with a high fraction of magnetite oxidizes more mercury. The oxidation of Hg^0 by Wyodak ash may have been due to the high content of unburned carbon, not to any inorganic constituent.

Galbreath et al. [30] found that α-Fe_2O_3 promoted mercury oxidation in bench-scale experiments, but did not significantly affect mercury speciation in real flue gases produced from burning sub-bituminous and lignite coals. Meanwhile, γ-Fe_2O_3 exhibited significant catalytic activity toward Hg^0 oxidation. However, the study of Abad-Valle et al. [31] showed that in the range of fly ashes studied, iron species do not affect the retention of mercury and do not play any role in heterogeneous mercury oxidation. Iron species in magnetospheres mainly include Fe_3O_4, α-Fe_2O_3, γ-Fe_2O_3, Fe^{2+}-

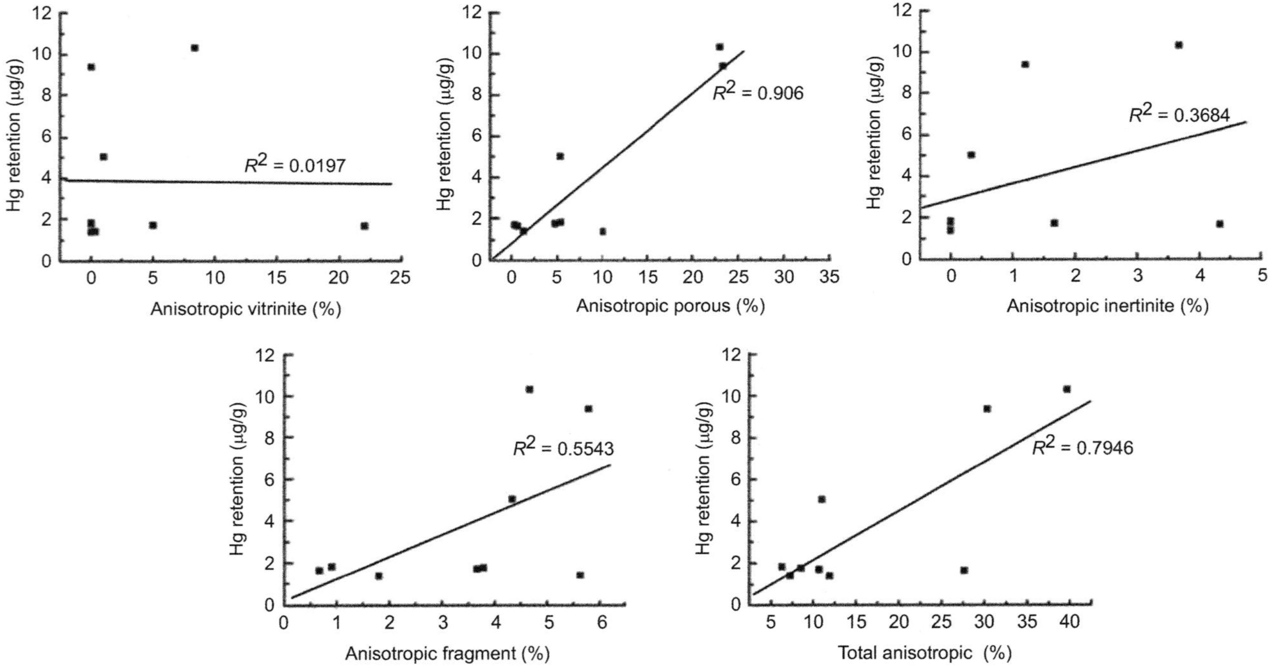

Figure 8.27 Relation between anisotropic carbons and mercury capture capacity.
From Zhao Y, Zhang J, Liu J, Diaz-Somoano M, Abad-Valle P, Martinez-Tarazona M, Zheng C. Experimental study on fly ash capture mercury in flue gas. Chinese Sci Bull 2010;53:976–83.

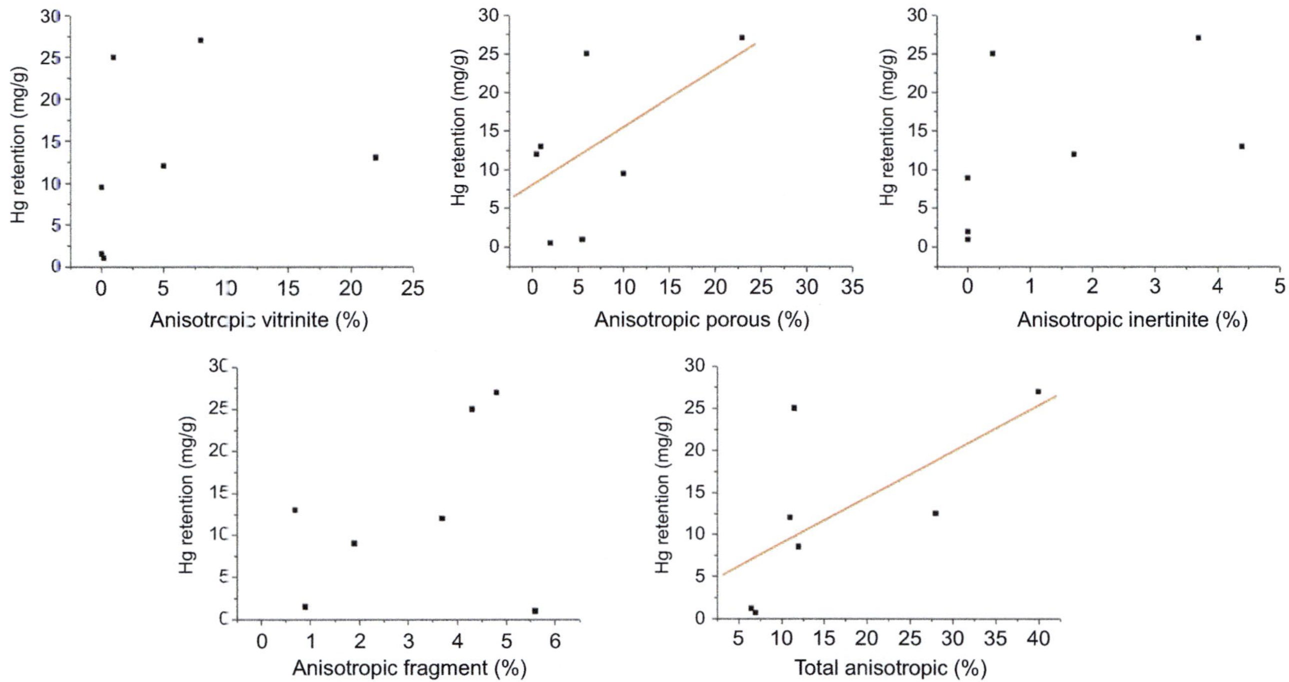

Figure 8.28 Relation between isotropic carbons and mercury capture capacity.
From Zhao Y, Zhang J, Liu J, Diaz-Somoano M, Abad-Valle P, Martinez-Tarazona M, Zheng C. Experimental study on fly ash capture mercury in flue gas. Chinese Sci Bull 2010;53:976–83.

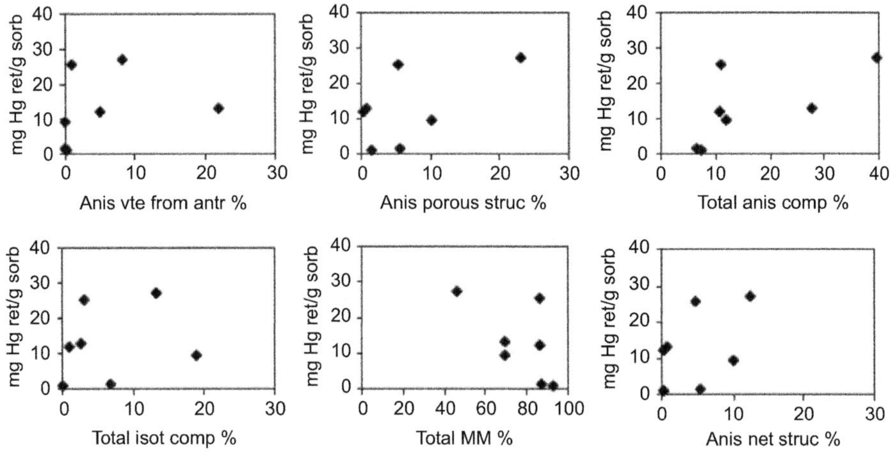

Figure 8.29 Anisotropic components of vitrinite from anthracite, anisotropic porous structures, total anisotropic components, total isotropic components, total mineral matter (MM), and anisotropic network structures versus mercury retention from Hg^0 evaporation.
From López-Antón M, Abad-Valle P, Díaz-Somoano M, Suárez-Ruiz I, Martínez-Tarazona M. The influence of carbon particle type in fly ashes on mercury adsorption. Fuel 2009;88(7): 1194–200.

silicate, and Fe^{3+}-silicate, while Fe_3O_4 is the dominant iron-bearing mineral [32]. The various physical–chemical characteristics of magnetospheres would result in the difference for Hg^0 removal capacity.

Yang et al. [33] found that, with the increase in iron content, the Hg^0 removal capacity of magnetospheres changed with a complex nonmonotonic character ($r = 0.65$), as shown in Fig. 8.32(a). For example, STM has a higher iron content than BJM, HSM, and SHM; however, η_T is much lower than these three samples. In particular, EZM samples with relatively high iron content (42.8 wt%) reached a minimum η_T (13.8%). However, if STM and EZM samples were excluded, the Hg^0 removal capacity generally increased with increasing iron content ($r = 0.91$), as shown in Fig. 8.32(b). This indicates that iron content is not the only factor determining the catalytic activity of the magnetic layer.

Various iron species presented in magnetospheres mainly include Fe_3O_4, α-Fe_2O_3, γ-Fe_2O_3, Fe^{2+}-silicate, Fe^{3+}-silicate, and FeSi [32]. The content of each iron species varies significantly from different magnetospheres. Among the iron species in magnetospheres, the Fe_3O_4, α-Fe_2O_3, and γ-Fe_2O_3 were considered as the active species for Hg^0 oxidation [29,30,34,35]. Therefore it can be inferred that in addition to iron content, Hg^0 removal ability also depends on the iron species. To understand the active species in Hg^0 removal ability, the relationship between the percentage of the forego-ing active substances and the iron content in the magnetic layer was first studied. As shown in Fig. 8.33, in addition to the STM and EZM samples, the total percentage of

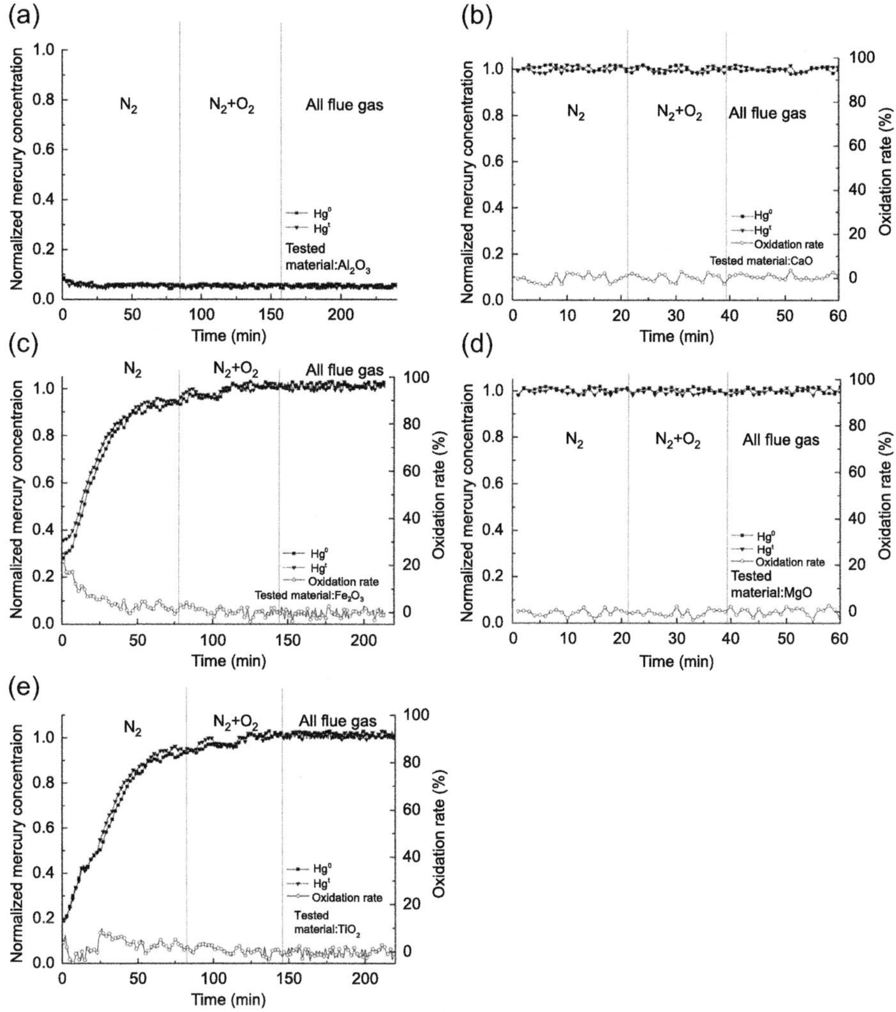

Figure 8.30 Mercury oxidation and adsorption of the simulated fly ash compositions: (a) Al_2O_3; (b) CaO; (c) Fe_2O_3; (d) MgO; and (e) TiO_2.
From Wang F, Wang S, Meng Y, Zhang L, Wu Q, Hao J. Mechanisms and roles of fly ash compositions on the adsorption and oxidation of mercury in flue gas from coal combustion. Fuel 2016;163:232–9.

iron lanthanum (Fe_3O_4 and γ-Fe_2O_3) and hematite (α-Fe_2O_3) increased as the iron content increased. However, the content of iron and hematite in STM and EZM is much lower than in other samples. Therefore a nonmonotonic change in Hg^0 removal capacity of iron content is understandable. Despite the relatively high iron content in STM and EZM, the lack of active material results in relatively low Hg^0 removal activity.

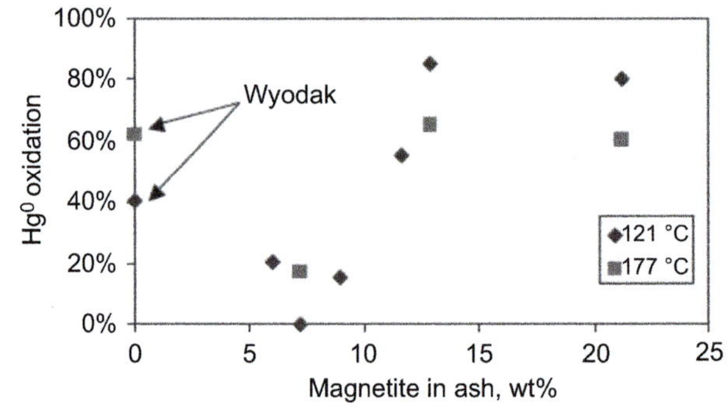

Figure 8.31 Effect of magnetite content of bituminous ash samples on oxidation of Hg^0 at 121 and 177°C.
From Dunham G, DeWall R, Senior C. Fixed-bed studies of the interactions between mercury and coal combustion fly ash. Fuel Process Technol 2003;82(2–3):197–213.

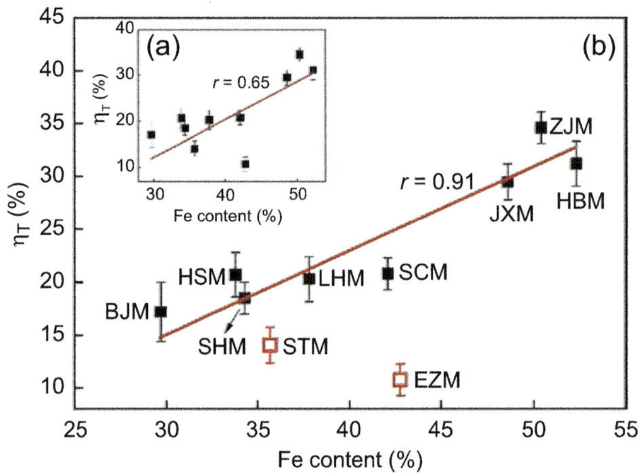

Figure 8.32 Relationship between iron content of magnetospheres and mercury removal capacity.
From Yang J, Zhao Y, Zhang S, Liu H, Chang L, Ma S, Zhang J, Zheng C. Mercury removal from flue gas by magnetospheres present in fly ash: Role of iron species and modification by HF. Fuel Process Technol 2017;167:263–70.

8.2.2.5 *Modification of fly ash for Hg removal*

Zhang et al. [36] investigated the effectiveness of fly ash modified with $CaCl_2$, $CaBr_2$, and HBr for absorbing mercury. As shown in Fig. 8.34, the mercury adsorption efficiency of modified fly ash is significantly improved as compared with unmodified fly ash. Fly ash modified by HBr exhibits the greatest adsorption capacity among the three additives. Under experimental conditions, the adsorption efficiencies of the two HBr-modified fly ashes (A and B) were 2.4 and 6.7 times greater for >200 mesh than for 80–200 mesh (Fig. 8.35). The improved properties of the modified fly ash are due to physical and chemical adsorption.

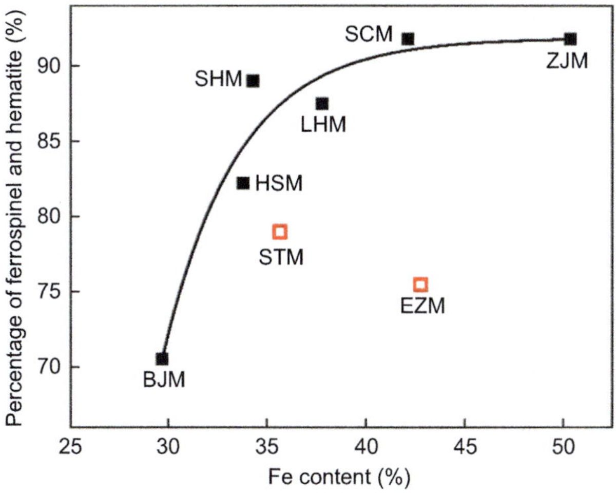

Figure 8.33 Dependence of ferrospinel and hematite percentage on iron content of magnetospheres.
From Yang J, Zhao Y, Zhang S, Liu H, Chang L, Ma S, Zhang J, Zheng C. Mercury removal from flue gas by magnetospheres present in fly ash: Role of iron species and modification by HF. Fuel Process Technol 2017;167:263–70.

Gu et al. [37] evaluated the Hg^0 adsorption capacity of fly ash modified with NH_4Br by different methods, as summarized in Table 8.6.

The bromine loading amount of the five modified samples is shown in Fig. 8.36. Mechanochemical methods have the best bromine recovery compared to the two other methods. Other methods require drying and ion exchange as well as filtration. Therefore bromine will be lost in these processes. Since the mechanochemical method saves transportation costs and the time of the preparation process, this method is suitable for full-scale production. Thus considering engineering applications, mechanochemical methods are the best way to produce modified fly ash. Fig. 8.37 shows that Hg^0 removal efficiency is significantly increased with increasing bromine loading amount.

8.2.2.6 Industrial application of fly ash for Hg removal

Wang et al. [38] conducted full-scale injection and adsorption experiments at a 300 MW plant. Fig. 8.38 shows a schematic diagram for the removal of mercury from injection and adsorption in a power plant. The fly ash was delivered to the site in bags and then put into the absorbent feeder. The adsorbent was then fed into the duct by air blower. The gaseous Hg^0 was adsorbed on fly ash and captured by the ESP. Thus the gaseous Hg^0 was removed effectively. In field experiments, the distance between the point of sorbent injection and ESP is very short, and the residence time of flue gas in this section is only about 1 s.

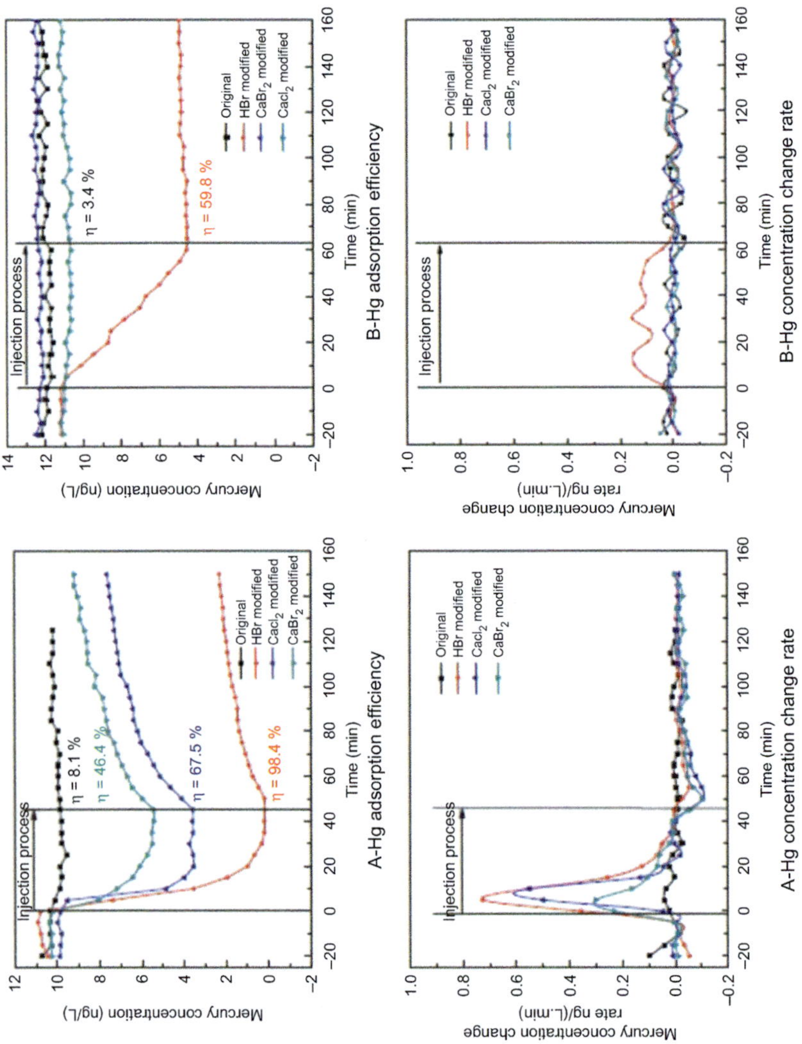

Figure 8.34 Hg adsorption ability of modified fly ash.
From Zhang Y, Duan W, Liu Z, Cao Y. Effects of modified fly ash on mercury adsorption ability in an entrained-flow reactor. Fuel 2014;128:274–80.

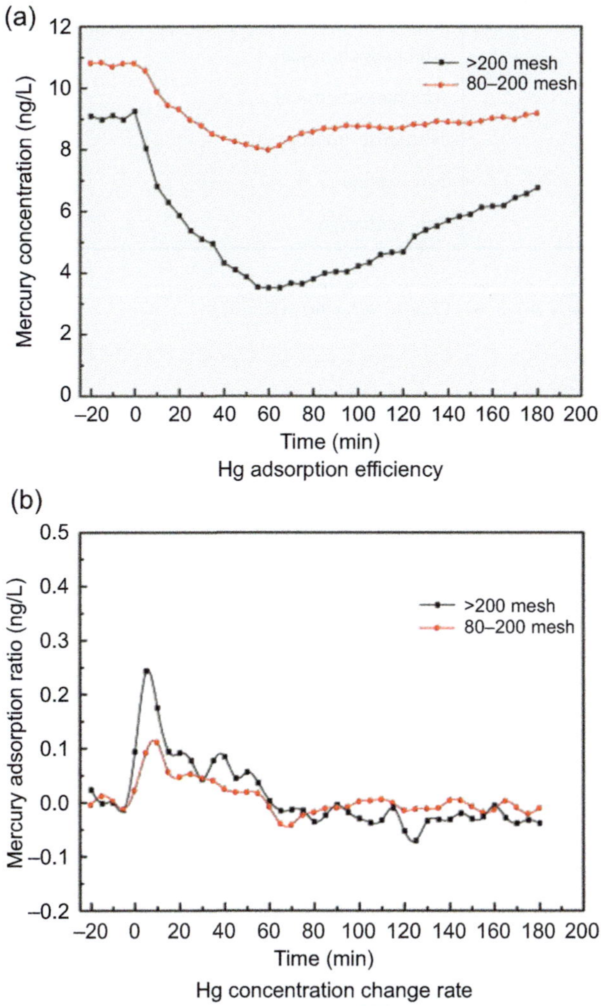

Figure 8.35 Effect of particle size on Hg adsorption ability of HBr-modified fly ash A. From Zhang Y, Duan W, Liu Z, Cao Y. Effects of modified fly ash on mercury adsorption ability in an entrained-flow reactor. Fuel 2014;128:274–80.

Fig. 8.39 shows mercury concentration tested by the 30B method and the CEM mercury measurement instrument on the same day. In the process of injecting the adsorbent, as the mass ratio of modified ash to mercury increases, the mercury concentration in flue gas decreases. Although there is no significant change in oxidized mercury, modified sorbent injection can reduce Hg^0 by 40%–50%. Combined with denitrification, dust removal, and desulfurization, the comprehensive removal rate can reach 75%–90%.

Table 8.6 Summary of modifications of fly ash samples.

Sample	Modified method	$m_{Br}/m_{FA}/\%$
1#	Unmodified	N/A
2#	Mechanochemical	0.15
3#	Mechanochemical	0.30
4#	Isometric impregnation	0.41
5#	Ion exchange	1.63
6#	Ion exchange	4.08

From Gu Y, Zhang Y, Lin L, Xu H, Orndorff W, Pan W. Evaluation of elemental mercury adsorption by fly ash modified with ammonium bromide. J Therm Anal Calorim 2015;119(3):1663–72.

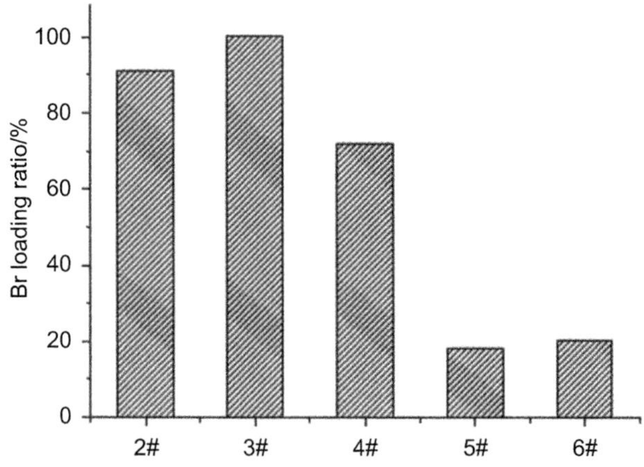

Figure 8.36 Bromine loading ratio of modified fly ash samples.
From Gu Y, Zhang Y, Lin L, Xu H, Orndorff W, Pan W. Evaluation of elemental mercury adsorption by fly ash modified with ammonium bromide. J Therm Anal Calorim 2015;119(3): 1663–72.

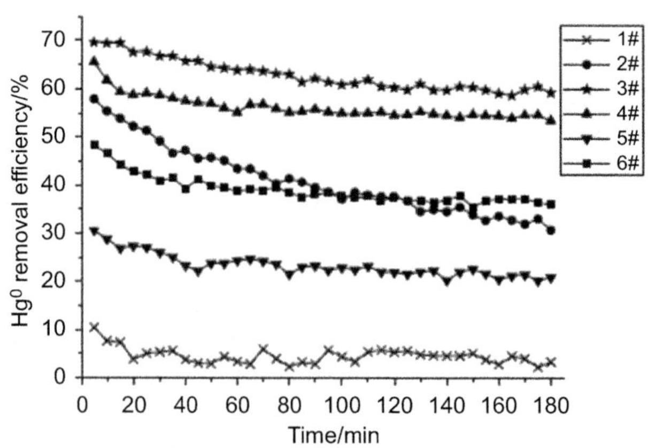

Figure 8.37 Hg^0 removal efficiency of fly ash samples.
From Gu Y, Zhang Y, Lin L, Xu H, Orndorff W, Pan W. Evaluation of elemental mercury adsorption by fly ash modified with ammonium bromide. J Therm Anal Calorim 2015;119(3): 1663–72.

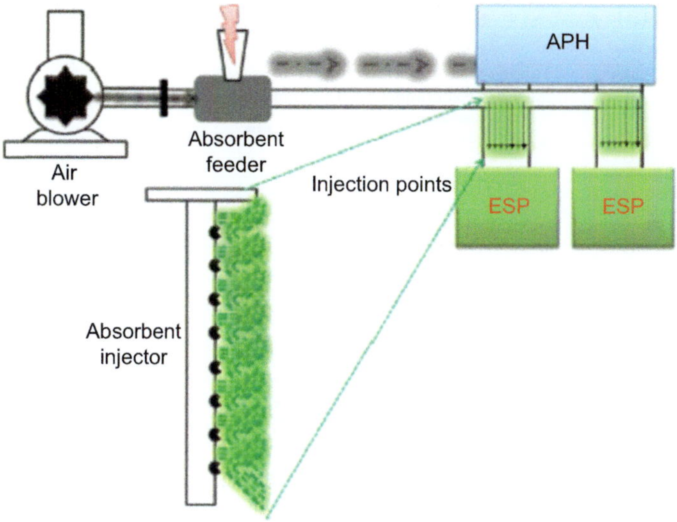

Figure 8.38 Schematic of injection setup at the 300 MW plant *ESP*, Electrostatic precipitator. *APH*, air pre-heater.
From Wang S, Zhang Y, Gu Y, Wang J, Zhang Y, Cao Y, Pan W. Using modified fly ash for mercury emissions control for coal-fired power plant applications in China. Fuel 2016;181: 1230–37.

8.2.2.7 Recyclable modified magnetosphere catalyst from fly ash for Hg removal

To remove Hg^0 from coal combustion flue gas and eliminate secondary pollution of spent sorbent, a novel magnetic catalyst based on $CuCl_2$-modified magnetospheres from fly ash ($CuCl_2$-MF [magnetospheres]) was developed [39–41]. The optimum loading of 6% $CuCl_2$ catalyst has the highest Hg^0 removal rate (90.6%) at 150°C. Chlorine-rich coordination acts as an active adsorption site for Hg^0, while chlorine-free coordination is inactive for Hg^0 removal. The removal of Hg^0 by the $CuCl_2$-MF catalyst in the presence of HCl showed good resistance to SO_2 (Fig. 8.40).

A particular feature of the magnetic ball catalyst is magnetism, which makes it possible to separate the spent catalyst from the fly ash for recycling. The management method of mercury-containing spent catalyst is proposed, which can be realized in two steps. First, the spent catalyst is captured by the ESP together with the fly ash after injecting the flue gas; the second step is to separate the spent catalyst by magnetic separation. In practical applications, the newly generated magnetospheres in fly ash during combustion would dilute the catalyst in the overall magnetospheres isolated after mercury capture. To eliminate the dilution of the newly generated magnetic layer, the catalyst can be injected before the last electric field of the ESP, as shown in Fig. 8.41. In this case, the spent catalyst is mixed with a minimum amount of fly ash, and any adverse effects regarding the newly generated magnetic layer will be minimized.

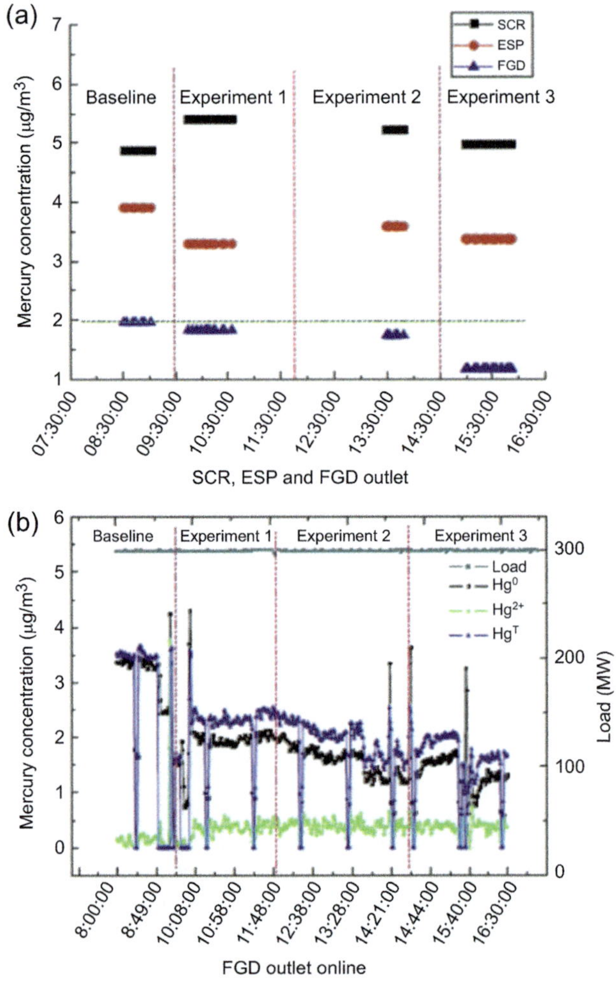

Figure 8.39 Hg concentration change before and after injection. *ESP*, Electrostatic precipitator; *FDG*, flue gas desulfurization; *SCR*, selective catalytic reduction. *CEM*, continuous emission monitoring.
From Wang S, Zhang Y, Gu Y, Wang J, Zhang Y, Cao Y, Pan W. Using modified fly ash for mercury emissions control for coal-fired power plant applications in China. Fuel 2016;181: 1230–7.

The interaction between Hg^0 and $CuCl_2$ with the participation of O_2 and/or HCl follows a three-step mechanism [40]: (1) the reduction of $CuCl_2$ to CuCl for the interaction with Hg^0, (2) the reoxidation of CuCl for the interaction with O_2 forming an intermediate copper oxygen chloride species, and (3) the rechlorination of oxychloride species resulting in the restoration of $CuCl_2$, as shown in Fig. 8.42.

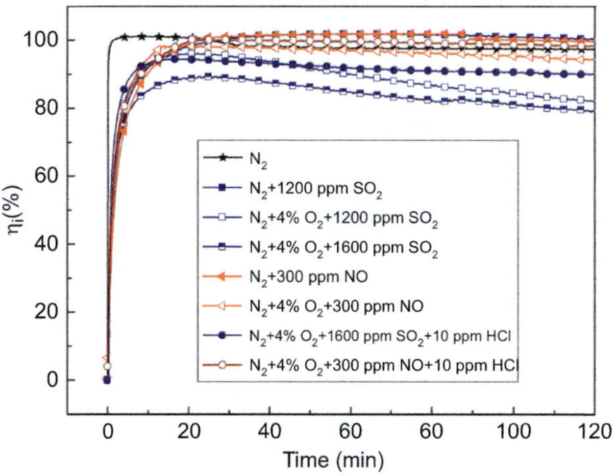

Figure 8.40 Hg removal performance under different atmospheres.
From Yang J, Zhao Y, Zhang J, Zheng C. Removal of elemental mercury from flue gas by recyclable $CuCl_2$ modified magnetospheres catalyst from fly ash. Part 1. Catalyst characterization and performance evaluation. Fuel 2016;164:419–28.

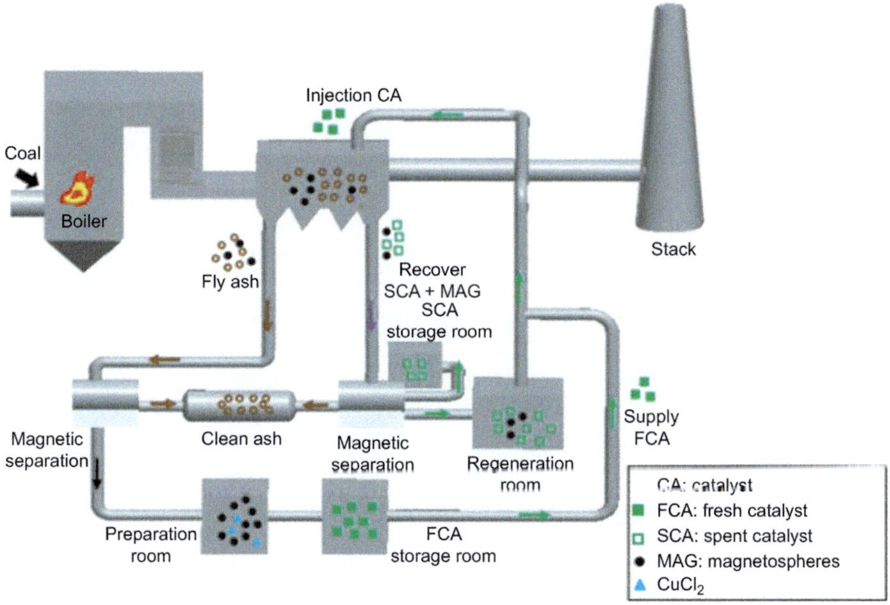

Figure 8.41 Schematic diagram of the utilization of a magnetosphere catalyst.
From Yang J, Zhao Y, Zhang J, Zheng C. Removal of elemental mercury from flue gas by recyclable $CuCl_2$ modified magnetospheres catalyst from fly ash. Part 1. Catalyst characterization and performance evaluation. Fuel 2016;164:419–28.

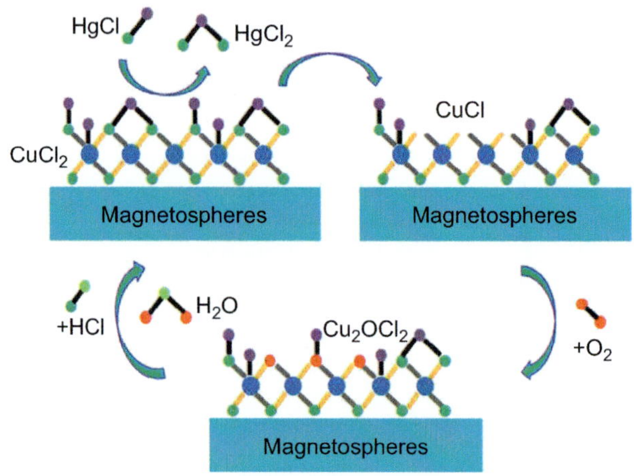

Figure 8.42 Schematic representation of the reaction mechanism for Hg^0 removal over the $CuCl_2$-MF catalyst with the participation of O_2 and/or HCl.
From Yang J, Zhao Y, Zhang J, Zheng C. Removal of elemental mercury from flue gas by recyclable CuCl2 modified magnetospheres catalyst from fly ash. Part 3. Regeneration performance in realistic flue gas atmosphere. Fuel 2016;173:1–7.

8.2.3 Hg removal by calcium-based adsorbents

An alternative to using AC is to use a calcium-based sorbent. The oxide-rich Ca-based adsorbent can effectively remove Hg^0 and Hg^{2+} at 80°C. SO_2 can increase the Hg removal rate of $Ca(OH)_2$ to 15%–20% and increase the Hg adsorption amount by more than 50% [42]. Similarly, SO_2 can also increase the Hg removal efficiency of CaO from 20% to 34%. This may help capture Hg due to the formation of an active site between the SO_2- and Ca-based adsorbents. Wang et al. [43] studied the vapor-phase Hg^0 adsorption performance of two Ca-based sorbents: $Ca(OH)_2$ and Mn^{x+}/$Ca(OH)_2$. At the beginning of the experiment, approximately 20% of the mercury in the flue gas was captured by $Ca(OH)_2$, but there was no adsorption capacity after 110 min. Impregnation with Mn^{x+} slightly increased the removal rate of mercury (Fig. 8.43). Manganese could promote the oxidation of Hg^0. $KMnO_4$-modified $Ca(OH)_2$ can oxidize a large amount of Hg^0 to Hg^{2+} in the presence of SO_2 or HCl, adsorbing more than 50% of total mercury.

8.2.4 Hg removal by mineral adsorbents

Mineral adsorbents have many advantages, such as numerous sources, large deposits, low cost, and unique structures with adsorption. Mineral adsorbents are promising candidates for industrial application in coal-fired power plants. In general, raw minerals have limited mercury removal capabilities. However, after modification, their Hg^0 removal performance has been improved to varying degrees. Ding et al. [44] studied the mercury removal performance of three types of natural minerals: bentonite (Ben), mordenite (Mor), and attapulgite (Atp). Several chemical accelerators, such as $CuCl_2$, $NaClO_3$, KBr, and KI, are used to increase the mercury removal capacity of the original adsorbent. As shown in Fig. 8.44, the average Hg^0 removal efficiency of the

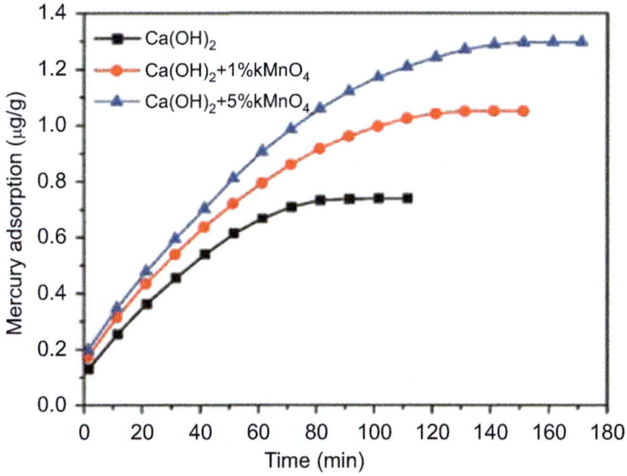

Figure 8.43 Comparison of the total mercury adsorption by $Ca(OH)_2$ and $Ca(OH)_2$ impregnated with $KMnO_4$.
From Wang Y, Duan Y. Effect of manganese ions on the structure of $Ca(OH)_2$ and mercury adsorption performance of $Mn_{x+}/Ca(OH)_2$ composites. Energy Fuels 2011;25(4):1553–8.

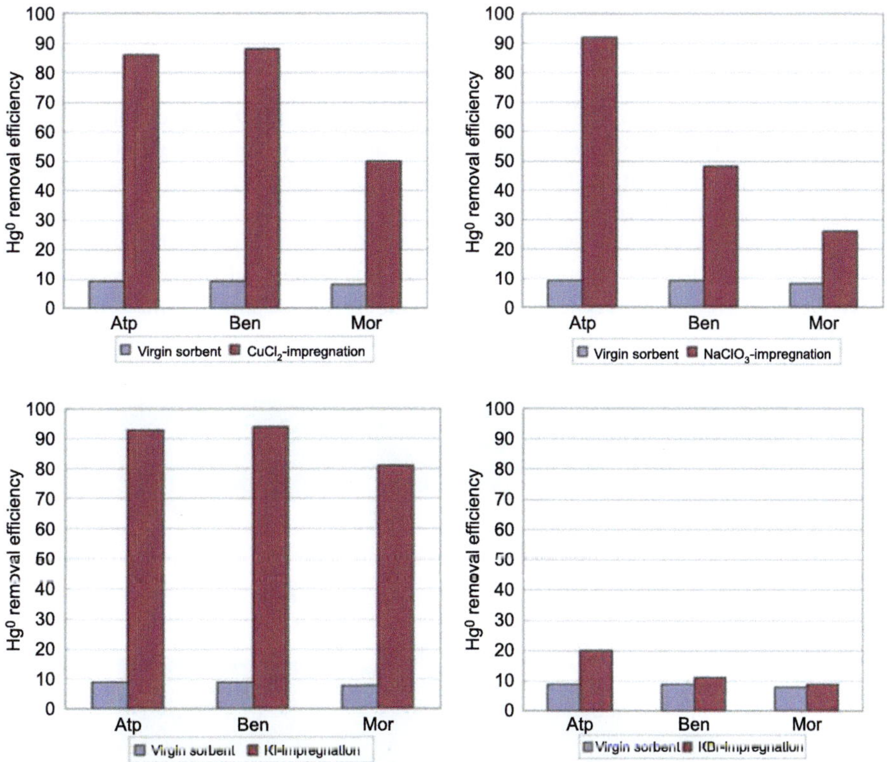

Figure 8.44 Hg^0 removal efficiency of raw and $CuCl_2$-, $NaClO_3$-, KI-, and KBr-impregnated sorbents.
From Ding F, Zhao Y, Mi L, Li H, Li Y, Zhang J. Removal of gas-phase elemental mercury in flue gas by inorganic chemically promoted natural mineral sorbents. Ind Eng Chem Res 2012; 51(7):3039–47.

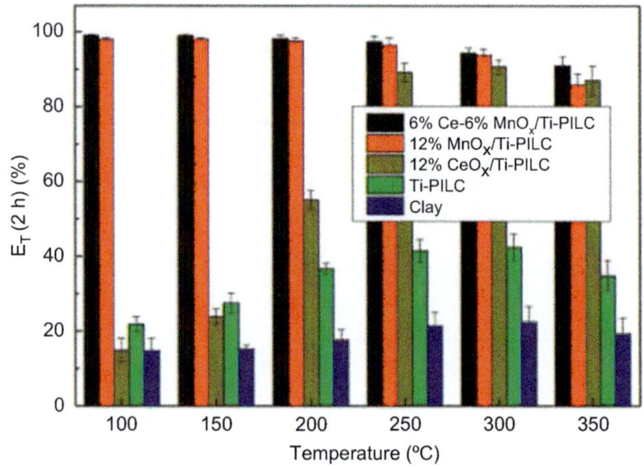

Figure 8.45 Total Hg^0 removal efficiency of the catalysts at 2 h.
From He C, Shen B, Chen J, Cai J. Adsorption and oxidation of elemental mercury over Ce-MnO_x/Ti-PILCs. Environ Sci Technol 2014;48(14):7891–8.

original mineral is about 10%. The average Hg^0 removal efficiency of $CuCl_2$-impregnated Atp (Cu-Atp) and $CuCl_2$-impregnated Ben (Cu-Ben) at 120°C was about 90%, respectively. The average Hg^0 removal efficiency of $NaClO_3$-impregnated Atp at 120°C exceeds 90%, while its performance is limited by the temperature. For the KI-impregnated adsorbent, as the temperature is raised from 70 to 150°C, the efficiency is steadily increased.

He et al. [45] investigated a series of innovative Ce–Mn/Ti-pillared clay (Ce–Mn/Ti-PILC) catalysts for Hg^0 capture. The 6% Ce–6% MnO_x/Ti-PILC catalyst showed above 90% Hg^0 removal efficiency at a wide reaction temperature window of 100–350°C (Fig. 8.45). Fig. 8.46 shows the mercury removal mechanism of 6%

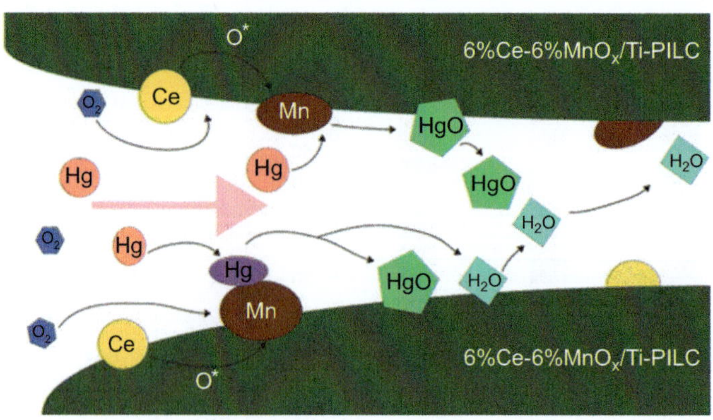

Figure 8.46 Illustration of mercury adsorption over 6%Ce–6%MnO_x/Ti-PILC.
From He C, Shen B, Chen J, Cai J. Adsorption and oxidation of elemental mercury over Ce-MnO_x/Ti-PILCs. Environ Sci Technol 2014;48(14):7891–8.

Ce−6% MnO_x/Ti-PILC. Adsorption played a dominant role in Hg^0 removal at the first reaction stage. As the reaction progresses, Hg^0 oxidation ability is improved. Both hydroxyl and lattice oxygen on the catalyst surface participate in Hg^0 oxidation. At low temperatures (150°C), the hydroxyl oxygen and lattice oxygen from $Ce^{4+} \rightarrow Ce^{3+}$ and $Mn^{3+} \rightarrow Mn^{2+}$ cause Hg^0 oxidation; at high temperatures (250°C), hydroxyl oxygen and lattice oxygen from $Mn^{4+} \rightarrow Mn^{3+}$ were responsible for Hg^0 oxidation.

8.2.5 Hg removal by noble metals

Noble metals, including Pd, Au, Pt, and Ag, have been investigated as potential Hg oxidation catalysts. Hrdlicka et al. [46] studied the mercury oxidation performance in flue gas using gold and palladium catalysts on fabric filters. Au/TiO_2 and Pd/Al_2O_3 produce 40%−60% and 50%−80% of Hg^0 oxidation, respectively, as shown in Table 8.7. The Pd/Al_2O_3 catalyst was selected for a 19 kW pilot-scale burner. As shown in Table 8.8, the total mercury oxidation rate of the Pd/Al_2O_3 filter is higher than 90%. Each coal produces 5−10 μg/m^3 of total mercury in flue gas, while the Hg^0 concentration decreases to less than 1 μg/m^3 when passing the catalyst.

Gold is efficient for adsorbing mercury due to the formation of amalgam. Rodríguez-Pérez et al. [47] studied the mercury retention performance of a commercial AC (Norit RB3) impregnated with gold nanoparticles (Ag NPs). AC (RB3) was impregnated with different gold contents using polyvinyl alcohol and tetrakis (hydroxymethyl)-ruthenium chloride (THPC) methods, respectively. Although the mercury

Table 8.7 Total mercury oxidation for filter type Ry805 (uncoated and with all three catalysts)[a].

	Gas composition (ppm)				Total mercury oxidation (%)			
Sample	Cl_2	HCl	SO_2	NO	uncoated	TiO_2	Au/TiO_2	Pd/Al_2O_3
1	10				3	20	60	57
2		50			20	0	9	4
3			1000		33	0	11	5
4				100	9	0	10	49
5	10	50			0	17	53	10
6	10		1000		37	5	65	84
7	10			100	0	23	60	10
8		50	1000		30	0	14	45
9		50		100	30	26	41	58
10			1000	100	33	19	15	75
11	10	50	1000	100	37	21	62	75

[a]Baseline gas was 4% O_2, 10% H_2O, 20−30 μg/m^3 Hg^0, and balance N_2.
From Hrdlicka JA, Seames WS, Mann MD, Muggli DS, Horabik CA. Mercury oxidation in flue gas using gold and palladium catalysts on fabric filters. Environ Sci Technol 2008;42:6677−82.

Table 8.8 Total mercury oxidation and ash carbon content for each coal type tested in small pilot-scale testing[a].

	Total mercury oxidation (%)			
	uncoated filter		Filter with Pd/Al_2O_3	
	No Hg^0 doping	With Hg^0 doping	With Hg^0 doping	Carbon in ash (%)
Illinois No. 6	95	95		10
Eagle Butte	90	68	93	17
Falkirk	90	65	92	5

[a]Hg^0 doping is 10 μg/m^3. All of the tests were performed for 6 h.
From Hrdlicka JA, Seames WS, Mann MD, Muggli DS, Horabik CA. Mercury oxidation in flue gas using gold and palladium catalysts on fabric filters. Environ Sci Technol 2008;42:6677−82.

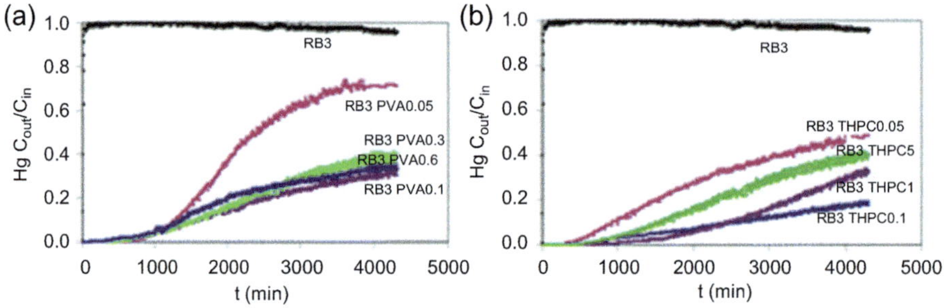

Figure 8.47 Curves of mercury adsorption in raw RB3 and RB3 samples with different gold contents, obtained by (a) polyvinyl alcohol and (b) tetrakis (hydroxymethyl)-ruthenium chloride.
From Rodriguez-Perez J, Lopez-Anton M, Diaz-Somoano M, Garcia R, Martinez-Tarazona M. Development of gold nanoparticle-doped activated carbon sorbent for elemental mercury. Energy Fuels 2011;25(5):2022−27.

retention capacity of these adsorbents for these two methods is similar, THPC is more efficient and in some cases 80% of Hg^0 removal efficiency was obtained (Fig. 8.47). The use of THPC is a promising method for the dropwise addition of gold (0.1%) to prepare a carbonaceous adsorbent for mercury capture.

Izquierdo et al. [48] developed regenerable mercury sorbents based on Au deposition on structured carbon-based monoliths. The adsorbent with an Au bulk content of less than 0.1% retains nearly 30 mg Hg/g Au. However, the mercury capture capacity of the adsorbent is not related to the amount of Au supported (Fig. 8.48). The most important factor for determining Hg^0 retention appears to be Ag NP size, and a balance between the amount of Au supported and the size of the NPs should be found.

Ag could also efficiently capture Hg^0 by forming an Ag−Hg amalgam. Fig. 8.49 shows a novel adsorbent based on Ag-loaded SBA-15 for the removal of Hg^0 [49].

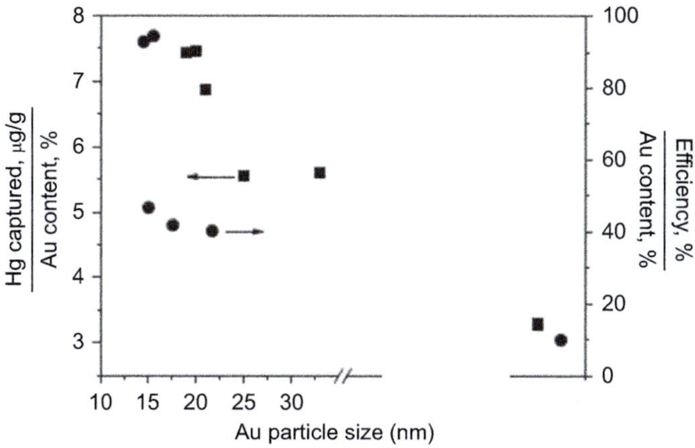

Figure 8.48 Relationship between Hg captured by the sorbents, the amount of Au, and its particle size (amount of Hg captured at 95% saturation; efficiency at 30% saturation).
From Izquierdo M, Ballestero D, Juan R, Garcia-Diez E, Rubio B, Ruiz C, Pino M. Tail-end Hg capture on Au/carbon-monolith regenerable sorbents. J Hazard Mater 2011;193:304–10.

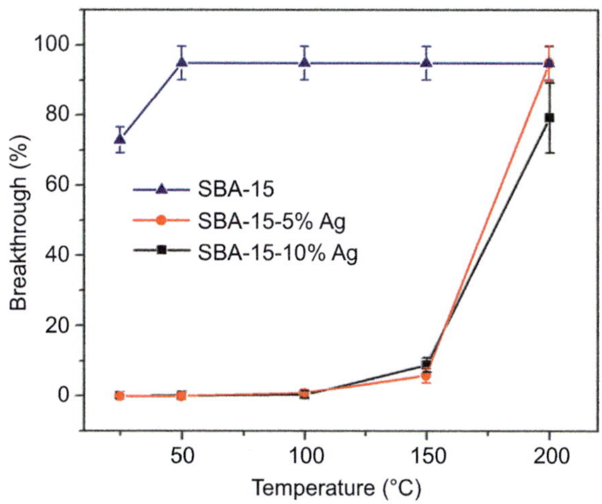

Figure 8.49 Mercury breakthrough results as a function of capture temperature for the samples (V_{Hg0} = 200 mL and $M_{adsorbent}$ = 30 mg).
From Xie Y, Yan B, Tian C, Liu Y, Liu Q, Zeng H. Efficient removal of elemental mercury (Hg^0) by SBA-15-Ag adsorbents. J Mater Chem A 2014;2(42):17,730–4.

The pure SBA-15 samples exhibit a negligible Hg^0 capturing capability. After the deposition of Ag NPs on SBA-15, more than 90% of mercury was captured at 150°C. This indicates promising potential applications on mercury emission control in coal-fired power plants. The possible adsorption mechanism could be the amalgamation of Hg^0 with Ag NPs.

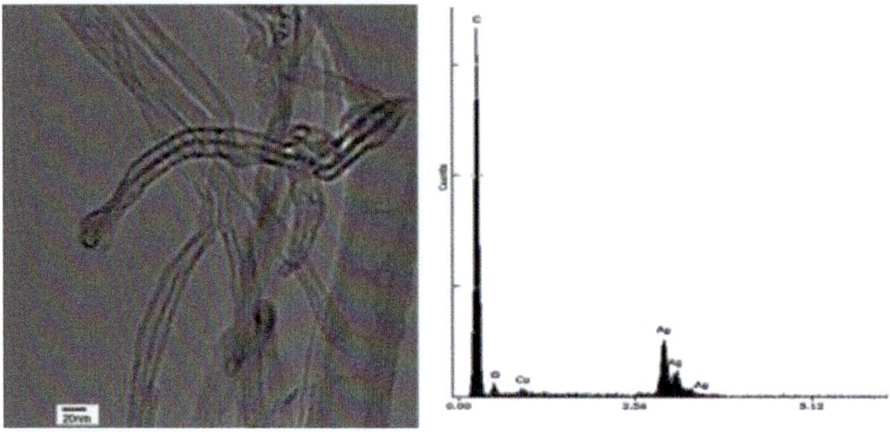

Figure 8.50 Characterization of the silver-loaded carbon nanotubes (Ag-CNTs) by transmission electron microscopy and energy-dispersive X-ray analysis.
From Luo G, Yao H, Xu M, Cui X, Chen W, Gupta R, Xu Z. Carbon nanotube-silver composite for mercury capture and analysis. Energy Fuels 2009;24(1):419–26.

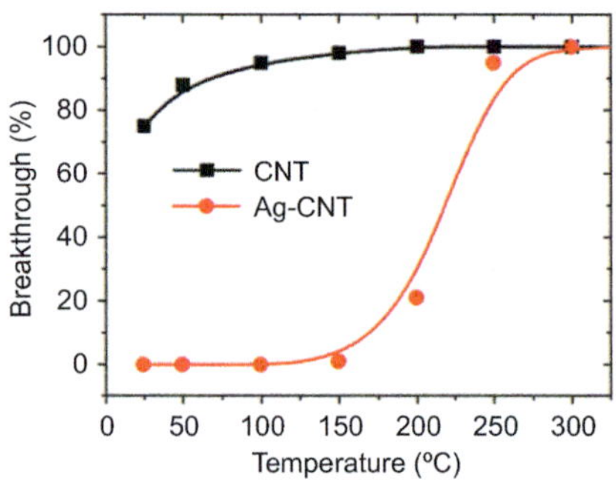

Figure 8.51 Mercury breakthrough tests of carbon nanotubes (CNTs) and silver-loaded carbon nanotubes (Ag-CNTs).
From Luo G, Yao H, Xu M, Cui X, Chen W, Gupta R, Xu Z. Carbon nanotube-silver composite for mercury capture and analysis. Energy Fuels 2009;24(1):419–26.

Luo et al. [50] analyzed the adsorption characteristics of mercury on carbon nanotubes (CNTs) and silver-loaded carbon nanotubes (Ag-CNTs) (Figs. 8.50 and 8.51). Even at a room temperature of 25°C, CNTs do not capture mercury with high efficiency and the penetration rate is 75%. In contrast, Ag-CNTs exhibit extremely high mercury capture capacity even at 150°C. Thus Ag-CNTs are particularly suitable for capturing mercury in coal-fired flue gas. The penetration temperature of Ag-CNTs for mercury adsorption is 300°C.

8.2.6 Hg removal by metal oxides

In recent years, transition metal oxide catalysts have been extensively studied to develop effective Hg^0 oxidation technologies. For higher Hg^0 oxidation efficiency, the presence of oxidant is essential for metal oxide catalysts. So, catalytic oxidation of Hg^0 using gaseous O_2 in the flue gas as the oxidant is a simple and economical method for Hg^0 control.

8.2.6.1 Iron-based catalyst

Among the available nanosized transition metal oxides catalysts, the nano-Fe_2O_3 catalyst has the advantages of strong toxicity resistance and high catalytic activity, and is an especially attractive candidate in certain special applications. Kong et al. [51] studied the heterogeneous oxidation of gas-phase Hg^0 by nano-Fe_2O_3. More than 40% of total mercury was oxidized at 300°C. However, the test temperature at 400°C caused sintering of the nano-catalyst, which led to lower Hg^0 oxidation efficiency. Hg^0 could be oxidized by active oxygen and lattice oxygen on the nano-Fe_2O_3 surface.

Yang's group developed a series of Fe-spinel catalysts, including $(Fe_{3-x}Ti_x)_{1-\delta}O_4$, Mn/γ-$Fe_2O_3$, $(Fe_{3-x}Mn_x)_{1-\delta}O_4$, $(Fe_2Ti_xMn_{1-x})_{1-\delta}O_4$, $(Fe_2Ti_{0.6}V_{0.4})_{1-\delta}O_4$, and so on. For $(Fe_{3-x}Ti_x)_{1-\delta}O_4$ $x = 0$, 0.2, 0.5, and 0.8 [52–59]. The cation vacancies were the active adsorption sites for Hg^0, and Mn^{4+} acted as the oxidizing agents. With the increase in Mn content, the percents of Mn^{4+} cations and cation vacancies on the surface increased. As a result, Hg^0 capture by $(Fe_{3-x}Ti_i)_{1-\delta}O_4$ was obviously promoted with the increase in Mn content. $(Fe_{2.2}Ti_{0.8})_{1-\delta}O_4$ showed excellent capacity for Hg^0 capture; the mercury capture capacity was above 1.5 mg/g at 100–300°C (Table 8.9).

A stoichiometric nanosized Mn–Fe spinel ($Fe_{2.2}Mn_{0.8}O_4$) was synthesized using a coprecipitation method [55]. After heat treatment in air at 400°C, chemical inhomogeneities caused by the difference in oxidation kinetics between Fe^{2+} and Mn^{2+}/Mn^{3+} were observed (Fig. 8.52). Mn, particularly Mn^{4+}, is concentrated on the particles' surface. In addition, the surface cation vacancy increased significantly due to the enrichment of Mn^{4+} cations. As a result, the capacity of $(Fe_{2.2}Mn_{0.8})_{1-\delta}O_4$-400 for Hg^0 capture was much better than that of MnO_x/γ-Fe_2O_3, $(Fe_{2.2}Mn_{0.8})_{1-\delta}O_4$-200, and $Fe_{2.2}Mn_{0.8}O_4$ (Fig. 8.53).

Table 8.9 Capacity of $(Fe_{3-x}Mn_x)_{1-\delta}O_4$ for elemental mercury capture (mg/g).

$(Fe_{3-x}Mn_x)_{1-\delta}O_4$	100°C	150°C	200°C	250°C	300°C
$x = 0$	<0.20	<0.20	0.26	0.44	0.34
$x = 0.2$	1.92	1.80	1.60	2.20	0.84
$x = 0.5$	2.90	2.92	2.42	4.26	1.74
$x = 0.8$	2.86	3.20	4.44	5.10	1.04
$x = 0.8$ with SO_2	2.38	1.92	1.72	2.48	0.96
$x = 0.8$ with SO_2 and HCl	2.01	1.92	2.07	1.54	2.21

From Yang S, Yan N, Guo Y, Wu D, He H, Qu Z, Li J, Zhou Q, Jia J. Gaseous elemental mercury capture from flue gas using magnetic nanosized $(Fe_{3-x}Mn_x)_{1\text{-delta}}O_4$. Environ Sci Technol 2011;45(4):1540–6.

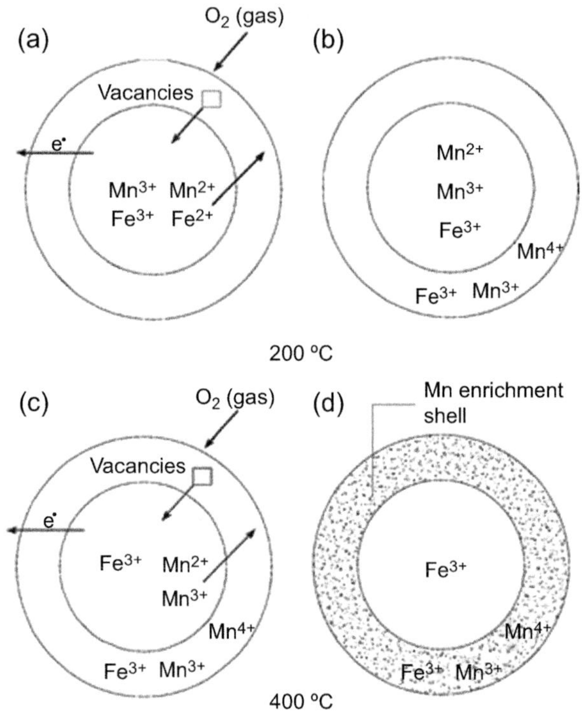

Figure 8.52 Oxidation schemes of $Fe_{2.2}Mn_{0.8}O_4$ at 200 and 400°C.
From Yang S, Guo Y, Yan N, Wu D, He H, Qu Z, Jia J. Elemental Mercury Capture from Flue Gas by Magnetic Mn–Fe Spinel: Effect of Chemical Heterogeneity. Ind Eng Chem Res 2011; 50(16):9650–56.

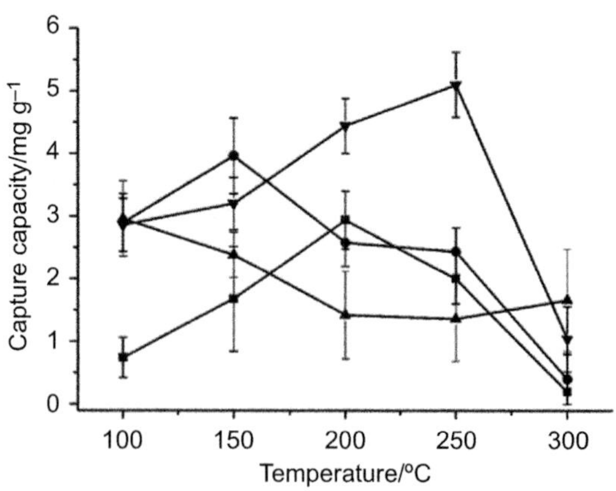

Figure 8.53 Capacity for elemental mercury capture: ■, $Fe_{2.2}Mn_{0.8}O_4$; ●, $(Fe_{2.2}Mn_{0.8})_{1-\delta}O_4$ 200; ▼, $(Fe_{2.2}Mn_{0.8})_{1-\delta}O_4$ 400; ▲, $MnO_x/\gamma\text{-}Fe_2O_3$.
From Yang S, Guo Y, Yan N, Wu D, He H, Qu Z, Jia J. Elemental Mercury Capture from Flue Gas by Magnetic Mn–Fe Spinel: Effect of Chemical Heterogeneity. Ind Eng Chem Res 2011; 50(16):9650–56.

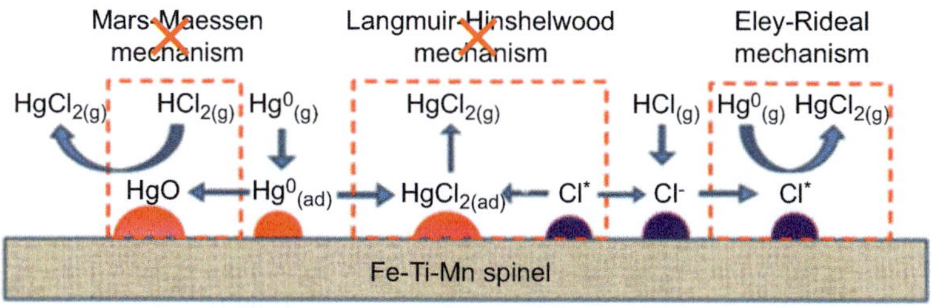

Figure 8.54 Illustration of the mercury adsorption mechanism of Fe—Ti—Mn spinel. From Xiong S, Xiao X, Huang N, Dang H, Liao Y, Zou S, Yang S. Elemental Mercury Oxidation over Fe—Ti—Mn Spinel: Performance, Mechanism, and Reaction Kinetics. Environ Sci Technol 2017;51(1):531—539.

Titanium (Ti) was incorporated into nonstoichiometric Mn—Fe spinel to improve its performance for Hg^0 capture [57]. Although the amount of Mn^{4+} on $(Fe_2Ti_xMn_{1-x})_{1-\delta}O_4$ is smaller than that on $(Fe_{3-x}Mn_x)_{1-\delta}O_4$, the number of available vacancies for Hg^0 oxidation is significantly increased. As a result, Hg^0 capture of Mn—Fe spinel is generally promoted by incorporation of Ti. The Hg^0 oxidation on Fe—Ti—Mn spinel follows the Eley—Rideal mechanism (i.e., the reaction of gaseous Hg^0 with adsorbed HCl), and the oxidation rate of Hg^0 mainly depends on the surface $Cl^{\bullet}$ concentration [54], as shown in Fig. 8.54.

8.2.6.2 Manganese-based catalysts

The MnO_x-based catalyst has been extensively studied for the low temperature (<200°C) selective catalytic reduction (SCR) of NO_x. Cimino et al. [60] studied the performance of two Mn-based catalysts for simultaneous SCR of NO_x with NH_3 and Hg capture. The effect of carrier type (TiO_2 on Al_2O_3) on NO_x conversion and Hg capture efficiency was investigated. The Mn-based adsorbent on TiO_2 showed better Hg capture performance than that on Al_2O_3 (Fig. 8.55). The most likely cause is the higher proportion of Mn^{4+} sites.

MnO_x/Al_2O_3 presents excellent Hg^0 adsorption performance in the absence of HCl, and its optimal adsorption temperature is about 623 K (Fig. 8.56) [61]. However, when HCl or chlorine (Cl_2) is introduced into the flue gas, the catalytic oxidation of Hg^0 becomes dominant. Above 90%, Hg^0 removal efficiency was obtained in the presence of 20 ppm of HCl or 2 ppm of Cl_2 (Fig. 8.57). Sulfur dioxide (SO_2) has an inhibitory effect on Hg^0 adsorption on the catalyst, but its inhibitory effect was offset by Cl_2 (Fig. 8.58).

Doping molybdenum (Mo) could significantly improve Hg^0 oxidation over Mn-based catalysts [61], the activity of which was even better than the precious metal catalyst Pd/α-Al_2O_3 (Fig. 8.59). In addition, the Mo-doped catalyst exhibits excellent sulfur resistance at lower temperatures, and the catalytic oxidation efficiency of Mo(0.03)—Mn/α-Al_2O_3 in the presence of 500 ppm of SO_2 exceeds 95%, and is about 48% for the unmodified catalyst (Fig. 8.60).

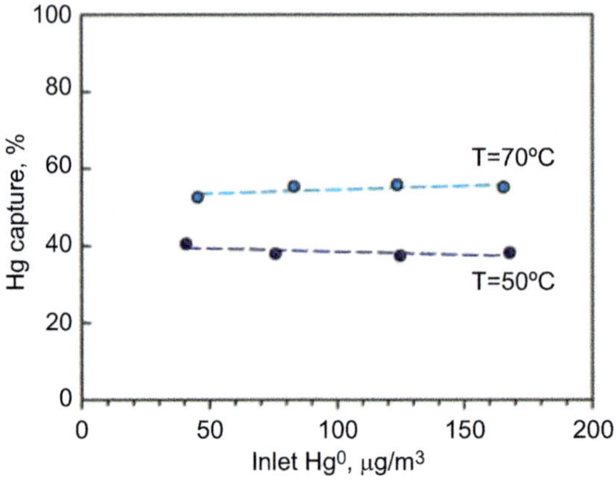

Figure 8.55 Effect of the inlet Hg^0 concentration in air on the mercury removal efficiency over an Mn/T catalyst at 50 and 70°C. Qt = 78 Sl/h; sample weight Mn/T = 10 mg.
From Cimino S, Scala F. Removal of elemental mercury by MnO_x catalysts supported on TiO_2 or Al_2O_3. Ind Eng Chem Res 2015;55(18):5133–8.

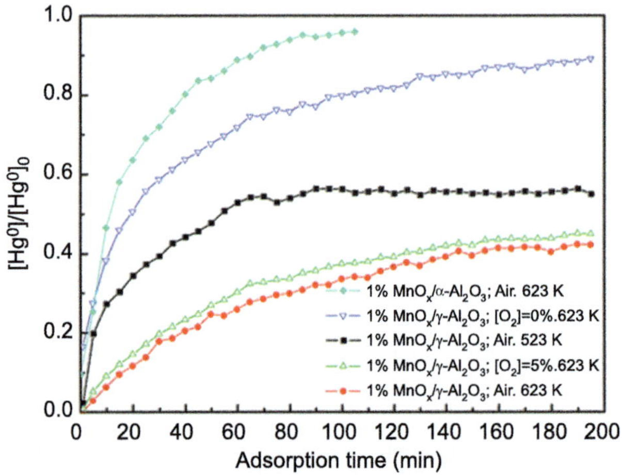

Figure 8.56 Breakthrough curves of Hg^0 across the catalysts. Air was used as the balance gas, and the packed volume of the catalyst was 3.2 mL (corresponding to 3.6 g of $MnO_x/R{-}Al_2O_3$ or 2.4 g of $MnO_x/\gamma{-}Al_2O_3$) if not indicated particularly.
From Qiao S, Chen J, Li J, Qu Z, Liu P, Yan N, Jia J. Adsorption and catalytic oxidation of gaseous elemental mercury in flue gas over MnOx/Alumina. Ind Eng Chem Res 2009;48: 3317–22.

8.2.6.3 Copper-based catalysts

Copper-based catalysts exhibit excellent Hg^0 oxidation activity [62]. As shown in Fig. 8.61, the Hg^0 and Hg^{2+} concentrations at the reactor inlet were 5.8 μg/m^3 and 1.1 μg/m^3, respectively. After passing through the catalyst, the Hg^0 and Hg^{2+}

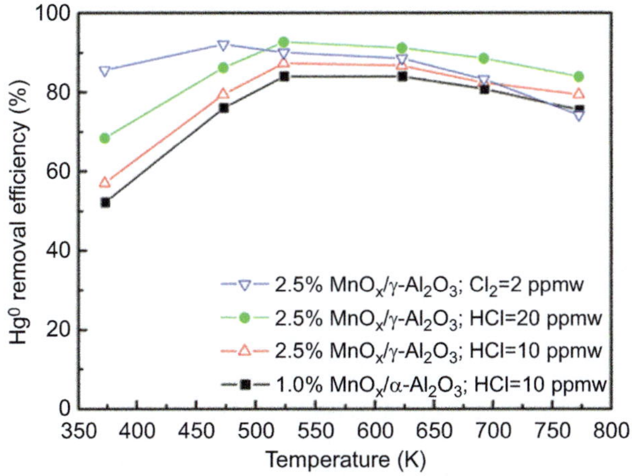

Figure 8.57 Removal efficiency of Hg^0 in the presence of HCl or Cl_2 under various temperatures.
From Qiao S, Chen J, Li J, Qu Z, Liu P, Yan N, Jia J. Adsorption and catalytic oxidation of gaseous elemental mercury in flue gas over MnOx/Alumina. Ind Eng Chem Res 2009;48:3317–22.

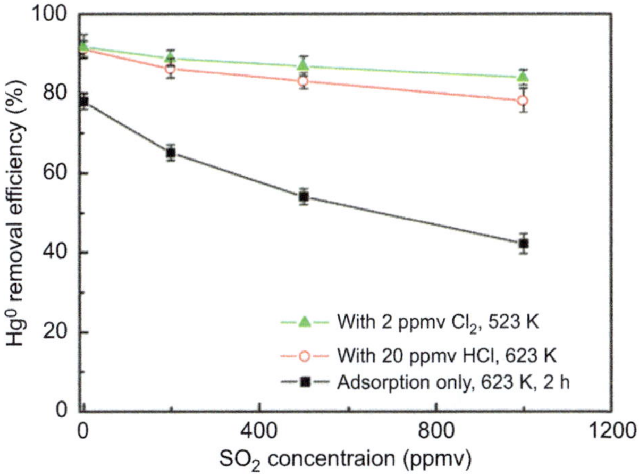

Figure 8.58 Effect of SO_2 on Hg^0 removal efficiency at various conditions. The catalyst was $MnO_x/\gamma\text{-}Al_2O_3$ with 2.5% Mn content.
From Qiao S, Chen J, Li J, Qu Z, Liu P, Yan N, Jia J. Adsorption and catalytic oxidation of gaseous elemental mercury in flue gas over MnOx/Alumina. Ind Eng Chem Res 2009;48: 3317–22.

concentrations were decreased to 0.9 μg/m^3 and 6.5 μg/m^3, respectively. Thus 85% of Hg^0 is oxidized to Hg^{2+} by CuO NPs in the actual coal combustion gas.

Fig. 8.62 shows the effect of particle size on Hg^0 oxidation at 423 K. The Hg^0 oxidation efficiency was 80% over the CuO NPs between 50 and 90 nm, but at 620 nm CuO NPs, the conversion at 473 K was 20%. The reduced conversion at

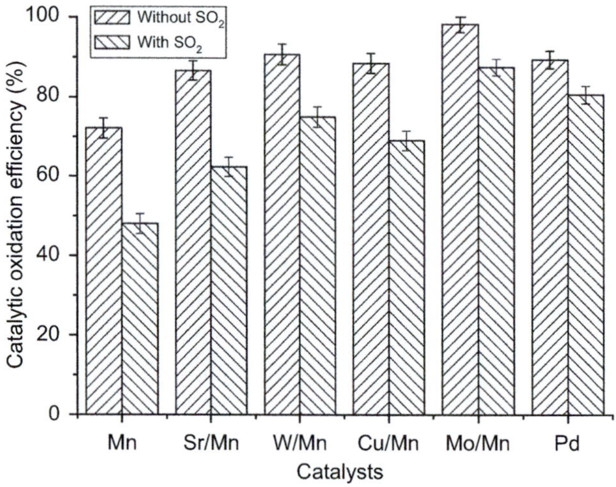

Figure 8.59 Comparison of the Hg^0 catalytic oxidation efficiencies with and without SO_2 for the various doped catalysts at 423 K (20 ppm HCl).
From Qiao S, Chen J, Li J, Qu Z, Liu P, Yan N, Jia J. Adsorption and catalytic oxidation of gaseous elemental mercury in flue gas over MnOx/Alumina. Ind Eng Chem Res 2009;48: 3317–22.

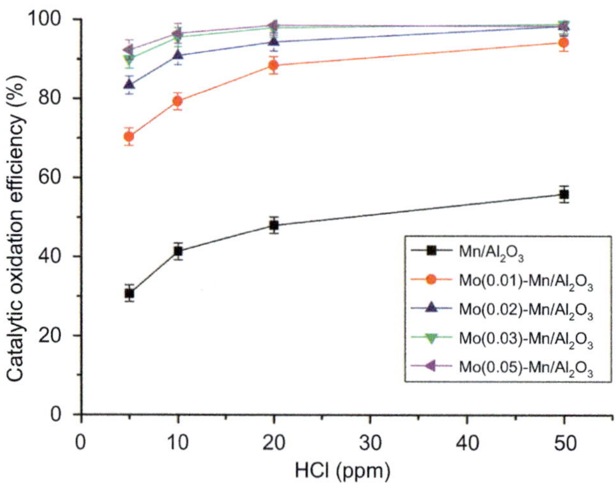

Figure 8.60 Effect of the doped Mo content on the Hg^0 oxidation efficiency at various HCl concentrations and at 423 K (with 500 ppm SO_2).
From Qiao S, Chen J, Li J, Qu Z, Liu P, Yan N, Jia J. Adsorption and catalytic oxidation of gaseous elemental mercury in flue gas over MnOx/Alumina. Ind Eng Chem Res 2009;48: 3317–22.

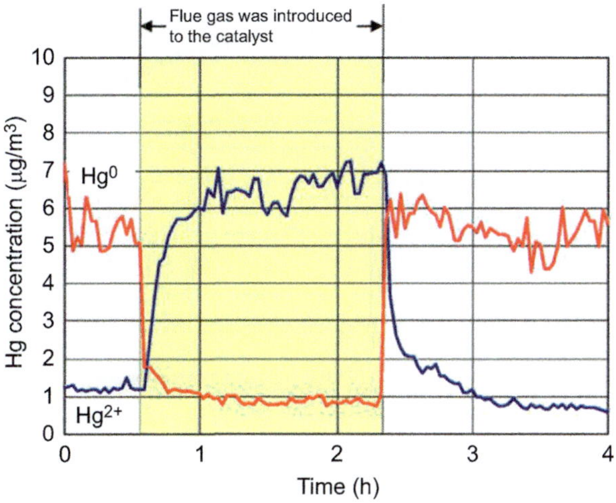

Figure 8.61 Mercury oxidation with CuO nanoparticles in actual coal combustion gas.
From Yamaguchi A, Akiho H, Ito S. Mercury oxidation by copper oxides in combustion flue gases. Powder Technol 2008;180(1−2):222−6.

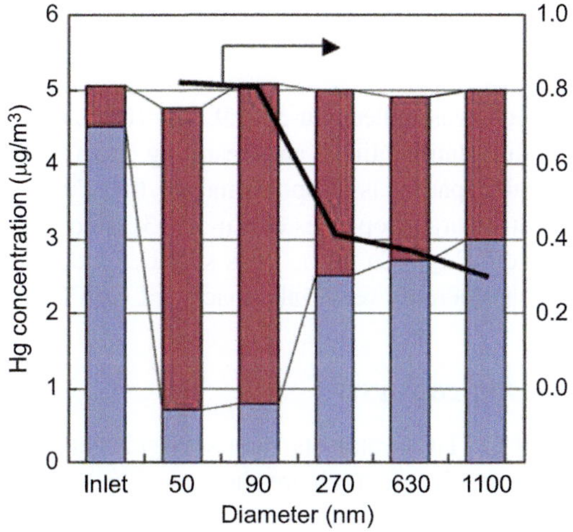

Figure 8.62 Influence of particle size on mercury oxidation at 423 K.
From Yamaguchi A, Akiho H, Ito S. Mercury oxidation by copper oxides in combustion flue gases. Powder Technol 2008;180(1−2):222−6.

473 K is caused by sintering of particles, but the decrease in conversion in the range 363−473 K may be caused by an increase in temperature rather than a decrease in particle size.

Hg^0 removal on copper-coated porous carbonaceous (PC) materials (Cu/PC) was investigated by Kim et al. [63]. Cu coating was performed on the PC surface by

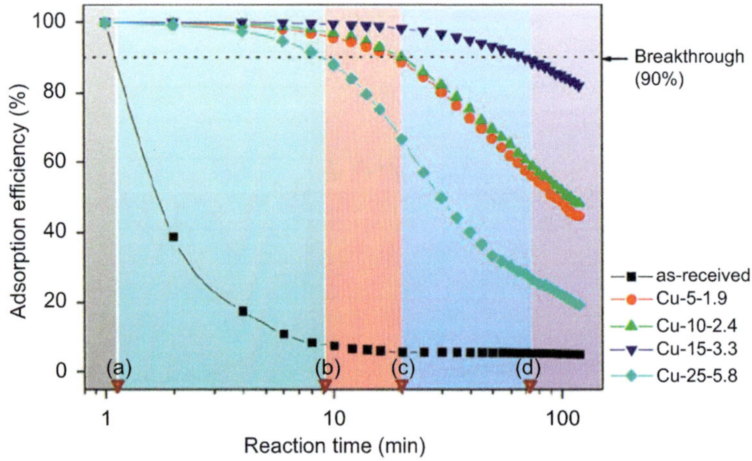

Figure 8.63 Hg^0 removal efficiency of the Cu/PC as a function of the plating time: (a) breakthrough time for the as-received sample, (b) breakthrough time for Cu-25−5.8, (c) breakthrough times for Cu-5−1.9 and Cu-10−2.4, (d) breakthrough time for Cu-15−3.3. Breakthrough means 90% filter performance for elemental mercury.
From Kim B, Bae K, Park S. Elemental mercury vapor adsorption of copper-coated porous carbonaceous materials. Micropor Mesopor Mat 2012;163:270−5.

electroless plating, which was called Cu-5−1.9, Cu-10−2.4, Cu-15−3.3, and Cu-25−5.8 according to the plating, time, and resulting metal content. As shown in Fig. 8.63, Hg^0 removal capacity is proportional to the Cu content reaching Cu-15−3.3. Although the structural properties of Cu-15−3.3 and Cu-25−5.8 are similar, Cu-25−5.8 has the lowest Cu_2O/Cu ratio. Thus surface oxidation level and total Cu content can determine the mercury removal capacity of Cu/PC.

8.2.6.4 *Cobalt-based catalysts*

Liu et al. [64] prepared Co/TiO_2 catalysts by a sol−gel method and employed gas-phase Hg^0 oxidation, the results of which are shown in Fig. 8.64. The optimal loading of Co was 7.5%, which yielded more than 90% Hg^0 oxidation efficiency within the temperature range 120−330°C. The high activity was mainly attributed to the enrichment of well-dispersed Co_3O_4. Oxygen performed a key role in the mercury oxidation process, while HCl could corrode the oxidized mercury and release it to the gas phase.

Mei et al. [65−67] studied the mercury removal performance of AC-Co, $CuCoO_4$/Al_2O_3, $MnCoO_4$/Al_2O_3, ($CuCoO_4$ + NH_4Cl)/Al_2O_3, and ($CuCoO_4$ + NH_4Br)/Al_2O_3. As shown in Fig. 8.65, when the Co_3O_4 loading value is 20%, AC-Co achieves the best mercury removal performance. At 350°C, the mercury removal efficiency was 66%. However, AC-Co is only suitable for situations without SO_2. Hg^0 oxidation ability and reusability of $Cu_xCo_{3-x}O_4$ were studied with the intention of improving the SO_2 antipoisoning ability of Co_3O_4 [67]. With the continuous increase of x from 0.75 to 2.25, $Cu_xCo_{3-x}O4$'s SO_2 antipoisoning ability increased.

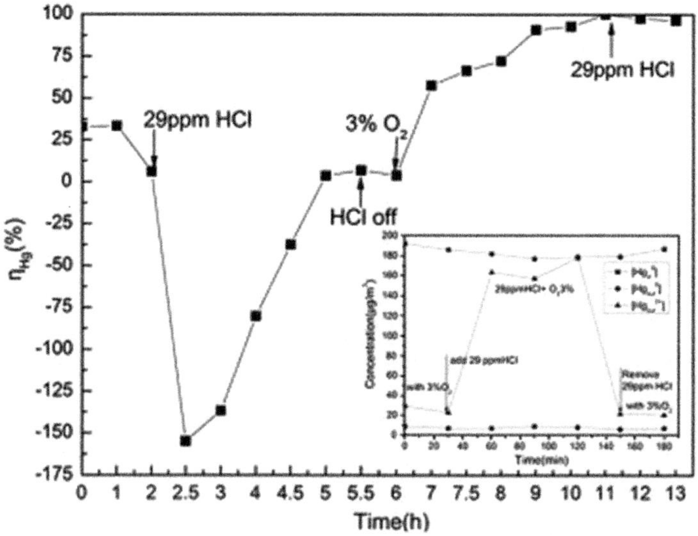

Figure 8.64 Effect of oxygen and hydrochloric on mercury oxidation efficiency. Condition: Hg^0] = 180 μg/m^3, temperature = 150°C, balanced gas = N_2, flow rate = 700 mL/min, GHSV = 105,000 h^{-1}.
From Liu Y, Wang Y, Wang H, Wu Z. Catalytic oxidation of gas-phase mercury over Co/TiO_2 catalysts prepared by sol−gel method. Catal Commun 2011;12(14):1291−4.

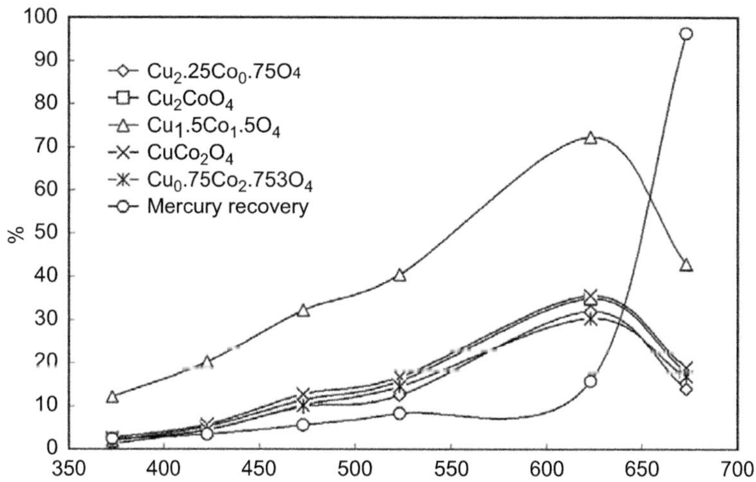

Figure 8.65 Hg^0 oxidation ability versus temperature of different $Cu_xCo_{3-x}O_4$ samples under condition 1 and the mercury recovery ratio of spent $Cu_{1.5}Co_{1.5}O_4$ under different temperatures.
From Mei Z, Shen, Z, Zhao Q, Yuan T, Zhang Y, Xiang F, Wang W. Removing and recovering gas-phase elemental mercury by $Cu_{(x)}Co_{(3-x)}O_{(4)}$ (0.75 < or = x < or = 2.25) in the presence of sulfur compounds. Chemosphere 2008;70(8):1399−404.

The gas-phase Hg^0 oxidation capacities of Al_2O_3 (AL) loaded with $CuCoO_4$ (AL-C), $CuCoO_4 + NH_4Cl$ (AL-CCl), and $CuCoO_4 + NH_4Br$ (AL-CBr) were investigated to produce more cost-effective adsorbents for controlling Hg^0 emissions during coal combustion [66]. NH_4Cl or NH_4Br-doped AL-C showed much higher Hg^0 oxidizing capacity than AL-C, and the best NH_4Cl or NH_4Br adulteration value was 30%. AL-C, AL-CCl, and AL-CBr have excellent SO_2 resistance to poisoning.

To remove Hg^0 in coal combustion flue gas and eliminate secondary mercury pollution of the spent catalyst, a new regenerable magnetic catalyst based on cobalt oxide-loaded magnetospheres from fly ash (Co-MF) was developed [68]. The use of this novel renewable magnetic catalyst for mercury removal is shown schematically in Fig. 8.66. The magnetic layer is first separated from the fly ash to synthesize the catalyst, which is then injected into the flue gas to remove mercury. The spent catalyst is then collected by a particle control device and recovered again from the fly ash using magnetic separation. The spent adsorbent is regenerated and recycled after heat treatment at a suitable temperature.

The best supported catalyst with 5.8% cobalt species achieved about 95% Hg^0 removal efficiency at 150°C under a simulated flue atmosphere. O_2 can enhance the Hg^0 removal activity of the magnetic layer catalyst by the Mars–Maessen mechanism. SO_2 showed an inhibitory effect on Hg^0 removal ability. A lower concentration of NO can increase the removal efficiency of Hg^0. However, when the NO concentration was increased to 300 ppm, a slight inhibition of NO was observed. An Hg^0 removal

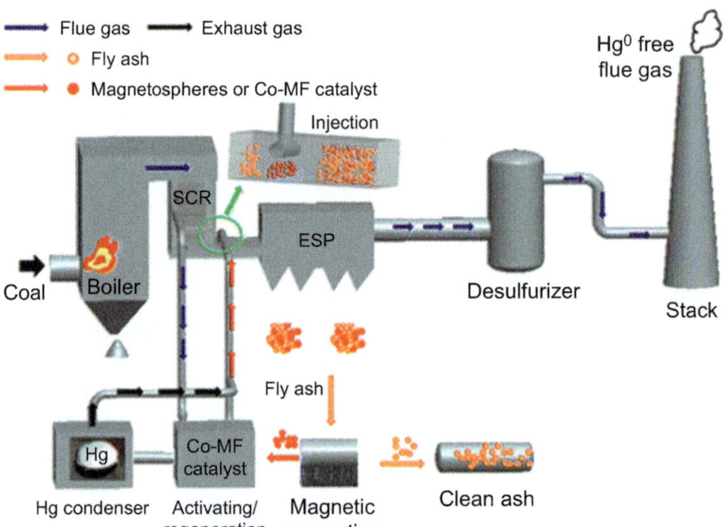

Figure 8.66 Schematic of regenerable magnetosphere catalyst for mercury emission control from coal-fired power plants. *ESP*, Electrostatic precipitator; *SCR*, selective catalytic reduction. From Yang J, Zhao Y, Zhang J, Zheng C. Regenerable cobalt oxide loaded magnetosphere catalyst from fly ash for mercury removal in coal combustion flue gas. Environ Sci Technol 2014;48(24):14,837–43.

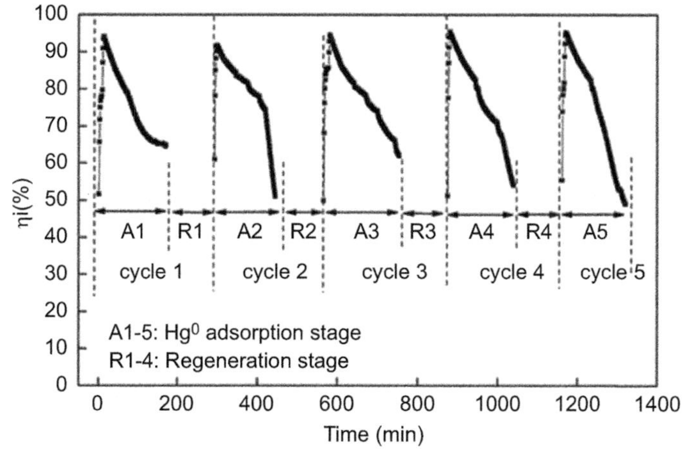

Figure 8.67 Hg^0 removal capacity of Co-MF catalyst over five oxidation−regeneration cycles. From Yang J, Zhao Y, Zhang J, Zheng C. Regenerable cobalt oxide loaded magnetosphere catalyst from fly ash for mercury removal in coal combustion flue gas. Environ Sci Technol 2014;48(24):14,837−43.

efficiency of greater than 95.5% was obtained in the presence of 10 ppm of HCl due to the formation of active chlorine species on the surface. H_2O has a serious inhibitory effect on Hg^0 removal efficiency. A repeated oxidation−regeneration cycle demonstrated that the spent Co-MF catalyst was efficiently regenerated by heat treatment at 400°C for 2 h (Fig. 8.67).

8.3 Adsorbents for capturing as in coal-derived flue gas

8.3.1 As removal by activated carbon

López-Antón et al. [69] studied the capacity of different ACs (Norit RBHG3, Norit RB3, CA) for retaining arsenic from coal combustion and gasification gases. Fig. 8.68 shows the As retention capacity under different conditions. The retained element percentage (%E) and the other maximum retention capacities (MRCs) are listed in Table 8.10. Although the theoretical data predicted different arsenic compositions in the combustion and gasification atmospheres, it can be observed that As retention in the two atmospheres is similar. Thermodynamic data at an equilibrium at 120°C show that As−Ca/Fe reactions are theoretically possible (reactions 8.1−8.8):

$$6CaO + As_4O_{10}(g) = 2Ca_3(AsO_4)_2 \quad K = 1.14 \times 10^{150} \tag{8.1}$$

$$6CaCO_3 + As_4O_{10}(g) = 2Ca_3(AsO_4)_2 + 6CO_2(g) \quad K = 1.07 \times 10^{58} \tag{8.2}$$

$$2CaO + As_4(g) + 6H_2O(g) = 2Ca(AsO_2)_2 + 6H_2(g) \quad K = 2.80 \times 10^{40} \tag{8.3}$$

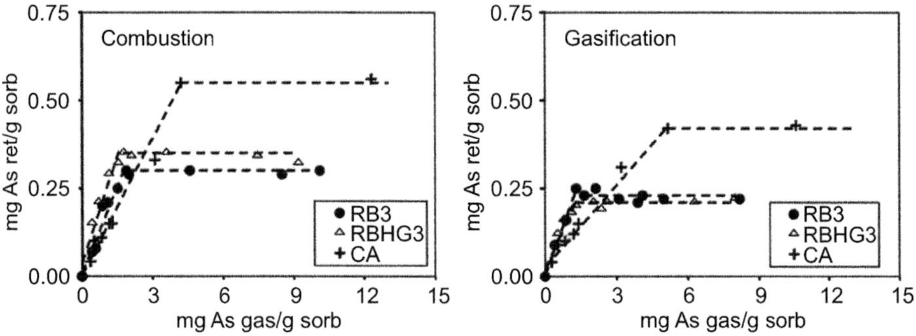

Figure 8.68 Arsenic retention in activated carbons in combustion and gasification atmospheres at the same particle size.
From López-Antón M, Díaz-Somoano M, Fierro J, Martínez-Tarazona M. Retention of arsenic and selenium compounds present in coal combustion and gasification flue gases using activated carbons. Fuel Process Technol 2007;88(8):799–805.

Table 8.10 Retention of arsenic from As_2O_3 evaporation in combustion and gasification atmospheres at 120°C.

		Combustion		Gasification	
		MRC		MRC	
Sorbent	**Particle size (mm)**	**mg g^{-1}**	**%E**	**mg g^{-1}**	**%E**
RB3	3	0.29	7 ± 2	0.19	8 ± 2
	0.2–0.5	0.30	17 ± 2	0.23	20 ± 2
RBHG3	3	0.25	10 ± 2	0.21	6 ± 2
	0.2–0.5	0.35	26 ± 7	0.21	18 ± 3
CA	0.2–0.5	0.56	14 ± 3	0.43	13 ± 3

%E, Retained element percentage; *MRC*, maximum retention capacity.
From López-Antón M, Díaz-Somoano M, Fierro J, Martínez-Tarazona M. Retention of arsenic and selenium compounds present in coal combustion and gasification flue gases using activated carbons. Fuel Process Technol 2007;88(8):799–805.

$$2CaCO_3 + As_4(g) + 6H_2O(g) = 2Ca(AsO_2)_2 + 6H_2(g) + 2CO_2(g) \;\; K = 2.50 \times 10^{08} \tag{8.4}$$

$$2Fe_2O_3 + As_4O_{10}(g) = 4FeAsO_4 \;\; K = 9.29 \times 10^{33} \tag{8.5}$$

$$4FeS + As_4O_{10}(g) + 7O2(g) = 4FeAsO_4 + 4SO_2(g) \;\; K = 1.10 \times 10^{308} \tag{8.6}$$

$$4FeS + As_4(g) = 4FeAsS \;\; K = 3.98 \times 10^{14} \tag{8.7}$$

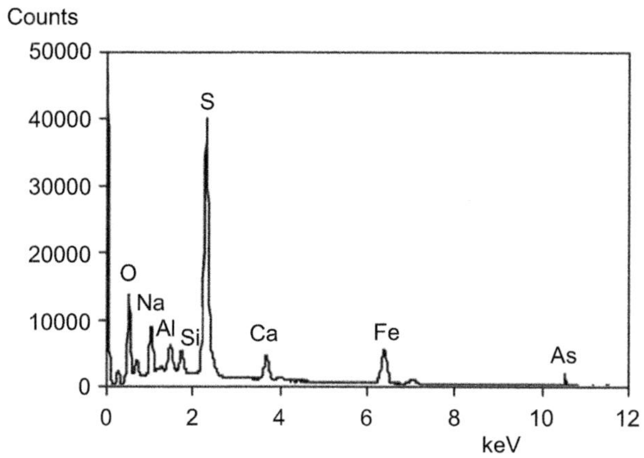

Figure 8.69 Energy-dispersive X-ray analysis of postretention activated carbon CA obtained in the combustion atmosphere.
From López-Antón M, Díaz-Somoano M, Fierro J, Martínez-Tarazona M. Retention of arsenic and selenium compounds present in coal combustion and gasification flue gases using activated carbons. Fuel Process Technol 2007;88(8):799–805.

$$4FeS + As_4S_4(g) + 4H_2(g) = 4FeAsO_4 + 4H2S(g) \quad K = 6.00 \times 10^{13} \tag{8.8}$$

Scanning electron microscopy/energy-dispersive X-ray (SEM–EDX) analysis demonstrated the presence of relatively high As quantities in particles that have high proportions of Fe (Fig. 8.69). The theoretically predicted reactions are more favorable in oxidant conditions, which would explain the slightly higher MRC value obtained in the combustion atmosphere (Table 8.10). Moreover, the smaller particle size of the sorbent is favorable for As retention (Table 8.10).

8.3.2 As removal by fly ash

López-Antón et al. [70] evaluated the As retention capacities of different fly ashes. The %E and MRC are listed in Table 8.11. The arsenic retention capacity is similar in both combustion and gasification atmospheres (Fig. 8.70). Unburned materials in different fly ashes did not have a significant effect on As retention. As the unburned carbon content increased, only a small increase in MRC and efficiency was observed. The MRC shows a higher value in the CTP, the unburned carbon content of which is smallest. Although the high surface area due to the increase in unburned carbon content favors As retention, the carbon content hardly increases retention in the same fly ash.

Fig. 8.71 shows the relationships between As and Ca content in CTA obtained from combustion and gasification atmospheres. Similar relationships were obtained between As and Fe. Moreover, an analysis by SEM/EDX in the CTP origin sample confirmed the presence of relatively high As content in particles that have high proportions of Fe (Fig. 8.72). Possible Ca and/or Fe interactions with As are shown in Eqs. (8.9)–(8.16).

Table 8.11 Retention of arsenic in the sand and fly ashes in the two gas atmospheres at 120°C.

	Combustion		Gasification	
	MRC		MRC	
Sorbent	**mg g^{-1}**	**%E**	**mg g^{-1}**	**%E**
Sand	0.04	0.05	0.05	0.06
CTA orig.	2.8 ± 0.1	12 ± 4	2.1 ± 0.1	12 ± 4
CTA >150	3.5 ± 0.1	13 ± 6	2.9 ± 0.1	15 ± 6
CTA >150 agl.	4.2 ± 0.1	18 ± 9	3.8 ± 0.1	18 ± 5
CTP orig.	5.3 ± 0.1	17 ± 6	4.5 ± 0.1	21 ± 5

%E, retained element percentage; *MRC*, maximum retention capacity.
From López-Antón M, Díaz-Somoano M, Spears D, Martínez-Tarazona M. Arsenic and selenium capture by fly ashes at low temperature. Environ Sci Technol 2006;40(12):3947–51.

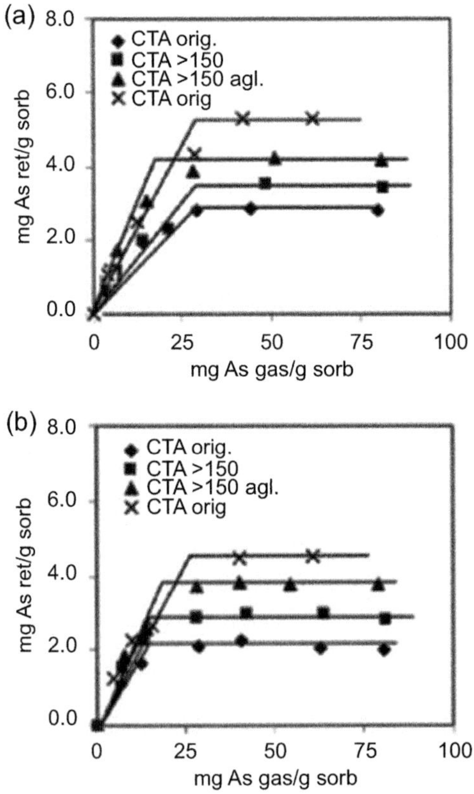

Figure 8.70 Arsenic retention in fly ashes in (a) the combustion and (b) the gasification atmosphere.
From López-Antón M, Díaz-Somoano M, Spears D, Martínez-Tarazona M. Arsenic and selenium capture by fly ashes at low temperature. Environ Sci Technol 2006;40(12):3947–51.

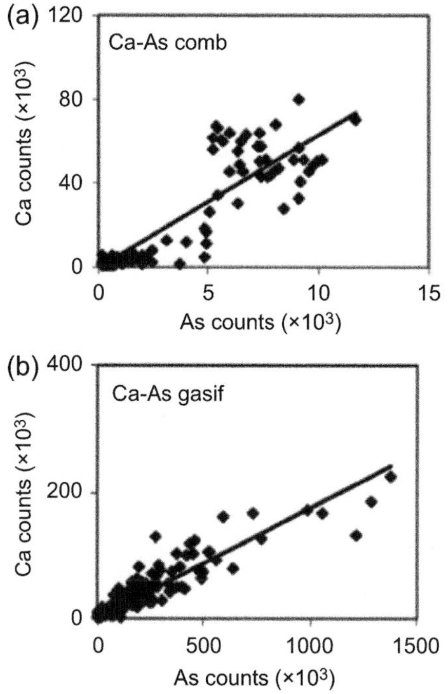

Figure 8.71 Relationships observed by laser ablation inductively coupled plasma mass spectrometry between arsenic and calcium in the (a) combustion and (b) gasification atmospheres.
From López-Antón M, Díaz-Somoano M, Spears D, Martínez-Tarazona M. Arsenic and selenium capture by fly ashes at low temperature. Environ Sci Technol 2006;40(12):3947–51.

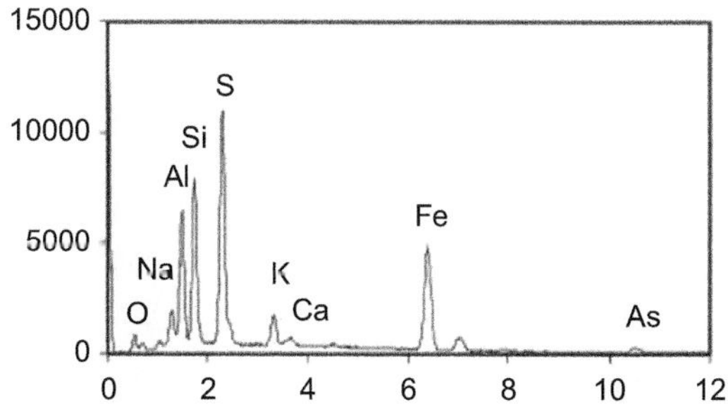

Figure 8.72 Energy-dispersive X-ray analysis of postretention fly ash CTP obtained in the gasification atmosphere.
From López-Antón M, Díaz-Somoano M, Spears D, Martínez-Tarazona M. Arsenic and selenium capture by fly ashes at low temperature. Environ Sci Technol 2006;40(12):3947–51.

$$6CaO + As_4O_{10}(g) = 2Ca_3(AsO_4)_2 \quad K = 1.14 \times 10^{150} \tag{8.9}$$

$$6CaCO_3 + As_4O_{10}(g) = 2Ca_3(AsO_4)_2 + 6CO_2(g) \quad K = 1.07 \times 10^{58} \tag{8.10}$$

$$2CaO + As_4(g) + 6H_2O(g) = 2Ca(AsO_2)_2 + 6H_2(g) \quad K = 2.80 \times 10^{40} \tag{8.11}$$

$$2CaCO_3 + As_4(g) + 6H_2O(g) = 2Ca(AsO_2)_2 + 6H_2(g) + 2CO_2(g) \quad K = 2.50 \times 10^{08} \tag{8.12}$$

$$2Fe_2O_3 + As_4O_{10}(g) = 4FeAsO_4 \quad K = 9.29 \times 10^{33} \tag{8.13}$$

$$4FeS + As_4O_{10}(g) + 7O2(g) = 4FeAsO_4 + 4SO_2(g) \quad K = 1.10 \times 10^{308} \tag{8.14}$$

$$4FeS + As_4(g) = 4FeAsS \quad K = 3.98 \times 10^{14} \tag{8.15}$$

$$4FeS + As_4S_4(g) + 4H_2(g) = 4FeAsO_4 + 4H2S(g) \quad K = 6.00 \times 10^{13} \tag{8.16}$$

The amount and location of calcium in a solid matrix are important for understanding the As/Ca interaction [71]. Fig. 8.73 shows that the ratio of Ca/SiO_2 affects the interaction between Ca and As oxide in the solid matrix. CaO is the most effective solid reactant, while calcium monosilicate is the least effective. As the Ca/SiO_2 ratio decreases, the interaction with the As_4O_6 vapor weakens. Figs. 8.73 and 8.74 show that the carrier gas played no significant role in the amount of As reacted with each solid reactant. All three solids capture arsenic at the same level.

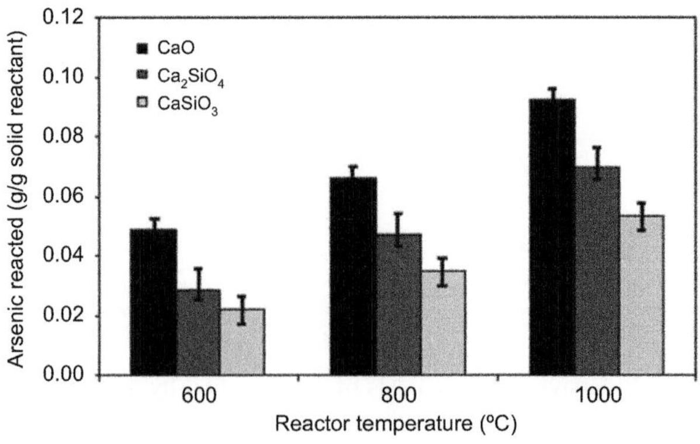

Figure 8.73 Extent of arsenic reaction with CaO, $2CaO \cdot SiO_2$, and $CaO \cdot SiO_2$ for 1 h at 13 ppm $As_4O_{6(g)}$ vapor concentration in air. Basis: mass of arsenic (as As) per unit mass of indicated solid.
From Sterling R, Helble J. Reaction of arsenic vapor species with fly ash compounds: kinetics and speciation of the reaction with calcium silicates. Chemosphere 2003;51(10):1111–9.

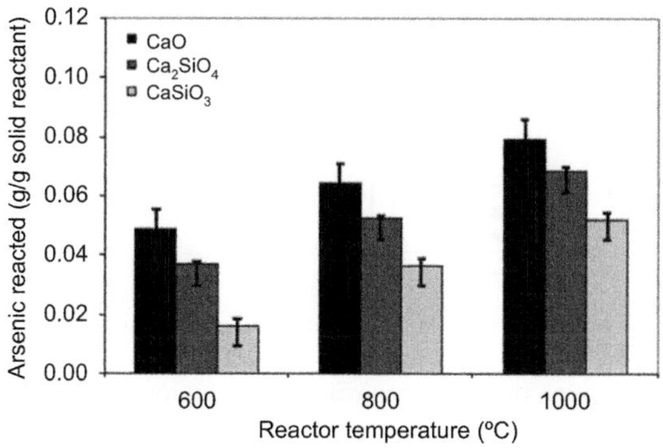

Figure 8.74 Extent of arsenic reaction with CaO, $2CaO \cdot SiO_2$, and $CaO \cdot SiO_2$ for 1 h at 13 ppm $As_4O_{6(g)}$ vapor concentration in N_2. Basis: mass of arsenic (as As) per unit mass of indicated solid.
From Sterling R, Helble J. Reaction of arsenic vapor species with fly ash compounds: kinetics and speciation of the reaction with calcium silicates. Chemosphere 2003;51(10):1111–9.

8.3.3 As removal by calcium-based sorbent

Arsenic oxide can interact with CaO to form different calcium arsenates: $CaAs_2O_6$, $Ca_2As_2O_7$ (dicalcium arsenate), and $Ca_3As_2O_8$ (tricalcium orthoate). The thermal stability of these species increases in the order $CaAs_2O_6$ (750°C) < $Ca_2As_2O_7$ (950°C) < $Ca_3As_2O_8$ (1400°C). Fig. 8.75 shows the amount of arsenic captured at

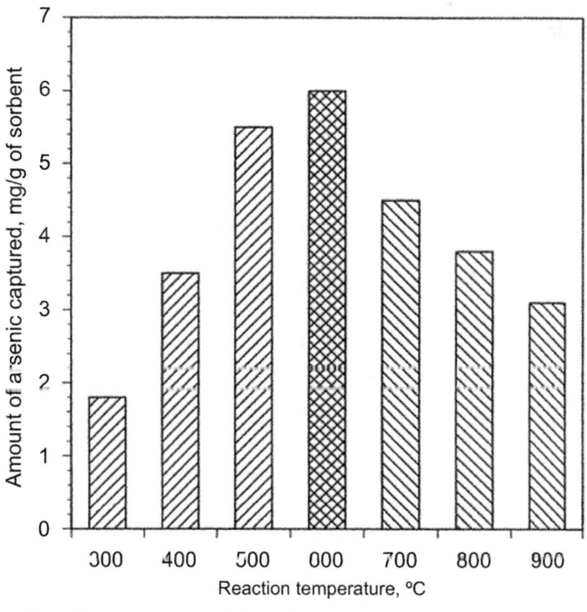

Figure 8.75 Arsenic oxide capture by CaO in the medium- and high-temperature range. As_2O_3 concentration: 14 ppm in air. Reaction time: 10 min.
From Jadhav R, Fan L. Capture of gas-phase arsenic oxide by lime: kinetic and mechanistic studies. Environ Sci Technol 2001;35(4):794–9.

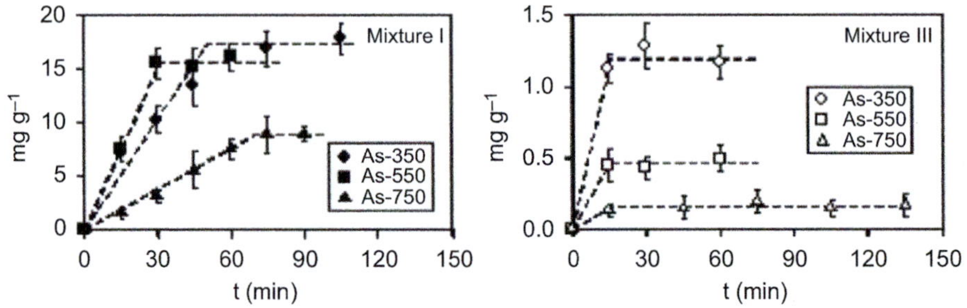

Figure 8.76 Effect of sorption time on arsenic and selenium retention on limestone in different conditions.
From Diaz-Somoano M, Martinez-Tarazona MR. Retention of arsenic and selenium compounds using limestone in a coal gasification flue gas. Environ Sci Technol 2004;38:899–903.

300–900°C when CaO is exposed to 14 ppm of arsenic oxide in air [72]. The capture capacity increases in the temperature range 300–600°C and gradually decreases when the temperature exceeds 600°C. The interaction between the two components was found to be temperature dependent. $Ca_3As_2O_8$ was a product below 600°C, while $Ca_2As_2O_7$ was the dominant form in the range 700–900°C. The optimal capture temperature of arsenic oxide occurred in the range 500–600°C.

Diaz-Somoano et al. [73] studied the retention of arsenic compounds from coal gasification flue gas using limestone. When H_2S did not exist in the atmosphere, the arsenic retained capacity can reach 17 mg/g, while the presence of H_2S results in a significant decrease in arsenic retention (Fig. 8.76). Sorbent characterization, thermal stability, and water solubility tests indicate that chemical reaction is one of the mechanisms for capturing arsenic, and $Ca(AsO_2)_2$ is the main compound formed.

Arsenic could be effectively captured by CaO at 573–1323 K [74]. Fig. 8.77 shows that, with the increase in retention time from 1 to 30 min, more arsenic is captured at

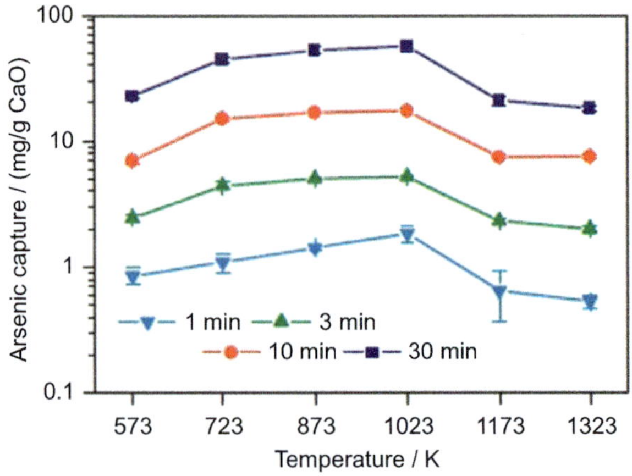

Figure 8.77 Arsenic capture by CaO at temperatures from 573 to 1323 K.
From Chen D, Hu H, Xu Z, Liu H, Cao J, Shen J, Yao H. Findings of proper temperatures for arsenic capture by CaO in the simulated flue gas with and without SO_2. Chem Eng J 2015;267: 201–6.

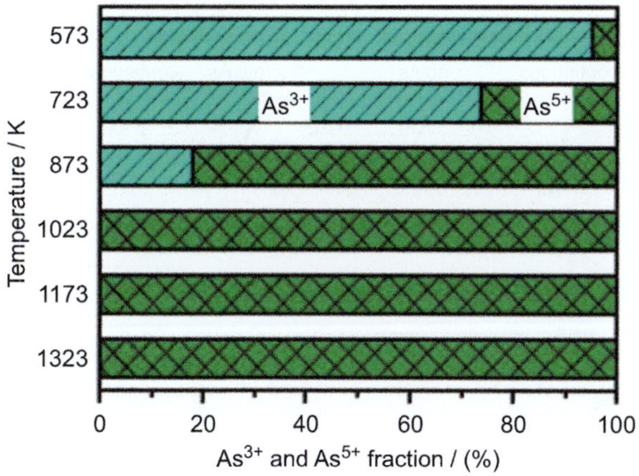

Figure 8.78 Arsenic speciation in the adsorption products.
From Chen D, Hu H, Xu Z, Liu H, Cao J, Shen J, Yao H. Findings of proper temperatures for arsenic capture by CaO in the simulated flue gas with and without SO_2. Chem Eng J 2015;267: 201–6.

each temperature. This suggests that there were enough active sites for capturing As in CaO within 30 min. The adsorption temperature significantly affected the As capture capacity. As the temperature increased from 573 to 723 K, the As capture amount increased significantly, and because the temperature further increased to 1023 K, the amount of arsenic captured increased slightly. However, at higher temperatures (1173 and 1323 K), CaO captures less arsenic.

The As capture mechanism by CaO is very complicated. Arsenic vapor in the CaO particles can be stabilized by condensation, physical adsorption, and chemical oxidation. Fig. 8.78 shows that the captured arsenic is primarily in the form of As^{3+} at 573 (94.9%) and 723 K (73.9%), which is primarily attributed to direct condensation and/or physical adsorption. When the temperature increased from 573 to 723 K, the physical adsorption of AsOO(g) by CaO was promoted. Even at 873 K, 17.8% of the arsenic in the product was still determined by physical adsorption. With a further increase in temperature, the chemical oxidation of As_2O_3(g) was promoted. In the reaction product obtained at 1023–1323 K, all of the arsenic was present as the As^{5+} form.

8.3.4 As removal by metal and metal oxides

The adsorption of arsine by Cu–Pd alloys was studied using a high-throughput composition spread alloy film sample library [75]. As shown in Fig. 8.79, when the Cu content (xCu) was below 0.2 and above 0.95, particulate As-rich layers form on the Cu-Pd alloy surfaces. In contrast, over a wide range of intermediate compositions, xCu = 0.20–0.75, As does not interact with the alloy surface. This suggests that mixed Cu–Pd arsenides are not thermodynamically favored and that the rate of alloy component segregation to form single arsenides is slow under these exposure

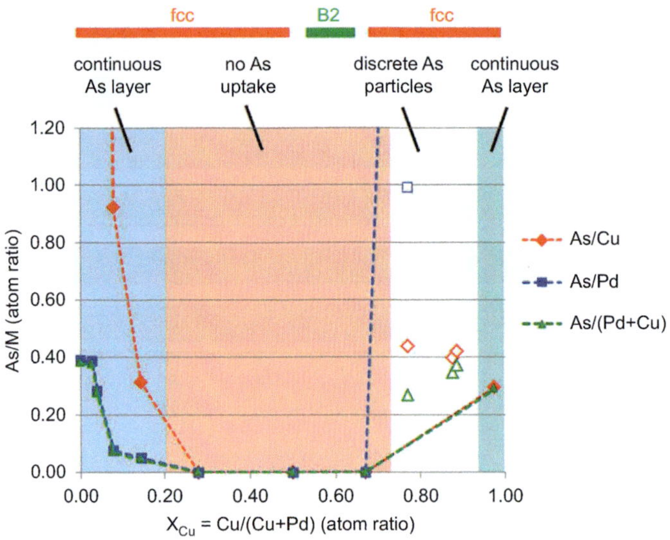

Figure 8.79 Arsenic uptake as a function of alloy composition. The bar at the top shows alloy phase behavior at 523°C.
From Uffalussy K, Miller J, Howard B, Stanko D, Yin C, Granite E. Arsenic adsorption on copper–palladium alloy films. Ind Eng Chem Res 2014;53(18):7821–7.

conditions. At alloy compositions between $x\text{Cu} = 0.75$ and $x\text{Cu} = 0.95$, As interacts with the alloy surface to form discrete As-containing particles on an As-free background. No correlation between As uptake pattern and alloy structure is observed in this study.

Poulston et al. [76] investigated the high temperature removal of AsH_3 from gasified coal using 5 wt% Pd on alumina sorbents. Arsenic uptake was found to be substantially linear with exposure time and significantly higher than unpromoted alumina beads. Although the adsorbent is less likely to saturate under this load, the arsenic loading exceeds 7% (Fig. 8.80). As the arsenic loading on the adsorbent increased, the $PdAs_2$ phase was identified in the electron-probe microanalysis and X-ray power diffraction (XRD) patterns (Figs. 8.81 and 8.82). The potential effect of arsenic adsorption on palladium on the catalyst, electrode, and membrane can be used to separate hydrogen from the fuel gas.

Zhang et al. [77] studied the gaseous As_2O_3 adsorption capacity of CaO, Fe_2O_3, and Al_2O_3. As the temperature increases at 600–900°C, the amount of arsenic adsorbed by Fe_2O_3, CaO, and Al_2O_3 decreases (Fig. 8.83). The Fe_2O_3 exhibited better temperature adaptability than CaO and Al_2O_3. In contrast, the concentration of arsenic in Al_2O_3 was significantly dependent on the temperature. Fe_2O_3 has the best adsorption performance, while Al_2O_3 has the worst adsorption performance. The amount of adsorption increases as the inlet gaseous arsenic concentration increases. The higher adsorption temperature slows the increasing rate. The inlet concentration has a greater effect on Fe_2O_3 than the other adsorbents (Figs. 8.84 and 8.85). Within the range of arsenic inlet

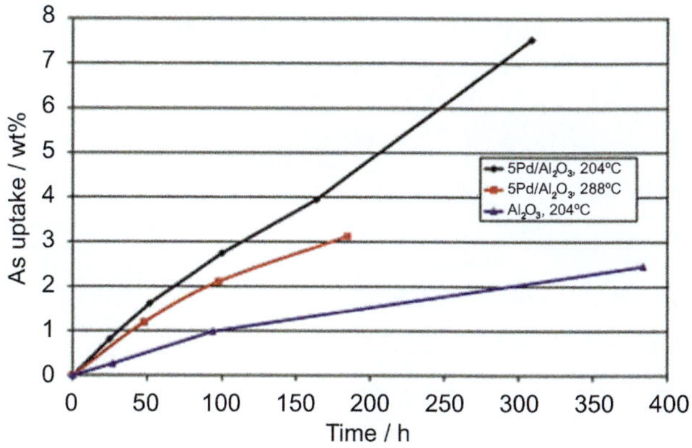

Figure 8.80 Arsine uptake as a function of AsH_3 exposure time on both Al_2O_3 and 5 wt% Pd/ Al_2O_3 at 204 and 288°C.
From Poulston S, Granite E, Pennline H, Hamilton H, Smith A. Palladium based sorbents for high temperature arsine removal from fuel gas. Fuel 2011;90(10):3118–21.

concentrations studied, the adsorption efficiency remained essentially unchanged and no sorbent saturation occurred.

Metal oxide mixtures based on iron, titanium, or zinc oxides (zinc ferrites and zinc titanates) were tested for arsenic retention [78]. The study was carried out in a laboratory-scale fixed-bed reactor at 550°C. As shown in Fig. 8.86, a metal oxide mixture containing zinc ferrite has proven to be a suitable adsorbent with an arsenic retention capacity of 21 mg/g. The results indicate that the arsenic compound can remain in the adsorbent together with the sulfur compound and desorb during the regeneration of the adsorbent.

8.4 Adsorbents for capturing F in coal-derived flue gas

Defluorination technology can be divided into flue gas defluorination and combustion fluorine fixation. Flue gas defluorination refers to the removal of HF by leaching or spraying alkaline fluorinating agent in the flue before the precipitator. Combustion of fluorine refers to the incorporation of fluorinating agent into the coal or spraying into the combustion chamber. A fluorine-fixing agent removes HF during combustion. The advantage of burning fluorine is that the direct addition of a fluorine-fixing agent in coal is a simple process and does not require huge equipment investment and operating costs of flue gas defluorination; the solid fluorine product should not cause secondary pollution to water source and soil.

Qi et al. [79] reported the test results of fluorine retention in alkaline-earth metal compound combustion to develop a highly efficient fluoride retention additive. At a combustion temperature of 1000°C, the defluorination rate is between 32.75% and 87.61%, and the range of variation is large. The effect of fluorine removal is in a descending order of defluorination rate: $BaCO_3 > SrCO_3 > Ca(OH)_2 > CaCO_3 > CaO > Ba(OH)_2 \cdot 8H_2O > Mg(OH)_2 > MgCO_3$. The results

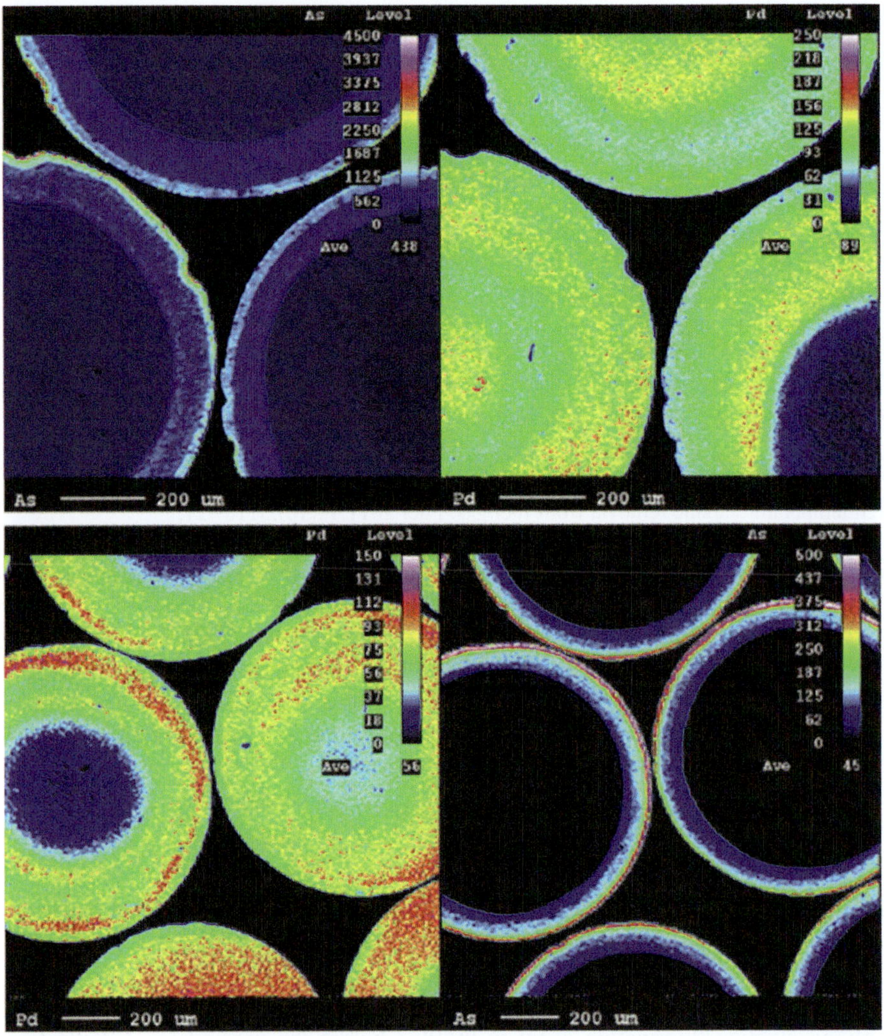

Figure 8.81 Electron-probe microanalysis of Pd/Al_2O_3 beads after exposure to AsH_3 at (*top*) 204°C with As on the left and Pd on the right and (*bottom*) 288°C with Pd on the left and As on the right.
From Poulston S, Granite E, Pennline H, Hamilton H, Smith A. Palladium based sorbents for high temperature arsine removal from fuel gas. Fuel 2011;90(10):3118–21.

show that the defluorination rate of the compound increases gradually with the increase in the atomic number of the group IIA elements, indicating that the alkaline earth metal fluoride with a large atomic number has good thermal stability. For calcium compounds, $CaCO_3$ and $Ca(OH)_2$ have similar fluorine fixation effects, which are significantly better than CaO. If analyzed according to the foregoing rules, the effect of $Ba(OH)_2$ on fluorine fixation should be close to that of $BaCO_3$, but the defluorination

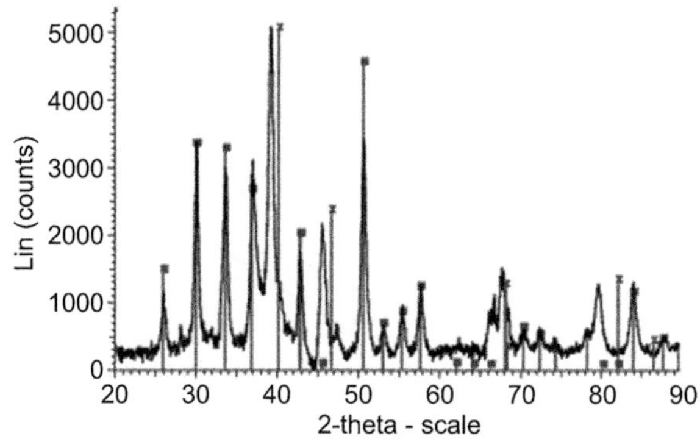

Figure 8.82 X-ray power diffraction pattern of $5Pd/Al_2O_3$ after exposure to AsH_3 in syngas for 309 h at 204°C.
From Poulston S, Granite E, Pennline H, Hamilton H, Smith A. Palladium based sorbents for high temperature arsine removal from fuel gas. Fuel 2011;90(10):3118–21.

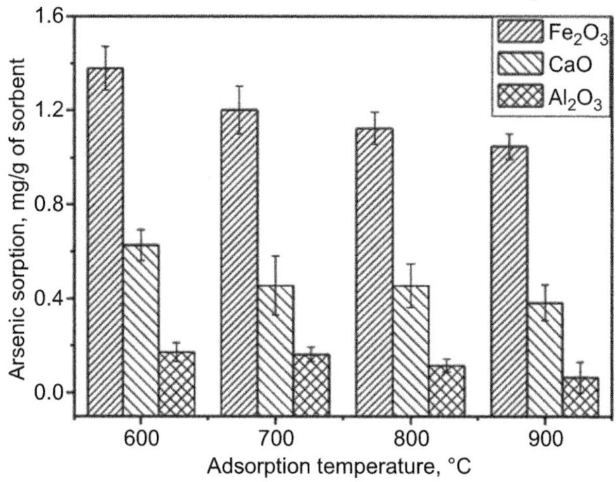

Figure 8.83 Amount of arsenic captured by various metal oxide adsorbents at different temperatures.
From Zhang Y, Wang C, Li W, Liu H, Zhang Y, Hack P, Pan W. Removal of gas-phase As_2O_3 by metal oxide adsorbents: Effects of experimental conditions and evaluation of adsorption mechanism. Energy Fuels 2015;29(10):6578–85.

rate of the experimental results is only 42.84%, which is much lower than the defluorination rate of $BaCO_3$ at 87.61%; this is due to the laboratory. The $Ba(OH)_2$ reagent is $Ba(OH)_2 \cdot 8H_2O$, and the dehydration of crystal water promotes the fluorine-fixing reaction to the reverse reaction direction, that is, the hydrolysis reaction of the fluorine-fixing product reduces the fluorine-fixing effect. When the combustion temperature is raised to 1200°C, the order of the fluorine fixation effect of each absorbent is still consistent with 1000°C. For a certain absorbent, the effect of fluorine fixation is

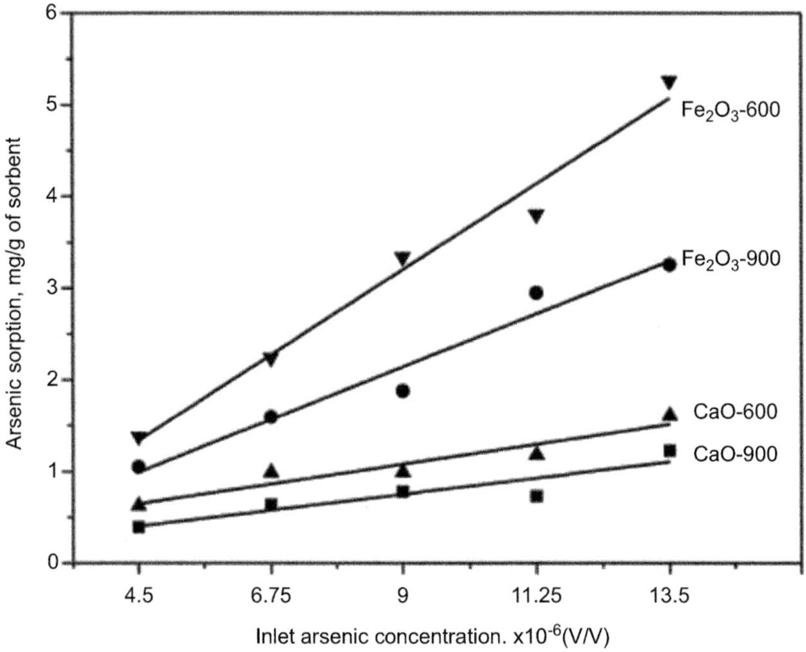

Figure 8.84 Effect of gas-phase arsenic concentration on As_2O_3 capture at different temperatures.
From Zhang Y, Wang C, Li W, Liu H, Zhang Y, Hack P, Pan W. Removal of gas-phase As_2O_3 by metal oxide adsorbents: Effects of experimental conditions and evaluation of adsorption mechanism. Energy Fuels 2015;29(10):6578–85.

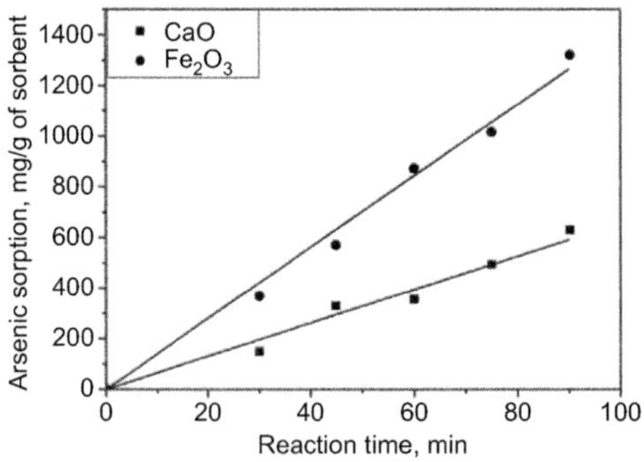

Figure 8.85 Rate of reaction of CaO and Fe_2O_3 with 4.5 ppm As_2O_3(g) in dry air at 600°C.
From Zhang Y, Wang C, Li W, Liu H, Zhang Y, Hack P, Pan W. Removal of gas-phase As_2O_3 by metal oxide adsorbents: Effects of experimental conditions and evaluation of adsorption mechanism. Energy Fuels 2015;29(10):6578–85.

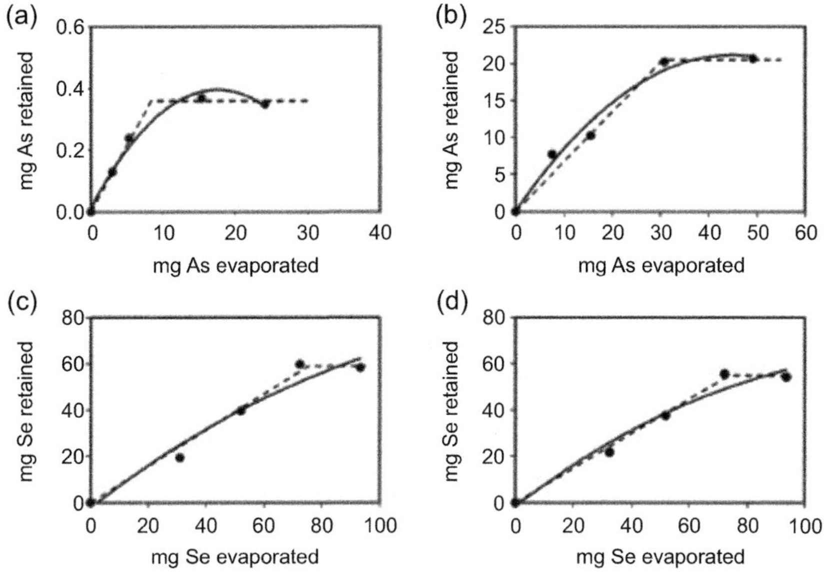

Figure 8.86 Results obtained during retention experiments for arsenic with (a) ZT-6 and (b) ZFT-11, and for selenium with (c) ZT-6 and (d) ZFT-11 (ZT-6: m(ZnO): m(TiO_2) = 0.8:1.0, ZFT-11: m(ZnO): m(Fe_2O_3): m(TiO_2) = 1.0:0.5:0.5).
From Diaz-Somoano M, López-Antón M, Martínez-Tarazona M. Retention of arsenic and selenium during hot gas desulfurization using metal oxide sorbents. Energy Fuels 2004;18(5): 1238–42.

significantly lower than that at 1000°C, and the defluorination rate is only 6.13–77.25%; however, the effect of reducing the fluorine fixation effect is different. The overall trend is the atomic number of elements with the IIA group. The increase in the fluorine-fixing effect of the compound is decreased, indicating that the alkaline-earth metal fluoride having a large atomic number has good thermal stability. The effect of the fluorine fixation of each absorbent decreases with the increase in the combustion temperature, indicating that the fluorine-fixing product alkaline-earth metal fluoride has undergone hydrolysis reaction.

The CaO–HF reaction is the main chemical reaction process of calcium-based fluorine retention. Wang et al. [80] established an unreacted shrinking core dynamic model of CaO–HF to reveal the fluorine retention mechanism. As the reaction temperature and reaction time increase, the conversion of CaO increases. However, when the temperature is 1073–1273 K, the reaction temperature does not have much influence on the CaO conversion rate. As the HF concentration increases, the CaO conversion rate increases. As the initial particle size of CaO decreases, the CaO concentration increases significantly, that is, the initial particle size of CaO has a significant effect on the fluorine retention rate.

An iron–aluminum–cerium (Fe–Al–Ce) hydroxide adsorbent with a high F adsorption capacity was investigated using X-ray photoelectron spectroscopy (XPS) and ^{19}F magic-angle spinning nuclear resonance (^{19}F MAS NMR) [81]. The XPS O1s peak shows a quantitative ligand exchange relationship between the metal-

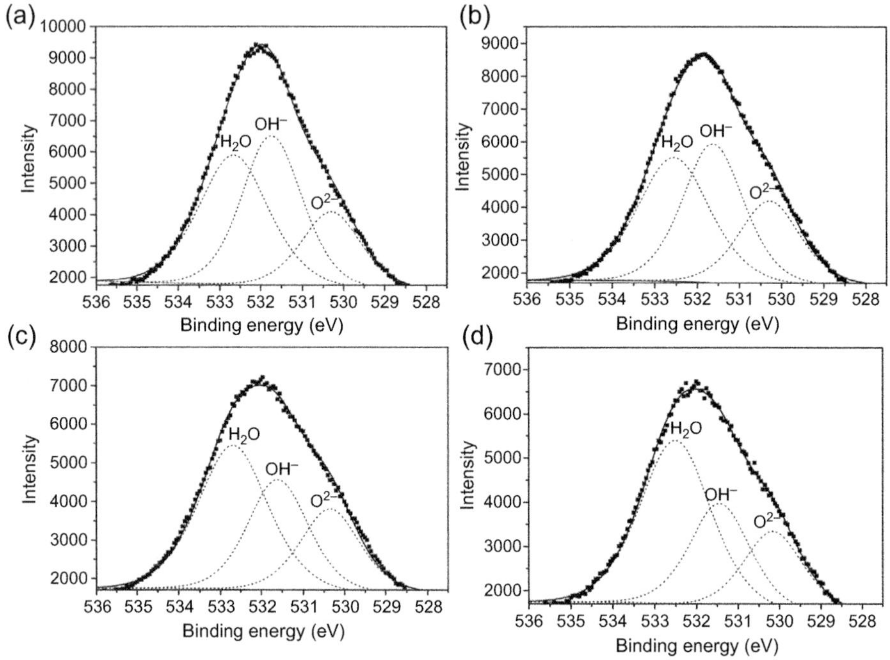

Figure 8.87 (a) O1s spectra of Fe—Al—Ce, (b) Fe—Al—Ce—F (10 mg/g), (c) Fe—Al—Ce—F (80 mg/g), and (d) Fe—Al—Ce—F (138 mg/g).
From Wu X, Zhang Y, Dou X, Zhao B, Yang M. Fluoride adsorption on an Fe—Al—Ce trimetal hydrous oxide: characterization of adsorption sites and adsorbed fluorine complex species. Chem Eng J 2013;223:364—70.

hydroxyl (—OH) group and the F ion (Fig. 8.87). The XPS F1s peak shows that F ions can replace all three metal—OH groups (Fig. 8.88). Ce—OH is the preferred adsorption site at low F load (10 mg/g), while Al—F is the most abundant complex species with increased F load (80 and 138 mg/g). In addition, Fig. 8.89 shows that Al_3—F, Ce_x—Fy, and Fe_x—Fy were identified by ^{19}F MAS NMR analysis at 20 and 80 mg/g F loading, while AlF_3 was considered to be the major species of high F load (138 mg/g).

8.5 Adsorbents for capturing Se in coal-derived flue gas

8.5.1 Se removal by activated carbon

López-Antón et al. [69] studied the capture of Se in ACs (Norit RBHG3, Norit RB3, CA). Fig. 8.90 shows the results of Se retention experiments under different conditions. As shown, higher MRC and adsorption kinetics were obtained in the gasification atmosphere. XPS demonstrated that the selenium species in the gas phase are different in both atmospheres (Fig. 8.91). Se^{4+} and Se^{2-} are the species on the adsorbent surface after exposure to combustion and gasification atmospheres, respectively.

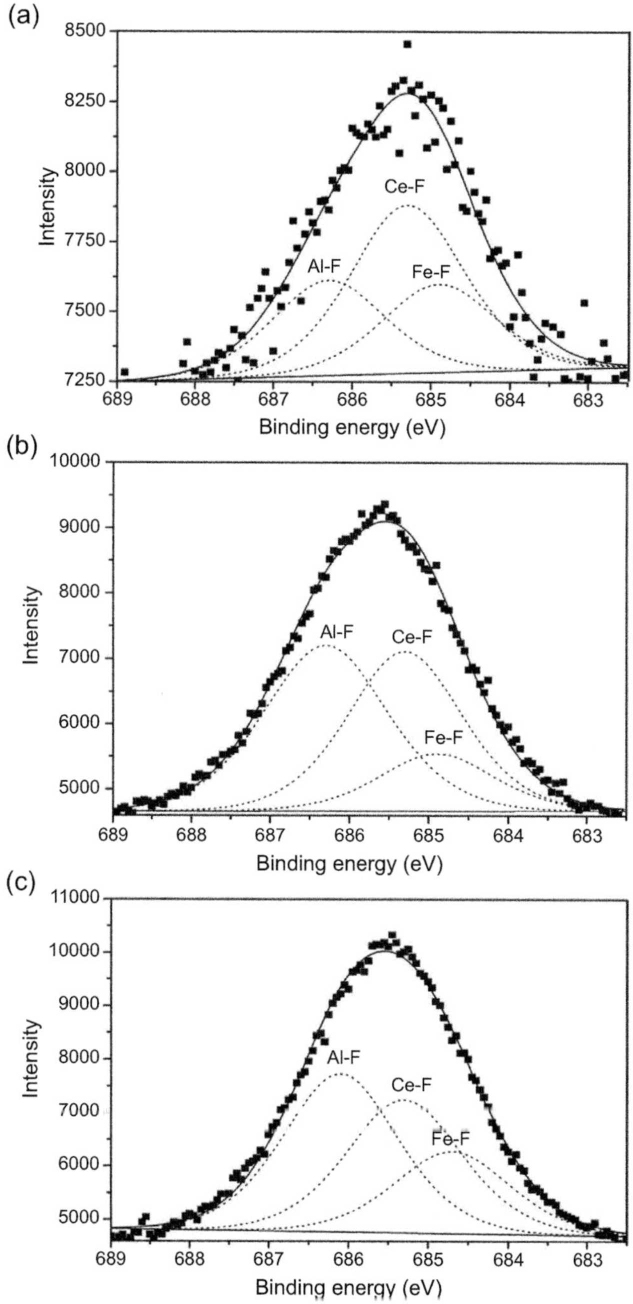

Figure 8.88 (a) F1s spectra of Fe–Al–Ce–F (10 mg/g), (b) Fe–Al–Ce–F (80 mg/g), and (c) Fe–Al–Ce–F (138 mg/g).
From Wu X, Zhang Y, Dou X, Zhao B, Yang M. Fluoride adsorption on an Fe–Al–Ce trimetal hydrous oxide: characterization of adsorption sites and adsorbed fluorine complex species. Chem Eng J 2013;223:364–70.

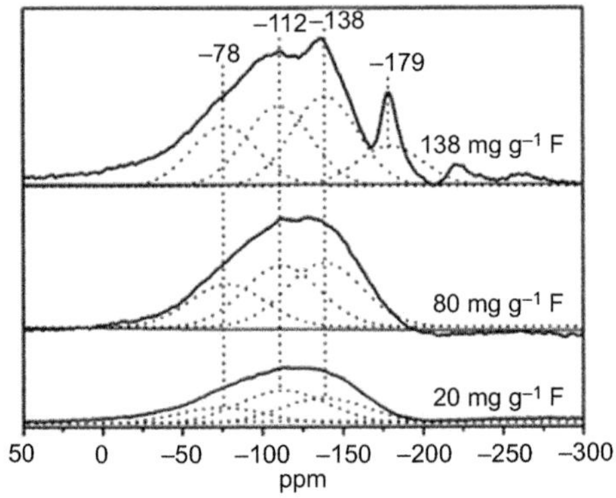

Figure 8.89 ^{19}F magic-angle spinning nuclear resonance spectra of Fe−Al−Ce adsorbents at different fluoride loads. The *dotted curves* represent simulated spectra.
From Wu X, Zhang Y, Dou X, Zhao B, Yang M. Fluoride adsorption on an Fe−Al−Ce trimetal hydrous oxide: characterization of adsorption sites and adsorbed fluorine complex species. Chem Eng J 2013;223:364−70.

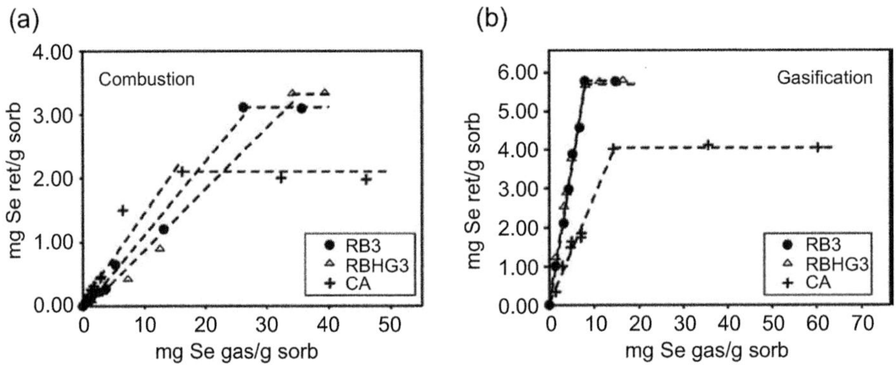

Figure 8.90 Selenium retention in activated carbons in (a) combustion and (b) gasification atmospheres at the same particle size.
From López-Antón M, Díaz-Somoano M, Fierro J, Martínez-Tarazona M. Retention of arsenic and selenium compounds present in coal combustion and gasification flue gases using activated carbons. Fuel Process Technol 2007;88(8):799−805.

8.5.2 Se removal by fly ash

López-Antón et al. [70] studied selenium capture by fly ashes at low temperature. Table 8.12 shows that, under a combustion atmosphere, a considerable amount of condensation is retained. The amount retained in the sand is 30% of the amount retained when the bed contains sand + fly ash. The MRC is slightly higher in the gasification atmosphere. There was no significant difference in the kinetics of the adsorption process under two atmospheres. When compared with the retained capacity of CTA that has different unburned carbon contents, no significant change was observed. No Ca−Se or Fe−Se association was found in CTA or CTP fly ash in a combustion

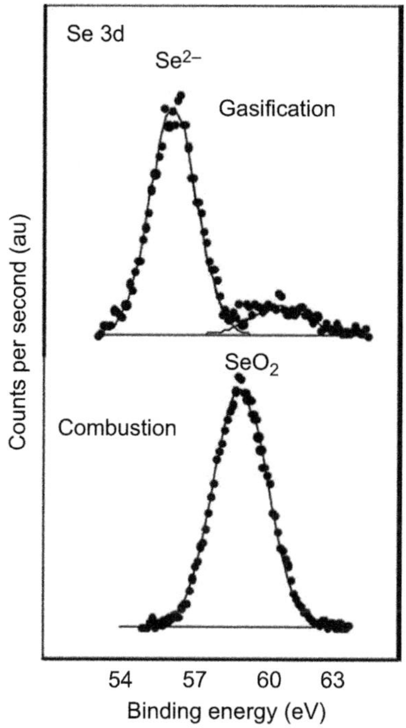

Figure 8.91 X-ray photoelectron spectroscopy spectrogram of RB3 activated carbon in combustion and gasification atmospheres.
From López-Antón M, Díaz-Somoano M, Fierro J, Martínez-Tarazona M. Retention of arsenic and selenium compounds present in coal combustion and gasification flue gases using activated carbons. Fuel Process Technol 2007;88(8):799–805.

Table 8.12 Retention of selenium in the sand and fly ashes in the two gas atmospheres at 120°C.

	Combustion		Gasification	
	MRC		MRC	
Sorbent	mg g^{-1}	%E	mg g^{-1}	%E
Sand	4.7	9.7	0.06	0.08
CTA orig.	15.6 ± 0.3	27 ± 6	17.4 ± 0.3	36 ± 4
CTA >150	16.5 ± 0.3	26 ± 4	18.4 ± 0.3	36 ± 2
CTA >150 agl.	17.7 ± 0.2	31 ± 4	19.3 ± 0.3	37 ± 4
CTP orig.	17.8 ± 0.3	38 ± 5	21.5 ± 0.4	35 ± 6

%E, Retained element percentage; *MRC*, maximum retention capacity.
From López-Antón M, Díaz-Somoano M, Spears D, Martínez-Tarazona M. Arsenic and selenium capture by fly ashes at low temperature. Environ Sci Technol 2006;40(12):3947–51.

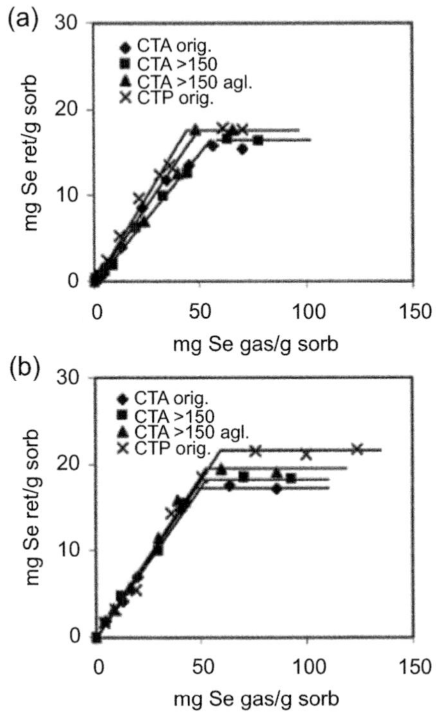

Figure 8.92 Selenium retention in fly ashes in (a) combustion and (b) gasification atmospheres. From López-Antón M, Díaz-Somoano M, Spears D, Martínez-Tarazona M. Arsenic and selenium capture by fly ashes at low temperature. Environ Sci Technol 2006;40(12):3947–51.

atmosphere (Fig. 8.92). However, under a gasification atmosphere, the relationship between Se, Ca, and Fe was observed (Fig. 8.93). The following reaction may be the cause of Ca–Se and Fe–Se interaction at 120°C.

$$CaO + H_2Se(g) = CaSe + H_2O(g) \quad K = 4.07$$

$$FeS + H_2Se(g) = FeSe + H_2S(g) \quad K = 8.97 \times 10^2$$

López-Antón et al. [82] studied the selenium content in each fly ash fraction. Se concentrations were significantly varied in different samples. To assess the relationship between Se and unburned carbon, the selenium content of the fraction separated from the ash was compared to the LOI value. Selenium performed differently in each sample studied. In CTL, Se is present in the fraction of the highest unburned carbon (40%). In the CTA sample, the highest Se concentration was present in the portion with an LOI value of 22.1%, while the lowest Se concentration was present in the portion with an LOI value of 22.4% (Fig. 8.94). In the size fraction of the CTES fly ash, the maximum selenium concentration was found to be part of the intermediate

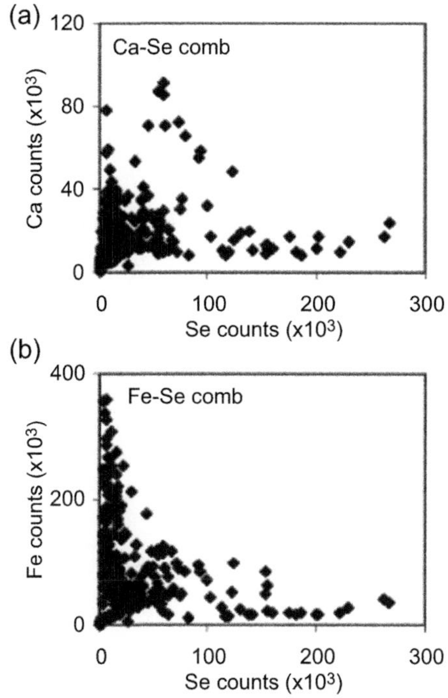

Figure 8.93 Relationships observed by laser ablation inductively coupled plasma mass spectrometry between (a) selenium and calcium and (b) selenium and iron in the combustion atmosphere.
From López-Antón M, Díaz-Somoano M, Spears D, Martínez-Tarazona M. Arsenic and selenium capture by fly ashes at low temperature. Environ Sci Technol 2006;40(12):3947−51.

LOI value (20%−30%). Finally, the fraction with 9.5% LOI showed the highest selenium content (Fig. 8.95), and no correlation was found for the unburned content. These findings indicate that there is no relationship between unburned carbon and mercury or selenium capture in the studied fly ash.

8.5.3 Se removal by calcium-based sorbents

Lime and limestone are extensively employed for the capture of sulfur during coal processing. Moreover, the Ca-based sorbents also demonstrated good retention characteristics for selenium during combustion. Diaz-Somoano et al. [73] studied the selenium removal performance of limestone in coal gasification. As shown in Fig. 8.96 and Table 8.13, the MRC of selenium depends on the temperature and gas composition. At 550 and 750°C, a similar Se retention capacity was obtained between 50 and 65 mg/g in both atmospheres, while the maximum efficiency was achieved at 750°C (92 wt%) in the absence of H_2S.

Pure CaO was used for Se capture from Se-rich stone coal combustion flue gases at 800°C [83]. As shown in Fig. 8.97, when the CaO dosage increased from 0.10 to

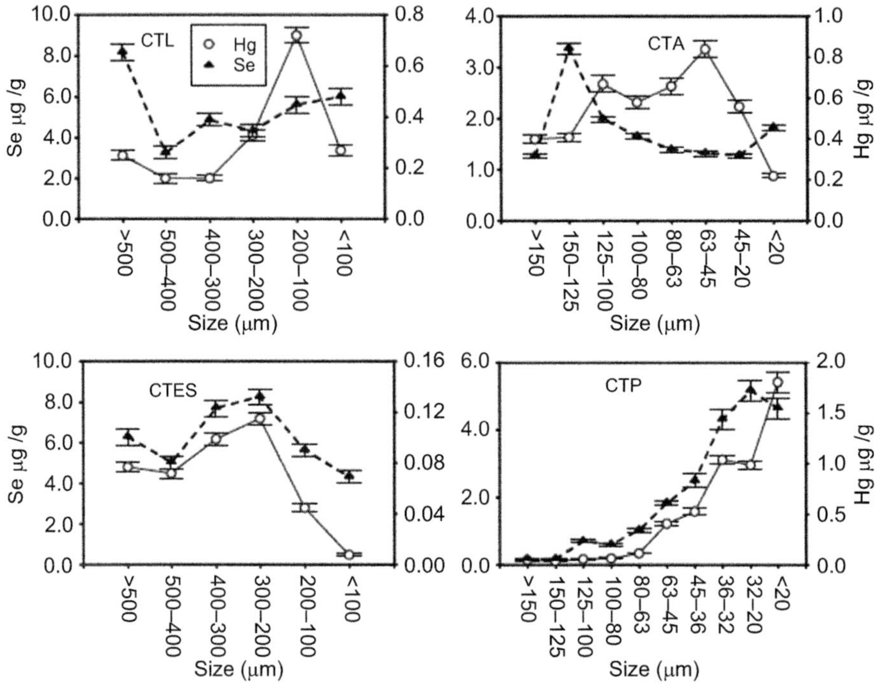

Figure 8.94 Selenium content in the size fractions of different fly ashes. (CTA, CTL, CTES: the fly ash originated from power station-burned high-rank coals, bituminous coals, and sub-bituminous coals. CTP: the fly ash originated from an fluidized-bed combustion (FBC) plant burning a mixture of bituminous coal and coal wastes of high calorific value.)
From López-Antón M, Díaz-Somoano M, Abad-Valle P, Martínez-Tarazona M. Mercury and selenium retention in fly ashes: Influence of unburned particle content. Fuel 2007;86(14): 2064–70.

0.50 g, the adsorption efficiency significantly increased from 46.43% to 65.49%. However, a further increase in CaO quality results in only a slight additional increase in adsorption efficiency. Since the reaction between CaO and SeO_2 to produce $CaSeO_3$ is exothermic, the local temperature of the adsorbent during combustion may increase with the increasing number of reactants. This could accelerate the decomposition of $CaSeO_3$. Moreover, the available surface area of the adsorbent may be smooth because the amount of adsorbent increases beyond a certain mass. When the CaO mass was 4.00 g, the highest efficiency achieved was about 67.40%. This indicates that CaO acts as an effective adsorbent for capturing Se from the flue gas produced by the combustion of selenite-rich coal.

8.5.4 Se removal by noble metal and metal oxides

Pd sorbents supported on alumina were prepared by different methods for removing hydrogen selenide from simulated fuel gas [84]. The dispersion and location of Pd

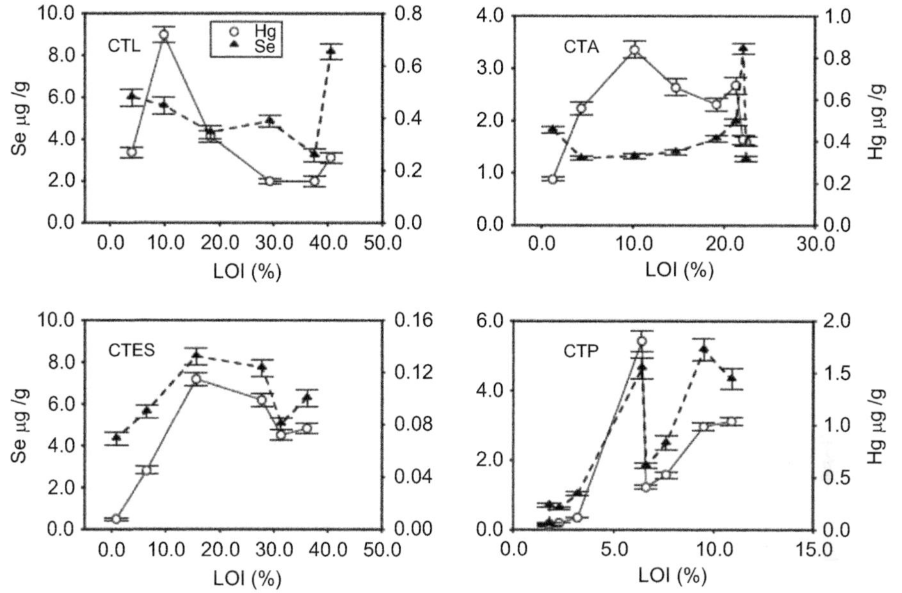

Figure 8.95 Unburned content (loss on ignition [LOI]) versus selenium content in different fly ashes.

From López-Antón M, Díaz-Somoano M, Abad-Valle P, Martínez-Tarazona M. Mercury and selenium retention in fly ashes: Influence of unburned particle content. Fuel 2007;86(14): 2064–70.

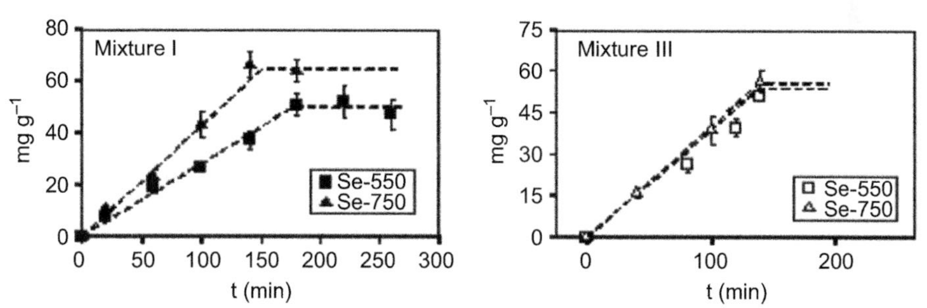

Figure 8.96 Effect of sorption time on selenium retention on limestone in different conditions.
From Diaz-Somoano M, Martinez-Tarazona M. Retention of arsenic and selenium compounds using limestone in a coal gasification flue gas. Environ Sci Technol 2004;38(3):899–903.

in alumina-supported sorbents influenced the performance of the sorbents (Fig. 8.98). The standard Pd/Al_2O_3 (8.5% Pd/Al_2O_3) adsorbent shows the best penetration performance relative to hydrogen selenide, because the Pd in the adsorbent is the least susceptible to interaction with H_2S. Although the performance of the Pd/Al_2O_3-son (sonochemical preparation method) adsorbent is the worst, it should be noted that its Pd loading is only 40% of the other two adsorbents. It is difficult to standardize

Table 8.13 Maximum retention capacities (MRCs) and efficiencies (E) obtained in the retention experiments.

		Mixture I (without H_2S)		Mixture III (with H_2S)	
		MRC		MRC	
Element	*T* (°C)	mg g^{-1}	%E	mg g^{-1}	%E
As	350	17.3 ± 0.6	44	1.19 ± 0.10	15
	550	15.7 ± 0.5	95	0.46 ± 0.03	5.8
Se	750	8.92 ± 0.1	22	0.16 ± 0.02	1.7
	550	49.9 ± 2.5	55	53.5 ± 2.4	70
	750	64.8 ± 1.6	92	55.6 ± 1.0	76

From Diaz-Somoano M, Martinez-Tarazona M. Retention of arsenic and selenium compounds using limestone in a coal gasification flue gas. Environ Sci Technol 2004;38(3):899–903.

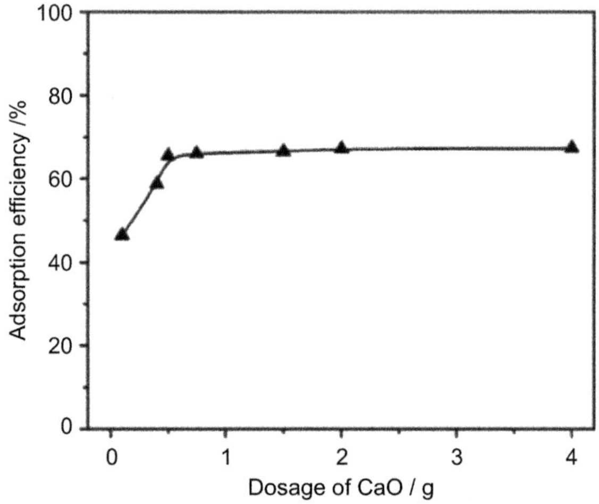

Figure 8.97 Se adsorption efficiency of CaO. Quantity of Se-rich stone coal: 5.00 g, reaction temperature: 800°C, reaction time: 90 min, flow rate of air: 100 mL min^{-1}
From Xu S, Shuai Q, Huang Y, Bao Z, Hu S. Se capture by a CaO–ZnO composite sorbent during the combustion of Se-rich stone coal. Energy Fuels 2013;27(11):6880–6.

the breakthrough performance value by wt% Pd, but attempting to do so on a time scale basis will make the performance of the Pd/Al_2O_3-sub adsorbent the same as that of the boron-modified adsorbent.

Metal oxide mixtures based on iron, titanium, or zinc oxides (zinc ferrites and zinc titanates) were tested for selenium retention [78]. As shown in Fig. 8.99, a selenium retention capacity of 56 mg/g was obtained in a metal oxide mixture containing zinc

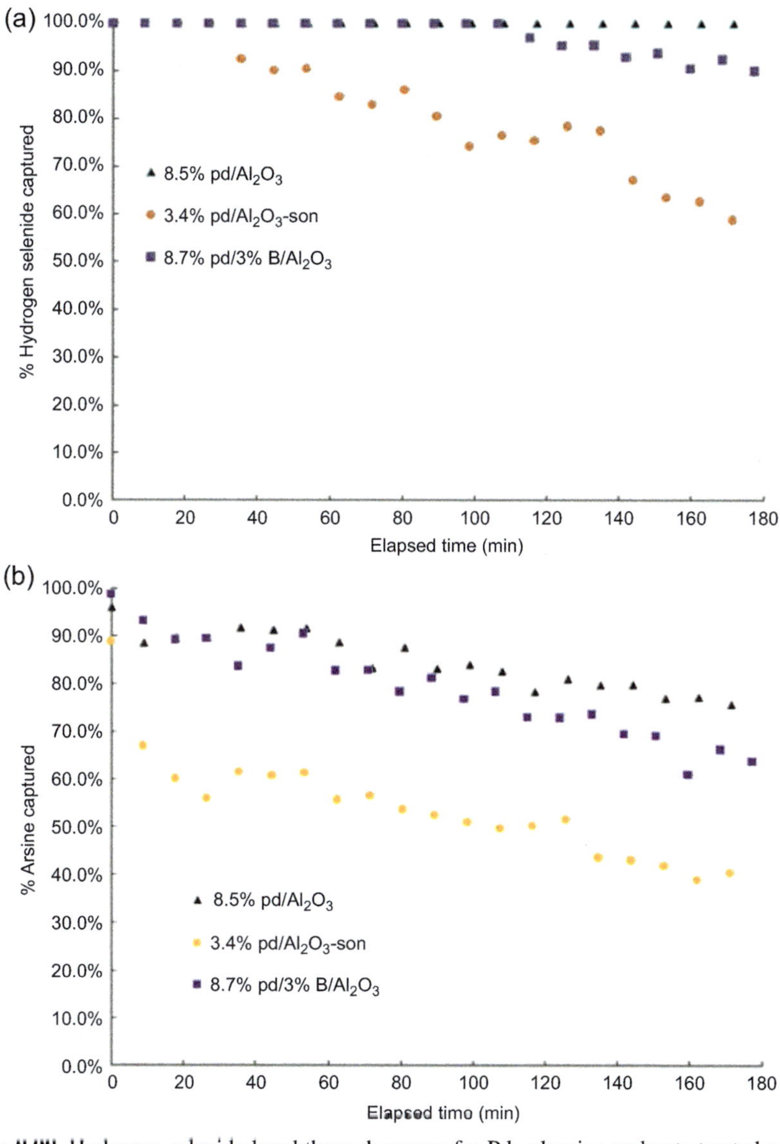

Figure 8.98 Hydrogen selenide breakthrough curves for Pd—alumina sorbents treated with synthetic fuel gas.
From Baltrus J, Granite E, Rupp E, Stanko D, Howard B, Pennline H. Effect of palladium dispersion on the capture of toxic components from fuel gas by palladium-alumina sorbents. Fuel 2011;90(5):1992–98.

titanate. A metal oxide mixture containing zinc ferrite exhibits a selenium retention capacity of 55 mg/g. Table 8.14 shows that selenium retention is similar in the two metal oxide mixtures with an MRC value close to 55 mg/g. A similar selenium retention efficiency close to 80% was found in both adsorbents.

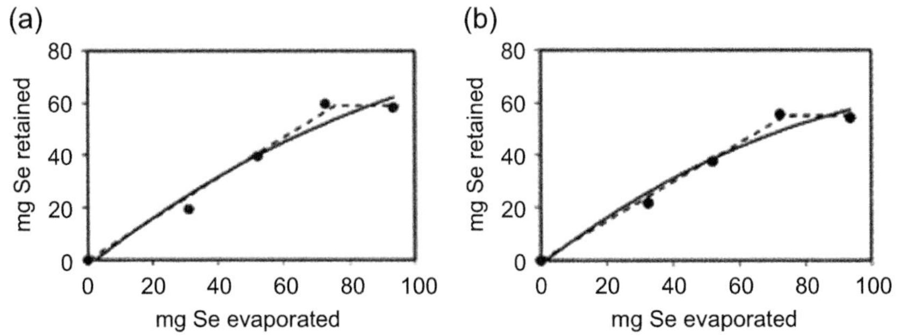

Figure 8.99 Results obtained during retention experiments for selenium with (a) ZT-6 and (b) ZFT-11.
From Diaz-Somoano M, López-Antón M, Martínez-Tarazona M. Retention of arsenic and selenium during hot gas desulfurization using metal oxide sorbents. Energy Fuels 2004;18(5): 1238–42.

Table 8.14 Maximum retention capacities (MRCs) and efficiencies (E) obtained.

	MRC ($mg\ g^{-1}$)		E (%)	
Sorbent	**As**	**Se**	**As**	**Se**
ZT-6	0.35 ± 0.04	56.4 ± 2.42	2	83
ZFT-11	20.5 ± 1.43	55.0 ± 3.12	66	77

From Diaz-Somoano M, López-Antón M, Martínez-Tarazona M. Retention of arsenic and selenium during hot gas desulfurization using metal oxide sorbents. Energy Fuels 2004;18(5):1238–42.

$$Zn_2TiO_4 + H_2Se(g) + H_2S(g) \rightarrow ZnSe + ZnS + TiO_2 + 2H_2O(g)$$

$$2ZnFe_2O_4 + H_2Se(g) + 5H_2S(g) + 2H_2(g) \rightarrow ZnSe + ZnS + 4FeS + 8H_2O(g)$$

A calcium-based sorbent was produced by combining CaO and nanosized metal oxides such as ZnO, Al_2O_3, and Fe_3O_4 to capture Se from Se-rich coal stone [83]. As shown in Fig. 8.100, the adsorption efficiency of all adsorbents first increases as the temperature increases from 700 to 800°C, and then decreases as the temperature further increases to 1000°C. The maximum efficiency observed at 800°C may be due to a compromise between a relatively weak adsorption reaction between Se and the adsorbent at low temperatures and some desorption of Se at high temperatures. Adding nano-Al_2O_3 to CaO reduces the adsorption efficiency, while nano-ZnO- and nano-Fe_3O_4-based composite adsorbents have higher adsorption efficiency than pure CaO. However, pure nano-ZnO exhibits limited adsorption capacity, while both nano-Fe_3O_4 and nano-Al_2O_3 exhibit very small adsorptions. This may be attributed to the highest capacity of the three composite adsorbents studied by CaO–ZnO composites. The maximum adsorption efficiency obtained with CaO–ZnO was 90.60% (Fig. 8.101).

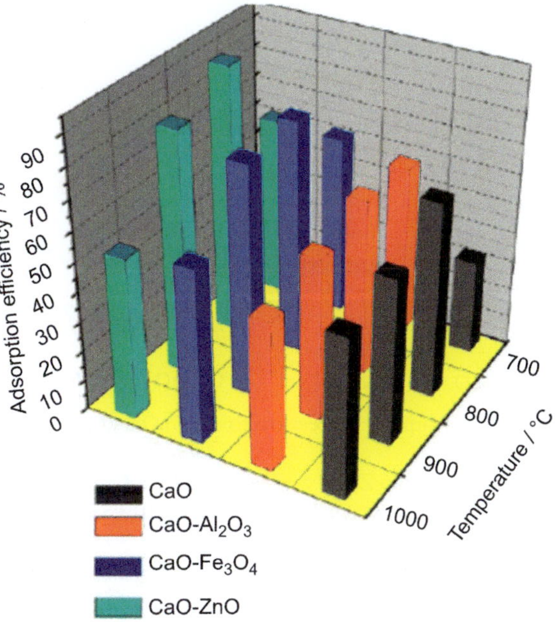

Figure 8.100 Se adsorption efficiency of calcium-based composite sorbents. Quantity of Se-rich stone coal: 5.00 g, quantity of CaO: 0.50 g, quantity of nanomaterial: 0.15 g, particle size of nanomaterials: 60 nm, reaction time: 90 min, flow rate of air: 100 mL min^{-1}

From Xu S, Shuai Q, Huang Y, Bao Z, Hu S. Se capture by a CaO–ZnO composite sorbent during the combustion of Se-rich stone coal. Energy Fuels 2013;27(11):6880–6886.

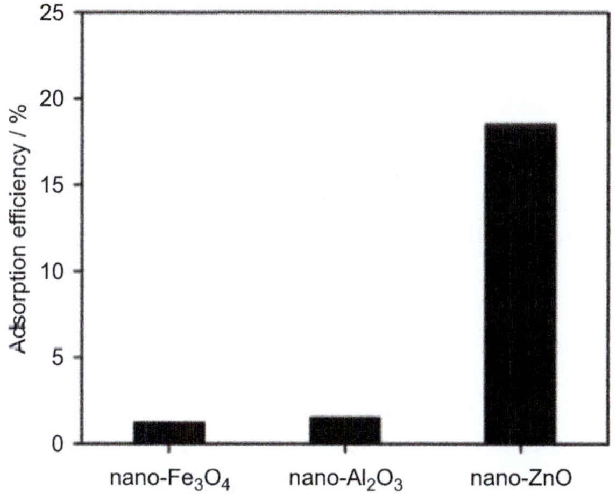

Figure 8.101 Se adsorption efficiency of pure nanomaterials. Quantity of Se-rich stone coal: 5.00 g, quantity of nanomaterial: 0.50 g, particle size of nanomaterial: 60 nm, reaction time: 90 min, reaction temperature: 800°C, flow rate of air: 100 mL min^{-1}.

From Xu S, Shuai Q, Huang Y, Bao Z, Hu S. Se capture by a CaO–ZnO composite sorbent during the combustion of Se-rich stone coal. Energy Fuels 2013;27(11):6880–6.

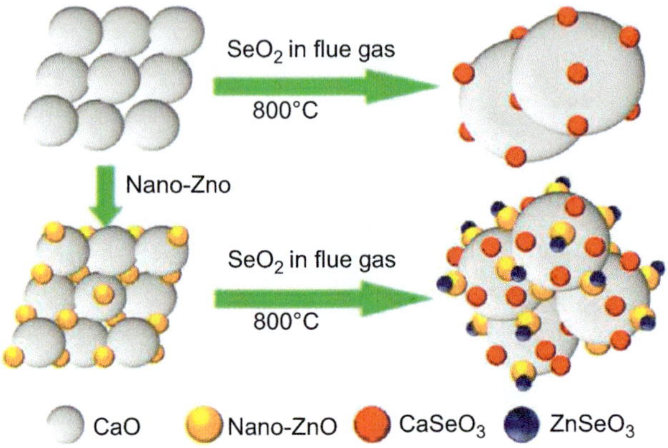

Figure 8.102 Proposed Se adsorption process over pure CaO and CaO−ZnO composite sorbents.
From Xu S, Shuai Q, Huang Y, Bao Z, Hu S. Se capture by a CaO−ZnO composite sorbent during the combustion of Se-rich stone coal. Energy Fuels 2013;27(11):6880−6.

As shown in Fig. 8.102, gaseous SeO_2 is initially physically adsorbed on the pure CaO surface and then reacted with CaO to form $CaSeO_3$. The improvement in adsorption activity obtained by introducing nano-ZnO may be attributed to the following aspects. Nano-ZnO itself can adsorb SeO_2 to form $ZnSeO_3$, which helps to improve the adsorption capacity. Another possibility is that the increased reaction interface leads to the addition of highly dispersed nano-ZnO to the CaO surface.

8.6 Adsorbents for capturing Cr in coal-derived flue gas

ACFs were applied to control Cr in incineration flue gas [85]. Three ACFs, various adsorption temperatures (150, 250, and 300°C), and ACF weights were experimentally determined. The effects of ACF type and ACF weight on solid-state Cr removal were insignificant, while the effect of the ACF type on Cr removal was negligible. The boiling point of various Cr metal compounds may exceed the incineration temperature of 800°C, and the concentration of Cr in fly ash and the gas phase may be lower than that of bottom ash. Therefore changes in the ACF surface structure do not significantly affect the removal of Cr, as shown in Fig. 8.103.

Fig. 8.104 shows solid metal removal efficiency at various adsorption temperatures. The mixed metal physically adsorbed or coagulated on the fly ash may evaporate and absorb at 300°C. In addition, the metal condensed on the ACF surface at 150°C blocks most of the pores on the ACF surface, reducing the chance of metal entering the pores and reducing the metal concentration physically adsorbed on the ACF. Therefore, the removal rate of the mixed metal can be up to 250°C.

The effect of ACF weight on Cr removal is negligible (Fig. 8.105). Most of the Cr metal compounds are distributed in the bottom ash and the concentration of gaseous Cr is low. Therefore the adsorption capacity of a single ACF is sufficient to adsorb Cr in the flue gas.

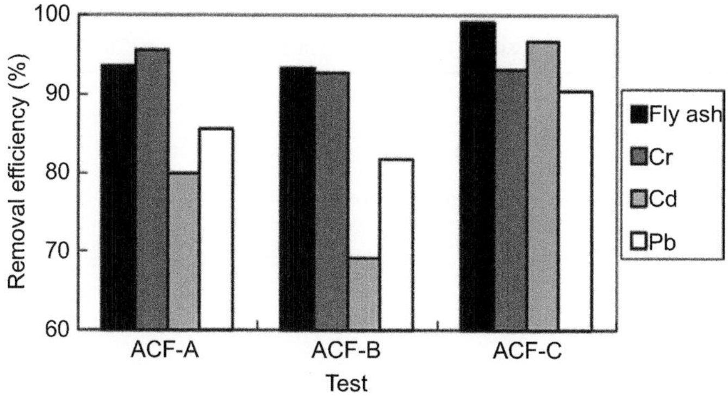

Figure 8.103 Effects of three activated carbon fibers (ACFs) on the removal efficiency of solid-state metals (Cr, Cd, and Pb): Run 1 (ACF-A), Run 2 (ACF-B), and Run 3 (ACF-C).
From Liu Z. Control of heavy metals during incineration using activated carbon fibers. J Hazard Mater 2007;142(1−2):506−11.

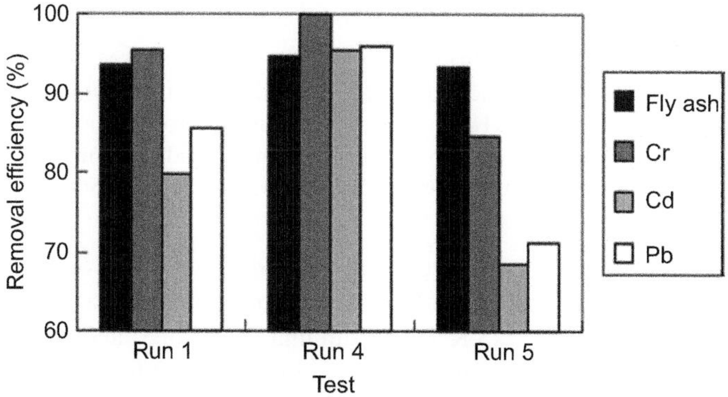

Figure 8.104 Influences of adsorption temperature on fly ash and solid-state metals (Cr, Cd, and Pb) removal: Run 1 (150°C), Run 4 (250°C), and Run 5 (300°C).
From Liu Z. Control of heavy metals during incineration using activated carbon fibers. J Hazard Mater 2007;142(1−2):506−11.

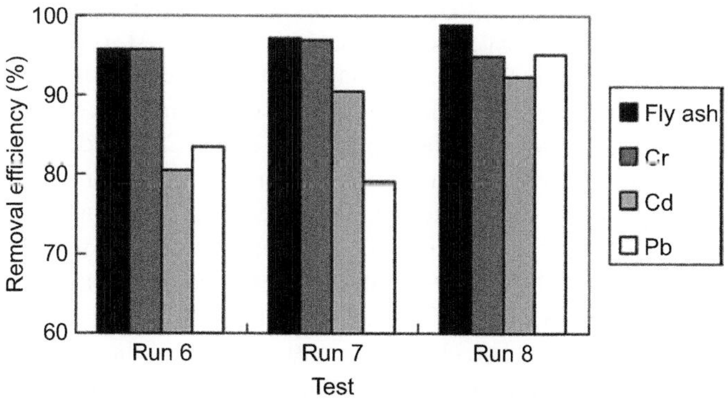

Figure 8.105 Effects of the weight of ACFs on the removal efficiency of fly ash and solid-state metals (Cr, Cd, and Pb): Run 6 (1.12 g), Run 7 (1.64 g), and Run 8 (2.28 g).
From Liu Z. Control of heavy metals during incineration using activated carbon fibers. J Hazard Mater 2007;142(1−2):506−11.

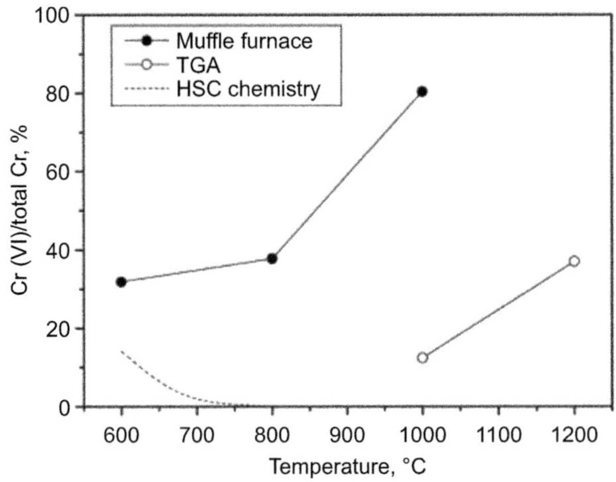

Figure 8.106 Cr(VI) fractions in ashing products of pure oxide mixture of CaO and Cr_2O_3 in a muffle furnace and thermogravimetric analysis (TGA) at different temperatures.
From Chen J, Jiao F, Zhang L, Yao H, Ninomiya Y. Use of synchrotron XANES and Cr-doped coal to further confirm the vaporization of organically bound Cr and the formation of chromium (VI) during coal oxy-fuel combustion. Environ Sci Technol 2012;46(6):3567−73.

For studying the absorbed mechanics and absorbed effectiveness on variant objects, TE capture experiments with absorbents were performed in a laboratory-scale fluidized-bed combustor at 1223 K [86]. Alumina, limestone, and kaolinite served as solid absorbents with three kinds of coal, including Pingdingshan bituminous coal, Jiaozuo anthracite coal, and Liupanshui lean coal. Experimental results indicated that all three absorbents were effective for chromium (especially for bituminous coal). As shown in Fig. 8.106, CaO was effective for capturing Cr(VI)-bearing species, particularly oxychloride vapors at 1000°C, forming $CaCrO_4$ and $Ca_3(CrO_4)_2$ via direct stabilization of Cr(VI) oxychloride vapor by CaO particle or indirect oxidation of Cr(III) via the initial formation of Ca chromite [87]. Adding CaO into coal also confirmed the capture of Cr as Cr(III) by CaO at 600°C. At 1000°C, the use of CaO retained 84.77% of Cr in the ash, resulting in an increase of about 20% compared to the Cr-doped coal only at 1000°C. In addition, the Cr(VI) fraction in the ash reached 16.9%, which was 4.9% relative to the original ash of the Cr-doped coal.

To understand the partition and emission of chromium in the combustion process, Zhao et al. [88] performed combustion experiments using a typical high-chromium coal and a highly volatile bituminous coal (HN coal) adding chromate in a bench-scale drop tube furnace. Chromium retained from the combustion of high Cr coal by six adsorbents (Al_2O_3, CaO, Fe_2O_3, zeolite, bentonite, and bauxite) was investigated in a fixed-bed reactor. The chromium removal rate of the six adsorbents for high Cr coal combustion is shown in Fig. 8.107. Among the adsorbents, the removal rate of bauxite is the highest, and the removal rate of chromium in high-chromium coal combustion is about 92%, mainly due to its maximum BET surface area. However, the BET surface area is not the only factor controlling chromium retention.

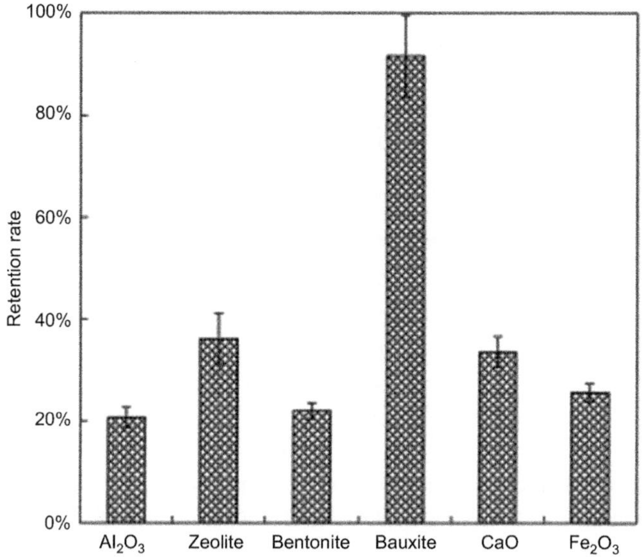

Figure 8.107 Removal rates of chromium by different sorbents.
From Zhao Y, Zhang J, Zheng C. Release and removal using sorbents of chromium from a high-Cr lignite in Shenbei coalfield, China Fuel 2013;109:86–93.

Zoelite with the smallest BET surface area can capture about 36% of chromium in hot gases. The presence of sodium in the zeolite may be the main cause of high chromium removal. CaO can react with chromium to form $CaCrO_3$ or $CaCr_2O_4$. The adsorbent Fe_2O_3 retains less chromium than CaO, and alumina and bentonite show the lowest retention.

XRD results show that CrO_3 can release chromium in the form of Cr_2O_3, and CaO reacts easily with the Cr compound to form chromate (Fig. 8.108). After interaction at 1000°C, a certain amount of calcium chromate ($Ca_5Cr_3O_{12}$) and aluminum chromate (Al_3CrO_6) were identified in the mixture product. $FeCrO_4$ from the reaction of Fe_2O_3 and chromium oxide was found in the mixture only after cofiring at 1200°C (Fig. 8.109). These results indicate that chromium in coal can react with solid adsorbents to form new chromium compounds.

8.7 Summary

Adsorbent injection is considered a promising approach to effectively control TE emissions from coal-derived flue gas. Various adsorbents have been developed to capture TEs, especially Hg. Adsorbents such as AC, fly ash, calcium-based adsorbents, metal oxides, and mineral adsorbents have been used to capture TEs in flue gases. AC injection has good Hg removal performance and has been commercially used in many coal-fired power plants. Fly ash has good adsorption properties for As and Hg. Removal performance can be widely affected by physicochemical properties and reaction

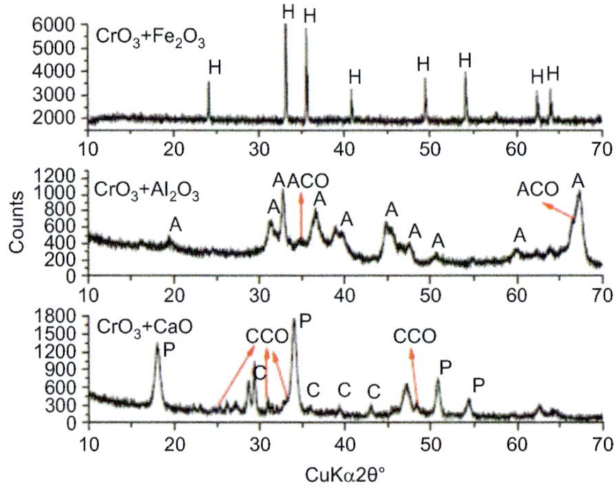

Figure 8.108 X-ray diffraction patterns of mixture products of chromium oxide and sorbents under a temperature of 1000°C (H: Hematite; A: Alumina; P: Portlandite; C: Calcite; ACO: Al3CrO6; CCO: $Ca_5Cr_3O_{13}H$).
From Zhao Y, Zhang J, Zheng C. Release and removal using sorbents of chromium from a high-Cr lignite in Shenbei coalfield, China Fuel 2013;109:86–93.

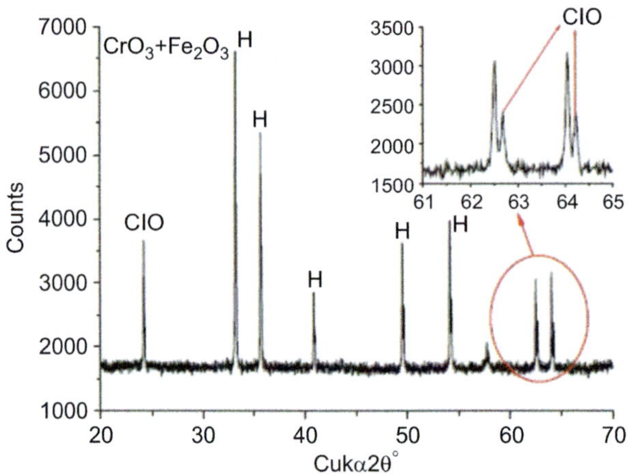

Figure 8.109 X-ray diffraction patterns of mixture products of chromium oxide and Fe_2O_3 under a temperature of 1200°C (H: Hematite, CIO: $FeCr_2O_4$).
From Zhao Y, Zhang J, Zheng C. Release and removal using sorbents of chromium from a high-Cr lignite in Shenbei coalfield, China Fuel 2013;109:86–93.

conditions. F and Hg can be removed by mixing the calcium-based adsorbent and the halogen compound with coal, respectively, during the combustion of coal. Metal oxides such as CaO, Fe_2O_3, and Al_2O_3 can also capture As, Se, and Cr. Among these mineral adsorbents, bauxite, kaolinite, and zeolite are usually used to capture Cr.

Adsorbent injection is a promising technology for removing mercury, but further research is needed to understand the competitive adsorption mechanism between mercury and other TEs to develop adsorbents for the simultaneous removal of mercury and other TEs.

Acknowledgments

This project was supported by the National Natural Science Foundation of China (51776227), Natural Science Foundation of Hunan Province, China (2018JJ1039, 2018JJ3675).

References

[1] Wu S, Yang W, Zhou J, Wang H, Xie Z. Effects of properties of activated carbon on its activity for mercury removal and mercury desorption from used activated carbons. Energy Fuels 2015;29(3):1946−50.
[2] Hsi H, Tsai C, Lin K. Impact of surface functional groups, water vapor, and flue gas components on mercury adsorption and oxidation by sulfur-impregnated activated carbons. Energy Fuels 2014;28(5):3300−9.
[3] Min H, Ahmad T, Lee S. Mercury adsorption characteristics as dependent upon the physical properties of activated carbon. Energy Fuels 2016;31(1):724−9.
[4] Li Y, Lee C, Gullett B. Importance of activated carbon's oxygen surface functional groups on elemental mercury adsorption. Fuel 2003;82(4):451−7.
[5] Diamantopoulou I, Skodras G, Sakellaropoulos G. Sorption of mercury by activated carbon in the presence of flue gas components. Fuel Process Technol 2010;91(2):158−63.
[6] Li Y, Lee C, Gullett B. The effect of activated carbon surface moisture on low temperature mercury adsorption. Carbon 2002;40(1):65−72.
[7] Uddin M, Yamada T, Ochiai R, Sasaoka E. Role of SO_2 for elemental mercury removal from coal combustion flue gas by activated carbon. Energy Fuels 2008;22(4):2284−9.
[8] Sjostrom S, Durham M, Bustard C, Martin C. Activated carbon injection for mercury control: Overview. Fuel 2010;89(6):1320−2.
[9] Zhuang Y, Martin C, Pavlish J, Botha F. Cobenefit of SO_3 reduction on mercury capture with activated carbon in coal flue gas. Fuel 2011;90(10):2998−3006.
[10] Sano A, Takaoka M, Shiota K. Vapor-phase elemental mercury adsorption by activated carbon co-impregnated with sulfur and chlorine. Chem Eng J 2017;315:598−607.
[11] Yao Y, Velpari V, Economy J. Design of sulfur treated activated carbon fibers for gas phase elemental mercury removal. Fuel 2014;116:560−5.
[12] Zeng H, Jin F, Guo J. Removal of elemental mercury from coal combustion flue gas by chloride-impregnated activated carbon. Fuel 2004;83(1):143−6.
[13] Zhang B, Zeng X, Xu P, Chen J, Xu Y, Luo G, Xu M, Yao H. Using the novel method of nonthermal plasma to add Cl active sites on activated carbon for removal of mercury from flue Gas. Environ Sci Technol 2016;50(21):11837−43.
[14] Sasmaz E, Kirchofer A, Jew A, Saha A, Abram D, Jaramillo T, Wilcox J. Mercury chemistry on brominated activated carbon. Fuel 2012;99:188−96.
[15] Zhou Q, Duan Y, Zhu C, She M, Zhang J, Yao T. In-flight mercury removal and cobenefit of SO_2 and NO reduction by NH_4Br impregnated activated carbon injection in an entrained flow reactor. Energy Fuels 2015;29(12):8118−25.

[16] Lee S, Seo Y, Jurng J, Lee T. Removal of gas-phase elemental mercury by iodine- and chlorine-impregnated activated carbons. Atmos Environ 2004;38(29):4887–93.
[17] Zhong L, Li W, Zhang Y, Norris P, Cao Y, Pan W. Kinetic studies of mercury adsorption in activated carbon modified by iodine steam vapor deposition method. Fuel 2017;188: 343–51.
[18] Lee S, Lee J, Keener T. Mercury oxidation and adsorption characteristics of chemically promoted activated carbon sorbents. Fuel Process Technol 2009;90(10):1314–8.
[19] Ghorishi S, Keeney R, Serre S, Gullett B, Jozewicz W. Development of a Cl-impregnated activated carbon for entrained-flow capture of elemental mercury. Environ Sci Technol 2002;36(20):4454–9.
[20] Zhou Q, Duan Y, Hong Y, Zhu C, She M, Zhang J, Wei H. Experimental and kinetic studies of gas-phase mercury adsorption by raw and bromine modified activated carbon. Fuel Process Technol 2015;134:325–32.
[21] Zhang Y, Zhao L, Guo R, Wang J, Cao Y, Orndorff W, Pan W. Influences of NO on mercury adsorption characteristics for HBr modified fly ash. Int J Coal Geol 2017;170: 77–83.
[22] He P, Zhang X, Peng X, Wu J, Jiang X. Effect of fly ash composition on the retention of mercury in coal-combustion flue gas. Fuel Process Technol 2016;142:6–12.
[23] Zhao Y, Zhang J, Liu J, Díaz-Somoano M, Abad-Valle P, Martínez-Tarazona M, Zheng C. Experimental study on fly ash capture mercury in flue gas. Sci China Technol Sci 2010; 53(4):976–83.
[24] Abad-Valle P, Lopez-Anton M, Diaz-Somoano M, Martinez-Tarazona M. The role of unburned carbon concentrates from fly ashes in the oxidation and retention of mercury. Chem Eng J 2011;174(1):86–92.
[25] Goodarzi F, Hower J. Classification of carbon in Canadian fly ashes and their implications in the capture of mercury. Fuel 2008;87(10–11):1949–57.
[26] Hower J, Senior C, Suuberg E, Hurt R, Wilcox J, Olson E. Mercury capture by native fly ash carbons in coal-fired power plants. Prog Energy Combust 2010;36(4):510–29.
[27] López-Antón M, Abad-Valle P, Díaz-Somoano M, Suárez-Ruiz I, Martínez-Tarazona M. The influence of carbon particle type in fly ashes on mercury adsorption. Fuel 2009;88(7): 1194–200.
[28] Wang F, Wang S, Meng Y, Zhang L, Wu Q, Hao J. Mechanisms and roles of fly ash compositions on the adsorption and oxidation of mercury in flue gas from coal combustion. Fuel 2016;163:232–9.
[29] Dunham G, DeWall R, Senior C. Fixed-bed studies of the interactions between mercury and coal combustion fly ash. Fuel Process Technol 2003;82(2–3):197–213.
[30] Galbreath K, Zygarlicke C, Tibbetts J, Schulz R, Dunham G. Effects of NO_x, α-Fe_2O_3, γ-Fe_2O_3, and HCl on mercury transformations in a 7-kW coal combustion system. Fuel Process Technol 2005;86(4):429–48.
[31] Abad-Valle P, Lopez-Anton M, Diaz-Somoano M, Juan R, Rubio B, Garcia J, Martinez-Tarazona M. Influence of iron species present in fly ashes on mercury retention and oxidation. Fuel 2011;90(8):2808–11.
[32] Yang J, Zhao Y, Zyryanov V, Zhang J, Zheng C. Physical–chemical characteristics and elements enrichment of magnetospheres from coal fly ashes. Fuel 2014;135:15–26.
[33] Yang J, Zhao Y, Zhang S, Liu H, Chang L, Ma S, Zhang J, Zheng C. Mercury removal from flue gas by magnetospheres present in fly ash: role of iron species and modification by HF. Fuel Process Technol 2017;167:263–70.
[34] Guo P, Guo X, Zheng C. Roles of γ-Fe_2O_3 in fly ash for mercury removal: results of density functional theory study. Appl Surf Sci 2010;256(23):6991–6.

[35] Jung J, Geatches D, Lee K, Aboud S, Brown G, Wilcox J. First-principles investigation of mercury adsorption on the α-Fe_2O_3(1102) surface. J Phys Chem C 2015;119(47):26512–8.
[36] Zhang Y, Duan W, Liu Z, Cao Y. Effects of modified fly ash on mercury adsorption ability in an entrained-flow reactor. Fuel 2014;128:274–80.
[37] Gu Y, Zhang Y, Lin L, Xu H, Orndorff W, Pan W. Evaluation of elemental mercury adsorption by fly ash modified with ammonium bromide. J Therm Anal Calorim 2015; 119(3):1663–72.
[38] Wang S, Zhang Y, Gu Y, Wang J, Zhang Y, Cao Y, Pan W. Using modified fly ash for mercury emissions control for coal-fired power plant applications in China. Fuel 2016;181: 1230–7.
[39] Yang J, Zhao Y, Zhang J, Zheng C. Removal of elemental mercury from flue gas by recyclable $CuCl_2$ modified magnetospheres catalyst from fly ash. Part 1. Catalyst characterization and performance evaluation. Fuel 2016;164:419–28.
[40] Yang J, Zhao Y, Zhang J, Zheng C. Removal of elemental mercury from flue gas by recyclable $CuCl_2$ modified magnetospheres catalyst from fly ash. Part 2. Identification of involved reaction mechanism. Fuel 2016;167:366–74.
[41] Yang J, Zhao Y, Zhang J, Zheng C. Removal of elemental mercury from flue gas by recyclable $CuCl_2$ modified magnetospheres catalyst from fly ash. Part 3. Regeneration performance in realistic flue gas atmosphere. Fuel 2016;173:1–7.
[42] Wang Y, Duan Y, Huang Z, Meng S, Yang L, Zhao C. Vapor-phase elemental mercury adsorption by $Ca(OH)_2$ impregnated with MnO_2 and Ag in fixed-bed system. Asia Pac J Chem Eng 2009;5(3):479–87.
[43] Wang Y, Duan Y. Effect of manganese ions on the structure of $Ca(OH)_2$ and mercury adsorption performance of $Mn_{x+}/Ca(OH)_2$ composites. Energy Fuels 2011;25(4):1553–8.
[44] Ding F, Zhao Y, Mi L, Li H, Li Y, Zhang J. Removal of gas-phase elemental mercury in flue gas by inorganic chemically promoted natural mineral sorbents. Ind Eng Chem Res 2012;51(7):3039–47.
[45] He C, Shen B, Chen J, Cai J. Adsorption and oxidation of elemental mercury over Ce-MnO_x/Ti-PILCs. Environ Sci Technol 2014;48(14):7891–8.
[46] Hrdlicka JA, Seames WS, Mann MD, Muggli DS, Horabik CA. Mercury oxidation in flue gas using gold and palladium catalysts on fabric filters. Environ Sci Technol 2008;42: 6677–82.
[47] Rodriguez-Perez J, Lopez-Anton M, Diaz-Somoano M, Garcia R, Martinez-Tarazona M. Development of gold nanoparticle-doped activated carbon sorbent for elemental mercury. Energy Fuels 2011;25(5):2022–7.
[48] Izquierdo M, Ballestero D, Juan R, Garcia-Diez E, Rubio B, Ruiz C, Pino M. Tail-end Hg capture on Au/carbon-monolith regenerable sorbents. J Hazard Mater 2011;193:304–10.
[49] Xie Y, Yan B, Tian C, Liu Y, Liu Q, Zeng H. Efficient removal of elemental mercury (Hg^0) by SBA-15-Ag adsorbents. J Mater Chem A 2014;2(42):17730–4.
[50] Luo G, Yao H, Xu M, Cui X, Chen W, Gupta R, Xu Z. Carbon nanotube-silver composite for mercury capture and analysis. Energy Fuels 2009;24(1):419–26.
[51] Kong F, Qiu J, Liu H, Zhao R, Ai Z. Catalytic oxidation of gas-phase elemental mercury by nano-Fe_2O_3. J Environ Sci China 2011;23(4):699–704.
[52] Dang H, Liao Y, Ng T, Huang G, Xiong S, Xiao X, Yang S, Wong P. The simultaneous centralized control of elemental mercury emission and deep desulfurization from the flue gas using magnetic Mn–Fe spinel as a co-benefit of the wet electrostatic precipitator. Fuel Process Technol 2016;142:345–51.

[53] Liao Y, Xiong S, Dang H, Xiao X, Yang S, Wong P. The centralized control of elemental mercury emission from the flue gas by a magnetic rengenerable Fe-Ti-Mn spinel. J Hazard Mater 2015;299:740−6.
[54] Xiong S, Xiao X, Huang N, Dang H, Liao Y, Zou S, Yang S. Elemental mercury oxidation over Fe-Ti-Mn spinel: performance, mechanism, and reaction kinetics. Environ Sci Technol 2017;51(1):531−9.
[55] Yang S, Guo Y, Yan N, Wu D, He H, Qu Z, Jia J. Elemental mercury capture from flue gas by magnetic Mn−Fe spinel: effect of chemical heterogeneity. Ind Eng Chem Res 2011; 50(16):9650−6.
[56] Yang S, Guo Y, Yan N, Wu D, He H, Qu Z, Yang C, Zhou Q, Jia J. Nanosized cation-deficient Fe-Ti spinel: a novel magnetic sorbent for elemental mercury capture from flue gas. ACS Appl Mater Interfaces 2011;3(2):209−17.
[57] Yang S, Guo Y, Yan N, Wu D, He H, Xie J, Qu Z, Jia J. Remarkable effect of the incorporation of titanium on the catalytic activity and SO_2 poisoning resistance of magnetic Mn−Fe spinel for elemental mercury capture. Appl Catal B Environ 2011;101(3−4): 698−708.
[58] Yang S, Guo Y, Yan N, Wu D, He H, Xie J, Qu Z, Yang C, Jia J. A novel multi-functional magnetic Fe-Ti-V spinel catalyst for elemental mercury capture and callback from flue gas. Chem Commun 2010;46(44):8377−9.
[59] Yang S, Yan N, Guo Y, Wu D, He H, Qu Z, Li J, Zhou Q, Jia J. Gaseous elemental mercury capture from flue gas using magnetic nanosized $(Fe_{3-x}Mn_x)_{1-delta}O_4$. Environ Sci Technol 2011;45(4):1540−6.
[60] Cimino S, Scala F. Removal of elemental mercury by MnO_x catalysts supported on TiO_2 or Al_2O_3. Ind Eng Chem Res 2015;55(18):5133−8.
[61] Qiao S, Chen J, Li J, Qu Z, Liu P, Yan N, Jia J. Adsorption and catalytic oxidation of gaseous elemental mercury in flue gas over MnOx/Alumina. Ind Eng Chem Res 2009;48: 3317−22.
[62] Yamaguchi A, Akiho H, Ito S. Mercury oxidation by copper oxides in combustion flue gases. Powder Technol 2008;180(1−2):222−6.
[63] Kim B, Bae K, Park S. Elemental mercury vapor adsorption of copper-coated porous carbonaceous materials. Microporous Mesoporous Mater 2012;163:270−5.
[64] Liu Y, Wang Y, Wang H, Wu Z. Catalytic oxidation of gas-phase mercury over Co/TiO_2 catalysts prepared by sol−gel method. Catal Commun 2011;12(14):1291−4.
[65] Mei Z, Shen Z, Zhao Q, Wang W, Zhang Y. Removal and recovery of gas-phase element mercury by metal oxide-loaded activated carbon. J Hazard Mater 2008;152(2):721−9.
[66] Mei Z, Shen Z, Mei Z, Zhang Y, Xiang F, Chen J, Wang W. The effect of N-doping and halide-doping on the activity of $CuCoO_4$ for the oxidation of elemental mercury. Appl Catal B Environ 2008;78(1−2):112−9.
[67] Mei Z, Shen Z, Zhao Q, Yuan T, Zhang Y, Xiang F, Wang W. Removing and recovering gas-phase elemental mercury by $Cu_{(x)}Co_{(3-x)}O_{(4)}$ (0.75< or = x < or =2.25) in the presence of sulphur compounds. Chemosphere 2008;70(8):1399−404.
[68] Yang J, Zhao Y, Zhang J, Zheng C. Regenerable cobalt oxide loaded magnetosphere catalyst from fly ash for mercury removal in coal combustion flue gas. Environ Sci Technol 2014;48(24):14837−43.
[69] López-Antón M, Díaz-Somoano M, Fierro J, Martínez-Tarazona M. Retention of arsenic and selenium compounds present in coal combustion and gasification flue gases using activated carbons. Fuel Process Technol 2007;88(8):799−805.
[70] López-Antón M, Díaz-Somoano M, Spears D, Martínez-Tarazona M. Arsenic and selenium capture by fly ashes at low temperature. Environ Sci Technol 2006;40(12):3947−51.

[71] Sterling R, Helble J. Reaction of arsenic vapor species with fly ash compounds: kinetics and speciation of the reaction with calcium silicates. Chemosphere 2003;51(10):1111–9.
[72] Jadhav R, Fan L. Capture of gas-phase arsenic oxide by lime: kinetic and mechanistic studies. Environ Sci Technol 2001;35(4):794–9.
[73] Diaz-Somoano M, Martinez-Tarazona MR. Retention of arsenic and selenium compounds using limestone in a coal gasification flue gas. Environ Sci Technol 2004;38(3):899–903.
[74] Chen D, Hu H, Xu Z, Liu H, Cao J, Shen J, Yao H. Findings of proper temperatures for arsenic capture by CaO in the simulated flue gas with and without SO_2. Chem Eng J 2015; 267:201–6.
[75] Uffalussy K, Miller J, Howard B, Stanko D, Yin C, Granite E. Arsenic adsorption on copper–palladium alloy films. Ind Eng Chem Res 2014;53(18):7821–7.
[76] Poulston S, Granite E, Pennline H, Hamilton H, Smith A. Palladium based sorbents for high temperature arsine removal from fuel gas. Fuel 2011;90(10):3118–21.
[77] Zhang Y, Wang C, Li W, Liu H, Zhang Y, Hack P, Pan W. Removal of gas-phase As_2O_3 by metal oxide adsorbents: effects of experimental conditions and evaluation of adsorption mechanism. Energy Fuels 2015;29(10):6578–85.
[78] Diaz-Somoano M, López-Antón M, Martínez-Tarazona M. Retention of arsenic and selenium during hot gas desulfurization using metal oxide sorbents. Energy Fuels 2004; 18(5):1238–42.
[79] Qi Q, Y G, Zhang H, Liu J, Cen K. Experimental research on fluorine retention of alkaline-earth metal compounds during coal combustion. J Liaoning Tech Univ 2006;25:801–3 [In Chinese].
[80] Wang H, Qi Q, Wang J, Liu J, Zhou J, Cen K. Kinetic mechanics of the reaction between CaO and HF during coal combustion. J China Coal Soc 2014;39(1):161–5 [In Chinese].
[81] Wu X, Zhang Y, Dou X, Zhao B, Yang M. Fluoride adsorption on an Fe–Al–Ce trimetal hydrous oxide: characterization of adsorption sites and adsorbed fluorine complex species. Chem Eng J 2013;223:364–70.
[82] López-Antón M, Díaz-Somoano M, Abad-Valle P, Martínez-Tarazona M. Mercury and selenium retention in fly ashes: influence of unburned particle content. Fuel 2007;86(14): 2064–70.
[83] Xu S, Shuai Q, Huang Y, Bao Z, Hu S. Se capture by a CaO-ZnO composite sorbent during the combustion of Se-rich stone coal. Energy Fuels 2013;27(11):6880–6.
[84] Baltrus J, Granite E, Rupp E, Stanko D, Howard B, Pennline H. Effect of palladium dispersion on the capture of toxic components from fuel gas by palladium-alumina sorbents. Fuel 2011;90(5):1992–8.
[85] Liu Z. Control of heavy metals during incineration using activated carbon fibers. J Hazard Mater 2007;142(1–2):506–11.
[86] Lu J, Yu L, Zhang J. Study of control with absorbents on trace elements during fluidized bed combustion. Proceedings of the CSEE 2004;3:187–92.
[87] Chen J, Jiao F, Zhang L, Yao H, Ninomiya Y. Use of synchrotron XANES and Cr-doped coal to further confirm the vaporization of organically bound Cr and the formation of chromium (VI) during coal oxy-fuel combustion. Environ Sci Technol 2012;46(6): 3567–73.
[88] Zhao Y, Zhang J, Zheng C. Release and removal using sorbents of chromium from a high-Cr lignite in Shenbei coalfield, China. Fuel 2013;109:86–93

Trace element resource recovery from coal and coal utilization by-products

9

Yongchun Zhao, Yuming Zhou, Junying Zhang, Chuguang Zheng
State Key Laboratory of Coal Combustion, Huazhong University of Science and Technology, Wuhan, China

9.1 Introduction

China is the largest producer and user of coal in the world. The high volume of coal usage in China has focused attention not only on the toxic trace elements that may be released from coal combustion but also on the valuable elements that may be present in the coal and associated ash [1]. Under some specific coalfield or geological conditions, valuable trace elements (e.g., germanium [Ge], gallium [Ga], uranium [U], and rare earth elements [REEs]) can be significantly enriched in coal and coal utilization by-products, equal to or even higher than their concentrations in many conventional types of ores [2]. The economic value of valuable trace elements in coal has been recognized for more than 100 years. Some metals (e.g., Ge, aluminum [Al], Ga) have been successfully extracted at an industrial scale from Chinese coals, and others have significant potential for such extraction. In China, Ge from Lincang, Yunnan Province, and Wulantuga, Inner Mongolia, has been studied in detail and is being used industrially [2]. Ga, as well as Al, is now also under study for extraction from coal combustion residues [2]. Major challenges remaining for coal scientists include the development of economic extraction methods from coal ash and the control of toxic elements released during the metal extraction process to protect human health and to avoid environmental pollution. Dai et al. have given a very comprehensive review of valuable elements in Chinese coals [1] and systematically summarized the average contents of valuable metals in common coals, as well as the enrichment origin and modes of occurrence of trace elements.

In this chapter, the enrichment of Ge, Ga, U, and REEs in coal and by-products is systematically reviewed, and the extraction methods for these elements from coal and by-products are also summarized. This chapter, together with previous review articles, will provide significant insight into valuable element resource recovery from coal and coal utilization by-products.

Emission and Control of Trace Elements from Coal-Derived Gas Streams
https://doi.org/10.1016/B978-0-08-102591-8.00009-X

9.2 Enrichment of Ge in coal and its recovery from coal products

9.2.1 Enrichment of Ge in coal

Germanium is an element with high economic value and is used in novel and high-technological industrial applications, such as advanced electronic and optical devices [3]. The concentration of Ge in typical Ge-rich coal and by-products is listed in Table 9.1. The first Ge-rich coal deposit found in the former USSR was in the

Table 9.1 Concentrations of Ge in Ge-rich coals and their fly ashes (μg/g).

	Minimum	Maximum	Average	Samples	Source
World coal (brown)			2.0	84	[4]
World coal (hard)			2.2	127	[4]
Chinese coal	0.5	10	4	3084	[5]
Wulantuga coal	32	820	137	939	[6]
Wulantuga coal	22	1894	427	12	[7]
Wulantuga coal	23.3	1424	168	—	[8]
Wulantuga coal	45	1170	274	13	[9]
Lincang coal	25.5	2523	852	52	[8]
Lincang coal	737	2176	1642	26	[10]
Yimin coal	19	71	43	8	[11]
Yimin coal	<1	470			[17]
Angrensk coal			30		[12]
Tarbagataisk			53		
Rakovsk			230		
Pavlovsk (Spetzugli + Luzanovsk)			450		
Shkotovsk			1043		
Bikinsk (Cheremshovii + Phedos'evsk)			300		
Novikovsk (Central + Southern)			700		
Wulantuga fly ash			2820.2		[2]
Lincang fly ash			3901.7	20	[2,13]
Puertollano IGCC fly ash	194	420	283	8	[14]
Spetzugli fly ash			4906.1	12	[13]

Angren river valley (Uzbekistan) in the late 1950s. Seredin and Finkelman summarized the Ge-rich coal deposits in Russia and China in 2008; other similar deposits with higher Ge contents and higher total Ge resources than in the Angrensk deposit were found later in various regions of Russia as well as in China [12]. The Russian Far East and the Inner Mongolia Province of China are the regions where Ge-bearing coals are most prevalent, with inferred reserves of germanium estimated to be approximately 6000–7000 and 5600 metric tonnes (t), respectively [12]. Dai et al. have summarized the typical coal-hosted deposits in the world that are currently being mined as raw materials for Ge extraction: Lincang (Yunnan, China), Wulantuga (Inner Mongolia, China), Yimin (Inner Mongolia, China), and Spetzugli (Primorye, Russian Far East) [1,2,9] (Fig. 9.1). The concentration of Ge in the coal-hosted ore deposits is about 1900 μg/g, much higher than the average concentration for world lignite and hard coals.

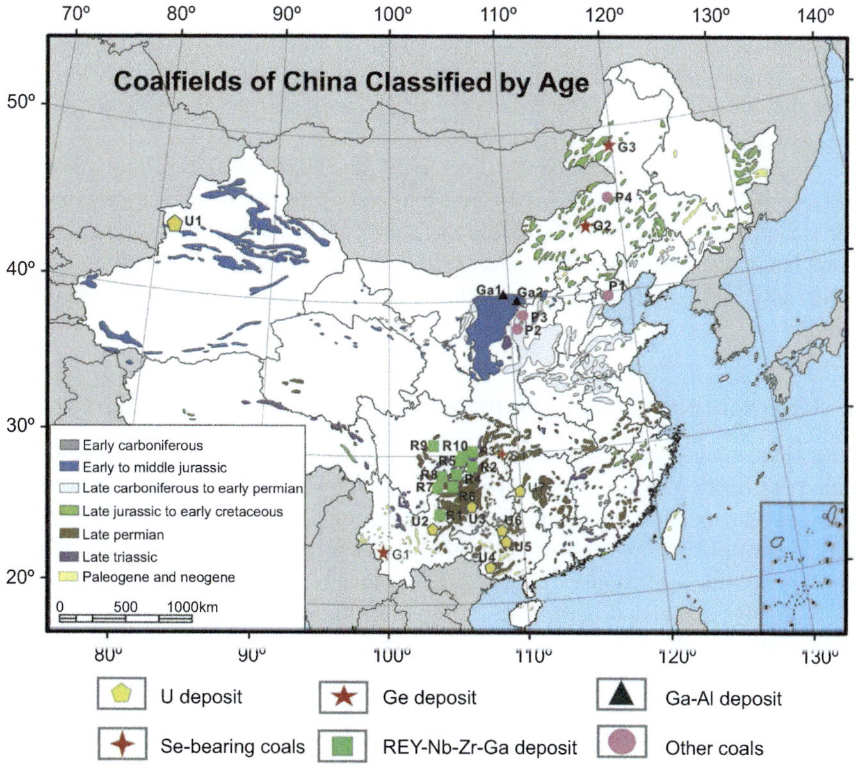

Figure 9.1 Distribution of coal-hosted ore deposits in China [1]. G1, Lincang; G2, Wulantuga; G3, Yimin; Ga1, Jungar; Ga2, Daqingshan; P1, Kailuan; P2, Pingshuo; P3, Datong; P4, Baiyinhua; R1, Xuanwei; R2, Zhongliangshan; R3, Huayingshan; R4, Guxu; R5, Moxinpo; R6, Zhenxiong; R7, Yiliang; R8, Luokan; R9, Xinde; R10, Sujiawan; U1, Yili; U2, Yanshan; U3, Guiding; U4, Fusui; U5, Heshan; U6, Yishan; U7, Chenxi. *REY*, rare earth elements and yttrium.

The Wulantuga high-Ge coal deposit is located in the Shengli coalfield, Inner Mongolia, northeastern China (Fig. 9.1). The Ge concentration in the coal ranges from 22 to 1894 μg/g [2,6]. Du et al. estimated that there was 1700 t of metallic Ge extractable from coal with Ge content >100 μg/g [6]. There is a negative correlation between the mean Ge content and the thickness of the coal seam, and the lower, upper, and middle parts of the coal seams are high in Ge [6]. A major organic affinity of Ge was determined [9,15], fine-grained Ge mineral was also found in these coals [7]. The designed capacity for Ge production in the recovery plant at the Wulantuga deposit is 100 t per year; various Ge compounds, including high-purity zone-refined Ge ingot (from 99.99999% to 99.999999%), are produced from the recovery plant [1].

The Lincang Neogene Ge-deposit is located in western Yunnan Province (Fig. 9.1) and was discovered in the 1950s while workers were prospecting for U [2]. The proven Ge reserves are 1000 t in two ore fields (Dazhai, 800 t; Zhongzhai, 200 t) of the Lincang Ge deposit [16], and the content of Ge in the deposit is about 25–2500 μg/g [2,8]. Production from the Lincang Ge recovery plant is between 39 and 47.6 t Ge per year [13]. Ge-bearing coals in both the Wulantuga and the Lincang deposits contain high concentrations of lithophilic elements (Wulantuga, W, Be, U, and Cs; Lincang, W, Be, U, Cs, and Nb; they are typical compositions of felsic rocks) and highly volatile elements that are members of the epithermal or volcanogenic group (Wulantuga, Sb, As, Hg, and Tl; Lincang, Sb and As) [2].

The largest Ge-bearing coal deposit, not only in China but also in the world, is probably in the Yimin coalfield of Inner Mongolia, with an estimated resource of 4 kt Ge [17]. The content of Ge in these mineralized coals and partings varies greatly, from less than 1 to 470 μg/g, and is mostly in the range between 50 and 200 μg/g [3]. However, only limited information is available as of this writing about this unique deposit [2,17].

9.2.2 Ge recovery from coal products derived from high-Ge coal

Font et al. investigated eight typical Ge-enriched fly ashes (FAs) from the Puertollano integrated gasification with combined cycle (IGCC) power plant [14]. The Ge content in these coal gasification FAs can reach 10 times of that in the raw coal, with an average content of 283 μg/g. Up to 80% of Ge occurs in water-soluble GeO_2, GeS_2, and slightly soluble GeS phases in IGCC FA, and the remaining 20% of insoluble Ge is probably associated with Ge in aluminosilicates rather than with GeO_2 (Fig. 9.2) [14]. Dai et al. present a systematic investigation of the composition and modes of occurrence of elements in coal combustion products derived from three high-Ge coals, Lincang, Wulantuga, and Spetzugli. The Ge in the FAs is highly enriched, up to 4.66% (on an organic-free basis) (Fig. 9.3) [13]. Ge oxides are the major Ge carrier in the FAs. Other Ge-bearing mineral phases, however, were also identified, including glass, Ca ferrites, solid solutions of Ge in SiO_2, and probably elemental Ge or Ge(Ge–W) carbide, as well as previously unknown complex oxides, including (Ge, As)O_x, (Ge, As, Sb)O_x, (Ge, As, W)O_x, and (Ge, W)O_x. Some portion of the Ge occurs as adsorbed species in different types of unburnt carbon in the ash particles.

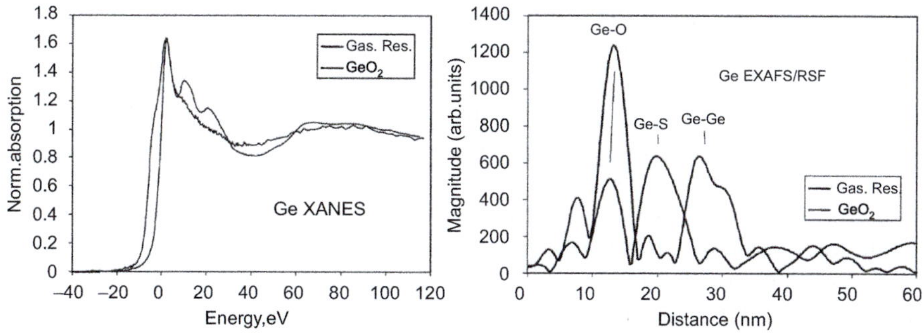

Figure 9.2 X-ray absorption near edge structure (*XANES*) (left) and extended X-ray absorption fine structure/radial structure function (*EXAFS/RSF*) (right) spectra for Ge in integrated gasification with combined cycle fly ashes [14]. *Gas. Res*, Gasification Residue.

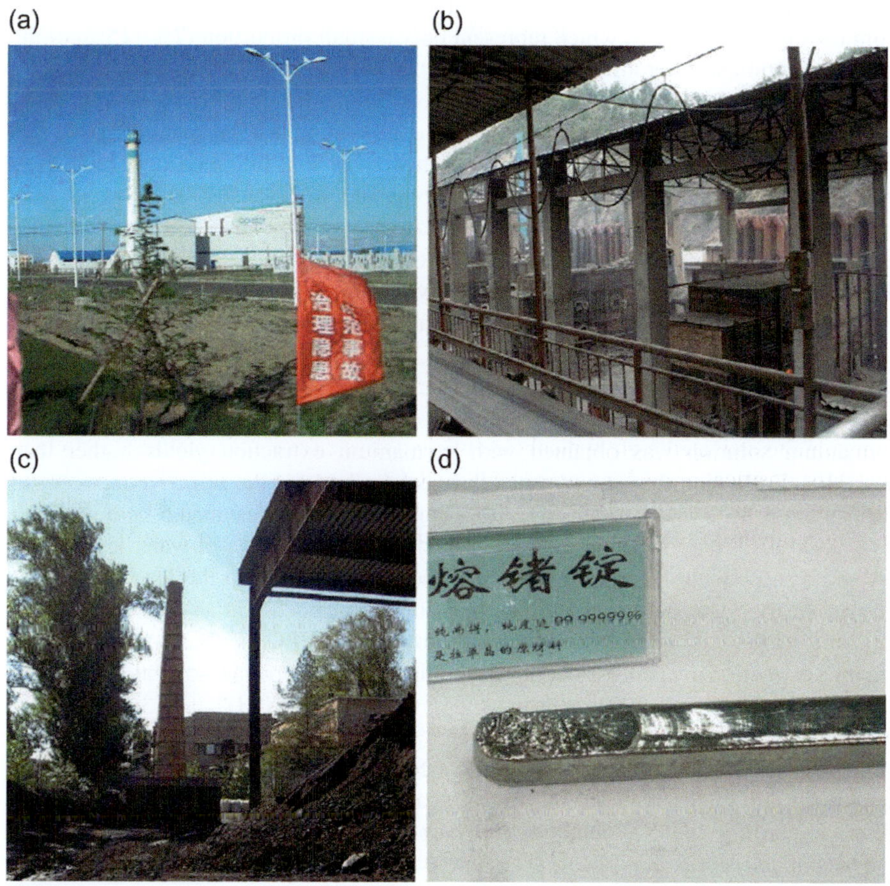

Figure 9.3 Power plants for combustion of Ge-bearing coal. (a) Wulantuga; (b) Lincang; (c) Spetzugli. (d) High-purity metallic germanium (99.99999%) produced by Xilinguole Tongli Ge Refinery Co., Ltd., from Wulantuga fly ash [13].

The designed capacity for Ge production in the recovery plant at the Wulantuga deposit is 100 t per year, which is close to the annual world consumption of Ge in 2010–11 (120 t) [13]. The coals from the mines are crushed and then are fed into vortex furnaces for combustion. The combustion conditions are strictly controlled to ensure that Ge is concentrated mostly in the FA collected by the baghouse filters, up to 14,870 μg/g [13]. The Ge production of the Lincang Ge recovery plant is 39–47.6 t per year. The Ge-rich coals are fed into chain conveyer furnaces for combustion, and Ge-rich FA is collected by baghouse fabric filters after its temperature has been decreased to 110°C–120°C by condensator cooling. The collecting efficiency of the FA is around 99.8%. Ge is concentrated mostly in the FA collected by the baghouse filters, up to 38,964 μg/g [13]. The ensured Ge reserves in the Spetzugli high-Ge coal deposit are about 1000 t. The designed capacity for Ge production is 21 t per year (http://www.geapplic.ru/en/). The coals from the mine are fed into a flare-layered boiler for combustion. The collection efficiency of the FA is similar to that for the Lincang deposit. Ge is also concentrated mostly in the FA collected by the baghouse filters, up to 24,600 μg/g [13]. In all three plants, however, Ge is extracted from the fine-grained FAs, which represent only a small proportion (7%–15%) of the total coal combustion products [13].

Ge-concentrated FA is a potential source of germanium, and many studies have focused on the recovery of germanium from coal FAs by several methods [18]. The first step of most of these processes is the leaching of FA, followed by a process to separate germanium from other elements contained in leachates, such as arsenic, molybdenum, nickel, antimony, vanadium, or zinc. Conventional processes are precipitation with tannin, distillation of $GeCl_4$, and solvent extraction (SX). SX appears to be the most attractive procedure due to the easy scale-up and operation control. In an exemplary study, a simple hydrometallurgical method for the selective recovery of germanium from FA generated in an IGCC process was presented. The method consisted of the leaching of FA with water and the subsequent concentration and selective separation of germanium by a solvent method. As a result, a high-purity germanium solution was obtained with germanium extraction yields higher than 90% [19]. Following this, a hydrometallurgical method for the selective recovery of germanium from FA was tested at a pilot plant scale with a designed 5 kg/h recovery (Fig. 9.4) [20]. The pilot plant flow sheet comprised a first stage of water leaching of FA and a subsequent selective recovery of the germanium from the leachate by SX. The SX method was based on Ge complexation with catechol in an aqueous solution followed by the extraction of the Ge–catechol complex $\left(Ge(C_6H_4O_2)_3{}^{2-}\right)$ with an organic extraction reagent (trioctylamine) diluted in an organic solvent (kerosene), the relevant reaction probably being $2(C_8H_{17})_3N_{(org)} + Ge(OH)^0_{4(aq)} + 3(C_6H_6O_2)_{(aq)} \leftrightarrow ((C_8H_{17})_3NH)_2Ge(C_6H_4O_2)_{3(org)} + 4H_2O_{(aq)}$, followed by the subsequent stripping of the organic extract. Under the operational conditions tested, approximately 50% of the germanium could be recovered from FA after a water extraction at room temperature. Regarding the SX method, the best operational conditions for obtaining a concentrated germanium-bearing solution practically free of impurities were as follows: extraction time 20 min; aqueous phase/organic phase volumetric ratio of 5; stripping with 1 M NaOH, stripping time 30 min; and stripping phase/organic phase

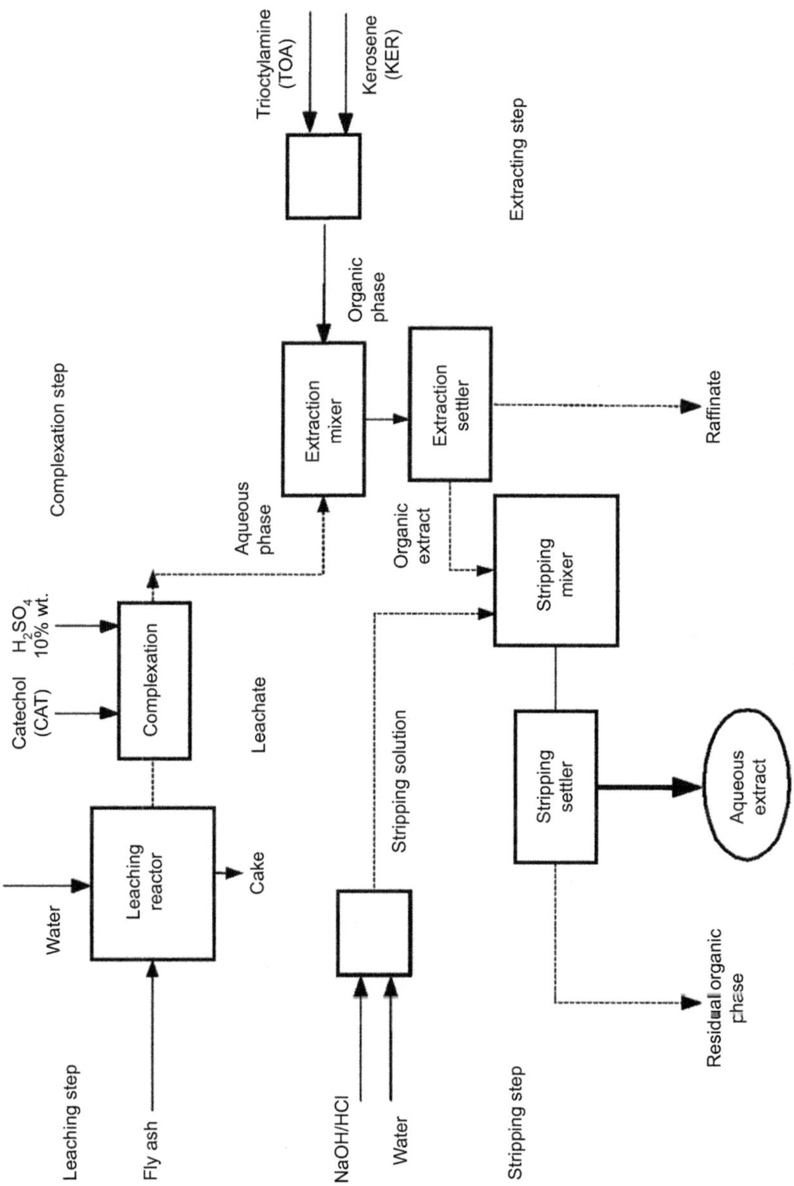

Figure 9.4 Schematic arrangement of the Ge-extraction pilot plant flow sheet [20].

volumetric ratio of 5. Ninety-five percent of the germanium was recovered from water leachates using those conditions [20].

In addition to the highly elevated Ge, it should also be noted that some elements (including Hg, As, Sb, Tl, U, and Be) enriched in Ge-bearing coals and their combustion products have high toxicity [9,13]. However, their negative influences on human health have rarely been studied. The health of workers in power plants and Ge-recovery factories is under a high threat of these toxic trace elements due to the following two conditions [9,13]: the high concentrations of the aforementioned toxic trace elements in FA together with Ge, because of the high volatility of Sb, As, Hg, and Tl under high-temperature conditions, and the organic modes of occurrence of U and Be in Ge-rich coals in the Wulantuga and Lincang deposits. The contents of Tl, Hg, Sb, and As in an FA sample collected by a baghouse filter from the Wulantuga Ge recovery factory were as high as 58, 67.5, 5160, and 9215 μg/g, respectively. Such fine particulate matter in FA with high concentrations of toxic trace elements represents a special danger, because they can penetrate into the lungs of people in contact with FA in the course of industrial activity.

9.3 Enrichment of Ga in coal and its recovery from coal products

9.3.1 Enrichment of Ga in coal and fly ash

Ga is typically a dispersed element and it is relatively rare for a Ga ore deposit to form [21]. The Ga contents in world coal are 5.5 and 6.0 μg/g, for brown coal and hard coal, respectively [4]. The Ga concentrations in Chinese coal and US coal are 6.55 and 5.7 μg/g, respectively [21]. Only a few Chinese coals contain >30 μg/g Ga, which is the industrial grade of Ga content in coal according to the Chinese standard. A Ga coal deposit was found in the Heidaigou Surface Mine of the Jungar coalfield, Inner Mongolia, in northern China (Fig. 9.1), with a weighted average of 45 μg/g Ga, 1.5 times the industrial grade content in coal. The proven reserves of Ga in this mine are estimated to be 54 kt, making it a super-large Ga ore deposit [2]. The management of the Heidaigou Surface Mine, which is owned by Shenhua Group Jungar Energy Co., Ltd., has built a pilot plant to develop technologies to extract the metals from combustion residues.

The concentration of Ga in Ga-rich coals and their FAs is listed in Table 9.2. The content of Ga could be up to 50−100 μg/g in coal and can be as high as several hundred micrograms per gram or even a few kilograms per tonne in coal FA, making Ga-rich coals a potential resource for Ga recovery as a by-product of coal combustion [12]. Ga contents were 0.1%−0.5% in the FA of thermoelectric power stations in Japan and 1.58% in England [12]. High Ga contents in coal from Kentucky, USA (150−900 μg/g in the ash) are generally confined to near-tonstein coal horizons and near-contact zones of coal beds [66]. In Primorsk Krai, the highest Ga contents are typical of Ge-bearing coals of the Rakovsk deposit (35−100 μg/g on a whole coal

Table 9.2 Concentration of Ga in Ga-rich coals and their fly ashes (μg/g).

	Minimum	Maximum	Average	Samples	Source
World coal (brown)			5.5	86	[4]
World coal (hard)			6.0	127	[4]
Chinese coal			6.55	2451	[2]
US coal			5.7	7565	[22]
Heidaigou Mine	11	76	45	12	[23]
Puertollano IGCC fly ash	149	320	198	8	[14]
Ontario Hydro, Lakeview fly ash			64.3		[24]
N.B. Power, Dalhousie Generating Station fly ash			55.0		
N.B. Power, Grand Lake Generating Station fly ash			86.8		
Saskatchewan Power, Poplar River fly ash			112		
Saskatchewan Power, Shand fly ash			120		
Trasaita Utilities fly ash, Sundance, Alberta			104		
Trasaita Utilities, Keephills Plant fly ash, Alberta			80		
Alberta Power Ltd, Sheerness Generating Station fly ash			62.1		
Wabamun Generating Plant fly ash, Alberta			66.6		
Ontario Hydro, Nanticoke fly ash			56.5		
Manitoba hydro, Brandon fly ash			64.6		
City Power plant fly ash, Changsha, China			37.5		
Hadera Power Plant fly ash, Israel			60.0		

IGCC, integrated gasification with combined cycle.

basis and 100−300 μg/g in the ash), which makes it possible to recover both Ge and Ga together. Moreover, anomalous Ga accumulations (100−400 μg/g in the ash) are found in some beds in the Kuzbass basin [12,67].

Fang and Gesser tested 13 different FAs from various part of Canada, China, and Israel. The Ga in the FA ranged from 37.5 to 120 μg/g [24]. And the concentration of Ga increased with decreasing FA particle size, which indicated that most of the Ga was condensed on the surface of the ash from the vapor phase during coal combustion. Font et al. also investigated the occurrence of Ga in typical FAs from the Puertollano IGCC power plant [14]. Its concentration in the coal gasification FAs could reach Ge contents of 10 times of that in the raw coal, with an average content of 283 μg/g. The speciation techniques suggest that Ga in IGCC FA mainly occurs as oxides (Fig. 9.5) [14]. Ga can occur as free oxide, Ga_2O_3, or Ga^{3+} substituting for Al^{3+} in the aluminosilicate matrix of IGCC FAs. And a small fraction of Ga may occur as a sulfide, probably substituting for Zn in sphalerite.

9.3.2 Ga recovery from coal products

Considerable effort has been devoted to the extraction of Ga from the FA generated by coal-fired power plants. Lu et al. present a systematic review of the extraction of Ga from coal FA [25]. Fang and Gesser developed a technique for Ga recovery from coal FA, in which a sample containing 107 μg/g Ga was first roasted at 500°C for 10 h [24]. Subsequently, it was leached at room temperature for 2 h using 2 M hydrochloric acid with an L/S ratio of 6. The equilibrium reactions in the acidic system of gallium−chloride−water are:

$$Ga_2O_3 + 8HCl \rightarrow 2HGaCl_4 + 3H_2O \quad (9.1)$$

$$Ga(OH)_3 + 4HCl \rightarrow HGaCl_4 + 3H_2O \quad (9.2)$$

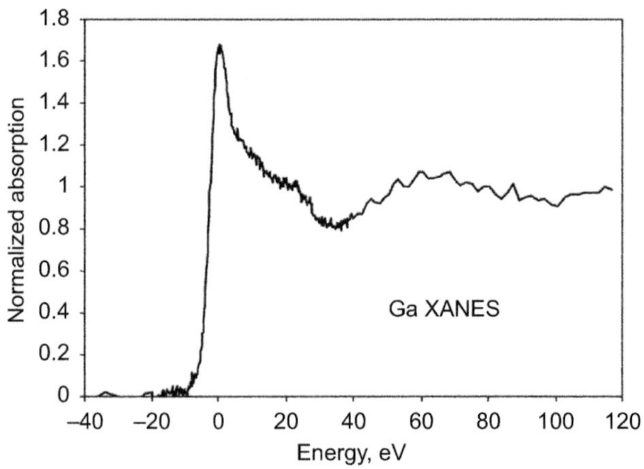

Figure 9.5 X-ray absorption near edge structure (*XANES*) spectrum for Ga in integrated gasification with combined cycle fly ash [14].

The results demonstrated that Ga in coal FA can be efficiently recovered by a hydrometallurgical process. More than 95% of the Ga in the ash could be leached into solution [24]. The proposed flow sheet is shown in Fig. 9.6, and consists of two-stage reverse leaching, removal of silica and calcium, reduction of Fe^{3+}, extraction of Ga by polyurethane foam, and purification and recovery of aluminum and iron. The open-cell polyurethane foam sponge proved to be an efficient medium for the SX of Ga from the complex FA leach solution.

Gutiérrez et al. [26] also studied Ga recovery from FA by an acid leaching and dual SX process [26]. The ash, which contained 110 μg/g Ga, was treated with 6 M hydrochloric acid. The leaching liquor (containing Ga as $[GaCl_4]^-$) was then mixed with Amberlite LA-2 (5 vol%) dissolved in kerosene to extract both Ga and Fe. Then, the loaded organic phase was washed with distilled water to strip Ga and Fe, and then Fe was precipitated from the washing solution with sodium hydroxide. The Ga remaining in solution was subsequently extracted selectively with a 10 vol% LIX 54 solution in kerosene. The Ga-loaded organic phase was stripped with a 0.5 M hydrochloric acid solution. Using this method, approximately 114 μg/g Ga was obtained from the stripped solution with 83% recovery.

In addition to the acid leaching process, the sodium hydroxide extraction method was also evaluated for Ga recovery from FA by Font et al. [27]. Eight FA samples

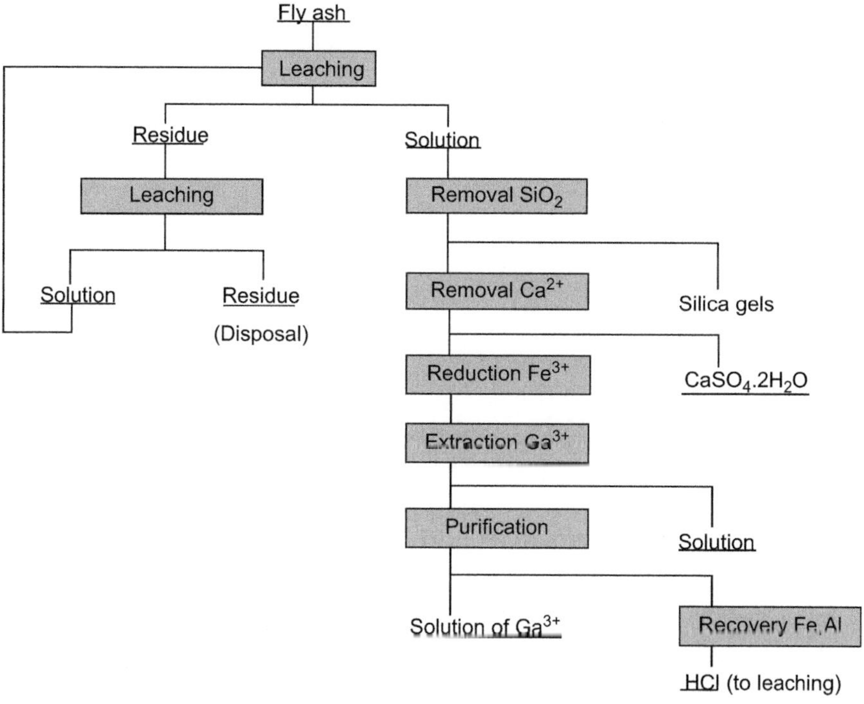

Figure 9.6 Proposed flow sheet for Ga recovered by a hydrometallurgical process for the treatment of coal fly ash [24].

were collected under different feed conditions from the Puertollano IGCC power plant for Ga extraction tests, in which the Ga content ranged from 149 to 320 μg/g. The optimal Ga extraction conditions were as follows: 25°C, 0.7–1 M NaOH, L/S ratio of 5 L/kg, and 6 h. Under these conditions, a Ga leaching rate of approximately 60%–86% was obtained, equivalent to 197–275 mg Ga per kilogram of FA. Recirculation of the leachates increased the initial Ga concentrations from 25–38 to 188–215 mg/L. The concentrated solution (containing Ga as $[Ga(OH)_4]^-$) was further treated by a two-stage carbonation process to recover Ga. Using the proposed method, the final recovery yield obtained for Ga reached 132–152 mg Ga per kilogram of FA, equivalent to 1582–1821 kg of Ga per year given the current annual production of the Puertollano IGCC plant (12,000 t).

A pilot plant producing 4000 t of alumina with a Ga recovery system was built in 2012 in Ordos, western China, by the Shenhua Energy Co., Ltd. [28]. The pilot plant was based on a previously developed laboratory-scale process for Ga recovery from a circulating fluidized-bed FA, as presented in Fig. 9.7 [68]. First, the raw FA was crushed (≤100 mesh) and the iron was removed through magnetic separation (iron level ≤1 wt%). Subsequently, the pretreated sample was dissolved in hydrochloric acid (20–37 wt%) to obtain a leachate (pH 1–3, Ga present as $[GaCl_4]^-$). Ga in the leachate was then adsorbed with a macroporous cationic resin. Subsequently, the eluent produced from washing the loaded resin was treated with a masking agent to mask ferric ions, and Ga was readsorbed on the resin from the purified solution. Then, the secondary eluent, generated from eluting the resin, was subjected to an alkalinity adjustment with a sodium hydroxide solution (180–245 g/L).Finally, the filtered Ga-rich solution was electrolyzed at 35–45°C under a current of 180–200 mA/L and a voltage of 4 V to produce Ga metal (99.9% purity). Through this process, the Ga adsorption efficiency and the iron removal rate were >96% and 80% in hydrochloric acid leachate, respectively. Although the alumina product passed the production appraisal in 2013 the plant operation had already been stopped because of the low extraction rate and the difficulty in the removal of impurities [28]. As of this writing, little progress has been made toward further developing this process and its application to industry.

9.4 Enrichment of U in coal and its recovery from coal products

9.4.1 Enrichment of U in coal and coal products

Uranium is one of the typical natural radionuclides with long half-lives in coal and ash. Several studies have been reported on the concentration and distribution of uranium in coal [1,2,10,12,22,30–34]. The concentrations of U in U-rich coals and their FAs are listed in Table 9.3. The US Geological Survey reported that the average uranium content in US coals is 2.1 μg/g, and the study of Ketris and Yudovich found that the

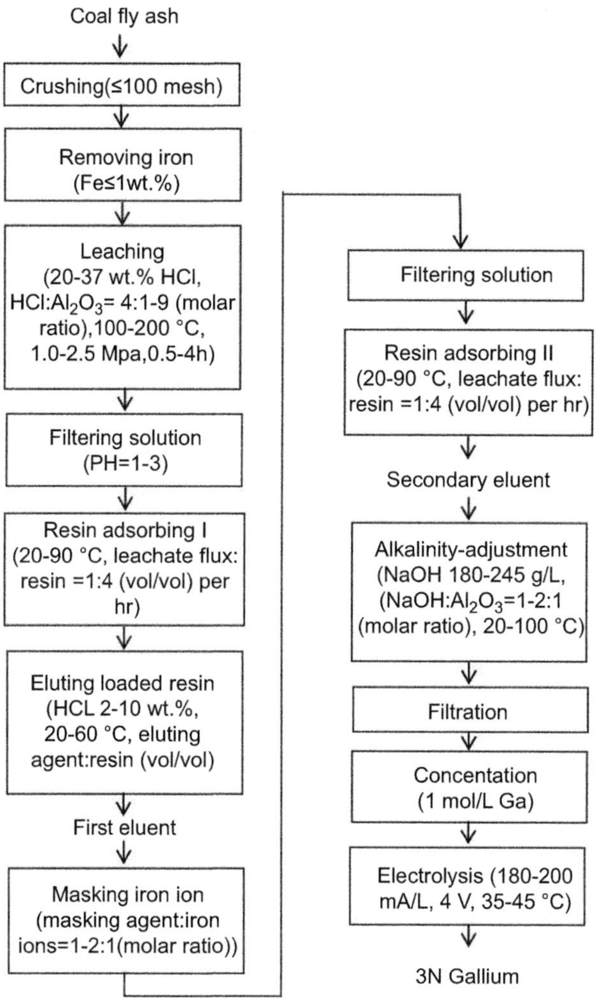

Figure 9.7 Schematic diagram of gallium recovery from fly ash [25,29].

average uranium content in world coals is about 1.9 and 2.9 μg/g for brown coal and hard coal, respectively [4]. Huang et al. reported that the average uranium content in Chinese coals is about 3 μg/g [15], which is close to the US and global average values, but up to 25,600 μg/g has been found in coal in some parts of China.

Uranium-bearing coals are probably most abundant in Middle Asia. Resources of the largest U deposits in the coal basins of this region are as high as 37,000 t (Koldzhatsk) and 60,000 t (Nizhneillisk), some smaller U deposits and occurrences in Russian coal basins are situated in the southern Urals (Donsk), Transbaikalia (Urtuisk), Kansk-Achinsk basin (Berezovsk open mine), and Primorski Krai (Rakovsk, Luzanovsk, etc.) [12].

Table 9.3 Concentration of U in U-rich coals and their fly ashes (μg/g).

	Minimum	Maximum	Average	Samples	Source
World coal (brown)			2.9	63	[4]
World coal (hard)			1.9	110	[4]
Chinese coal			3		[15]
Yili coal, Xinjiang	76	1835	519.8	6	[2]
Guiding coal, Guizhou	211	288	237.8	4	
Heshan, Guangxi	21.4	48.7	38.5	5	
Yanshan, Yunnan	111	178	153		
Late Permian coals, Western Guizhou	0.1	176	14	71	[34]
Chenxi, Hunan			48.7	1	[33]
Bangmai, Yunnan			71.5	1	[32]
Mengzi, Yunnan	0.14	141.5	36.8	223	[35]
Gange, Yunnan	0.1	315.8	66.3	115	[15]
Qiubei, Yunnan	38.3	64.7	55.7	5	
Early Permian coal, Datong, Shanxi	33	42	38.2		[31]
Late Carboniferous coal, Datong, Shanxi	5	82	28.2		
Neogene coal, Dazhai, Yunnan	9.6	130	52.5		
Miocene coal, Dazhai, Yunnan	1.1	640	56		
Luquan, Yunnan	22.6	47.2	34.1		
Puan, Guizhou	2.5	133	32.4		
Zhijin, Guizhou			49.6		
Yishan, Guangxi	35.0	123	71.7		
Sawabuqi, Xinjiang	210	520	365		
Shengli, Inner Mongolia	0.42	148	25.9		
Shiping, Sichuan	0.75	155	39.8		
Moxinpo, Chongqing	295	476	376		
Rongyang, Guizhou			70.5		
Mengwang coal, Lincang, Yunnan			112.56		[36]

Table 9.3 Continued

	Minimum	Maximum	Average	Samples	Source
South Texas lignite low-temperature ash	1130	4610	2970	4	[37]
Figueira coal power plant, Brazil	117	1190			[38,39]
Coal-fired power plant fly ash, China	5.5	13.6		2	[39]
4# power plant bottom ash, Ulaanbaatar, Mongolia	10.6	154	54.3	144	[40]
Lincang bottom ash, China			374		[41]

Coal with a uranium concentration of higher than 200 μg/g is regarded as a resource for industrial extraction [42]. Some coals in China with abnormally high uranium concentrations (100 times higher than the average of common Chinese coals) are classified as coal-hosted uranium deposits. Chen et al. summarized some coal-hosted uranium deposits [43], i.e., coals from the Yili and Tarim basins of Xinjiang Province, Bangmai basin of Yunnan Province, and Mabugang basin of Guangdong Province. In addition to the U-rich coals in the Yili coal basin of Xinjiang Province, which are closely situated to the large U deposits in Middle Asia, some smaller deposits of U-rich coals, including Tengchong, Yanshan, Jianshui, and Lincang of Yunnan Province; Guiding of Guizhou Province; and Heshan of Guangxi Province, have been found in southern China (Fig. 9.1) [2]. Chen et al. has summarized the significantly uranium-rich coals in China [31], which are also listed in Table 9.3. Significant uranium enrichment in coals also occurs in Shanxi, Xinjiang, Inner Mongolia, Sichuan, and Chongqing Provinces of China. The Yili basin is a Mesozoic−Cenozoic continental basin, derived from the Paleozoic interarc taphrogenic trough. The U-bearing strata are located in the lower Jurassic coal-bearing sedimentary sequences and include in situ leaching sandstone type as well as lignites. Regarding the uranium concentrations in coals of different ages, it decreases in the following order: Paleogene and Neogene > Late Permian > Late Triassic > Late Carboniferous and Early Permian > Late Jurassic and Early Cretaceous > Early and Middle Jurassic [31].

The modes of occurrence of the uranium in coal could significantly affect its behavior during combustion and also its extraction from coal products, because most of the uranium associated with minerals might be encapsulated by sintering of the ash. Considerable research has demonstrated that the uranium in coal is mainly associated with the organic matter [1,2,10,12,15,22,31,34], which is attributed to the exchange reactions occurring between H^+ ions of humic acid and uranyl ions dissolved in solution. In the uranium-rich coals, various uranium minerals are also found, such as autonite, carnotite, coffinite, brannerite, torbernite, uraninite, uranophane, pitchblende, zircon, thorite, samarskite, euxenite, and some phosphates [1,2,12,37,43]. The uranium in these minerals is mainly present as isomorphous substitutions in the crystal lattices of the minerals. Overall, with the exception of minor detectable uranium minerals, the organic matter and silicates are primarily the hosts of uranium in Chinese coals, although detailed information about speciation and structure is limited [43].

When coal is combusted in a coal-fired power plant, the uranium is concentrated in the by-products, i.e., slag, bottom ash, and FA. The by-products are commonly enriched in uranium 3- to 10-fold that of the feed coals [44]. However, only limited references about uranium content in by-products derived from coal have been found. Mohan prepared low-temperature ashes of a lignite sample from the Pawelek mine, Karnes County, Texas, USA. They found that uranium was concentrated in low-temperature ashes at 1130–4610 μg/g [37]. In the Parami study [45], uranium concentrations in FA range from 1.1 to 21.7 μg/g in the Philippines' coal-fired thermal power plants. In another study by Flues [38], in Brazil, uranium concentrations in the FA range from 117 to 1190 μg/g. Zhang et al. reported the uranium content and distribution in a coal-fired power plant in China, but the uranium content in coal by-products was quite low, only 0.29, 4.7, and 5.9 μg/g, for gypsum, bottom ash, and FA, respectively [39]. Yang et al. found that uranium is enriched in the magnetospheres (magnetic fractions of FAs) from Chinese and Russian power plants [46]; uranium concentrations in magnetospheres are about 1.2–1.5 times those in raw FA.

9.4.2 Uranium recovery from coal by-products

Uranium production from coal ash is technically feasible, and it was extracted from coal ash in the past in very specific historical situations [47]. Due to the obvious dangers in the generation of electricity in nuclear power plants and in the production of nuclear weapons, the public references on uranium extraction and recovery are limited. The Oak Ridge National Laboratory has conducted leaching behavior of lignite ash for uranium recovery since 1950s. The lignite was roasted in a muffle furnace for 2 h at 800°C, and the average uranium extraction was almost 80% under conditions of 5% H_2SO_4 at 80°C [48]. Nearly 80% of the uranium in coal ashes was leached out with H_2SO_4 after 14 days [49]. Sparton Resources, Inc. (2007), reported the extraction of U from coal FA. This process included acid leaching (H_2SO_4) and formation of yellowcake by passing the filtered leachate solution through a standard ion exchange resin, stripping the resin, and precipitating a U compound. This process is essentially similar to the U extraction method used by the primary U ore processing plants. However, this research is preliminary and requires further work to improve recovery [50].

Maslov et al. investigated the uranium recovery from coal ash dumps of Mongolia [40]. The ash samples were irradiated on an MT-25 microtron electron accelerator for 2 h and cooled over 1 h. Then, 1 g of irradiated ash was treated with various solvents at a phase ratio of 1:20 in 50-mL plastic test tubes. The mixture was stirred over 5–24 h; solid and liquid phases were separated by filtering the mixture through a nuclear filter. For radiochemical separation and isolation of uranium from the solution obtained after

Table 9.4 Results of uranium leaching in different modes [40].

Solvent	Temperature, °C	Treatment time, h	Chemical yield, %
H_2O	20	24	1.1
Na_2CO_3 (10%)	90	2	9.2
H_2SO_4 (45%)	90	2	45.4
8 M $HNO_3 + H_2O_2$ (10%)	90	2	53.5
8 M HNO_3 + HF (10%)	20	24	99

leaching the anion-exchange technique was used. From Table 9.4 it is seen that in the case of treatment of the ash with a mixture of 8 M HNO_3 + HF (10%), the level of uranium leaching reached 99%. Such a degree of recovery of uranium is due to the fact that the vitrified portion of ash, which contains at least 50% uranium and is formed as a result of thermal transformations of the mineral matter of coal, is destroyed in the presence of hydrofluoric acid. From the data presented in the table, it follows that no more than 1% of the uranium can be washed out from FAs by water under natural conditions. The leaching of uranium with a carbonate (10%) solution indicates the presence of uranium in the valence state of +6 in the ash. Elements such as Ag, Al, Ba, Ca, Cr, Cs, K, Mg, Mn, Na, Ni, Rb, Sc, Th, Ti, and Y, and REEs, are eluted from the resin with a 2 M HCl solution in the first two portions of the eluate. The third elution portion allocates more than 90% of the uranium [40] (Fig. 9.8).

Li's group investigated the leaching of uranium from lignite and coal ash [51,52]. They discussed the effects of the leaching conditions and oxidants on the uranium

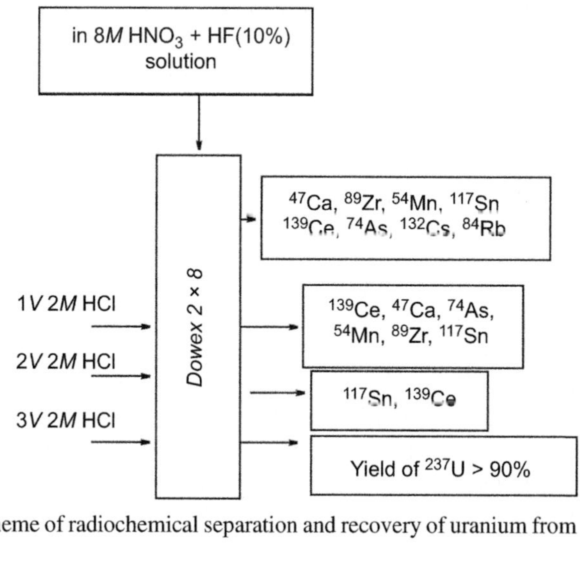

Figure 9.8 Scheme of radiochemical separation and recovery of uranium from the leachate [40].

leaching efficiency of the lignite and coal ash. The leaching of uranium from the lignite by a Na_2CO_3/$NaHCO_3$ mixture was efficient under the optimal solid/liquid and CO_3^{2-}/HCO_3^- ratios, which were 1:20 (g/mL) and 2:1, respectively. The five oxidants, H_2O_2, NaClO, $KMnO_4$, $K_2S_2O_8$, and O_2, can significantly enhance the leaching of uranium in the lignite. And the use of O_2 is the most efficient option, which increases the leaching efficiency up to 72.23% after a 12-h reaction. Tessier sequential extraction was used to divide the uranium in the coal ash into five different fractions. Leachable uranium mainly consisted of carbonates and Fe−Mn oxides, accounting for 50% of the total uranium in coal ash. The optimum experiment conditions for uranium leaching from coal ash were 0.3 M H_2SO_4, 1:20 g/L solid/liquid ratio, 12 h, and 60°C, under which the leaching efficiency was almost 45%. The oxidant had a positive effect on uranium leaching. When 4% $KMnO_4$ was added into 0.3 M H_2SO_4, the leaching efficiency was increased to 55% by its strong oxidation and ability to enhance ionic strength. Lei et al. found that roasting the coal ash with $CaCl_2$ enhanced the leaching of uranium by HNO_3; 87.3% of the total uranium can be recovered via the optimized roast−leach process [53]. Sun et al. also investigated uranium recovery by $CaCl_2$ roasting followed by HNO_3 leaching from a high-germanium coal bottom ash. The experimental results showed that uranium mainly exists incorporated into aluminosilicates (mainly glass phase) (59.1%) and Fe-oxides (34.9%). The extraction of uranium was achieved by Tessier sequential extraction, acid leaching, magnetic separation, and mechanical activation combined with alkali activation, as well as calcination with $CaCl_2$ followed by HNO_3 leaching. The results demonstrated that the optimal method of uranium extraction was calcination with $CaCl_2$ followed by leaching with HNO_3. By this method, the recovery of uranium can be as high as 95.8%.

9.5 Enrichment of rare earth elements in coal and their recovery from coal products

9.5.1 *Enrichment of rare earth elements in coal and coal products*

REEs include lanthanum (La), cerium (Ce), praseodymium (Pr), neodymium (Nd), promethium (Pm), samarium (Sm), europium (Eu), gadolinium (Gd), terbium (Tb), dysprosium (Dy), holmium (Ho), erbium (Er), thulium (Tm), ytterbium (Yb), and lutetium (Lu). The rare earths are also often considered to include the metals scandium (Sc) and yttrium (Y). REEs (or REYs if including Y) have attracted much attention in recent years, not only because the demand for these elements has grown rapidly, but also because of the supply crisis of 2010 and the price spike of 2011 [1,12,54−56]. Seredin and Dai present a systematic classification of these elements and their significance, sources, and problems [56]. REYs have now become a "cornerstone" for modern technologies, especially in alternative power and energy-saving applications. Although the crisis did not last for long, the short-lived alarm produced by export restrictions from China was sufficient to initiate a treasure hunt by way of

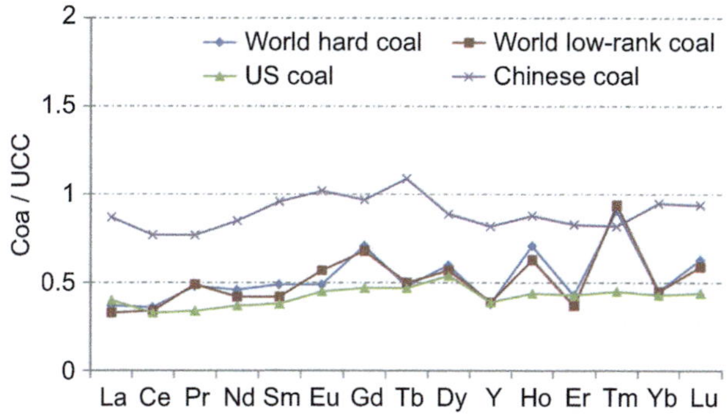

Figure 9.9 Distribution patterns of average values of rare earth elements and yttrium (REYs) in common US coals, Chinese coals, and world low-rank and hard coals, normalized to average REY concentrations in the upper continental crust (*UCC*) [54].

exploration for REE deposits all over the world. The price of some critical metals in 2011 increased about 10 times more than they were 1½ years before. In 2011, global production of REEs was approximately 132,000 t—95% of which was supplied by China.

The estimated average sum of REY concentrations in Chinese coals (138 μg/g [21]) is around twice the average sum of REY concentrations for world coals (68 μg/g [4]). The average REY content of US coals (62.1 μg/g [22]) is close to that of world coals. The distribution plots of REY averages for both Chinese and US coals are similar to those of the upper continental crust (Fig. 9.9).

The average REY content in world (404 μg/g) and US (517 μg/g) coal ashes is approximately 3 times higher than that in the upper continental crust (168.4 μg/g [57]) [56]. The REY oxides (REOs), which are generally used to estimate the abundance of the metals in ores, are 483 and 621 μg/g, respectively, after recalculation from the content of single elementary substance. These values are close to the REO content of some ion-adsorbed deposits (Longnan, 840 μg/g, and Huashan, 700 μg/g [58]) and large-tonnage and low-grade ores of hydrothermal origin (Round Top Mountain, USA, 690 μg/g; http://www.trer.com). Hence, if the average REY content in coal ashes, based on analyses of numerous samples, is comparable to that of some conventional ores, the wastes of coal combustion could and should be considered as possible sources of these metals [56].

Five types of elemental assemblage associated with REY enrichment in coal and coal-bearing strata in China have been observed and systematically discussed by Dai et al. [1]: REY-Zr(Hf)-Nb(Ta)-Ga, REY-U-V-Cr-Se-Re, REY-Nb(Ta)-Zr(Hf)-U, REY-U(Mo, Se), and REY-Al-Ga. As described by Dai et al. (2018), the coal-hosted REY-Zr(Hf)-Nb(Ta)-Ga ore deposits in China have been identified only in the southwestern part of the country, covering areas that include eastern Yunnan, western Guizhou, Chongqing, and southern Sichuan provinces. The REY-Zr(Hf)-Nb(Ta) deposits contain up to 0.1%–0.5% REO, 1%–3% (Zr, Hf)$_2$O$_5$, and 0.05%–0.1% (Nb, Ta)$_2$O$_5$ (for metalliferous coals, the rare metal concentrations are on an ash basis). A coal deposit with highly elevated concentrations of REY-U-V-Cr-Se-Re was found in the Moxinpo coalfield, Sichuan Province, and is highly enriched in REY oxides (12,390 μg/g), as well as U (917 μg/g), V_2O_5 (13,098 μg/g), Cr_2O_3 (8569 μg/g), Se (160 μg/g), and Re (6.21 μg/g) (all on an ash basis) [55].

The US Department of Energy/National Energy Technology Laboratory (DOE/NETL) performed an assessment of the amounts of REEs that are likely to be found in coal deposits and the mineral matter—particularly the clays—associated with these deposits [59]. The initial resource estimates indicate that 6 million t of REEs could potentially be produced from recoverable coal reserves in select western state coal basins in Montana, Wyoming, Colorado, Utah, New Mexico, and Arizona. Similar estimates indicate that 4.9 million t are potentially available from among the coal deposits found in Pennsylvania, West Virginia, Kentucky, and Virginia. These estimates represent total resources, with REE contents in associated mineral matter over a cutoff grade of 500 μg/g, and are based on estimates of recoverable coal reserves in these regions found elsewhere.

As REYs are nonvolatile elements, their distribution patterns in an average coal ash should be the same as those in average coal. Knowing how the REYs are distributed in coal ash is essential to devising effective approaches for REY extraction from this potential resource. Kolker et al. used the Stanford−USGS SHRIMP-RG ion microprobe to determine grain-scale REE partitioning in a suite of 19 US and international coal FA samples [60]. Results confirm the occurrence of REEs in aluminosilicate glasses consistent with previous results that showed a strong positive correlation between Al and REE content in a larger group of US and international coal FA samples. Those results suggest that extraction of REE from the aluminosilicate glass fraction in coal ash will yield a significant portion of the REE present and, therefore, this fraction should be targeted in ongoing efforts to recover REE from coal FA.

9.5.2 Rare earth elements recovery from coal and its by-products

There are several methods developed for the recovery of rare elements from coal FA. Zhang et al. systematically reported promising recovery methods of REEs from coal and coal by-products [61]. The chlorination method is favored due to selectivity and high reactivity [50]. The necessary apparatus required for this process is very expensive. This technique can be used to recover only certain rare elements when other complex recovery processes are not suitable. From this standpoint, Matyes et al. reported a nondestructive digestion process for the recovery of rare elements [62], including Lanthanides and Actinides and group V and VI elements that are adsorbed onto the surfaces of coal combustion residues. This technique includes: (1) reducing the grain size of the combustion residues from 0.1 to 1 mm, (2) digestion of the residues with water, (3) digestion of the residues with a dilute aqueous alkali solution (1.5%−8%), (4) digestion of the residues with a dilute aqueous acid solution (0.5%−5%) at a temperature below the boiling point of water, (5) separating the solid and liquid phases after each digestion, and (6) isolating the metals from the liquid phases. This study shows that the mild digestion agents used dissolve an unexpectedly high proportion of rare metals present in the by-products and that increasing the heat and residence time lowers the availability of the metals. This process is not appropriate for the recovery of individual elements from the mixed solution.

Joshi et al. demonstrated the recovery of REEs from coal FA [63]. In this method, first the coal FA was treated with a mineral acid (HNO_3) at 90°C to form a more concentrated mineral acid solution. This solution is mixed with an organic solution (tributyl phosphate and kerosene) that includes the rare earth salts. The organic solution is mixed with water to form an aqueous solution. REEs were substantially recovered from the aqueous solution by using an ion-exchange process. In a continuation-in-part of the aforementioned patent, US Patent No. 0139871 (2015) was carried out using the same method [64], but here the coal FA was sorted into a substantially magnetic and a nonmagnetic fraction. This indicates that magnetic treatment can maximize the efficiency of extracting REEs from coal FA. However, the extraction efficiency of these methods was not reported. Furthermore, the ion-exchange method can remove other metals from the solution. This method needs to be optimized to improve the yields for REEs. Rozelle et al. examined two US coal by-products for ion extraction, using ammonium sulfate, an ionic liquid, and a deep eutectic solvent as lixiviants, and showed that extraction of REEs in each case produced high recoveries of REYs to the solution. Although the REEs are known to occur in low concentrations in US coals and coal products, the data of Rozelle et al. suggest that at least in the cases of the materials examined, US coal by-products may be technically suitable as REE ores [65]. A report by the DOE and NETL to the US Congress indicated that a preliminary reserve base has been established in two key coal-producing regions in the United States, and these two regions alone have potential reserves in the millions of tonnes. However, the key to unlocking this potential reserve base for US REE production from coal and coal by-products is the improvement of separation and recovery technologies.

9.6 Summary

Through the studies of the past decades, the valuable trace elements in coal and coal by-products have attracted more and more attention even though their average content in coal is not high, as in ore sources. Several specific Ge-, Ga-, U-, and REE enriched coal fields have been found and investigated broadly. Although some technical problem still exist, the value and significance of coal and coal by-products as the source of Ge, Ga, U, and REE have been recognized.

Extraction of valuable trace elements from coal ashes greatly differs among coals and regions, and also varies for different kinds of elements. There has not been a unified extraction method applied for valuable trace elements from coal and coal by-products. Different extraction methods, which depend on different combustion conditions (e.g., combustion temperature, category of raw coal, furnace type) and modes of occurrence of the trace elements in raw coals, would lead to different leaching efficiencies. Therefore, to effectively extract valuable trace elements from coal and coal ashes, there are still many problems that need be resolved, such as:

1. identification of the sources of coal and coal by-products with highest concentrations of valuable trace elements;
2. following the detailed understanding of the form and occurrence of valuable trace elements in coal and coal by-products, which will be very helpful for the recovery technologies;
3. design, development, and testing of various separation technologies to recover valuable trace elements from coal and coal by-products; several fundamental studies need to be performed for developing and optimizing a process for economic and environmentally friendly recovery;
4. large-scale demonstration plants for valuable trace elements recovery need to be deployed to test the techniques;
5. the serious human health problems caused by the process of valuable trace elements recovery need to be focused on and paid more attention, because most coal or coal by-products enriched in valuable trace elements also contain higher concentrations of toxic trace elements.

Acknowledgments

This project was supported by the National Key Technologies R&D Program (grant no. 2018YFB0605104.

References

[1] Dai SF, Yan XY, Ward CR, Hower JC, Zhao L, Wang XB, et al. Valuable elements in Chinese coals: a review. Int Geol Rev 2018;60:590–620.
[2] Dai SF, Ren DY, Chou CL, Finkelman RB, Seredin VV, Zhou YP. Geochemistry of trace elements in Chinese coals: a review of abundances, genetic types, impacts on human health, and industrial utilization. Int J Coal Geol 2012;94:3–21.
[3] Li J, Zhuang XG, Querol X, Font O, Izquierdo M, Wang ZM. New data on mineralogy and geochemistry of high-Ge coals in the Yimin coalfield, Inner Mongolia, China. Int J Coal Geol 2014;125:10–21.
[4] Ketris MP, Yudovich YE. Estimations of Clarkes for Carbonaceous biolithes: world averages for trace element contents in black shales and coals. Int J Coal Geol 2009;78: 135–48.
[5] Wang T, Huang W, Yan D, Tang X. Progress of research on mineralization mode of large coal-Ge deposits in China: coal-Ge deposit in wulantuga of Inner Mongolia and Lincang of Yunan. Earth Sci Front 2016;23:113–23.
[6] Du G, Zhuang X, Querol X, Izquierdo M, Alastuey A, Moreno T, et al. Ge distribution in the Wulantuga high-germanium coal deposit in the Shengli coalfield, Inner Mongolia, northeastern China. Int J Coal Geol 2009;78:16–26.
[7] Zhuang X, Querol X, Alastuey A, Juan R, Plana F, Lopez-Soler A, et al. Geochemistry and mineralogy of the Cretaceous Wulantuga high-germanium coal deposit in Shengli coal field, Inner Mongolia, Northeastern China. Int J Coal Geol 2006;66:119–36.

[8] Qi H, Hu R, Su W, Qi L, Feng J. Continental hydrothermal sedimentary siliceous rock and genesis of superlarge germanium (Ge) deposit hosted in coal: a study from the Lincang Ge deposit, Yunnan, China. Sci China Earth Sci 2004;47:973–84.
[9] Dai S, Wang X, Seredin VV, Hower JC, Ward CR, O'Keefe JMK, et al. Petrology, mineralogy, and geochemistry of the Ge-rich coal from the wulantuga ge ore deposit, Inner Mongolia, China: New data and genetic implications. Int J Coal Geol 2012;90–91:72–99.
[10] Dai SF, Wang PP, Ward CR, Tang YG, Song XL, Jiang JH, et al. Elemental and mineralogical anomalies in the coal-hosted Ge ore deposit of Lincang, Yunnan, Southwestern China: key role of N-2-CO_2-mixed hydrothermal solutions. Int J Coal Geol 2015;152:19–46.
[11] Li J, Zhuang XG, Querol X, Font O, Moreno N, Zhou JB. Environmental geochemistry of the feed coals and their combustion by-products from two coal-fired power plants in xinjiang Province, Northwest China. Fuel 2012;95:446–56.
[12] Seredin VV, Finkelman RB. Metalliferous coals: a review of the main genetic and geochemical types. Int J Coal Geol 2008;76:253–89.
[13] Dai S, Seredin VV, Ward CR, Jiang J, Hower JC, Song X, et al. Composition and modes of occurrence of minerals and elements in coal combustion products derived from high-Ge coals. Int J Coal Geol 2014;121:79–97.
[14] Font O, Querol X, Huggins FE, Chimenos JM, Fernández AI, Burgos S, et al. Speciation of major and selected trace elements in IGCC fly ash. Fuel 2005;84:1364–71.
[15] Huang W, Wan H, Finkelman R, Tang X, Zhao Z. Distribution of uranium in the main coalfields of China. Energy Explor Exploit 2012;30:819–36.
[16] Hu R-Z, Qi H-W, Zhou M-F, Su W-C, Bi X-W, Peng J-T, et al. Geological and geochemical constraints on the origin of the giant Lincang coal seam-hosted germanium deposit, Yunnan, SW China: a review. Ore Geol Rev 2009;36:221–34.
[17] Wu W, Mo R, Wang Z. Occurrence features and geological work of germanium resource in Yimin coal field, Inner Mongolia. Inn. Mong. Geol. 2002;1:27–30.
[18] Kursun Unver I, Terzi M. Distribution of trace elements in coal and coal fly ash and their recovery with mineral processing practices: a review. J Mining Environ 2018;9:641–55.
[19] Arroyo F, Fernandez-Pereira C. Hydrometallurgical recovery of germanium from coal gasification fly ash. solvent extraction method. Ind Eng Chem Res 2008;47:3186–91.
[20] Arroyo F, Fernandez-Pereira C, Olivares J, Coca P. Hydrometallurgical recovery of germanium from coal gasification fly ash: pilot plant scale evaluation. Ind Eng Chem Res 2009;48:3573–9.
[21] Dai S, Li D, Chou C-L, Zhao L, Zhang Y, Ren D, et al. Mineralogy and geochemistry of boehmite-rich coals: new insights from the Haerwusu surface mine, jungar coalfield, inner Mongolia, China. Int J Coal Geol 2008;74:185–202.
[22] Finkelman RB. Trace and minor elements in coal. In: Engel MH, Macko SA, editors. Organic geochemistry. New York: Plenum Press; 1993. p. 593–607.
[23] Dai SF, Ren DY, Chou CL, Li SS, Jiang YF. Mineralogy and geochemistry of the No. 6 coal (Pennsylvanian) in the junger coalfield, Ordos basin, China. Int J Coal Geol 2006;66: 253–70.
[24] Fang Z, Gesser HD. Recovery of gallium from coal fly ash. Hydrometallurgy 1996;41: 187–200.
[25] Lu F, Xiao T, Lin J, Ning Z, Long Q, Xiao L, et al. Resources and extraction of gallium: a review. Hydrometallurgy 2017;174:105–15.
[26] Gutiérrez B, Pazos C, Coca J. Recovery of gallium from coal fly ash by a dual reactive extraction process. Waste Manag Res 1997;15(4):371–82.
[27] Font O, Querol X, Juan R, Casado R, Ruiz CR, López-Soler Á, et al. Recovery of gallium and vanadium from gasification fly ash. J Hazard Mater 2007;139:413–23.

[28] Ding J, Ma S, Shen S, Xie Z, Zheng S, Zhang Y. Research and industrialization progress of recovering alumina from fly ash: a concise review. Waste Manag 2017;60:375–87.
[29] Zou P, Wei C, Jiang Y, Han J, Guo Z, Gu D, et al., Extraction of gallium from coal ash by crushing coal ash, performing wet magnetic separation, adding hydrochloric acid, filling liquid in macropore cation resin column, eluting, adding screening agent, concentrating, and electrolyzing. China Shenhua Energy Co Ltd (Shgr-C) Shenhua Zhunger Energy Co Ltd (Shgr-C).
[30] Duan P, Wang W, Sang S, Qian F, Shao P, Zhao X. Partitioning of hazardous elements during preparation of high-uranium coal from Rongyang, Guizhou, China. J Geochem Explor 2018;185:81–92.
[31] Chen J, Chen P, Yao D, Huang W, Tang S, Wang K, et al. Abundance, distribution, and modes of occurrence of uranium in Chinese coals. Minerals 2017;7.
[32] Tang X, Huang W. Trace elements in Chinese coals. Beijing: The Commercial Press; 2004.
[33] Li W. Uranium distribution in high-sulfur coal and stone coals of China. Coal Anal Utilization 1986:27–37.
[34] Dai S, Ren D, Tang Y, Yue M, Hao L. Concentration and distribution of elements in late permian coals from western guizhou province, China. Int J Coal Geol 2005;61:119–37.
[35] Xi W. Uranium concentrations of coals in yunnan province. China Coal Geology 1992;4:32.
[36] Yang Z, Li Y, Lou Q, Liu D, Ning Y, Yang S, et al. Release of uranium and other trace elements from coal ash by $(NH_4)_2SO_4$ activation of amorphous phase. Fuel 2019;239: 774–85.
[37] Mohan MS, Ilger JD, Zingaro RA. Speciation of uranium in a South Texas lignite: additional evidence for a mixed mode of occurrence. Energy Fuels 1991;5:568–73.
[38] Flues M, Camargo IMC, Silva PSC, Mazzilli BP. Radioactivity of coal and ashes from Figueira coal power plant in Brazil. J Radioanal Nucl Chem 2006;270:597–602.
[39] Zhang Y, Shi M, Wang J, Yao J, Cao Y, Romero CE, et al. Occurrence of uranium in Chinese coals and its emissions from coal-fired power plants. Fuel 2016;166:404–9.
[40] Maslov OD, Tserenpil S, Norov N, Gustova MV, Filippov MF, Belov AG, et al. Uranium recovery from coal ash dumps of Mongolia. Solid Fuel Chem 2010;44:433–8.
[41] Sun Y, Qi G, Lei X, Xu H, Wang Y. Extraction of uranium in bottom ash derived from high-germanium coals. Procedia Environ Sci 2016;31:589–97.
[42] Dai S, Ren D. Concentration and the sequential chemical extraction procedures of U and Th in the Paleozoic coals from the Ordos Basin. J China Coal Soc 2004;29:56–60.
[43] Chen J, Chen P, Yao D, Huang W, Tang S, Wang K, et al. Geochemistry of uranium in Chinese coals and the emission inventory of coal-fired power plants in China. Int Geol Rev 2018;60:621–37.
[44] Hower J, Dai S, Eskenazy G. Distribution of uranium and other radionuclides in coal and coal combustion products, with discussion of occurrences of combustion products in Kentucky power plants. Coal Combust Gasif Prod 2016;8:44–53.
[45] Parami VK, Sahoo SK, Yonehara H, Takeda S, Quirit LL. Accurate determination of naturally occurring radionuclides in Philippine coal-fired thermal power plants using inductively coupled plasma mass spectrometry and γ-spectroscopy. Microchem J 2010;95: 181–5.
[46] Yang JP, Zhao YC, Zyryanov V, Zhang JY, Zheng CG. Physical-chemical characteristics and elements enrichment of magnetospheres from coal fly ashes. Fuel 2014;135:15–26.
[47] Monnet A, Percebois J, Gabriel S. Assessing the potential production of uranium from coal-ash milling in the long term. Resour Pol 2015;45:173–82.
[48] Hurst FJ. Recovery of uranium from lignites. Hydrometallurgy 1981;7:265–87.

[49] Paul M, Seferinoğlu M, Ayçık GA, Sandström Å, Smith ML, Paul J. Acid leaching of ash and coal: time dependence and trace element occurrences. Int J Miner Process 2006;79: 27−41.
[50] Sahoo PK, Kim K, Powell MA, Equeenuddin SM. Recovery of metals and other beneficial products from coal fly ash: a sustainable approach for fly ash management. Int J Coal Sci Technol 2016;3:267−83.
[51] Yang Z, Li Y, Ning Y, Yang S, Tang Y, Zhang Y, et al. Effects of oxidant and particle size on uranium leaching from coal ash. J Radioanal Nucl Chem 2018;317:801−10.
[52] Ning Y, Li Y, Zhang Y, Tang P. Study on the oxidative leaching of uranium from the lignite in the CO_{3}^{2-}/HCO_{3}^{-} system. Geofluids 2017:8 pages. Article ID 7130971. https://doi.org/10.1155/2017/7130971.
[53] Lei X, Qi G, Sun Y, Xu H, Wang Y. Removal of uranium and gross radioactivity from coal bottom ash by $CaCl_2$ roasting followed by HNO_3 leaching. J Hazard Mater 2014;276: 346−52.
[54] Dai S, Graham IT, Ward CR. A review of anomalous rare earth elements and yttrium in coal. Int J Coal Geol 2016;159:82−95.
[55] Dai SF, Xie PP, Jia SH, Ward CR, Hower JC, Yan XY, et al. Enrichment of U-Re-V-Cr-Se and rare earth elements in the late Permian coals of the moxinpo coalfield, Chongqing, China: genetic implications from geochemical and mineralogical data. Ore Geol Rev 2017; 80:1−17.
[56] Seredin VV, Dai S. Coal deposits as potential alternative sources for lanthanides and yttrium. Int J Coal Geol 2012;94:67−93.
[57] Taylor SR, McLennan SM. The continental crust: its composition and evolution. 1985.
[58] Bao Z, Zhao Z. Geochemistry of mineralization with exchangeable REY in the weathering crusts of granitic rocks in South China. Ore Geol Rev 2008;33:519−35.
[59] DOE/NETL. Report on rare earth elements from coal and coal byproducts. Washington, DC: United States Department of Energy; 2017. p. 43.
[60] Kolker A, Scott C, Hower JC, Vazquez JA, Lopano CL, Dai S. Distribution of rare earth elements in coal combustion fly ash, determined by SHRIMP-RG ion microprobe. Int J Coal Geol 2017;184:1−10.
[61] Zhang W, Rezaee M, Bhagavatula A, Li Y, Groppo J, Honaker R. A review of the occurrence and promising recovery methods of rare earth elements from coal and coal by-products. Int J Coal Prep Util 2015;35:295−330.
[62] Matyas B, Gerber P, Solymos A, Kaszanitzky F, Panto G, Leffler J. Process for recovering rare metals from the combustion residue of coal by digestion. US; 1987.
[63] Joshi PB, Preda DV, Skyler DA, Tsinberg A, Green DB, Marinelli WJ. Recovery of rare earth elements and compounds from coal ash. United States: Physical Sciences, inc.; 2013.
[64] Joshi PB, Preda DV, Skyler DA, Scherer A, Green BD, Marinelli WJ. Recovery of rare earth elements and compounds from coal ash. US; 2015.
[65] Rozelle PL, Khadilkar AB, Pulati N, Soundarrajan N, Klima MS, Mosser MM, et al. A study on removal of rare earth elements from U.S. Coal byproducts by ion exchange. Metall Mater Trans 2016;3:6−17.
[66] Hower JC, Ruppert LF, Eble CF. Lanthanide, yttrium, and zironium anomalies in the fire clay coal bed, Eastern Kentucky. Int. J. Coal Geol 1999;39:141−53.
[67] Nifantov BF. Valuable and toxic elements in coals, vol. II. Geoinformmark, Moscow (in Russian): Coal Resources of Russia; 2003. p. 77−91.
[68] Gu D, Jiang Y, Wei C, Han J, Deli YU, Guo Z, Guo C. Patent and Trademark Office. Washington, DC: U.S: U.S. Patent No. 8,728,296; 2014.

Index

Note: 'Page numbers followed by "f" indicate figures and "t" indicate tables'.

G

CPI Antony Rowe
Eastbourne, UK
July 19, 2019